Surveying

TENTH EDITION

Francis H. Moffitt
University of California, Berkeley

John D. Bossler
The Ohio State University

 ADDISON-WESLEY

An imprint of Addison Wesley Longman, Inc.

Menlo Park, California • Reading, Masschusetts • New York • Harlow, England
Berkeley, California • Don Mills, Ontario • Sydney • Bonn • Amsterdam • Tokyo • Mexico City

Acquisitions Editor: Michael Slaughter
Assistant Editor: Susan Slater
Executive Editor: Chuck Iossi
Production Editor: Diane Williams
Copy Editor: Barbara Conway
Art Supervisor: Kevin Berry
Cover Design: Yvo Riezebos
Cover Photo: Tony Stone Image/Sean Ellis

Surveying, Tenth Edition

Library of Congress Cataloging-in-Publication Data

Moffitt, Francis H.
      Surveying / Francis H. Moffitt, John D. Bossler.—10th ed.
         p.   cm.
      Includes index.
      ISBN 0-673-99752-9
      1. Surveying. I. Bossler, John D. II. Title.
TA545.M74 1998
526.9—dc21

96-53010
CIP

1 2 3 4 5 6 7 8 9 10—MA—01 00 99 98 97

Preface

Surveying is essentially the science of measurement of the spatial positions of the features on or near the surface of the earth. The practice of surveying is changing rapidly because of the availability of electronically controlled instruments, automatic capture and rapid transfer of field measurements, widespread adoption of the Global Positioning System (GPS), and the proliferation of various Geographic and Land Information Systems (GIS/LIS). Nevertheless, the principles underlying the new technology remain fairly constant. For example, the basic equations and computing techniques presented in this book are used to develop the computer programs employed by practicing surveyors. Software is available for practically any situation encountered by surveyors, including coordinate geometry, traverse calculations, adjustment computations, directions from astronomical observations, road design and layout, earthwork calculations, and stakeout data.

The authors have attempted to present the material underlying these foregoing advances in such a way that the student can learn the principles of surveying by self-study. References at the ends of the chapters are intended to expand the knowledge acquired through study of the basic principles found in the book.

Several major revisions have been made since the previous edition. Moreover, every chapter has been brought up to date. The information on inertial surveying has been dropped completely. Chapter 10 is devoted to GPS, thus separating traditional control surveying from this expanding technology and providing a simple but comprehensive explanation of GPS. Emphasis is placed on the NAD 83 as contrasted with NAD 27, as the redefinitions of the control networks in North America are now complete and the new datum is being utilized at a rapidly growing rate. To enhance the student's understanding of the material in Chapter 11 we provided a computer program (TRANSFOR) to compute state plane coordinates. The examples provided with the program are the same as those given in the text. This program is available on the Addison-Wesley ftp site, ftp.aw.com/cseng/authors/moffitt. (If you are unable to access this material please contact your local Addison Wesley Longman representative.)

Chapter 20 on Geographic Information Systems has been added, along with a problem set and a complete and extensive bibliography. Appendix A has been shortened considerably by removing the sections on the adjustment of triangulation.

We have worked hard to make this book readable and easy to understand while at the same time presenting current material that will be valuable to the modern surveyors.

The following publications are highly recommended to any student contemplating a career in surveying and mapping:

ACSM Bulletin, published by the American Congress on Surveying and Mapping, 5410 Grosvenor Lane, Bethesda, MD 20814 (301)493-0200.

Geodetical Information Magazine, published by the Geodetical Information and Trading Center, G.I.T.C.b.v., Villa Novalaan 14,8531 HS Lemmer, The Netherlands, +31(0)-5146-1854.

GeoInfo Systems, published by Advanstar Communications, 859 Willamette Street, Eugene, OR 97400 (541)343-1200.

GIS World, published by GIS World Inc., 155 E. Boardwalk Drive, Suite 250, Fort Collins, CO 80525 (970)223-4848.

GPS World, published by GPS World Corporation, 859 Willamette Street, Eugene, OR 97440 (541)343-1200.

Point of Beginning, better known as *P.O.B.,* published by Business News Publishing Company, P.O. Box 7069, 755 West Big Beaver Road, Suite 1000, Troy, MI 48007 (810)362-3700.

Professional Surveyor, published by Professional Surveyors Publishing Company, Inc., Suite 501, 2300 Ninth Street South, Arlington, VA 22204-2320 (703)892-0733.

Surveying and Land Information Systems, published by the American Congress on Surveying and Mapping.

Technical Papers of the Annual and Fall Conventions of the American Congress on Surveying and Mapping.

The majority of the articles listed in the bibliographies are taken from the publications listed above.

The authors wish to express their thanks to Mr. Edward McKay and Mr. James Stem of the National Geodetic Survey for their generous help in providing information for Chapters 9 and 11. We wish to thank a number of students and faculty at The Ohio State University and other professionals in the field who have reviewed portions of this edition of the book and have provided helpful comments to the authors. We are indebted to them. We must provide special thanks to the following: Mr. Bishnu Phuyal at The Ohio State University for checking a number of examples, Dr. George Dedes, formerly of The Ohio State University and now with Topcon-Geo Comp, Ltd., and Dr. Richard Langley of the University of New Brunswick (along with special thanks to GPS World) for their review and contribution to the GPS chapter. Thanks to Dr. Ayman Habib of The Ohio State University, who created the program TRANSFOR by modifying the NGS plane coordinate computation working out the examples in the book using that software. Thanks to Mr. James Saunders and Ms. Mary Tabler of The Ohio State University Center for

Mapping for reviewing and adding to the chapter on GIS. Thanks to Dr. Charles Toth of The Ohio State University Center for Mapping for his comments on the chapter on photogrammetry. We wish to offer very special thanks to Mr. David Cunningham of The Ohio State University Center for Mapping for his unique contributions in applying his rare combination of artistic ability and computer skills in developing many of the new figures (graphical illustrations) that appear in this edition. Finally, the authors wish to thank the instrument manufacturers who generously furnished the updated photographs used in this edition.

Francis H. Moffitt
John D. Bossler

Contents

Chapter 3   Leveling   61

Chapter 4   The Measurement of Angles   128

Chapter 5   Random Errors   177

Chapter 6   Field Operations with the Transit or Theodolite   213

Chapter 7   Direction of a Line   230

Chapter 8   Traverse Surveys and Computations   241

Chapter 9   Horizontal Control Networks   317

Chapter 10   The Global Positioning System   349

Chapter 11   State Plane Coordinate and Universal Transverse Mercator Systems   369

Chapter 12 Practical Astronomy 432

Chapter 13 Horizontal and Vertical Curves 466

Chapter 14   Tacheometry   506

Chapter 15   Topographic Surveys   528

Chapter 16   Photogrammetry   550

Chapter 17 Earthwork 590

Chapter 18 United States Public Land Surveys 627

Chapter 19 Municipal and Subdivision Surveys 655

Chapter 20   Geographic Information Systems   672

Appendix A   Adjustment of Elementary Surveying Measurements by the Method of Least Squares   688

Appendix B   The Adjustment of Instruments   719

Appendix C   Tables   726

1 Introduction

1-1 Surveying

The purpose of surveying is to locate the positions of points on or near the surface of the earth. Some surveys involve the measurement of distances and angles for the following reasons: (1) to determine horizontal positions of arbitrary points on the earth's surface, (2) to determine elevations of arbitrary points above or below a reference surface, such as mean sea level, (3) to determine the configuration of the ground, (4) to determine the directions of lines, (5) to determine the lengths of lines, (6) to determine the positions of boundary lines, and (7) to determine the areas of tracts bounded by given lines. Such measurements are data-gathering measurements.

In other surveys it is required to lay off distances, angles, and grade lines to locate construction lines for buildings, bridges, highways, and other engineering works, and to establish the positions of boundary lines on the ground. These distances and angles constitute layout measurements.

A survey made to establish the horizontal or vertical positions of arbitrary points is known as a *control survey*. A survey made to determine the lengths and directions of boundary lines and the area of the tract bounded by these lines, or a survey made to establish the positions of boundary lines on the ground, is termed a *cadastral, land, boundary,* or *property survey*. A survey conducted to determine the configuration of the ground is termed a *topographic survey*. The determination of the configuration of the bottom of a body of water is a *hydrographic survey*. Surveys executed to locate or lay out engineering works are known as *construction surveys*. A survey performed by means of aerial photography is called an *aerial survey* or a *photogrammetric survey*.

The successful execution of a survey depends on surveying instruments of a rather high degree of precision and refinement, and also on the proper use and handling of these instruments in the field. All surveys involve some computations, which may be made directly in the field, or performed in the office, or in both places.

Some types of surveys require very few computations, whereas others involve lengthy and tedious computations. In the study of surveying therefore, the student not only must become familiar with the field operation techniques but must also learn the mathematics applied in surveying computations.

1-2 Basic Definitions

To gain a clear understanding of the procedures for making surveying measurements on the earth's surface, it is necessary to be familiar with the meanings of certain basic terms. The terms discussed here have reference to the actual figure of the earth.

An *oblate spheroid,* also called an *ellipsoid of revolution,* is a solid obtained by rotating an ellipse on (or around) its shorter axis. It is helpful to construct an idealized figure of the earth, which is usually an oblate spheroid where the earth's rotational axis serves as the shorter or minor axis. Because of its relief, the earth's surface is not a true spheroid. However, an imaginary surface representing a mean sea level extending over its entire surface very nearly approximates a spheroid. This imaginary surface is used as the figure on which surveys of large extent are computed.

A *vertical line* at any point on the earth's surface is the line that follows the direction of gravity at that point. It is the direction that a string will assume if a weight is attached to the string and the string is suspended freely at the point. At a given point there is only one vertical line. The earth's center of gravity cannot be considered to be located at its geometric center, because vertical lines passing through several different points on the surface of the earth do not intersect in that point. In fact, all vertical lines do not intersect in any common point. A vertical line is not necessarily normal to the surface of the earth, nor even to the idealized spheroid. The angle between the vertical line and the normal to the spheroid at a point is called the *deflection of the vertical.*

A *horizontal line* at a point is any line that is perpendicular to the vertical line at the point. At any point there are an unlimited number of horizontal lines.

A *horizontal plane* at a point is the plane that is perpendicular to the vertical line at the point. There is only one horizontal plane through a given point.

A *vertical plane* at a point is any plane that contains the vertical line at the point. There are an unlimited number of vertical planes at a given point.

A *level surface* is a continuous surface that is at all points perpendicular to the direction of gravity. It is exemplified by the surface of a large body of water at complete rest (unaffected by tidal action).

A *horizontal distance* between two given points is the distance between the points projected onto a horizontal plane. The horizontal plane, however, can be defined at only one point. For a survey the reference point may be taken as any one of the several points of the survey.

A *horizontal angle* is an angle measured in a horizontal plane between two vertical planes. In surveying this definition is effective only at the point at which the measurement is made or at any point vertically above or below it.

A *vertical angle* is an angle measured in a vertical plane. By convention, if the angle is measured upward from a horizontal line or plane, it is referred to as a *plus* or *positive* vertical angle, and also as an *elevation* angle. If the angle is measured downward, it is referred to as a *minus* or *negative* vertical angle, and also as a *depression* angle.

A *zenith angle* is also an angle measured in a vertical plane, except that, unlike a vertical angle, the zenith angle is measured down from the upward direction of the plumb line. Obviously the zenith angle is equal to 90° minus the vertical angle.

The *elevation* of a point is its vertical distance above or below a given reference level surface (see Section 3-1).

The *difference in elevation* between two points is the vertical distance between the two level surfaces containing the two points.

Plane surveying is that branch of surveying wherein all distances and horizontal angles are assumed to be projected onto one horizontal plane. A single reference plane may be selected for a survey where the survey is of limited extent. For the most part this book deals with plane surveying.

Geodetic surveying (sometimes referred to as *control surveying*) is that branch of surveying wherein all distances and horizontal angles are projected onto the surface of the reference spheroid that represents mean sea level on the earth.

The surveying operation of leveling takes into account the curvature of the spheroidal surface in both plane and geodetic surveying. The leveling operation determines vertical distances and hence elevations and also differences of elevation.

1-3 Units of Measurement

In the United States the linear unit most commonly used at the present time is the foot, and the unit of area is the acre, which is 43,560 ft^2. In most other countries throughout the world, distances are expressed in meters. The meter is also used by the National Geodetic Survey of the United States Department of Commerce as well as other federal and state agencies in the United States engaged in establishing control. However, the published results of some of these control survey operations are given in both units, or are available to the user in both units.

On all U.S. government land surveys, the unit of length is the Gunter's chain, which is 66 ft long and is divided into 100 links, each of which is 0.66 ft or 7.92 in. long. A chain, therefore, equals $\frac{1}{80}$ mile. This is a convenient unit where areas are to be expressed in acres, since 1 acre = 10 square chains. A distance of 2 chains 18 links can also be written as 2.18 chains. Any distance in chains can be readily converted into feet, if desired, by multiplying by 66.

In the Southwest portions of the United States that were influenced by the Spanish, another unit, known as the vara, has been used. A vara is about 33 in. long. The exact length varies slightly in different sections of the Southwest, where the lengths of property boundaries are frequently expressed in this unit.

For purposes of computation and plotting, decimal subdivisions of linear units are the most convenient. Most linear distances are therefore expressed in feet and tenths, hundredths, and thousandths of a foot. The principal exception to this practice is in the layout work on a construction job, where the plans of the structures are dimensioned in feet and inches. Tapes are obtainable graduated either decimally or in feet and inches.

Volumes are expressed in either cubic feet or cubic yards.

Angles are measured in degrees (°), minutes (′), and seconds (″). One circumference = 360°; 1° = 60′; 1′ = 60″. In astronomical work some angles are expressed in hours (^{h}), minutes (^{m}), and seconds (^{s}). Since one circumference = 24^h = 360°, it follows that 1^h = 15° and 1° = $\frac{1}{15}^h$ = 4^m (see Section 12-5).

Although some surveying instruments that measure angles in the sexagesimal system are graduated in degrees, minutes, and seconds, it is usually necessary, in computations with hand calculators, to convert to degrees and decimal degrees (unless the calculator has provision for this conversion) to compute the trigonometric functions of the angles, and then if necessary to convert back to degrees, minutes, and seconds. The number of decimal places to be retained in the decimal degree is a function of the least reading of the instrument or of the given angle. Equivalents of the decimal part of a degree are

$$0.00001° = 0.036″$$
$$0.0001° = 0.36″$$
$$0.001° = 3.6″$$
$$0.01° = 36″$$
$$0.1° = 6′$$

Thus, if the angle is given to the nearest minute, two decimal places must be retained; if the angle is given to the nearest second, four decimal places must be retained.

EXAMPLE 1-1   Convert 153° 43′ 17.2″ to decimal form.

Solution:
$$153° \ 43′ \ 17.2″ = 153° \left(43\frac{17.2}{60} \right)' = 153° \ 43.287′$$
$$153° \ 43.287′ = \left(153\frac{43.287}{60} \right)° = 153.72144°$$

which is also written 153°.72144.

EXAMPLE 1-2   Convert 24.4652° to degrees, minutes, and seconds.*

Solution:
$$0.4652° \times 60 = 27.912′$$
$$0.912′ \times 60 = 54.7″$$

The value of the conversion is thus 24° 27′ 54.7″

---

*Many handheld calculators have provision for converting from degrees, minutes, and seconds to decimal degrees and vice versa using conversion keystrokes.

To help visualize the physical size of small dimensions, consider that 0.01 ft is very nearly equal to $\frac{1}{8}$ in., and that 0.10 ft is therefore about $1\frac{1}{4}$ in. Also, the sine or tangent of 1′ of arc is approximately 0.0003 (three zeroes 3). Thus 1′ of arc subtends about 0.03 ft or $\frac{3}{8}$ in. in 100 ft. At 1000 ft, 1′ of arc subtends 0.30 or $3\frac{1}{2}$ in.

The sine or tangent of 1″ of arc is approximately 0.000005 (five zeroes 5). Thus at 1000 ft, 1″ subtends 0.005 ft or about $\frac{1}{16}$ in. Another relationship is embodied in the expression "a second is a foot in 40 miles." Taking the radius of the earth to be 4000 miles, a second of arc at the center of the earth subtends approximately 100 ft on the earth's surface.

Quite frequently it becomes necessary to convert a small angle from its arc or radian value to its value expressed in seconds, and vice versa. A unit radian is the angle subtended by an arc of a circle having a length equal to the circle's radius. Thus 2π rad = 360°; 1 rad = 57° 17′ 44.8″ and 0.01745 rad = 1°. We find from trigonometry that the sine, the tangent, and the arc or radian value of 1″ are all equal within the limits of computational practicality. As a consequence of this, and designating θ as a small angle,

$$\text{arc } \theta \simeq \sin \theta \simeq \tan \theta$$

If the small angle is expressed in seconds, its arc, sine, or tangent can be determined by the following relationships:

$$\text{arc } \theta = \theta'' \times \text{arc } 1'' \qquad \text{(exact)}$$
$$\sin \theta = \theta'' \times \sin 1'' \qquad \text{(approximate)}$$
$$\tan \theta = \theta'' \times \tan 1'' \qquad \text{(approximate)}$$

The value of the arc, sine, and tangent of 1″ is 0.0000048481 to 10 decimal places. If the arc or radian value of a small angle is known, or if the sine or tangent is known, then its value in seconds can be determined by rearranging the three preceding expressions to give

$$\theta'' = \frac{\text{arc } \theta}{\text{arc } 1''} = \frac{\text{arc } \theta}{48,481 \times 10^{-10}} = 206,265 \text{ arc } \theta$$

$$\theta'' = \frac{\sin \theta}{\sin 1''} = \frac{\tan \theta}{\tan 1''}$$

EXAMPLE 1-3   A target 4 in. wide is placed on the side of a building 1600 ft away from a survey point. What angle is subtended at the point between the left and right edges of the target?

Solution: The target width is 0.333 ft and thus the arc of the small angle is 0.333/1600 = 0.000208. The angle in seconds is then 0.000208/48,481 × 10⁻¹⁰ = 42.9″ or 0.000208 × 206,265 = 42.9″

1-4   Metric Equivalents

The metric system, used extensively throughout the world, is finding a more wide-spread adoption in the United States. This adoption has come about primarily to facilitate exchange of scientific and technical data with countries using the metric system. It obviously facilitates the exporting and importing of goods. The metric system is also called the *International System of Units* and is abbreviated SI.

The basis of distance measurement in the metric system is the meter, which is defined as 1,650,763.73 wavelengths of orange-red krypton gas with an atomic weight of 86 and at a specified energy level in the spectrum, or as the distance light travels, in a vacuum, in 1/299,792,458 second with time measured by a cesium-133 atomic clock. This natural standard was adopted in lieu of a man-made physical standard, such as the International Prototype Meter Bar, because of the ever-present danger that the latter might be destroyed by accident or by a hostile act.

The meter is more commonly subdivided into the following units:

$$1 \text{ centimeter (cm)} \quad = 0.01 \text{ meter}$$
$$1 \text{ millimeter (mm)} \quad = 0.001 \text{ meter}$$
$$1 \text{ micrometer (μm)} \quad = 0.001 \text{ mm} = 10^{-6} \text{ meter}$$
$$1 \text{ nanometer (nm)} \quad = 0.001 \text{ μm} = 10^{-9} \text{ meter}$$
$$1 \text{ angström (Å)} \quad = 0.1 \text{ nm} = 10^{-10} \text{ meter}$$

It was not until 1959 that the technical and scientific agencies of the United States and the United Kingdom adopted the following equivalence:

$$1 \text{ U.S. inch} \quad = 2.54 \text{ centimeters}$$
$$1 \text{ British inch} = 2.54 \text{ centimeters}$$
$$1 \text{ U.S. foot} \quad = 30.48 \text{ centimeters}$$

This unit (1 foot = 30.48 cm) is called the *international foot.*

The previous U.S. equivalent was 1 meter = 39.37 U.S. inches, or 1 U.S. inch = 2.540005 + centimeters. Since plane and geodetic surveys may involve very long distances, sometimes on the order of 200 miles, and since results of important nationwide surveys going back more than 50 years are still used, inconsistencies would be introduced if the 1959 equivalence was adopted for surveying operations. Therefore the previous equivalence of 1 meter = 39.37 U.S. inches is used to define the *U.S. survey foot.* Thus 1 survey foot = 12/39.37 meter. This differs from the new definition of the foot (1 ft = 30.48 cm) by about two parts per million (2 ppm). To get an appreciation of this difference, consider that in a distance of 100,000 meters, which is slightly more than 60 miles, the difference between the two values is 0.66 ft or 8 in.

The kilometer (km) is 1000 m = 3280.833 ft, which is approximately $\frac{5}{8}$ of a mile. One mile is approximately 1.6 km.

The unit of area in many countries that use the metric system is the *hectare* (ha), which is 10,000 m². Thus one hectare is approximately $2\frac{1}{2}$ acres.

Volumes are expressed in cubic meters (m³), which is approximately 1.3 cubic yards (yd³).

1-5 Centesimal System

Many surveying and mapping instruments with angular scales are graduated in the centesimal system. In this system one circumference is divided into 400 divisions called *gon*. These divisions are also referred to as *grads*(^{g}). The relationship between the gon, the grad, and the degree are determined by the following equivalences:

$$
\begin{array}{lllll}
400 \text{ gon} & = 400^g & & = 360° & = 360° \; 00' \; 00'' \\
1 \text{ gon} & = 1^g & & = 0.9° & = 0° \; 54' \; 00'' \\
0.1 \text{ gon} & = 0.1^g & = 10^c & = 0.09° & = 0° \; 05' \; 24'' \\
10 \text{ mgon} & = 0.01^g & = 1^c & = 0.009° & = 0° \; 00' \; 32.4'' \\
1 \text{ mgon} & = 0.001^g & = 10^{cc} & = 0.0009° & = 0° \; 00' \; 03.2'' \\
0.1 \text{ mgon} & = 0.0001^g & = 1^{cc} & = 0.00009° & = 0° \; 00' \; 00.32''
\end{array}
$$

An angle expressed as 324.4625^g (or 324.4625 gon) could also be expressed as 324^{g}46^{c}25^{cc}, but this latter notation would be much more awkward and entirely unnecessary. However, it is found in some past literature in surveying and photogrammetry as well as in other scientific literature. If trigonometric tables based on centesimal units are available, many computations are greatly simplified.

To convert an angle given in gon to its equivalent value in degrees, multiply the gon value by 0.9.

EXAMPLE 1-4   Convert 324.4625 gon to degrees, minutes, and seconds.

Solution: Multiplying by 0.9 gives 324.4625 × 0.9 = 292.01625°. Conversion from decimal degrees to degrees, minutes, and seconds then follows.

$$
\begin{array}{ll}
0.01625° \times 60 = 0.975' & \text{(minutes)} \\
0.975 \;\; \times 60 = 58.5'' & \text{(seconds)}
\end{array}
$$

The value of the conversion is thus 292° 00′ 58.5″.

When converting from degrees, minutes, and seconds to gon, the angle must be converted to decimal degrees and the results then divided by 0.9.

EXAMPLE 1-5   Convert 142° 22′ · 14.5″ to gon.

Solution:

$$142° \; 22' \; 14.5'' = 142° \left(22\frac{14.5}{60} \right) = 142° \; 22.2417'*$$

$$142° \; 22.2417' = 142 \left(\frac{22.2417}{60} \right)° = 142.37069°$$

$$142.37069° = \left(\frac{142.37069}{0.9} \right) \text{gon} = 158.18966 \text{ gon}$$

---

*See footnote on page 4.

1-6   Tables

Appendix C contains three tables that are useful in surveying computations. Table C-1 is useful for converting an angle given in degrees, minutes, and seconds into its radian value.

EXAMPLE 1-6   Determine the radian value of 37° 14′ 53.6″.

Solution: From Table C-1

37°	= 0.645 77 182
14′	= 0.004 07 243
53″	= 0.000 25 695
0.6″	= 0.000 00 291
37° 14′ 53.6″	= 0.650 10 411 radian

This calculation can be made quite easily using a handheld calculator by first converting 37° 14′ 53.6″ to decimal degrees and then multiplying by $\pi/180$. ◆

EXAMPLE 1-7   Compute the length of a circular arc of radius 2500 m that is subtended by an angle of 37° 14′ 53.6″.

Solution: The length of the arc is equal to the radius multiplied by the subtended angle in radians. Thus

$$\text{arc length} = 2500 \times 0.650\ 10\ 411 = 1625.260 \text{ m}$$ ◆

Tables C-2 and C-3 give trigonometric formulas for the solution of right triangles and oblique triangles, respectively.

1-7   Field Notes

When a survey is performed for the purpose of gathering data, the field notes become the record of the survey. If the notes have been carelessly recorded and documented, falsified, lost, or made incorrect in any way, the survey or a portion of the survey is rendered useless. Defective notes result in tremendous waste of both time and money. Furthermore, it will become obvious that, no matter how carefully the field measurements are made, the survey as a whole may be useless if some of those measurements are not recorded or if the meaning of any record is ambiguous.

Some surveying instruments (see Section 4-24) are capable of automatic storage of field measurements. However, they do not completely preclude the necessity of recording these data in the field notes because they are usually accompanied by field sketches and diagrams.

The keeping of neat, accurate, complete field notes is an exacting task. Although several systems of note keeping are in general use, certain principles apply to all. The aim is to make the clearest possible notes with the least expenditure of time and effort. Detailed examples of forms of notes suited to the principal surveying operations will be given later.

All field notes should show when, where, for what purpose, and by whom the survey was made. The signature of the recorder should appear. Habits are formed by constant repetition. One of the best habits is to make each page of notes complete. The field notes of a surveyor or engineer are often presented as evidence in court cases, and such things as time and place and members of the field party may be of extreme importance.

A hard pencil—4H or harder—should be used to prevent smearing. The notebook should be of good quality, since it is subjected to hard usage. No erasures should be made, because such notes will be under suspicion of having been altered. If an error is made, a line should be drawn through the incorrect value and the new value should be inserted above. In some organizations the notes are kept in ink, but this is rather inadvisable unless a waterproof ink is used or unless there is no possible chance of the notes becoming wet.

Clear, plain figures should be used, and the notes should be lettered rather than written. The record should be made in the field book at the time the work is done, and not on scraps of paper to be copied into the book later. Copied notes are not original notes, and there are too many chances for making mistakes in copying or of losing some of the scraps. All field computations should appear in the book so that possible mistakes can be detected later.

It should always be remembered that the notes are frequently used by other than the person who makes them. For this reason they should be so clear that there can be no possible chance of misinterpreting them.

If the numbers stamped on the field equipment are recorded, it may be possible to explain at some later date apparent discrepancies that were caused by some imperfection in the equipment. Thus, for example, errors in linear measurements can sometimes be explained by a broken tape being improperly repaired.

1-8   Data Collectors

The time required to enter survey data into a field book by hand and then, if necessary, to convert it to computer-readable form is extensive. Fortunately, equipment manufacturers have developed a variety of easy-to-use data collectors. All of these allow for manual entry of data, which takes about the same amount of time as it does to record the data in a field book. However, once collected, the data can be downloaded directly into a computer in a matter of minutes. Data collectors are also available as a part of total station instruments (see Section 4-24) and can be used to automatically record data collected by the total station.

Data collectors are generally able to perform calculations and other analysis; they are in fact computers. In many cases therefore the computations can be carried out in the field, and hence a great deal of time can be saved by eliminating numerous trips to check observations and so on.

Each data collector has a few unique features in its operation. However, each is fairly simple to use and can be learned quickly. Fig. 1-1 shows one such device.

FIGURE 1-1 Data collector. Courtesy of Pentax Corp.

Data collectors are useful in accomplishing the full range of surveying tasks, such as leveling and boundary and topographic surveys.

P.O.B., published by Business News Publishing Company prints the results of data collector surveys every three or four years. These surveys list the essential information on all data collectors on the current market.

1-9   Significant Figures and Rounding Off

It is common practice in the United States to use the recorded figures (digits) to imply the degree of accuracy of the measurement. The customary practice is to record the measured digits that are certain and the observer's best estimate of the value of the finest unit of measurement. No finer digits should be recorded. For example, a distance measured with a tape whose finest graduations are 0.01 ft and recorded as 99.98 ft has four *significant figures.* In this example three digits are certain but the last, 8, is estimated. The maximum round-off error is ±0.005 ft.

It is important that this procedure be followed since a recording of 99.980 could imply that the measurement was accurate to ±0.0005. Also, recording the value 100.0 ft "wastes" the precision of the measurement itself. The trailing digit 0 sometimes causes confusion, and occasionally a bar is placed over the last significant figure. For example, 99.9$\bar{8}$ has four significant figures, 10$\bar{0}$.0 has three significant figures, and so on. In this book and in common practice, the last 0 is considered significant.

All measurements are rounded off in some sense. Rounding is the process of dropping of digits that, for example, appear when eight digits are shown on a calculator but only two are significant. The normal convention is as follows:

- When the digit to be dropped is less than 5, the number is written without the digit. Therefore 117.573 becomes 117.57.
- When the digit to be dropped is exactly 5, the nearest even number is used. Therefore 117.575 becomes 117.58 and 117.565 becomes 117.56.
- When the digit to be dropped is more than 5, the preceding digit is increased by 1. Therefore 117.578 becomes 117.58.

When performing computations it is important to use an appropriate number of significant figures so that the computations themselves do not lessen the accuracy of the measurements and consequently the result. Likewise, it is misleading to carry more digits than is warranted. For example, if the area of a parcel is created by adding three separate smaller parcels together, with each sub-area having been determined by a different surveying technique, the result can be stated in a misleading fashion as shown:

Parcel A	3.684	acres
Parcel B	6.01	acres
Parcel C	11.0	acres
Total	= 20.694	acres
Correct answer	= 20.7	acres

The accuracy of the resulting area is probably no better than ±0.05 acres, because Parcel C's area was determined only to this accuracy.

Likewise, when multiplying and dividing, the number of significant figures in the answer is equal to the least number of significant figures in the terms of the product or quotient. For example, $117.58 \times 6.1 = 717.24$, but the answer should be shown as 720 unless both quantities were stated as exact numbers.

It is intuitive that when using a calculator or computer, round-off error occurs. Therefore it is advisable to use more decimal places in the computations than are significant. The number of places required is a function of the number of computations, but in plane surveying eight places will usually suffice. As with any such "rule," there are numerous exceptions.

1-10 Errors and Mistakes

The value of a distance or an angle obtained by field measurements is never exactly the true value, except by chance. The measured value approaches the true value as the number and size of errors in the measurements become increasingly small. An error is the difference between the true value of a quantity and the measured value of the same quantity. Errors result from instrumental imperfections, personal limitations, and natural conditions affecting the measurements. Examples of instrumental errors are (1) a tape that is actually longer or shorter than its indicated length; (2) errors in the graduations of the circles of an engineer's transit; and (3) a defect in the calibration of an electronic distance measuring device. Examples of personal limitations are the observer's inability to bisect a target or read a vernier exactly, inability to maintain a steady tension on the end of a tape, and failure to keep a level bubble centered at the instant at which a leveling observation is taken. Examples of natural conditions affecting a measurement are wind and temperature, and pressure and humidity changes causing the refraction of the direction or distance between two points.

An error is either a *systematic error* or a *random error.* A systematic error is one for which the magnitude and algebraic sign can theoretically be determined. If a tape is found to measure 99.94 ft between the 0-ft mark and the 100-ft mark when compared with a standard, then the full tape length introduces a systematic error of +0.06 ft each time it is used to measure the distance between two given points. If a tape is used at a temperature other than that at which it was compared with a standard, then the amount by which the nonstandard temperature increases or decreases the length of the tape can be computed from known characteristics of the material of which the tape is made.

A random error is one for which the magnitude and sign cannot be predicted. It can be plus or minus. Random errors tend to be small and tend to distribute themselves equally on both sides of zero. If an observer reads and records a value of, say, 6.242 ft when the better value is 6.243 ft, a random error of – 0.001 ft has been introduced. When an individual is holding a signal on which an instrument man is sighting, failure to hold the signal directly over the proper point will cause a random error of unknown size and algebraic sign in the measured angle. If, however, he *fixes* the signal eccentrically, the resultant error will be systematic.

Random errors tend to grow proportional to the square root of the number of them, but systematic errors grow directly proportional to the number of them.

A *mistake* is not an error but is a blunder on the part of the observer. Examples of mistakes are failure to record each full tape length in taping, misreading a tape, interchanging figures, and forgetting to level an instrument before taking an observation. Mistakes are avoided by exercising care in making measurements, by checking readings, by making check measurements, and to a great extent by common sense and judgment. If, for example, a leveling rod is read and the reading is recorded as 7.13 ft, whereas the levelman knows that this is absurd since he is very nearly at the top of a 14-ft rod, then he is exercising common sense in suspecting a mistake.

The subject of random errors is considered in more detail in Chapter 5. Systematic errors and methods for their elimination are discussed in the appropriate sections throughout the book.

1-11 Accuracy and Precision

Since surveying is after all a measurement science, it is necessary to distinguish between the two terms *accuracy* and *precision* that, if not understood, can cause needless confusion. The accuracy of a measurement is an indication of how close it is to the true value of the quantity that has been measured. To obtain an accurate measurement, the measuring instrument must have been calibrated by comparison with a standard. This allows for the elimination of systematic errors.

The precision of a measurement has to do with the refinement used in taking the measurement, the quality (but not necessarily the accuracy) of an instrument, the repeatability of the measurement, and the finest or least count of the measuring device.

As an illustration of the difference between these two terms, suppose that two different taping parties (see Chapter 2) measure the same line five times, each using a different 100-ft tape. The first party reports the following measurements: 736.80, 736.70, 736.75, 736.85, and 736.65 ft. The second party reports the following measurements: 736.42, 736.40, 736.40, 736.42, and 736.41 ft. Further suppose that the correct or true length of the line is 736.72 ft. Obviously, from an examination of the spread of the results, the measurements reported by the second party are more precise. However, those reported by the first party are more accurate because they tend to group around the true value. Thus the tape used by the second party has some kind of systematic error that has not been accounted for in the reported measurements.

PROBLEMS

1-1. How many hectares are contained in a rectangular field that measures 352.25 by 196.80 m?

1-2. How many acres are contained in the area given in Problem 1-1?

1-3. How many square feet are contained in the area given in Problem 1-1?

1-4. A distance of 40 chains 2 links is shown on a map between two boundary markers. What is the corresponding length in feet?

1-5. What is the length of the line in Problem 1-4 in meters?

1-6. Based on the equivalence 1 U.S. foot = 30.48 cm, how many feet are contained in a line that measures 27,542.331 m?

1-7. How many survey feet are contained in the line of Problem 1-6?

1-8. Convert the following decimal degrees to their corresponding values in the sexagesimal system: **(a)** 26.9°, **(b)** 196.23°, **(c)** 63.464°, **(d)** 312.1546°, **(e)** 19.52496°.

1-9. Convert the angles of Problem 1-8 to their equivalent grad values.

1-10. Convert the following angles to decimal degree form: **(a)** 16° 37′, **(b)** 254° 16′ 42″, **(c)** 96° 52′ 14.3″, **(d)** 35° 47′ 16.82″, **(e)** 174° 49′ 16.834″.

1-11. Convert the angles given in Problem 1-10 to their grad values.

1-12. Convert the following angles to decimal degrees: **(a)** 16^g, **(b)** 26.2^g, **(c)** 274.56^g, **(d)** 394.163^g, **(e)** 27.9444^g, **(f)** 37.46205^g.

1-13. Convert the angles given in Problem 1-12 to their values in the sexagesimal system.

1-14. Convert the following volumes to cubic meters: **(a)** 14,955 yd^3, **(b)** 129,590 yd^3, **(c)** 2545.6 yd^3, **(d)** 18,192.66 yd^3.

1-15. Convert the following volumes to cubic yards: **(a)** 546 m^3, **(b)** 2850.2 m^3, **(c)** 1485.66 m^3, **(d)** 42,455 m^3.

1-16. The sides of a triangle measure 1040.25, 1855.24, and 1318.18 m. Compute the three angles in the triangle expressed to the nearest 0.0001^g.

1-17. The sides of a triangle measure 302.55, 1696.20, and 1714.36 ft. Compute the three angles in the triangle (nearest second).

1-18. What is the area of the triangle of Problem 1-16 in hectares?

1-19. What is the area of the triangle of Problem 1-17 in acres?

1-20. Two sides and the included angle of a triangle are 1810.462 m, 1462.82 m, and 35.6362^g, respectively. Compute the length of the remaining side (nearest millimeter) and the two remaining angles (nearest 0.0001^g).

1-21. Two sides and the included angle of a triangle are 545.156 m, 1395.832 m, and 15° 22′ 46″, respectively. Compute the length of the remaining side (nearest millimeter) and the remaining two angles (nearest second).

1-22. Compute the area of the triangle in Problem 1-20 in hectares and in acres.

1-23. Compute the area of the triangle in Problem 1-21 in hectares and in acres.

1-24. In triangle ABC, $A = 27° 14′ 52″$, $B = 52° 35′ 44″$, and side $AB = c = 385.462$ ft. Compute sides a and b (nearest 0.001 ft).

1-25. The radius of a circle is 350.000 m. What arc length is subtended by a central angle of 125.4652^g (nearest millimeter).

1-26. The arc of a small angle is 0.0000621458. What is the angle in seconds (nearest 0.01″)?

1-27. What is the angle in Problem 1-26 in grads (nearest 0.00001^g)?

BIBLIOGRAPHY   1994 Data Collector Survey. *P.O.B.* 19:6.

Carver, G. P. 1993. Metrication: an economic wake-up call for surveyors and mappers. *P.O.B.* 19:1.

Greulich, G. H. 1982. Metric standards in subdivision and zoning. *Surveying and Mapping* 42:257.

Kennedy, G. J. 1984. Effective systems for control surveys. *Technical Papers Fall Conv. Amer. Cong. Surv. & Mapp.* p. 39.

Lane, A. 1992. Metric mandate. *Professional Surveyor* 12:2.

Strasser, Georg. 1975. The toise, the yard and the meter—the struggle for a universal unit of length. *Surveying and Mapping* 35:25.

Ward, H. O. 1989. Are field books obsolete? *ACSM Bull. 122, Amer. Cong. Surv. & Mapp.* p. 31.

2 Measurement of Horizontal Distances

2-1 Horizontal Distances

One of the basic operations of surveying is the determination of the distance between two points on the surface of the earth. In surveys of limited extent the distance between two points at different elevations is reduced to its equivalent horizontal distance either by the procedure used to make the measurement or by computing the horizontal distance from a measured slope distance. Distances are measured by scaling from a map, by pacing, by using an odometer, tacheometry, or electronic distance meters (EDMs), or by taping. This chapter will emphasize the use of the tape and the EDMs.

2-2 Pacing

Where approximate results are satisfactory, distances can be obtained by pacing. A person can best determine the length of his pace by walking over a line of known length several times, maintaining a natural walking stride. No particular advantage is obtained by developing a pace of, say, 3 ft. The natural stride is reproducible from day to day, whereas an artificial pace is not. The number of paces can be counted with a tally register or by the use of a pedometer, which is carried like a watch in a vertical position in the pocket.

2-3 Odometer

The odometer of a motor vehicle will give fairly reliable distances along highways, provided that the odometer is periodically checked against a known distance. This method can often be used to advantage on preliminary surveys where precise distances are not necessary. The odometer is used to obtain distances for writing descriptions of the locations of survey control points and markers.

2-4 Tacheometry

Distances can be measured indirectly by the use of optical surveying instruments in conjunction with measuring bars or rods. The measurements are performed quite rapidly and are sufficiently accurate for many types of survey operations. Their use, however, is limited to providing check measurements in high-accuracy surveys, in which the measurements are performed either by direct taping or by EDMs. Methods and operations used in this kind of measurement are referred to generally as *tacheometry.*

The telescope of an engineer's transit or theodolite usually contains three horizontal cross hairs that are used for tacheometry. The top and bottom hairs are called *stadia hairs.* These are usually so spaced that each unit intercepted between them on a graduated rod held vertically at a point some distance from the theodolite represents a distance of 100 units from the theodolite. Most distances obtained in topographic surveys conducted in the field are determined by stadia measurements. Both horizontal and vertical distances can be obtained by means of stadia tacheometry. This method is described in Chapter 14.

The self-reducing telemeter, which is discussed in Chapter 14, is another form of tacheometric instrument and performs generally the same function as the stadia hairs of the theodolite.

A somewhat more accurate method of tacheometry than the use of the transit and its variations is that of *subtense.* This consists of setting up a short base at one end of the line to be measured and then measuring the small horizontal angle subtended by the base with a precise theodolite at the other end of the line.

The short base, consisting of either a graduated rod or a rod containing a target on each end, is oriented so that it is in a horizontal position and is also perpendicular to the line being measured. The rod is called a *subtense bar.* The distance is computed from the isosceles triangle formed by the known base on the rod and the sides of the opposite measured angle. The subtense bar is discussed in greater detail in Chapter 14.

2-5 Electronic Surveying

In recent years several ingenious electronic systems have been developed for the express purpose of measuring distances in surveying with a high degree of accuracy. They are based on the invariant velocity of light or electromagnetic waves in a vacuum. It is of interest to note that the first of these instruments, which is called the *geodimeter,** was the outgrowth of instrumentation developed to determine an accurate value of the velocity of light. The value adopted by the International Association of Geodesy in 1975 is 299,792.458 km/s, and it differs by only 0.4 km/s from that determined by geodimeter instrumentation in 1951.

A discussion of the use of the various EDM instruments for determining distances is presented following the sections that describe the use of the tape.

---

*Geodimeter is a registered trademark of AGA Aktiebolag.

This newer system (EDM system) has practically displaced the tape in modern surveys except for relatively short measurements and for layout work in construction surveys.

2-6 Chains

Two kinds of chains were formerly used in surveying, namely, the 100-ft engineer's chain and the 66-ft Gunter's chain. Each kind is divided into 100 links. A link of an engineer's chain is therefore 1 ft long, and a link of a Gunter's chain is only 0.66 ft or 7.92 in. long. Such chains are seldom, if ever, used at present. Although the 66-ft chain is still employed as a *unit* of length in the survey of the public lands and of some farm lands, the actual field measurements are made with steel tapes graduated in chains and links. If the distances recorded are in chains, a chain of 66 ft is implied.

2-7 Tapes

Steel tapes for most surveying operations are graduated in feet or meters together with decimal parts of these units. Their lengths vary from 50 to 300 ft and from 15 to 100 m, although the 100-ft and the 30-m tapes are the most common lengths. Lightweight tapes may be graduated to hundredths of a foot or to centimeters for their entire length. However, the usual foot tape is graduated to feet, with an end foot divided to hundredths of a foot. Metric tapes are usually graduated to decimeters throughout with an end decimeter divided to millimeters.

As most engineering and architectural plans show dimensions in feet and inches, a tape graduated in feet and inches is an advantage on construction work for layout purposes.

Tapes graduated in meters are used in most countries outside of the United States. These tapes are also used on most geodetic work in all countries.

Tapes of cloth, or of cloth containing threads of bronze or brass, are sometimes used where low precision is permissible and where a steel tape might be broken, as in cross sectioning for a railroad or a highway.

For extreme precision an invar tape, made of an alloy of steel and nickel, is used. The advantage of a tape of this material is that its coefficient of thermal expansion is about one-thirtieth that of steel, and hence its length is not so seriously affected by temperature changes. However, since such tapes are expensive and must be handled very carefully to prevent kinking, invar tapes are only used for precise work.

Tapes are calibrated by comparison with a standard that is maintained by the National Institute of Standards and Technology (NIST was previously known as the National Bureau of Standards) and by certain state, county, and city agencies. A few universities and state sections of the American Congress on Surveying and Mapping also maintain standard lengths by which tapes can be calibrated. The owner of the tape specifies under what conditions the tape should be calibrated, that is, what tension should be applied to the tape, and whether it should be supported throughout its length or whether it is to be supported only at its ends. The calibration report then gives the length of the tape under the specified conditions and at some standard temperature, usually 68°F or 20°C. Tapes can be compared with the standard baselines measured by the National Geodetic Survey.

2-8 Equipment Used for Taping

For the direct measurement of a line several hundred feet or meters long, the nominal equipment used consists of a 100-ft or 30-m steel tape, two plumb bobs, one or more line rods, a set of taping pins, and, if the ground is hilly, a hand level. These items are shown in Fig. 2-1. A spring balance, which can be seen to the extreme right in Fig. 2-3, is used to apply the desired pull or tension to the tape during the measurement. The tension is expressed in pounds or kilograms.

Line rods, also called range poles, are from $\frac{3}{8}$ in. to more than 1 in. in diameter and from 6 to 8 ft (2 m) long. They are pointed at one end and are painted with alternate bands of white and red. The rods are used to sight on and thus keep the forward and rear ends of the tape on the line that is being measured.

Taping pins are used to mark the positions of the ends of the tape on the ground while a measurement is in progress. A taping pin may be a heavy spike but is usually a piece of No. 10 wire that is 10 to 18 in. long, is sharpened at one end, and is provided with an eye at the other end. Pieces of colored cloth can be tied to the eyes to make the pins more visible in tall grass or weeds.

The hand level is used to keep the two ends of the tape in the same horizontal plane when a measurement is made over rough or sloping ground.

The plumb bobs are used to project a point on the ground up to the tape, or to project a point on the tape down to the ground.

Some tapes are kept on reels when not in use. But a metal tape must be entirely removed from the reel when any length other than a few feet or meters is to be measured. If such a tape is not supplied with thongs on both ends with which

FIGURE 2-1 Equipment used for taping.

to hold the tape, a taping pin can be slipped through the eye at the end of the tape and used as a handle. A tape that is thrown together in the form of a series of loops when not in use must be carefully unwrapped and checked for short kinks before it can be used for measurement. As long as a tape is stretched straight, it will stand any amount of tension that two people can apply. If kinked or looped, however, a very slight pull is sufficient to break it.

2-9   Measurements with Tape Horizontal

The horizontal distance between two points can be obtained with a tape either by keeping the tape horizontal or by measuring along the sloping ground and computing the horizontal distance. For extreme precision, such as is required in determining the length of a baseline in a triangulation system, the latter method is used. This method is also advantageous where steep slopes are encountered and it would be difficult to obtain the horizontal distance directly.

For moderate precision where the ground is level and fairly smooth, the tape can be stretched directly on the ground, and the ends of the tape lengths can be marked by taping pins or by scratches on a paved area. Where the ground is level but ground cover prevents laying the tape directly on the ground, both ends of the tape are held at the same distance above the ground by the forward tapeman and the rear tapeman. The tape is preferably held somewhere between knee height and waist height. The graduations on the tape are projected to the ground by means of the plumb bobs. The plumb-bob string is best held on the tape graduation by clamping it with the thumb, so that the length of the string can be altered easily if necessary (this can be seen in Fig. 2-4). When a tape is supported throughout its length on the ground and subjected to a given tension, a different value for the length of a line will be obtained than when the tape is supported only at the two ends and subjected to the same tension (see Section 2-18). Where fairly high accuracy is to be obtained, the method of support must be recorded in the field notes, provided different methods of support are used on one survey. Experienced tapemen generally obtain equivalent results by plumbing the ends of the tape over the marks or by having the tape supported on the ground.

When the ground is not level, either of two methods may be used. The first is to hold one end of the tape on the ground at the higher point, to raise the other end of the tape until it is level, either by estimation or with the aid of the hand level, and then to project the tape graduation over the lower point to the ground by means of a plumb bob. The other method is to measure directly on the slope as described in Sections 2-11 and 2-12. These methods are shown in Fig. 2-2.

For high precision, a taping tripod or taping buck must be used instead of a plumb bob. Such a tripod is shown in Fig. 2-3. Taping tripods are usually used in groups of three, the rear tripod then being carried to the forward position. A pencil mark is scribed at the forward tape graduation, and on the subsequent measurement the rear tape graduation is lined up with this mark in order to carry the measurement forward. Since taping is usually done on the slope when tripods are used, the elevations of the tops of the tripods must be determined at the same time the taping proceeds. The elevations, which are determined by leveling (Chap-

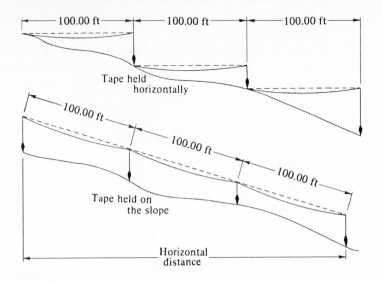

FIGURE 2-2 Taping over sloping ground using 100-ft tape.

FIGURE 2-3 Taping tripod.

ter 3), give the data necessary to reduce the slope distances to horizontal distances as discussed in Section 2-12.

The head tapeman carries the *zero* end of the tape and proceeds toward the far end of the line, stopping at a point approximately a tape length from the point of beginning. The rear tapeman lines in the forward end of the tape by sighting on the line rod at the far end of the line. Hand signals are used to bring the head tapeman on line. The rear tapeman takes a firm stance and holds the tape close to his body with one hand, either wrapping the thong around his hand as shown in Fig. 2-4 or holding a taping pin that has been slipped through the eye of the tape. Standing

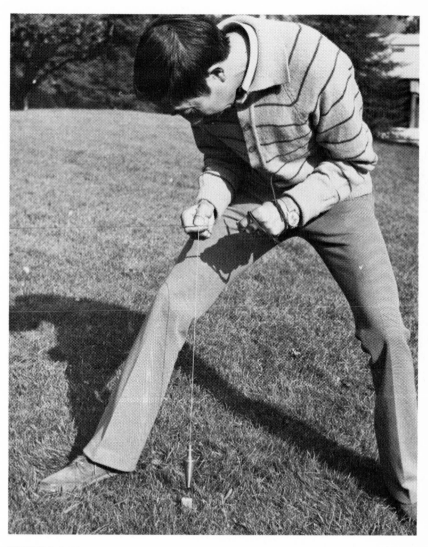

FIGURE 2-4 Plumbing over point.

to the side of the tape, he plumbs the end graduation over the point on the ground marking the start of the line. The tip of the plumb bob should be less than $\frac{1}{8}$ in. (about 3 mm) above the ground point.

The head tapeman applies the tension to be used, either by estimation or by means of a spring balance fastened to the zero or forward end of the tape. At approximately the correct position on the ground, he clears a small area where the taping pin will be set. After again applying the tension, the head tapeman waits for a vocal signal from the rear tapeman, indicating that the latter is on the rear point. When the plumb bob has steadied and its tip is less than $\frac{1}{4}$ in. or about 5 mm from the ground, the head tapeman dips the end of the tape slightly so that the plumb bob touches the ground. Then he, or a third member of the taping party, sets a taping pin or a surveying tack at the tip of the plumb bob to mark the end of the first full tape length, as shown in Fig. 2-5. The pin is set at right angles to the line and inclined at an angle of about 45° with the ground away from the side on which the rear tapeman will stand for the next measurement. The tape is then stretched out again to check the position of the pin. The notekeeper records the distance, 100.00 ft, or 30.000 m, in the field notes. The tape is advanced another tape length, and the entire process is repeated.

If the taping advances generally downhill, the head tapeman checks to see that the tape is horizontal by means of the hand level. If the taping advances generally uphill, the rear tapeman checks for level.

When the end of the line is reached, the distance between the last pin and the point at the end of the line will usually be a fractional part of a tape length. The rear tapeman holds the particular full-foot or decimeter graduation that will bring the subgraduations at the zero end of the tape over the point marking the end of the line. The head tapeman rolls the plumb-bob string along the subgraduations with his thumb until the tip of the plumb bob is directly over the ground point marking the end of the line.

Two types of end graduations of a tape that reads in feet are shown in Fig. 2-6. In view (a) the subgraduations are outside the zero mark, and the fractional part of a foot is added to the full number of feet. Hence the distance is 54 + 0.46 = 54.46 ft. This type is referred to as an *add* tape. In view (b) the subgraduations are between the zero and the 1-ft graduation, and the fractional part of a foot must be subtracted from the full number of feet. So the distance is 54 − 0.28 = 53.72 ft. This type is called a *cut* tape. Because of the variation in the type of end graduations, the rear tapeman must call out the actual foot mark he holds, and both the head tapeman and the notekeeper must agree that the value recorded in the field notes is the correct value.

Add tapes are more convenient to use than cut tapes simply because it is easier to add than to subtract the decimal part of the whole unit. The design of an add metric tape that is divided in decimeters throughout its length is shown in Fig. 2-7. The outside decimeter is further subdivided to centimeters and then to millimeters. In the illustration the rear tapeman holds the 14.7-m mark at the pin, and the head tapeman reads 0.072 m at the end of the line. The distance is thus 14.7 + 0.072 = 14.772 m.

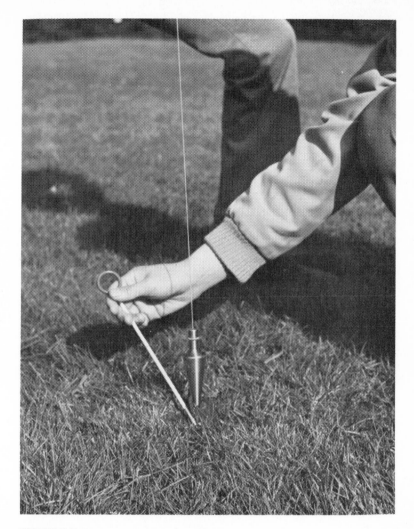

FIGURE 2-5 Setting taping pin to mark forward position of tape.

Distance = 54.46 ft

(a)

Distance = 53.72 ft

(b)

FIGURE 2-6 Graduations at end of foot-graduated tape.

FIGURE 2-7 Graduations at end of metric tape.

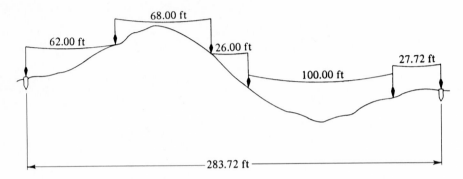

FIGURE 2-8 Breaking tape.

Where the slope is too steep to permit bringing the full length of the tape horizontal, the distance must be measured in partial tape lengths, as shown in Fig. 2-8. It is then necessary to enter a series of distances in the field notes. Some or all of them will be less than a full tape length. For a partial tape length, the head tapeman holds the zero end and the rear tapeman holds a convenient whole foot or decimeter mark that allows the selected length of tape to be horizontal. When the forward pin is set, this partial tape length is recorded in the field notes. The head tapeman then advances with the zero end of the tape, and the rear tapeman again picks up a convenient whole foot or decimeter mark and plumbs it over the pin. Each partial tape length is recorded as it is measured or as the forward pin is set. Fig. 2-9 illustrates the use of a device called a *tape clamp* for holding a tape at any place other than at an end.

If a tape clamp is not available, the rear tapeman must hold the tape in one hand in such a manner that it neither injures his hand nor damages the tape. At the same time he must be able to sustain a tension of between 10 and 20 lb, or between 5 and 10 kg. The technique shown in Fig. 2-10 is a satisfactory solution to this problem. The tape is held between the fleshy portion of the fingers and that of the palm. Enough friction is developed to sustain a tension upward of 25 to 30 lb (10 to 15 kg) without injury or discomfort to the tapeman. He must not turn his hand too sharply, however, otherwise the tape may become kinked.

FIGURE 2-9 Use of tape clamp.

All distances should be taped both forward and backward, to obtain a better value of the length of the line and to detect or avoid mistakes. When the backward measurement is made, the new positions of the pins should be completely independent of their previous positions. This practice eliminates the chance of repeating a mistake.

2-10   Tension

Most steel tapes are correct in length at a temperature of 68°F (20°C) when a tension of 10 to 12 lb or 5 kg is used and the tape is supported throughout the entire length. If this same tension is used when the tape is suspended from the two ends, the horizontal distance between the ends of the tape will be shorter than the nominal length. The amount of the shortening depends on the length and the weight of the tape. A light 100-ft tape weighs about 1 lb. Such a tape, when suspended from the two ends, would be shortened about 0.042 ft under a tension of 10 lb. A heavy 100-ft tape weighs about 3 lb and would be shortened about 0.375 ft under a tension of 10 lb. Some engineers attempt to eliminate this error by increasing the tension used. The tension for the light tape is then increased to about 18 lb. It is practically impossible to eliminate the error in the heavy tape by this method, as the tension would have to be increased to about 50 lb. Generally, less tension is used and a correction is applied to each measured length.

FIGURE 2-10 Method of holding tape when clamp is not available.

2-11   Reduction of Slope Measurements by Vertical Angles

Where slopes are considerable or where high accuracy is required, or where both these conditions occur, taped measurements are made directly along the slopes as stated in Section 2-9. The slope distances are then reduced to the corresponding horizontal distances by computations. To perform these computations, either the vertical angle of the slope measurement or the difference in elevation between the two ends of the tape or the line must be known.

In Fig. 2-11 a transit or theodolite (see Chapter 4) is set up over one end of the line to be measured. The inclination of the telescope is set so that the line of

FIGURE 2-11 Slope taping using transit or theodolite.

sight of the transit will be generally parallel with the slope of the ground. The instrumentman guides the two tapemen to keep the tape on line and on a constant slope. He then reads the vertical angle V or the zenith angle Z (see Section 4-19) to compute the horizontal distance. If the distance measured along the slope is s and the slope makes a vertical angle V with the horizontal, the corrected horizontal distance is then

$$H = s \cos V \tag{2-1}$$

EXAMPLE 2-1   A measurement is made along a line that is inclined by a vertical angle of 3° 22′ as measured using an engineer's transit. The slope measurement is 3236.86 ft. What is the corresponding horizontal distance?

Solution: Using Eq. (2-1),

$$H = 3236.86 \times 0.998274 = 3231.27 \text{ ft} \qquad \blacklozenge$$

EXAMPLE 2-2   A horizontal distance of 745.00 ft is to be established along a line that slopes at a vertical angle of 5° 10′. What slope distance should be measured off?

Solution: Using Eq. (2-1),

$$s = \frac{745.00}{0.99594} = 748.04 \text{ ft} \qquad \blacklozenge$$

The accuracy with which a vertical angle must be measured to reduce a slope distance to the corresponding horizontal distance will now be investigated. Accuracy in distances can be considered in one of two ways: an absolute accuracy and a relative accuracy. If a 10,000-ft distance is measured with an accuracy of a foot, then the implied absolute accuracy is 1 ft. The relative accuracy in this instance is 1 ft divided by 10,000 ft or 1/10,000 or 1 : 10,000. A statement of relative accuracy is preferred because it is independent of the lengths of lines in the survey.

Eq. (2-1) expresses the horizontal distance H in terms of the measured slope distance s and the angle of inclination V. To determine the accuracy with which the vertical angle must be measured to meet a given relative accuracy in the resulting horizontal distance, differentiate Eq. (2-1) with respect to V:

$$dH = -s \sin V \, dV \qquad (2\text{-}2)$$

in which dV is in radians. Disregarding the minus sign, the relative accuracy is

$$\frac{dH}{H} = \frac{s \sin V \, dV}{s \cos V} = \tan V \, dV \qquad (2\text{-}3)$$

The small angle dV can be computed for any desired relative accuracy and a particular slope angle.

EXAMPLE 2-3   To what accuracy must the slope angle of Example 2-1 be measured if the relative accuracy of the horizontal distance is to be 1/25,000?

Solution: By Eq. (2-3),

$$\tan V \, dV = \frac{1}{25,000} \text{ and } dV = \frac{1}{25,000 \tan 3° \, 22'}$$

Then (see Section 1-3)

$$dV'' = \frac{206,265}{25,000 \times 0.05833} = 140'' = 2' \, 20''$$

EXAMPLE 2-4   To what accuracy must the slope angle of Example 2-1 be measured if the horizontal distance is to be accurate to 0.02 ft?

Solution: By Eq. (2-2),

$$s \sin V \, dV = 0.02 \text{ ft} \text{ and } dV'' = \frac{0.02 \times 206,265}{3236.86 \times 0.05873} = 21.7''$$

The same results would be obtained by defining the relative accuracy as 0.02/3236.86 and then using Eq. (2-3) as shown in Example 2-3.

Eq. (2-3) can be used to compute a curve showing the necessary vertical angle accuracy required for different slope angles if the relative accuracy is known. Figure 2-12 shows three such curves for three different relative accuracies. An examination of these curves shows that 1′ accuracy in measuring vertical angles is adequate for 1 : 10,000 relative accuracy except for angles of more than 20°, which is fairly infrequent. On the other hand, to meet the 1 : 100,000 criterion, the accuracy of vertical angles must be better than 10″ for slopes of more than 10°. An investigation of the accuracy required in measuring slope angles becomes important when one considers the very high accuracies of distance measurement using EDMs (see Section 2-24). Poorly measured vertical angles degrade the overall accuracy of the distance measurements.

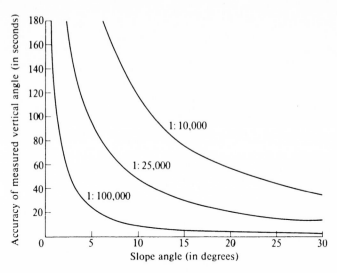

FIGURE 2-12 Accuracy required in measuring vertical angles when slope taping.

2-12　Reduction of Slope Measurements by Difference in Elevation

Measurements made on the slope can be reduced to their corresponding horizontal distances if the differences in elevation between the two ends of the tape have been measured by leveling (see Chapter 3). One method is to set a hub (a 1×1 in. or 2×2 in. stake driven flush with the ground) or a stake at each point marking the end of the tape after the taping pin has been pulled, and then to run a line of levels over these points to compute successive differences in elevation between the points. This technique assumes that the two tapemen hold their ends of the tape the same height above the ground, so that the measured difference in elevation is in fact the difference in elevation between the two ends of the tape.

In Fig. 2-13 s is the tape length along the sloping line, h is the difference in elevation between the two ends of the tape, and H is the corresponding horizontal distance to be determined. In this right-angled triangle

$$H = (s^2 - h^2)^{1/2} \tag{2-4}$$

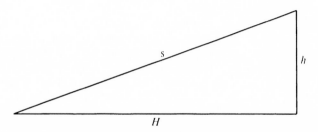

FIGURE 2-13 Reduction of slope distance using difference in elevation.

In precise measurement of baselines (see Section 9-4) or in high-order control surveys, in which the taping tripod described in Section 2-9 is employed, a level party either immediately precedes or follows the taping party to measure the elevations of the successive positions of the tripod tops before the tripods are moved forward along the line.

If a long sloping line is to be measured using the plumb bobs and taping pins, and if the ground slope is fairly uniform, the two tapemen can then measure parallel to the ground with the tape by holding the two ends an equal distance above the ground for each tape length. The difference in elevation between the two ends of the line is then measured using the level. However, if the slope of the line is irregular, hubs must be set at each tape length as discussed above. Each tape length is then reduced separately.

EXAMPLE 2-5   A line measures 1446.25 ft along a constant slope. The difference in elevation between the two ends of the line is 57.24 ft. What is the horizontal length of the line?

Solution: By Eq. (2-4),

$$H = (1446.25^2 - 57.24^2)^{\frac{1}{2}} = 1445.12 \text{ ft}$$

EXAMPLE 2-6   What slope distance must be laid out along a line that rises 6 ft/100 ft to establish a horizontal distance of 750 ft?

Solution: The difference in elevation between the two ends of the line is $6 \times 7.50 = 45.00$ ft. Then by Eq. (2-4), $s = (H^2 + h^2)^{1/2}$ or

$$s = (750^2 + 45^2)^{1/2} = 751.35 \text{ ft}$$

The difference C between the slope distance and the horizontal distance can be determined by expanding the right side of Eq. (2-4) by the binomial expansion to give

$$H = s - \frac{h^2}{2s} - \frac{h^4}{8s^3} - \cdots \tag{2-5}$$

Thus neglecting the last term in the above expansion gives

$$s - H = C = \frac{h^2}{2s} \tag{2-6}$$

An analysis of the accuracy required in determining the difference in elevation between the two ends of a tape in slope measurements can be made by differentiation of Eq. (2-6) with respect to h:

$$dC = \frac{h \, dh}{s}$$

If both sides are divided by s, the relative accuracy expression is obtained:

$$\frac{dC}{s} = \frac{h \, dh}{s^2} \tag{2-7}$$

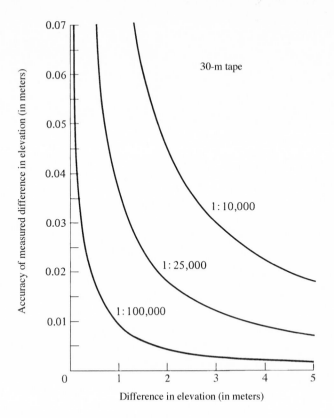

FIGURE 2-14 Accuracy required in measuring difference in elevation between two ends of a 30-m tape when slope taping.

EXAMPLE 2-7   A baseline is measured with a 50-m tape. The error in leveling over the tops of the taping posts is not to contribute more than one part in 100,000 error to the relative accuracy of the measurement. The largest difference in elevation between any two successive posts is 3.60 m. With what accuracy must this difference in elevation be determined?

Solution: By Eq. (2-7), the value of dC/s must be $1/100,000$ which in turn is $(h\,dh)/s^2$:

$$dh = \frac{s^2}{100,000h} = \frac{50^2}{100,000 \times 3.60} = 0.007 \text{ m}$$

Eq. (2-7) can be used to compute curves for a given tape length that show the necessary accuracy in determining elevations for various differences of elevation and for different accuracy ratios. The curves shown in Fig. 2-14 are for a 30-m tape at three different accuracy ratios.

2-13   Systematic Errors in Taping

The principal systematic errors in linear measurements made with a tape are (1) incorrect length of tape, (2) tape not horizontal, (3) fluctuations in the temperature of the tape, (4) incorrect tension or pull, (5) sag in the tape, (6) incorrect alignment, and (7) tape not straight.

2-14   Incorrect Length of Tape

At the time of its manufacture, a tape is graduated when under a tension of about 10 lb or 4.5 kg. A steel tape will maintain a constant length under a considerable amount of handling and abuse. This is not true, however, of an invar tape, which must be handled with great care. In either case, if a tape is compared with a standard length under specific conditions of temperature, tension, and method of support, the distance between the two end graduations will seldom equal the nominal length indicated by the graduation numbers. The correction to be applied to any measurement made with the tape to account for this discrepancy is called the *absolute correction C_a*, and is given by

$$C_a = \text{true length} - \text{nominal length} \qquad (2\text{-}8)$$

The true length is the value determined by calibration under specific conditions. The calibration or standard tension will range anywhere from 10 to 30 lb or 5 to 15 kg and is specified by the user of the tape. The calibration comparison can be made with the tape supported throughout its length, or supported only at the two ends, or supported at the two ends and at one or more intermediate points.

The absolute error $- C_a$ is usually assumed to be distributed uniformly throughout the length of the tape. Thus the absolute correction in a measured distance is directly proportional to the number and fractional parts of the tape used in making the measurement.

2-15   Tape Not Horizontal

If the tape is assumed to be horizontal but actually is inclined, an error is introduced. The amount of this error C can be computed from Eq. (2-6) or may be taken as $h^2/2s$. If one end of a 100-ft tape is 1.41 ft higher or lower than the other, the error will amount to 0.01 ft. It should be noted that the error is proportional to the square of the vertical distance. When one end is 2.82 ft above or below the other, the error increases to 0.04 ft.

Errors from this source are cumulative and may be considerable when measuring over hilly ground. The error can be kept at a minimum by using a hand level to determine when the tape is horizontal.

2-16 Changes in Temperature

The coefficient of thermal expansion of steel is about $0.0000065/°F$ or $1.15 \times 10^{-5}/°C$. As a difference of $15°F$ in temperature will cause a change in length of nearly 0.01 ft in a 100-ft tape, a measurement made with a tape on an extremely hot day in the summer cannot be expected to agree with one made during zero weather. A difference of $3°C$ changes the length of a 30-m tape by 1 mm.

If the temperature remains constant and is different from that at which the tape was standardized, the resulting error will be cumulative and in direct proportion to the number of tape lengths measured. When the field temperature differs considerably from the standard, the temperature should be determined and a correction should be applied to the measured length.

The amount of the correction for temperature is determined by the formula

$$C_t = L\alpha(T - T_s) \tag{2-9}$$

in which C_t is the correction to be applied, in feet or meters; L is the length of the tape actually used, in feet or meters; α is the coefficient of thermal expansion; T is the temperature at which the measurement is made; T_s is the temperature at which the tape was standardized.

Since the temperature of the tape may be considerably different from that of the surrounding air on a bright day, important measurements should be made on a cloudy day, or early in the morning, or late in the afternoon when the temperature of the air and the tape are similar.

The advantage of an invar tape over the steel tape is its very low coefficient of thermal expansion, which is $0.0000002/°F$ or $3.60 \times 10^{-7}/°C$. Because the effective temperature of a tape is very difficult to determine, even with the best thermometers, the invar tape will give better results than steel and should be used in surveys of high order. Even considering this advantage, however, the best results with an invar tape will be obtained on an overcast day or even at night.

The tape thermometer attached to the tape is fairly responsive to temperature changes. If taping is performed with the tape supported on the ground, the tape either takes on or gives off heat from or to the supporting material. Since different portions of the tape can be supported by different types of material—for example, blacktop pavement, grassy sod, concrete—the only place where the observed temperature is reasonably correct is that portion of the tape to which the thermometer is fastened. If the tape is suspended in air during the measurements, it reaches a state of thermal equilibrium in which the tape temperature is uniform throughout its length (except for spotty shaded areas). If the tape has been lying on the ground and then picked up for stretching, whoever reads the thermometer must wait for a minute or so for the tape to come to thermal equilibrium. Otherwise the effect of the tape being in contact with the ground will cause a considerable error in the observed temperature, sometimes as much as $30°F$ $(16°C)$.

2-17   Incorrect Tension

If a tape is used in the field under a tension different from the standard tension used in calibration, the tape will change length a slight amount according to the relationship between stress and strain. The amount to be added to or subtracted from the measured length to account for the difference in tension is a function of the measured length, the field tension, the standard tension, the cross-sectional area of the tape, and the modulus of elasticity of the material of which the tape is made. This is referred to as the *correction for pull* or tension and is given by

$$C_p = \frac{(P - P_s)L}{AE} \tag{2-10}$$

in which C_p is the correction per tape length, in feet or meters; P is the tension applied, in pounds or kilograms; P_s is the standard tension, in pounds or kilograms; L is the length in feet or meters; A is the cross-sectional area, in square inches or square centimeters; E is the modulus of elasticity of the steel. The modulus of elasticity of steel is about 28,000,000 to 30,000,000 psi (pounds per square inch) or 2,100,000 kg/cm².

A light tape, weighting about 1 lb, will have a cross-sectional area of about 0.0030 in². An increase of 10 lb in tension will stretch a 100-ft tape about 0.011 ft. The same increase with a heavy tape, weighing about 3 lb, will produce a change in length of but 0.004 ft.

On careful work, a spring balance is used for maintaining a constant tension. If the spring balance is not used, errors from incorrect tension will tend to be compensating, since the tensions applied to the tape may be either greater or less than the standard tension. The tendency, however, is usually to underpull rather than to exceed the standard tension.

2-18   Sag

If a tape has been standardized when supported throughout its entire length, a correction must be applied to every measurement that is made with the tape suspended from the two ends, or else the tension must be increased. A tape suspended in this way will take the form of a catenary, and the horizontal distance between the two ends will be less than when the tape is supported throughout its entire length. If the amount of the sag is such that the center of a 100-ft tape is about $7\frac{3}{8}$ in. below the two ends, the measured distance has been shortened by 0.01 ft. The correction for sag is given by the formula

$$C_s = n\frac{w^2 l^3}{24P^2} \tag{2-11}$$

in which C_s is the correction for sag, in feet or meters; n is the number of unsupported lengths; w is the weight per foot of tape, in pounds or kilograms per meter; l is the unsupported length, in feet or meters; P is the tension applied, in pounds or kilograms.

When a tape is suspended from the two ends and subjected to the standardized tension for the tape supported throughout its length, the observed distance is always greater than the true distance, and the error due to sag is constant.

If the total weight of the tape is W and the total length is L, then Eq. (2-11) can be expressed as

$$C_s = \frac{W^2 L}{24 P^2} \tag{2-12}$$

in which C_s is then the difference in distance between the ends of the tape when supported throughout and when supported only at the two ends.

Error from sag can be eliminated by applying the necessary corrections to the observed distances or, in the case of a light tape, by increasing the tension sufficiently to compensate for the effect of sag. When a tape having a total weight of W lb or kg is supported at the two ends, this required tension P_n, which is called the *normal tension* for the tape, may be found by solving the following formula by trial:

$$P_n = \frac{0.204 W \sqrt{AE}}{\sqrt{P_n - P_s}} \tag{2-13}$$

For a light tape the tension would be increased from 10 to about 18 lb. For a heavy tape the normal tension would be about 50 lb, which would be difficult to apply in practice.

EXAMPLE 2-8   A 30-m tape weighing 0.900 kg has a cross-sectional area of 0.0485 cm². The tape measures 30.000 m when supported throughout under a tension of 5 kg. The modulus of elasticity is $2.1 \times 10^6 \text{kg/cm}^2$. What tension is required to make the tape measure 30 m when supported only at the two ends?

Solution: Using Eq. (2-13), try 10 kg as a first approximation.

$$10 = \frac{0.204 \times 0.900 \sqrt{0.0485 \times 2,100,000}}{\sqrt{10 - 5}} = \frac{58.594}{\sqrt{5}} = 26.204 \text{ kg}$$

Try 18 kg.

$$18 = \frac{58.594}{\sqrt{13}} = 16.25 \text{ kg}$$

Try 17 kg.

$$17 = \frac{58.594}{\sqrt{12}} = 16.91 \text{ kg}$$

Try 16.95 kg.

$$16.95 = \frac{58.594}{\sqrt{11.95}} = 16.95 \text{ kg (checks)}$$

For extremely accurate work the tape is supported at enough points to render the error due to sag negligible; or the tape is standardized when it is supported in the manner in which it is to be used in the field.

2-19   Incorrect Alignment

If a field measurement comprises more than one tape length and the points marking the ends of the various lengths are not along a straight line, the measured length will be too great. The amount of error for any given tape length may be computed from Eq. (2-6), where h is the amount one end of the tape is off line. To produce an error of 0.01 ft in a 100-ft measurement, one end of the tape would have to be 1.41 ft off line. Under normal circumstances the rear tapeman should be able to keep the forward tapeman much closer to the true line than this, and the error from this source should be practically negligible. For measurements of high precision the tapemen can be kept on line with a theodolite.

2-20   Tape Not Straight

If the tape is not stretched straight, as when it is being bent either horizontally or vertically around trees or bushes or is blown by a strong wind, the measured length will be too great. The error is the least when the obstruction is near the middle of the tape. If the middle point of a 100-ft tape is 0.71 ft or about $8\frac{1}{2}$ in. off line, the resulting error in length is 0.01 ft. Errors from this source are cumulative.

2-21   Applying Corrections to Tape Measurements

The following examples illustrate the application of corrections to tape measurements.

EXAMPLE 2-9   A steel tape is standardized at 68°F under a tension of 15 lb when supported throughout its entire length, and the distance between the zero mark and the 100-ft mark is 99.98 ft. The tape weighs 0.013 lb/ft and has a cross-sectional area of 0.0040 in.² In the field this tape is used under a 15-lb tension and is supported at the ends only, and the temperature of the tape is recorded as 88°F throughout the measurement. A series of distances are measured with the tape held horizontal, and the observed distances are recorded as 100.00, 100.00, 100.00, and 57.22 ft. What is the total actual or true distance after corrections for systematic errors have been applied?

Solution: The total observed distance is 357.22 ft. The errors for incorrect length, for temperature, and for incorrect tension are all in direct proportion to the length of the measurement. Therefore the total correction for incorrect length is

$$-0.02 \times 3.57 = -0.071 \text{ ft}$$

and the total correction for temperature is

$$C_t = 357.22 \times 0.0000065 \times (88-68) = +0.046 \text{ ft}$$

The correction for incorrect tension is zero, since the standard pull was used when the measurements were made.

For computing the correction for sag, the three 100-ft measurements are treated together and the 57-ft measurement is treated separately. For the three 100-ft measurements, by Eq. (2-11),

$$C_s = 3 \times \frac{0.013^2 \times 100^3}{24 \times 15^2} = -0.094 \text{ ft}$$

For the 57-ft measurement,

$$C_s = \frac{0.013^2 \times 57.2^3}{24 \times 15^2} = -0.006 \text{ ft}$$

The total sag correction is -0.100 ft.

The total correction to be applied to the measured distance is $-0.071 + 0.046 - 0.100 = -0.125$ ft. To the nearest hundredth, it is -0.13 ft. So the corrected distance is 357.09 ft. ◆

EXAMPLE 2-10   A steel tape is standardized at 68°F under a tension of 15 lb when supported at the two ends only, and the distance between the zero mark and the 100-ft mark is 99.972 ft. The total weight of the tape is 2.80 lb, and its cross-sectional area is 0.0088 in.^{2} For steel E is assumed to be 30,000,000 psi. What is the distance between the zero and 100-ft marks when the tape is supported at the two ends only, the tension is 25 lb, and the standard temperature prevails?

Solution: First determine the distance when the tape is supported throughout and the tension is 15 lb. Then determine the distance for support throughout and a 25-lb pull. Finally, determine the distance when the tape is supported only at the two ends under a 25-lb pull. When the tape is supported throughout under a 15-lb pull, the distance between the end graduations will be greater than that with the tape supported only at the ends by the following amount:

$$C_s = \frac{0.028^2 \times 100^3}{24 \times 15^2} = 0.145 \text{ ft}$$

The distance for support throughout and a 15-lb pull is

$$99.972 + 0.145 = 100.117 \text{ ft}$$

When the tension is increased to 25 lb with the tape supported throughout, the additional increase in length is

$$C_p = \frac{(25 - 15) \times 100}{0.0088 \times 30,000,000} = 0.004 \text{ ft}$$

The distance for support throughout and a 25-lb pull is 100.121 ft.

Finally, when the support is taken away and the tape is supported only at the ends under a 25-lb pull, the distance will be less than that for support throughout. The correction is

$$C_s = \frac{0.028^2 \times 100^3}{24 \times 25^2} = 0.052 \text{ ft}$$

The distance between end graduations when the tape is supported at the ends only under a tension of 25 lb is 100.069 ft. ◆

EXAMPLE 2-11   A steel tape with a cross-sectional area of 0.039 cm^2 and weighing 0.909 kg has a length of 30.009 m between the zero and the 30-m marks when supported throughout at 20°C and subject to a tension of 5.5 kg. This tape is used to measure a distance along a uniform 4% grade and is supported throughout during the measurement. The tension applied is 12 kg. The temperature of the tape is 2°C. The measured slope distance is recorded as 381.615 m. What is the corrected horizontal distance?

Solution: All errors affecting this measurement are in direct proportion to the distance considered since there is no sag correction. For each tape length,

$$C_a = 30.009 - 30.000 = + 0.0090 \text{ m}$$
$$C_t = 30 \times 1.15 \times 10^{-5} (2 - 20) = -0.0062 \text{ m}$$
$$C_p = \frac{(12 - 5.5) \times 30}{0.039 \times 2.1 \times 10^6} = +0.0024 \text{ m}$$

The length of the tape as used in the field is thus 30.0000 + 0.0090 − 0.0062 + 0.0024 = 30.0052 m. The correct slope distance is thus

$$s = 381.615 + \frac{381.615}{30} \times 0.0052 = 381.681 \text{ m}$$

The slope angle is tan^{-1} 0.04 = 2° 17′, and the correct horizontal distance is 381.681 cos 2° 17′ = 381.376 m. ◆

2-22   Random Taping Errors

Random errors are introduced into taping measurements from several causes. Some of these are the following: (1) error in determining temperature of tape; (2) failure to apply the proper tension; (3) wind deflecting the plumb bob; (4) taping pin not set exactly where the plumb bob touched the ground; (5) inability of the tapeman to steady the plumb bob; (6) inability of the observer to estimate the last place in reading between graduations. Random errors and their effects on measurements will be discussed more fully in Chapter 5.

2-23   Specifications for Linear Measurements

In the preceding sections the various sources of error in linear measurements have been discussed. The principal sources of error in any tape length are those due to the incorrect length of the tape, variations in temperature and tension, incorrect determination of slope, and inaccurate marking of the tape end. These errors may either increase or decrease the length of the measurement. The total random error in any tape length may be assumed equal to the square root of the sum of the squares of the maximum individual errors.

Specifications for any desired accuracy can be prepared from a knowledge of these errors. Thus, if the actual error is not to exceed 1/5000, the following limits for the various errors in 100 ft would suffice:

- Length of tape to be known within 0.01 ft: Maximum error, ±0.01 ft.
- Temperature to be known within 10°F: Maximum error, ±0.006 ft.
- Tension to be known within 2 lb: Maximum error, ±0.001 ft.
- For slopes averaging 5%, vertical distance to be known within 0.2 ft: maximum error, ±0.01 ft.
- Tape lengths to be marked within 0.01 ft: maximum error, ±0.01 ft.
- Alignment errors to be eliminated by keeping tape on line by transit.
- Effect of sag to be eliminated by having tape standardized when suspended from the two ends: if the tension is known within 2 lb, the maximum sag error is about ±0.006 ft.

2-24   Electronic Distance Measuring Instruments (EDMs)

The electronic distance-measuring instruments referred to as EDMs in this text have practically replaced the tape for measuring all but relatively short distances. The instrument sends out a beam of light or high-frequency microwaves from one end of a line to be measured, and directs it toward the far end of the line. A reflector or transmitter-receiver at the far end reflects the light or microwaves back to the instrument where they are analyzed electronically to give the distance between the two points. The first such instrument was the geodimeter mentioned in Section 2-5. It became available to the general surveying and engineering profession in the early 1950s. The early geodimeter as well as all subsequent models employs a modulated light beam for determining distance. It was followed in the late 1950s by the tellurometer, an instrument that employs modulated microwaves for determining distances. Instruments similar to the tellurometer were introduced in subsequent years, and these along with the geodimeter became the standard instruments for measuring over long distances. The advantage of the microwave instruments is their operability in fog or moderate rain, day or night, as well as their generally longer range.

The development and perfection of small light-emitting diodes in the mid-1960s as well as a general miniaturization of electronics using solid-state components caused a revolution in the design of EDMs. This second generation of instruments is much more portable, takes less power to operate, and is much simpler to operate and read. However, these EDMs do not have the range that the earlier ones have.

A third generation of EDMs employing highly coherent laser light has been brought to perfection in recent years. This type of instrument has the distinct advantage of long-range, low-power requirements, and fairly good portability as well as ease of operation and readout.

Modern EDMs are fully automatic to such an extent that, after the instrument and its transponder (a light reflector or a microwave receiver-transmitter) have been set over the two ends of the line to be measured, the operator need only depress a button, and the slope distance is automatically displayed. More complete EDMs also have the capability of measuring horizontal and vertical or zenith angles as well as the slope distance (see Section 4-24). These instruments, referred to as *total station* instruments, display horizontal and vertical distances between the ends of the line, and horizontal and vertical or zenith angles to a survey point. These displayed outputs can be recorded on magnetic tape or stored in electronic field books for further calculations in a computer.

Because of the proliferation of EDMs available to the surveyor or engineer, no attempt is made in this textbook to catalog or describe all of them. Only the principles on which these instruments operate will be presented. The student is urged to consult the readings at the end of this chapter and to obtain descriptive brochures from instrument manufacturers to appreciate the wide variety of EDMs.

The classification of instruments can be made by their range capabilities, although this is somewhat arbitrary. A short-range EDM is one that is capable of measuring distances up to about 2 miles. This is not to say that all short-range instruments have a 2-mile range, because the upper range of some EDMs is only 1000 to 1500 m. The very short-range instruments, however, incorporate features such as automatic reading, light weight, very low power requirements, and adaptability to angle-measuring surveying instruments.

The instruments shown in Figs. 2-15 through 2-19 are all classed as short to intermediate range. The range can be extended by adding reflectors to the far end of the line to get a strong enough return signal at the longer distances. The light carrier for all of these instruments is infrared light in the 900- to 930-nm wavelength region. Note that this is out of the visible portion of the spectrum.

The instrument shown in Fig. 2-20 uses a laser light beam as the carrier. This is a typical long-range instrument, and under favorable atmospheric conditions can measure lines 40 miles long.

2-25   Measurement Principles of EDM Using Light Waves

In all EDMs using tungsten, mercury, laser, or infrared light as the carrier, a continuous beam of light is generated in the instrument. Before it enters the aiming optics and is directed to a reflector, such as shown in Fig. 2-21, that is placed at the other end of the line to be measured, this continuous beam is intensity modulated at a very high frequency. The modulation, in effect, chops the beam up into wavelengths that are a direct function of the modulating frequency. This wavelength is given by

$$\lambda = \frac{V_a}{f} \qquad (2\text{-}14)$$

FIGURE 2-15 Lietz RED mini 2 distance meter mounted atop a theodolite. Courtesy of Sokkia Corp. (Formerly Lietz Co.)

FIGURE 2-16 Topcon DM-S2 electronic distance meter. Courtesy of Topcon. Instrument Corporation of America.

in which λ is the wavelength of modulation, in meters; V_a is the velocity of the light through the atmosphere, in meters per second; f is the modulating frequency, in hertz. The value of V_a is a function of air temperature, pressure, and partial pressure of water vapor, as discussed in Section 2-27.

The intensity of the modulated light varies from zero at the beginning of each wavelength to a maximum at 90°, back to zero at 180°, to a second maximum at 270°, and back to zero at 360°. The distance between every other zero point is thus one full wavelength. The EDM therefore provides a measuring tape of light whose length is equal to the wavelength of the modulated light. For example, if the modulating frequency is 10 MHz and the velocity of the light is approximately 300,000 km/s, the modulated wavelength is about 30 m, or about 100 ft.

FIGURE 2-17 Zeiss Eldi 10 electronic distance meter attached to an electronic theodolite. Courtesy of Carl Zeiss, Inc.

In Fig. 2-22 the EDM is located to the left, at one end of a line to be measured, and the reflector R to the right, occupying the other end of the line. The reflector is a corner of a cube of glass in which the sides of the cube are perpendicular to one another within very close tolerances. This perpendicularity causes incoming light to be reflected internally and to emerge parallel with itself. The corner cube thus constitutes a retroreflector. An integral number of wavelengths plus a partial distance designated d make up the total distance from the EDM to the reflector back to the EDM. Note that if the reflector (or the EDM) were to be moved either forward or backward along the line by a distance of one-half wavelength, or any number of half wavelengths, the value of the partial distance d would be the same in each case. This partial distance is measured in the instrument by some type of a phase meter. The desired distance D between the two ends of the line is then given by

$$D = \tfrac{1}{2}(n\lambda + d) \tag{2-15}$$

in which n is the integral number of wavelengths in the double distance. One way in which this equation could be solved would be to have prior knowledge of the length of the double path to the nearest half wavelength, which requires that the length of the line be known to the nearest quarter wavelength. Since this is not

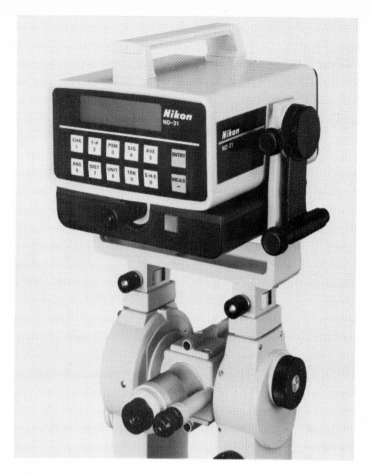

FIGURE 2-18 Nikon ND-31 electronic distance meter. Courtesy of Nikon, Inc.

practical, the ambiguity of n can be resolved by employing the multiple-frequency technique. If the measurement is made at one known frequency and it is then repeated using a slightly different frequency, two different values of d will be read on the phase meter. Knowing the two values of the wavelengths, two equations given by Eq. (2-15) can now be solved simultaneously to give the value of the unknown n and thus the desired distance D.

The multiple-frequency technique for resolving ambiguity is incorporated directly into the modern EDMs. One such system in common use is the decade-modulation technique described as follows. Assume that a modulation frequency of 15 MHz is set up in the instrument, resulting in a half wavelength of 10 m. Let a full sweep of the phase meter represent this 10-m distance. The phase-meter reading then gives the unit meter and decimal part of the meter in the measured distance from 0 to 9.999 m. In a distance of, say, 3485.276 m this frequency would give the 5.276 part. Switching down to 1.5 MHz, the half wavelength is now 100 m, which is resolved by the phase meter to give the tens of meters—in this instance 80 (8 tens). The next frequency is then 0.15 MHz, which in conjunction with the

FIGURE 2-19 Geodimeter 220 electronic distance meter. Courtesy of Geotronics of North America, Inc.

phase meter, gives the hundreds of meters, which in this instance is 400 (4 hundreds). Finally, a 15-kHz frequency will give the number of thousand meters in the distance, which in this instance is 3000 (3 thousands).

In most modern EDMs the different frequencies are sequentially changed in the instrument itself without intervention by the operator. The slope distance is then automatically displayed.

2-26   Measurement Principles of EDM Using Microwaves

The microwave instruments generate superhigh-frequency (SHF) or extremely high-frequency (EHF) electromagnetic waves in the range of 3 to 35 GHz as the carrier waves. These, in turn, are modulated to frequencies varying from 10 to 75 MHz, depending on the type of instrument. The length of the modulated wave is given by

FIGURE 2-20 K&E Rangemaster.

$$\lambda = \frac{V_r}{f} \qquad (2\text{-}16)$$

in which λ is the wavelength of modulation, in meters or feet; V_r is the velocity of the microwave through the atmosphere, in meters or feet per second; f is the modulating frequency, in hertz. The value of V_r depends on the prevailing temperature, atmospheric pressure, and the partial pressure of water vapor in the atmosphere, as discussed in Section 2-27. Two similar instruments are involved in measurement with the microwave instruments, one at each end of the line to be measured. These are referred to as the master and the remote. Observations are made at the master station, whereas the remote instrument, which must also have an operator, serves like a reflector in the light-generating EDMs.

FIGURE 2-21 Single and triple reflectors.

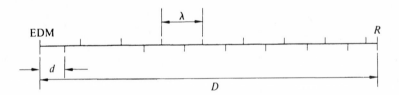

FIGURE 2-22 Measurement principle of EDM.

The operator at the master station selects a modulating frequency at which the microwaves are transmitted toward the remote instrument. He indicates to the remote operator by voice communication, which is built into the instrument, the transmitting frequency being used. The remote operator sets his instrument to correspond to that frequency. The signal is received by the remote instrument and is retransmitted back to the master with no delay. A phase meter at the master gives the total amount by which the outgoing and incoming signals are out of phase. This, in effect, gives the fractional or decimal part of the wavelength by which the double path from master to remote to master departs from an integral number of wavelengths. It is equivalent to distance d in Fig. 2-22.

If either the master or the remote instrument were to be moved toward or away from one another by one-half of the modulated wavelength, the phase meter would go through one complete cycle and display the same value as before. Thus the ambiguity that exists in the light-wave instruments is also present in the microwave instruments.

The technique for resolving the ambiguity in the number of whole wavelengths contained in the double distance is generally the same as that used in the light-generating EDM, that is, by the multiple-frequency technique.

2-27   Effect of Atmospheric Conditions on Wave Velocity

The conditions of the atmosphere that have an effect on the velocity of propagation of light and microwaves are the air temperature, the atmospheric pressure, and the relative humidity. The temperature and relative humidity, in turn, define the vapor pressure in the atmosphere. A knowledge of these conditions allows a determination of the *refractive index* of the air, which must be known to compute the velocity of light or microwaves under given meteorological conditions.

For light waves the refractive index n_g of standard air is given by

$$n_g = 1 + \left(287.604 + \frac{4.8864}{\lambda_c^2} + \frac{0.068}{\lambda_c^4}\right)10^{-6} \tag{2-17}$$

in which λ_c is the wavelength of the light carrier, in micrometers. For the kinds of light used in EDMs, the values of λ_c are as follows:

Carrier	$\lambda_c(\mu m)$
Mercury vapor	0.5500
Incandescent	0.5650
Red laser	0.6328
Infrared	0.900—0.930

The refractive index n_a for light waves at conditions departing from the standard air can then be computed by

$$n_a = 1 + \frac{0.359474(n_g - 1)p}{273.2 + t} - \frac{1.5026e \times 10^{-5}}{273.2 + t} \tag{2-18}$$

in which p is the atmospheric pressure, in millimeters of mercury (torr); t is the air temperature, in degrees Celsius; e is the vapor pressure, in torr.

For practically all distance measurements with light, the last term of Eq. (2-18) involving vapor pressures can be neglected since relative humidity has very little effect on light waves.

The velocity of the light wave in air, V_a, is related to the velocity of light in a vacuum by

$$V_a = \frac{c}{n_a} \qquad (2\text{-}19)$$

The value for c is 299,792.458 km/s.

EXAMPLE 2-12   The modulating frequency of a red laser beam is 24 MHz. This beam travels through the atmosphere, the temperature of which is 26°C and the atmospheric pressure 759 torr. What is the modulated wavelength of the light?

Solution: By Eq. (2-17), the refractive index of standard air for the laser carrier is

$$n_g = 1 + \left(287.604 + \frac{4.8864}{0.6328^2} + \frac{0.068}{0.6328^4}\right)10^{-6} = 1.0003002$$

By Eq. (2-18), the refractive index of the air under the given atmospheric conditions, neglecting the last term, is

$$n_a = 1 + \frac{0.359474(1.0003002 - 1) \times 759}{273.2 + 26} = 1.0002738$$

The velocity of laser light through this atmosphere is, by (Eq. 2-19),

$$V_a = \frac{299,792.5}{1.0002738} = 299,710.4 \text{ km/s}$$

Finally, the modulated wavelength is, by Eq. (2-14),

$$\lambda = \frac{299,710.4}{24 \times 10^6} = 0.01248793 \text{ km} = 12.4879 \text{ m} \qquad \blacklozenge$$

The effect of water vapor pressure, which can be neglected when working with light, is quite large when using microwave EDMs. Consequently, the relative humidity must be very carefully determined in the field at the time of measurement. A high-quality psychrometer that gives wet and dry bulb thermometer readings must be employed for the determination of vapor pressure.

The refractive index of microwaves n_r is given by

$$(n_r - 1)10^6 = \frac{103.49}{273.2 + t}(p - e) + \frac{86.26}{273.2 + t}\left(1 + \frac{5748}{273.2 + t}\right)e \qquad (2\text{-}20)$$

where p is the atmospheric pressure, in millimeters of mercury; e is the vapor pressure, in millimeters of mercury; t is the air temperature (dry bulb), in degrees Celsius.

The velocity of propagation of microwaves V_r through the atmosphere is then given by

$$V_r = \frac{c}{n_r} \qquad (2\text{-}21)$$

and the modulated wavelength is given by Eq. (2-16).

EXAMPLE 2-13   What is the wavelength in meters of microwaves modulated at a frequency of 10 MHz if the atmospheric pressure is 643 torr, temperature is 23.9°C, and vapor pressure is 3.5 torr?

Solution: By Eq. (2-20),

$$(n_r - 1)10^6 = \frac{103.49}{297.1}(643.0 - 3.5) + \frac{86.26}{297.1}\left(1 + \frac{5748}{297.1}\right)3.5 = 243.4$$

and $n_r = 1.0002434$. By Eq. (2-21),

$$V_r = \frac{299,792.5}{1.0002434} = 299,719.5 \text{ km/s}$$

Finally, by Eq. (2-16),

$$\lambda = \frac{299,719.5}{10 \times 10^6} = 0.02997195 \text{ km} = 29.97195 \text{ m}$$

The equations given for determining refractive index are presented in different forms in various articles and publications, and there are slight but insignificant variations in results obtained by the different expressions. These particular forms are presented here because they allow the reader to appreciate the relative significance of the temperature, pressure, and vapor pressure on both light and microwaves.

The effects of atmospheric conditions are handled in various ways by different EDM systems. The corrections are of little consequence for short distances using the light-generating EDMs. For longer lines an error of 10°C in the effective temperature of the beam path will introduce a relative error of 10 ppm, and an error of 25 mmHg in measuring atmospheric pressure will also introduce a relative error of 10 ppm. Corrections are either computed based on meteorological data determined at the time of measurement, or else the instrument circuitry is modified to take the atmospheric conditions into account by keying into the instrument a parts per million (ppm) correction.

In the case of the microwave EDMs, the partial pressure of water vapor obtained by the wet and dry bulb thermometer readings must be determined quite accurately. A 2-mm error in vapor pressure or a 1.5°C error in the difference between the dry and wet bulb temperatures will produce a relative error of about 10 ppm at normal temperatures. This relative error increases with an increase in air temperature. Assuming that the meteorological conditions have been measured satisfactorily, corrections to measured distances are easily made with the aid of the various charts, tables, or nomographs that are furnished with the instruments.

If an aneroid barometer (see Section 3-6) is used to determine atmospheric pressure, the elevation reading in feet or meters must be converted to the appropriate value of pressure in inches or millimeters of mercury. This conversion is made using charts furnished by the instrument maker. Fig. 2-23 shows a simplified chart that can be used to convert feet of elevation to inches of mercury. Altimeter dials used in surveying are usually graduated so that a reading of, say, 1000 ft will be obtained at sea level under standard atmospheric conditions. This is done to eliminate negative altimeter readings at or near sea level when the atmospheric pressure increases above standard. It is necessary to know this zero reading because it must be subtracted from the altimeter reading before using a chart such as that shown in Fig. 2-23.

2-28   Instrumental Errors in EDM

If a modern EDM is properly tuned, there will be very few sources of instrumental errors that need correcting. One error, referred to as the reflector constant, is caused by not having the effective center of the reflector plumbed over the far end of the line. This type of error is shown in Fig. 2-24 for a corner cube reflector. The distance through which the light travels in the glass cube during retroflection is $a + b + c$, which in turn is equal to $2t$. The distance t is measured from the face of the reflector to the corner of the glass cube. The equivalent air distance through which the light travels is $1.517 \times 2t$ on account of the refractive index of the glass. The effective corner of the cube is at R and represents the end of the line. If the plumb line passes vertically in front of point R, then an error c_R would be introduced into the measured length of the line. Then distance c_R would have to be subtracted in this instance. The reflector constant is a relatively small value, most commonly 30 to 40 mm. The reflector constant is effectively eliminated by advancing the electrical center of the EDM by a corresponding amount during manufacture.

Since reflectors are made and mounted by different manufacturers, the reflector constants of all of the reflectors used with any one given instrument may not all be the same. Thus it is necessary to determine a combined instrument-reflector constant C_I for each combination to be used by the surveyor. Once this constant has been determined, it is added to any measurement made with that particular combination. The constant C_I can be entered into most of the modern EDMs by the operator via a keyboard and is then automatically added to the measured slope distance.

Two methods can be employed to determine the combined constant. In the first a line as long as practicable is established and measured using a steel or invar tape. Because of the inherent accuracy of the EDMs, this line must be measured to a high degree of accuracy. If the known distance is then measured with the EDM, the distance corrected for meteorological conditions and the slope of the line should agree with the known (taped) distance. Otherwise

$$C_I = \text{taped distance} - \text{EDM measured distance} \tag{2-22}$$

FIGURE 2-23 Relationship between elevation in feet and atmospheric pressure in inches.

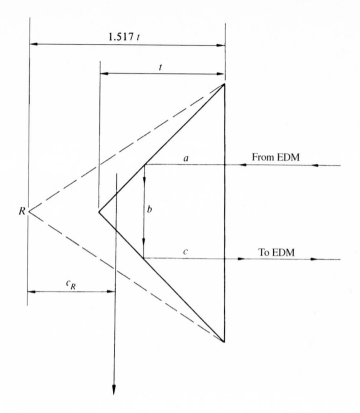

FIGURE 2-24 Reflector constant.

If a reliable baseline of known length is not available and if it is not feasible to measure a line, a second method can be employed to determine the instrument constant. In Fig. 2-25 three points A, B, and C are located on a straight line. The EDM occupies point A, and the distances AB and AC are measured. The EDM is moved to B, and the distance BC is measured. These three measurements are corrected for slope and meteorological conditions. The constant is then computed as (measured $AB + C_I$) + (measured $BC + C_I$) = measured $AC + C_I$ giving

$$C_I = \text{measured } AC - (\text{measured } AB + \text{measured } BC) \qquad (2\text{-}23)$$

These two methods for determining the combination instrument-reflector constant are also applied to microwave instruments. In this case the remote instrument is considered as the reflector.

FIGURE 2-25 Layout for determining combination instrument-reflector constant.

A determination of the value of C_I is made under the assumption that the instrument is properly tuned to give the correct modulation frequencies. An error in the frequency produces a scale error such as the incorrect tape length discussed in Section 2-14. For example, if the correct modulation frequency is 10 MHz and if the actual frequency departs from this by 100 Hz, a relative error of 10 ppm affects each measurement. The frequencies can be checked by means of a frequency counter, which can be obtained from an electronic laboratory or through the instrument manufacturer or service outlet. A frequency check should be performed at regular intervals, particularly if high-order surveys or surveys with very long lines are being performed. Alternatively, if the EDM is checked regularly against a known distance, applying corrections for instrument and reflector constants, meteorological conditions, and slope, a frequency shift can be detected.

2-29   Reduction of Slope Measurements in EDM

If the lines measured by EDMs are not too long, the slope distances obtained after the readings are corrected for meteorological and instrumental effects can be reduced to the corresponding horizontal distances by the methods outlined in Sections 2-11 and 2-12. To obtain the correct vertical angle used in the slope-reduction formulas, the height of the EDM above the ground station, the height of the reflector or remote EDM, the height of the instrument used to measure the vertical angle, and the height of the target on which the instrument was sighted when measuring the vertical angle must all be measured and recorded at the time of measurement.

In Fig. 2-26 the slope distance ER is measured with the EDM at E and the reflector at R. The vertical angle V is measured with the transit or theodolite at T to the target at P. In this instance the slope angle along the EDM line is larger than that along the line from T to P by the small angle ΔV.

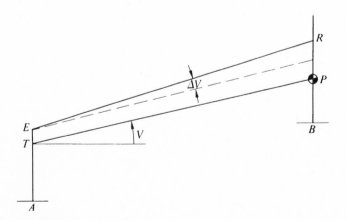

FIGURE 2-26 Difference between vertical angle of measured slope distance and vertical angle of line of sight.

Denoting AE as the height of the EDM, AT as the height of the theodolite, BR as the height of the reflector (or the remote instrument), and BP as the height of the target, the difference between the heights of the two instruments at A is $AE - AT$; and the difference between the heights of the reflector and the target is $BR - BP$. Note that $BR - BP$ is greater than $AE - AT$ in this instance. The difference, denoted ΔH, is then $(BR - BP) - (AE - AT)$ and is shown in Fig. 2-27. This value of ΔV in radians is closely approximated by $\Delta H \cos V/ER$, and $\Delta V'' = \Delta H \cos V/ER$ arc 1″. The vertical angle of the measured line is then $V' = V + \Delta V$.

EXAMPLE 2-14   A distance is measured along a slope with an EDM that, when corrected for meteorological conditions and instrumental constants, is 714.652 m. The EDM is 1.750 m above the ground, and the reflector is located 1.922 m above ground. A theodolite is used to measure a vertical angle of +4° 25′ 15″ to a target sitting 1.646 m above ground. The theodolite itself is 1.650 m above ground. What is the horizontal length of the line?

Solution: From the analysis given above, the situation is as shown in Fig. 2-26. The value of ΔH in Fig. 2-27 is

$$(1.922 - 1.646) - (1.750 - 1.650) = 0.276 - 0.100 = 0.176 \text{ m}$$

The small angle ΔV in seconds is then

$$\frac{0.176 \cos 4° 25′ 15″}{714.652 \times 0.485 \times 10^{-5}} = 51″$$

The slope of the measured line is then

$$+4° 25′ 15″ + 51″ = +4° 26′ 06″$$

The horizontal distance is computed using Eq. (2-1):

$$H = 714.652 \cos 4° 26′ 06″ = 714.652 \times 0.997006 = 712.512 \text{ m}$$

Many of the modern EDMs and the total station instruments discussed in Section 4-24 contain vertical angle-sensing elements that allow the horizontal and vertical distances to be displayed automatically. Some EDMs have provision for entering the vertical angles into a microprocessor contained in the instrument that computes and displays the horizontal and vertical distances from the measured

FIGURE 2-27 Reduction to correct vertical angle of slope measurement.

slope distance. The vertical angle is measured by a theodolite. In this case the target sighted on for measuring the vertical angle must be offset vertically from the reflector to maintain parallelism between the line of sight of the theodolite and the line between the EDM and the reflector.

If the slope distance obtained by EDM is to be reduced by difference in elevation between the two ends of the line, then the height of the EDM and the height of the reflector or remote EDM must be measured and recorded. In Fig. 2-28 the EDM is set up at A and the reflector at B. The height of the EDM is AE and that of the reflector is BR. The difference in elevation between A and B is h. But the difference in elevation between E and R along which the measurement is made is $h + \Delta h$. In this instance $\Delta h = BR - AE$. The corrected value of h is then used in Eq. (2-4) to obtain the horizontal distance from A to B.

If the measured lines are very long, the slope distance cannot be considered as the hypotenuse of a right triangle as assumed in short slope distances. When using a steel tape to make measurements, if the tape is held horizontal each time it is stretched out, the measurements are automatically made along a curved line that generally conforms with the curvature of the earth. This is shown ideally in Fig. 2-29 in which the curvature of the earth is exaggerated out of proportion to individual tape lengths. This curved distance, if reduced to sea level as explained in Chapter 9, is then called a *geodetic distance*.

When using an EDM, a single measurement is made of the entire line rather than in tape-length sections. Thus in Fig. 2-30 the EDM-measured slope distance is AB' and the desired distance is AB. The reduction required to compute AB is

FIGURE 2-28 Reduction to correct difference in elevation.

FIGURE 2-29 Taping along curved surface.

somewhat more complex than the simple expressions given by Eqs. (2-1) and (2-4). This topic is thoroughly discussed in Sections 9-11 and 9-12.

If a line measures 2 miles using the EDM and the vertical angle is 5°, the difference between the horizontal distance computed by Eq. (2-1) and that determined as outlined in Section 9-11 will be in error by no more than 1 part in 25,000. This relative accuracy is sufficient for all but high-order control surveys. Thus if the maximum range of the EDM is 2 miles, then the more elaborate slope-reduction calculations are needed only on lines that slope more than 5°. Figure 2-31 is a curve prepared on the basis of the slope reductions given in Section 9-11 for lines 10,000 ft or less in length in which the desired relative accuracy is not to drop below 1 part in 25,000. The curve indicates the maximum size of vertical slope angle as a function of distance. If the vertical angle exceeds this limit, then the result of Eq. (2-1) is not sufficiently accurate to give 1/25,000, and the more elaborate reduction of Section 9-11 must be employed. Most modern EDMs contain microprocessors which calculate the effect of curvature and refraction (see Section 3-2), thus eliminating this reduction when measuring long lines.

2-30 Layout Measurements Using EDM

In construction surveying, precalculated dimensions taken from plans and drawings must be laid out on the ground along well-defined construction lines. The steel tape is well suited to this type of surveying because the construction lines are usually rather short and the tape is convenient to handle. If the construction project is fairly extensive and involves the laying out of several long lines, however, the EDM can be employed to advantage. The technique to be used depends on the features of the instrument at hand.

If the EDM is designed to make a measurement along a direct line between it and a stationary reflector, which can be referred to as a static measurement, the technique to be used can take the following form. The approximate distance can be established within 1 or 2 ft using stadia hairs in the sighting telescope (see Sections 2-4 and 14-2) and two points are then set on a line a few feet in front of and behind the tentative point. The reflector is then set over the tentative point, and the distance is measured with the EDM. If there is significant slope, this distance is reduced to a horizontal (plan) distance using either Eq. (2-1) or (2-4). The presumably small difference between the measured distance and the distance indi-

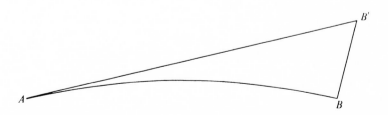

FIGURE 2-30 Measuring long line using EDM.

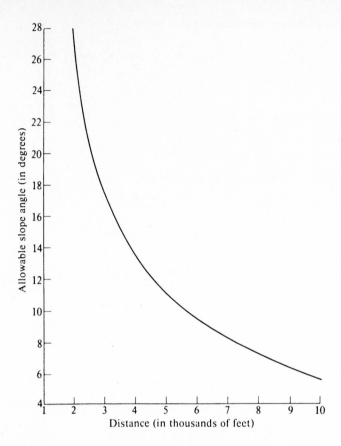

FIGURE 2-31 Allowable slope angle to permit using simple reduction equation to maintain relative accuracy of 1:25,000.

cated on the construction plans is then quickly laid off with a tape along the line defined by the two points previously set.

If the EDM is designed to follow a moving reflector to get the continuous slope distance, which can be referred to as a dynamic measurement, the positioning of the point is somewhat simplified. If the slope is significant, the correct slope distance to be laid off can be computed as shown in Example 2-2 or 2-6, in which the construction-plan distance is used for H in the examples. Alignment of the point can be made at the same time using the cross hairs of the sighting telescope.

The fully automatic EDM computes and displays horizontal distance from the slope distance measured with the instrument. After a tentative horizontal distance has been established, the reflector is moved forward or backward. Then continuous or near-continuous readout of horizontal distance is obtained. The point is set on line where the displayed distance is equal to the construction-plan distance.

2-31   Accuracy of EDM Measurements

At close range the accuracy of the EDM is limited by a constant uncertainty such as 1 cm, 5 mm, or 0.01 ft. As the measured distance is increased, this constant value becomes relatively inconsequential. Beyond, say, 500 to 1000 m, all of the currently available EDMs will give relative accuracies of better than 1 part in 25,000. This is very difficult to obtain by taping and requires taping tripods and very careful attention to systematic errors. On the other hand this accuracy is practically assured using the EDMs. The factors that limit both the relative and the absolute accuracies of EDM measurements are the meteorological conditions at the time of measurement. If these are known with sufficient accuracy, then all but the very short-range instruments are capable of relative accuracies of 1 part in 100,000 or better.

If very long lines must be measured with the maximum accuracy, then the uncertainty of the meteorological conditions along the entire beam path becomes important. For the majority of measurements in the short to intermediate range, meteorological measurements taken only at the instrument end of the line are sufficient to obtain the desired accuracy. This level of accuracy can be enhanced by taking a mean of the readings at both ends of the line. Improvements can further be made by sampling the meteorological conditions at intermediate points along the line, which of course is complicated usually by the necessity of elevating the meteorological instruments to considerable heights above the intervening terrain. The ultimate solution at present is to fly along the line and record the temperature and pressure profile all along the line. This technique has been employed in California along lines used to measure very small earthquake fault displacements over great distances. The various manufacturers usually state the accuracy as $\sigma = a + bd$, where d is the distance and a and b are constants for a particular instrument.

PROBLEMS

2-1. A surveyor paces a 100-ft length six times with the following results: $35\frac{1}{3}$, 34, $34\frac{2}{3}$, 35, $34\frac{2}{3}$, and 35 paces. How many paces must be stepped off to lay out a distance of 20 chains?

2-2. A surveyor paces a 50-m length five times with the following results: $56\frac{1}{2}$, 57, $56\frac{1}{2}$, 58, and 57 paces. How many paces must he step off to lay out a distance of 450.00 m?

2-3. A slope distance of 962.21 ft is measured between two points with a slope angle of 3° 16′. Compute the horizontal distance between the two points.

2-4. In Problem 2-3, if the vertical angle is in error by 2′, what error is produced in the horizontal distance?

2-5. A horizontal distance of 850.00 ft is to be laid out on a 2° 58′ slope. What slope distance must be laid out?

2-6. The difference in elevation between two points is 16.264 m. The measured slope distance is 343.516 m. Compute the horizontal distance.

2-7. A measurement of 148.264 m is made on a 4° 16′ slope. Compute the corresponding horizontal distance.

2-8. In problem 2-6, if the difference in elevation is 0.022 m, what error is produced in the horizontal distance?

2-9. A line was measured along sloping ground with a 30-m tape, and the following results were recorded:

Slope Distance (m)	Difference in Elevation (m)
30.000	1.792
30.000	0.930
18.520	0.966
30.000	3.075
12.422	0.660

What is the horizontal length of the line?

2-10. A tape that measures 99.96 ft between the zero and 100-ft mark is used to lay out foundation walls for a building 280.00×560.00 ft. What observed distances should be laid out?

2-11. A tape is calibrated and found to measure 100.04 ft between the 0- and 100-ft marks. What measurements should be laid out to establish a horizontal distance of 682.25 ft?

2-12. What distance on a 5% grade should be laid out with a tape that measures 30.010 m under field conditions if the horizontal distance is to be 430.000 m?

2-13. A 100-ft steel tape measures correctly when supported throughout its length under a tension of 10 lb and at a temperature of 72°F. It is used in the field at a tension of 18 lb and supported at the two ends only. The temperature throughout the measurement is 84°F. The measured length is 748.25 ft (the tape is suspended between the 48-ft mark and the zero end for the last measurement). The tape weighs 2.00 lb and has a cross-sectional area of 0.0060 in². Assuming that E is 28,000,000 psi for steel, what is the actual length of the line?

2-14. A 30-m tape weighs 12 g/m and has a cross section of 0.020 cm². It measures correctly when supported throughout under a tension of 8.5 kg and at a temperature of 20°C. When used in the field, the tape is supported at its two ends only, under a tension of 8.5 kg. The temperature is 13.5°C. What is the distance between the 0- and 30-m marks under these conditions?

2-15. A 100-ft tape is calibrated at 68°F and is found to measure 99.990 ft. A distance is measured as 515.68 ft at a temperature of 42°F. What is the correct distance?

2-16. A 100-ft steel tape weighs 1.80 lb and has a cross-sectional area of 0.0056 in². The tape measures 100.00 ft when supported throughout under a tension of 10 lb. Assume that $E = 28 \times 10^6$ psi. What tension, to the nearest $\frac{1}{4}$ lb, must be applied to overcome the effect of sag when the tape is supported at the two ends only?

2-17. A 50-m tape weighs 24 g/m and has a cross section of 0.038 cm². It measures 49.9862 m under a tension of 2.20 kg when supported at the two ends only. What does the tape measure if it is supported throughout under a tension of 6 kg? $E = 2.1 \times 10^6$ kg/cm².

2-18. With what accuracy must the difference in elevation between two ends of a 100-ft tape be known if the difference in elevation is 9.20 ft and the accuracy ratio is to be at least 1 : 10,000?

2-19. With what accuracy must a difference in elevation between two ends of a 30-m tape be known if the difference in elevation is 2.840 m and the accuracy ratio is to be at least 1 : 25,000?

2-20. The measured slope angle of a taped distance is 3° 54′. To what accuracy must the slope angle be measured if the relative accuracy is to be at least 1 : 20,000?

2-21. The measured slope angle of a measured slope distance of 342.535 m is 2° 24′. To what accuracy must the slope angle be measured if the horizontal distance is to be accurate to 5 mm?

2-22. Compute the refractive index of mercury vapor light at a temperature of 88°F and barometric pressure of 29.00 in.Hg. Neglect the effect of vapor pressure.

2-23. What is the refractive index of red laser light at a temperature of 20°C and barometric pressure of 725 mmHg? Neglect the effect of vapor pressure.

2-24. Microwaves are propagated through an atmosphere of 66°F, atmospheric pressure of 29.2 in.Hg, and vapor pressure of 0.51 in.Hg. If the modulating frequency is 30 MHz, what is the modulated wavelength?

2-25. What is the modulated wavelength of light of Problem 2-22 if the frequency of modulation is 30 MHz?

2-26. What is the modulated wavelength of light of Problem 2-23 if the frequency of modulation is 30 MHz?

2-27. Microwaves are modulated at a frequency of 75 MHz. They are propagated through an atmosphere at a temperature of 16°C, atmospheric pressure of 749 mmHg, and vapor pressure of 7.2 mmHg. What is the modulated wavelength?

2-28. Referring to Fig. 2-25, *AB* measures 796.16 ft; *BC* measures 423.25 ft; *AC* measures 1219.28 ft using a particular EDM reflector combination. A line measures 2946.22 ft with this instrument-reflector combination. What is the correct length of the line?

2-29. The height of an EDM set up at *A* is 5.32 ft. The height of the reflector set up at *B* is 4.30 ft. The height of the theodolite set up at *A* and used to measure a vertical angle is 5.00 ft. The height of the target at *B* on which the vertical angle sight is taken is 5.00 ft. The vertical angle is +4° 20′ 15″. The slope distance, after meteorological corrections, is 3451.55 ft. What is the horizontal distance between *A* and *B*?

2-30. The height of an EDM set up at *M* is 1.550 m. The height of the reflector set up at *P* is 1.400 m. The height of the theodolite set up at *M* used to measure the vertical angle is 1.600 m. The height of the target at *P* on which the vertical angle is sighted is 1.458 m. The slope distance, after meteorological corrections, is 975.26 m. The measured vertical angle is +3.2644^g. What is the horizontal distance between *M* and *N*?

BIBLIOGRAPHY 1993 EDM Survey, *P.O.B.* 19:2.

Bell, T. P. 1978. A practical approach to electronic distance measurement. *Surveying and Mapping* 38:335.

Buckner, R. B. 1970. Survey distance measurements and when to use what. *Proc. 30th Ann. Meet. Amer. Cong. Surv. & Mapp.* p. 1.

Carrol, C. L., Jr. 1974. Field comparison of steel surveyor's tapes. National Bureau of Standards Report, NBSIR 73-408. Washington, DC: U.S. Government Printing Office.

Critchfield, D. W. 1996. "Retroreflector Geometry." *Surveying and Land Information Systems* 55:4.

Hassan, E. B. 1986. Electromagnetic distance measurement along a path of non-linearly varying humidity. *Surveying and Mapping* 46:41.

House, S. L. 1982. Cyclic error in EDMI. *Surveying and Mapping* 42:281.

Kelly, M. L. 1979. Field calibration of electronic distance measuring devices. *Proc. Ann. Meet. Amer. Cong. Surv. & Mapp.* p. 425.

Kesler, J. M. 1973. EDM slope reduction and trigonometric leveling. *Surveying and Mapping* 33:61.

Love, J., and Spenser, J. F., Jr. 1981. Explanation and availability of calibration baseline data. *Tech. Papers Fall Conv. Amer. Cong. Surv. & Mapp.* p. 209.

Moffitt, F. H. 1958. The tape comparator at the University of California. *Surveying and Mapping* 18:441.

——1975. Calibration of EDMs for precision measurement. *Surveying and Mapping* 35:147.

Reilly, J. P. 1993 P.O.B. EDMI survey, *P.O.B.* 19:2.

Robertson, K. D. 1975. A method for reducing the index of refraction errors in length measurement. *Surveying and Mapping* 35:115.

Rosenberg, P. 1975. Corrections for EDM equipment mounted on surveying instruments. *Proc. Fall Conv. Amer. Cong. Surv. & Mapp.* p. 25.

Rüeger, J. M. 1985. Measurement and computation of cyclic errors of EDM instruments at different ranges. *Tech. Papers Fall Conv. Amer. Cong. Surv. & Mapp.* p. 94.

Townsend, A. R. 1985. Electronic field books: design, software, application. *Tech. Papers, 45th Ann. Meet. Amer. Cong. Surv. & Mapp.* p. 49.

Witte, B. U. 1982. Calibration of electro-optical rangefinders—experience gained and general remarks relative to calibration. *Surveying and Mapping* 42:151.

Yorczyk, R. A. 1985. The National Geodetic Survey's EDMI calibration baseline program. *Tech. Papers 45th Ann. Meet. Amer. Cong. Surv. & Mapp.* p. 32.

3 Leveling

3-1 Introduction

Leveling is the operation in surveying performed to determine and establish elevations of points, to determine differences in elevation between points, and to control grades in construction surveys. The elevation of a point has been defined as its vertical distance above or below a given reference level surface. The reference level surface used in the United States is the National Geodetic Vertical Datum of 1988. The NAVD 88 supersedes the National Geodetic Vertical Datum of 1929 (NGVD 29), which was the former official height reference (vertical datum) for the United States. NAVD 88 provides a modern, improved vertical datum for the United States, Canada, and Mexico. The NAVD 88 heights are the result of a mathematical (least squares) adjustment of the vertical control portion of the National Geodetic Reference System. Over 500,000 permanent benchmarks are included in the datum. The datum surface is an equipotential surface that passes through a point on the International Great Lakes Datum. The datum closely corresponds with mean sea level along the coasts of the United States.

A *benchmark* is a permanent or semipermanent physical mark of known elevation. It is set as a survey marker to provide a point of beginning for determining elevations of other points in a survey. A good benchmark is a bronze disk set either in the top of a concrete post or in the foundation of a structure. Specially designed marks used by the National Geodetic Survey (NGS) are the most stable. An NGS benchmark consists of a rod encased in grease inside a PVC pipe. Other locations for benchmarks are the top of a culvert headwall, the top of an anchor bolt, or the top of a spike driven into the base of a tree. The elevations of benchmarks are determined to varying degrees of accuracy by the field operations to be described in this chapter. Benchmarks established throughout the country by the NGS to a high order of accuracy define the North American Vertical Datum of 1988 (NAVD 88).

The basic instrument used in leveling is a spirit level that establishes a horizontal line of sight by means of a telescope fitted with a set of cross hairs and a level bubble. The level is described in later sections of this chapter. Other instruments used for determining vertical distances are the engineer's transit, the theodolite, the EDM, the aneroid barometer, the hand level, and the telescopic alidade. The use of the transit and the theodolite are explained in Chapters 4 and 14. A type of elevation can be determined by signals from the Global Positioning System (GPS) of satellites and will be discussed in Chapter 10.

3-2 Curvature and Refraction

The measurements involved in leveling take place in vertical planes. Consequently, the effect of earth curvature and atmospheric refraction must be taken into account because these effects occur in the vertical direction. They are allowed for either by the appropriate calculations or else by the measuring techniques designed to eliminate these effects.

To examine the effect of earth curvature, consider the amount by which the level surface passing through point A of Fig. 3-1 departs from a horizontal plane at the distance AB from the point of tangency. This departure is shown to be the vertical distance CB. If the effect of earth's curvature is designated c, and the distance from the point of tangency to the point in question is designated K, then the curvature effect can be shown to be

$$c = 0.667K^2 \qquad (3\text{-}1)$$

in which c is the curvature effect in feet and K is the distance in miles. This expression for earth curvature is based on a mean radius of the earth, or 3959 miles. It should be noted that the error due to curvature is proportional to the square of the distance from the point of tangency to the point in question. Thus, for a distance of 10 miles, the value of c is about 67 ft, whereas for a distance 100 ft, it diminishes to 0.00024 ft. In the metric system

$$c_M = 0.0785K_M^2 \qquad (3\text{-}2)$$

in which c_M is the curvature effect, in meters; K_M is the distance, in kilometers.

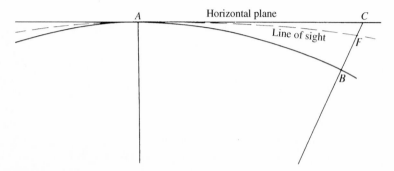

FIGURE 3-1 Curvature and refraction.

Rays of light passing through the earth's atmosphere in any direction other than vertical are refracted or bent from a straight path. This bending usually takes place in a direction toward the earth's surface under normal conditions of temperature and pressure gradients. Such bent rays of light tend to diminish the effect of curvature a slight amount, normally about 14% of the curvature effect. This effect is shown by the dashed line of sight in Fig. 3-1. The effect of both curvature and refraction is reduced to the distance *FB*. Thus under normal conditions the combined effect of curvature and refraction can be computed from one of the following:

$$(c + r) = 0.574K^2 \tag{3-3}$$

$$(c + r) = 0.0206M^2 \tag{3-4}$$

$$(c + r)_M = 0.0675K_M^2 \tag{3-5}$$

in which $(c + r)$ is the combined effect of curvature and refraction, in feet; $(c + r)_M$ is the combined effect of curvature and refraction, in meters; K is the distance, in miles; M is the distance, in thousands of feet; K_M is the distance, in kilometers.

When leveling with the spirit level, the line of sight is ordinarily no more than about 2 m or 6 ft above the ground throughout its length. Depending on the weather, the temperature gradient near the ground can be the reverse of what it is at say 20 ft above ground on account of the heat radiating from the ground. In this zone, therefore, the effect of atmospheric refraction is uncertain. Fortunately, however, because of the relatively short lines of sight that are encountered in spirit leveling and also because of balanced sight distance (see Section 3-29), this uncertainty is of minor consequence except in very precise leveling.

3-3 Trigonometric Leveling

The difference in elevation between two points can be determined by measuring the vertical angle of the line from one point to the other and then computing the difference in elevation from a knowledge of either the slope distance or the horizontal distance between the two points. In Fig. 3-2 *DE* is a horizontal line, and the

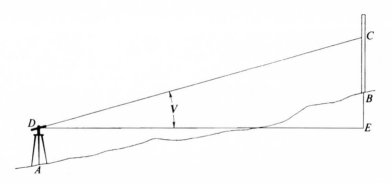

FIGURE 3-2 Trigonometric leveling.

vertical distance CE is $DC \sin V$ or $DE \tan V$. So the difference in elevation between A and B is $AD + CE - CB$, where AD is the height of the instrument above point A, and CB is the height of the line of sight above point B measured with a leveling rod or with some other type of graduated rod. This method of determining difference in elevation is limited to horizontal distances less than 1000 ft when moderate precision is sufficient, and to proportionately shorter distances as higher precision is desired.

Many of the EDMs discussed in Chapter 2 and the total station instruments discussed in Chapter 4 contain built-in microprocessors that compute the distance CE from the measured slope distance and the vertical angle. The vertical angle either is measured directly in some instruments or else must be entered into the keyboard of the EDM.

For distances beyond 1000 ft the effects of curvature and refraction must be considered and applied. The situation shown in Fig. 3-3 represents a line of intermediate length, up to perhaps 2 miles. In Fig. 3-3(a) the angle at E is shown to be 90° and is so assumed in this range. At 2 miles it is actually about 90° 01′ 44″. The

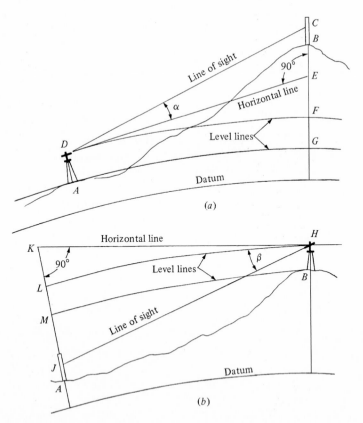

FIGURE 3-3 Curvature and refraction in trigonometric leveling.

vertical distance CE is equal to $DC \sin \alpha$ or $DE \tan \alpha$, depending on whether the slope distance obtained by EDM or horizontal distance from triangulation (see Chapter 10) is known. The vertical distance from a level line through D to point C is FC, which is $CE + EF$. But EF is the total effect of curvature and refraction, which, by Eq. (3-4), is $0.0206(DE/1000)^2$ or $0.0206(DC \cos \alpha /1000)^2$. The difference in elevation between A and B is then $AD + EF + CE - CB$. The values of AD and CB are recorded at the time of measurement.

To eliminate the uncertainty in the curvature and refraction correction, vertical-angle observations are made at both ends of the line as close in point of time as possible. This pair of observations is termed *reciprocal vertical-angle observation.* In Fig. 3-3(b) the angle at K is assumed to be 90°. The vertical distance JK is equal to $HJ \sin \beta$ or $HK \tan \beta$. The vertical distance from a level line through H to point J is JL, which is $JK - KL$. But KL is the total effect of curvature and refraction, which is $0.0206 (HK/1000)^2$ or $0.0206(HJ \cos \beta /1000)^2$. The difference in elevation from A to B is then $AJ + JK - KL - HB$. The values of AJ and HB are recorded at the time of measurement. The correct difference in elevation between the two ends of the line is then the mean of the two values computed both ways either with or without taking into account curvature and refraction.

The accuracy of the determination of difference in elevation over a long distance is basically a function of the uncertainty of the atmospheric refraction and of the accuracy of the vertical angles. The slope distance obtained with EDM will be so accurate that no appreciable error will be introduced from this source. The effect of uncertainty in atmospheric refraction is held to a minimum by reciprocal observations. Bearing in mind that a second is a foot in 40 miles, then in a distance of say 2 miles the error of difference in elevation can be held to 0.20 ft if the vertical-angle accuracy is about 4″, which is obtainable with the precise theodolites discussed in Chapter 4. This error is reduced in proportion to a reduction in distance. If reciprocal observations are taken every 1000 m and the accuracy of vertical-angle measurements is of the order of 3″ to 4″, the error can be held to $1\frac{1}{2}$ to 2 cm or about 0.05 to 0.07 ft.

When applying trigonometric leveling to very long lines, the slope distance is measured using the intermediate or long-range EDMs, and reciprocal vertical angles are measured using precise theodolites. If possible, the reciprocal angles should be made simultaneously to eliminate the refraction uncertainty. If this proves unfeasible, then more than one set should be observed at different times to average out the errors due to this uncertainty.

EXAMPLE 3-1   The slope distance between two mountain peaks determined by EDM measurement is 76,963.54 ft. The vertical angle at the lower of the two peaks to the upper peak is +3° 02′ 05″. The reciprocal vertical angle at the upper peak is – 3° 12′ 55″. The height of the instrument and the height of the target at the ends of the line are assumed to be equal. Compute the difference in elevation between the two peaks first by using only the single vertical angles and applying correction for curvature and refraction, and second by using the average obtained by both vertical angles.

Solution: The difference in elevation using only the single vertical angle α is

$$\Delta H = 76,963.54 \sin 3° 02' 05'' + 0.0206\left(\frac{76,963.54}{1000} \cos 3° 02' 05''\right)^2$$

$$76,963.54 \sin 3° 02' 05'' = 76,963.54 \times 0.05294113 = 4074.54$$

$$0.0206\left(\frac{76,963.54}{1000} \cos 3° 02' 05''\right)^2 = 0.0206 \times 5907 \qquad = +121.68$$

$$\Delta H = +4196.22 \text{ ft}$$

The difference in elevation using only the single vertical angle β is

$$\Delta H = 76,963.54 \sin 3° 12' 55'' - 0.0206[(76,963.54/1000) \cos 3° 12' 55'']^2$$

$$= +4316.71 - 121.64$$

$$= 4195.07 \text{ ft.}$$

The average from the two solutions is +4195.64 ft.

The difference in elevation without taking into account the correction for curvature and refraction, and using the average, is

$$76,963.54 \sin 3° 02' 05'' = 4074.54$$

$$76,963.54 \sin 3° 12' 55'' = 4316.71$$

$$\Delta H = \text{average} = +4195.63 \text{ ft}$$

3-4 Direct Differential Leveling

The purpose of differential leveling is to determine the difference of elevation between two points on the earth's surface. The most accurate method of determining differences of elevation is with the spirit level and a rod, as shown in Fig. 3-4. It is assumed that the elevation of point A is 976 ft and that it is desired to determine the elevation of point B. The level is set up, as described in Section 3-23, at some convenient point so that the instrument is higher than both A and B. A leveling rod is held vertically at point A, which may be on the top of a stake or on some solid object, and the telescope is directed toward the rod. The vertical distance from A to a horizontal plane can be read on the rod where the horizontal cross hair of the telescope appears to coincide. If the rod reading is 7.0 ft, the plane of the telescope is 7.0 ft above point A. The elevation of this horizontal plane is 976 + 7 = 983 ft. The leveling rod is next held vertically at B, and the telescope is directed toward

FIGURE 3-4 Direct differential leveling.

the rod. The vertical distance from B to the same horizontal plane is given by the rod reading with which the horizontal cross hair appears to coincide. If the rod reading at B is 3.0 ft, point B is 3.0 ft below this plane and the elevation of B is 983 − 3 = 980 ft. The elevation of the ground at the point at which the level is set up need not be considered.

The same result may be obtained by noting that the difference in elevation between A and B is 7 − 3 = 4 ft, and that B is higher than A. The elevation of B equals the elevation of A plus the difference of elevation between A and B, or 976 + 4 = 980 ft.

3-5   Stadia Leveling

Stadia leveling combines features of trigonometric leveling with those of direct differential leveling. In stadia leveling, vertical angles are read by using the transit or theodolite, and horizontal distances are determined at the same time by means of the stadia hairs mentioned in Section 2-4. As in direct differential leveling, the elevation of the ground at the point at which the instrument is located is of no concern in the process. Stadia leveling is a rapid means of leveling when moderate precision is sufficient. It is described in detail in Section 14-10.

3-6   Leveling with Aneroid Barometer

The fact that atmospheric pressure, and hence the reading of a barometer, decreases as the altitude increases is utilized in determining differences of elevation. Because of transportation difficulties, the mercurial barometer is not used for survey purposes. Instead, the aneroid barometer or altimeter is used in surveys in which errors of 5 to 10 ft are of no consequence. Altimeters vary in size from that of an ordinary watch to one that is 10 or 12 in. in diameter. The dials sometimes have two sets of graduations, namely, feet or meters of elevation and inches or millimeters of mercury. The smaller altimeters can be read by estimation to about 10 ft. A larger type, one of which is shown in Fig. 3-5, is much more sensitive, and differences of elevation of 2 or 3 ft can be detected.

An altimeter is, of course, subject to natural changes in atmospheric pressure due to weather, and it is also subject to effects of temperature and humidity. For this reason altimeters should be used in groups of three or more.

In Fig. 3-6 an altimeter is maintained at L, which is a point of known elevation and is designated the low base. A second altimeter, whose reading has been compared with the one kept at the low base, is taken to H, which is a second point of known elevation and is designated the high base. At regular intervals, say every 5 min, the altimeters at the low and high bases are read. A third altimeter, which is referred to as the field altimeter or roving altimeter, is initially compared to the low-base altimeter and then taken to various points whose elevations are to be established. At each point the reading of the roving altimeter and the time are recorded. The elevation of any point can be determined from the reading of the roving altimeter at the point and the readings of the altimeters at the high and low

FIGURE 3-5 Precision altimeter.

FIGURE 3-6 Two-base altimetry.

bases at that time. In order to maintain accuracy, limitations must be imposed on the elevation difference between the two bases and on the area covered by the roving altimeter. These limitations tend to eliminate the effects of changing atmospheric pressure due to the weather and also the effects of differences in temperature and humidity.

In the following table the readings of the three altimeters indexed at the low base are shown. The high-base altimeter is taken to the high base and the roving altimeter is taken to various field points.

TABLE 3-1 Two-Base Altimeter Readings.
Indexing on Low Base; Low-Base Altimeter 1180; High-Base Altimeter 1189; Roving Altimeter 1188

Time	Low-Base Reading (ft)	Low-Base Elevation (ft)	High-Base Reading (ft)	High-Base Elevation (ft)	Roving-Altimeter Reading (ft)
1:55 P.M.	1180	225	1637	658	1360
2:15 P.M.	1186	225	1650	658	1425
2:20 P.M.	1184	225	1646	658	1452

Before either the high-base readings or the roving-altimeter readings are used for computing elevations, they must be corrected for index error. Thus the high-base readings must be reduced by 9 ft and the roving-altimeter readings by 8 ft. It is possible to make all three altimeters read the same at the time of indexing by physically adjusting the pointers, but this practice is not recommended. The corrected readings are shown here for convenience.

Time	Low-Base Reading (ft)	Corrected High-Base Reading (ft)	Corrected Roving-Altimeter Reading (ft)	Difference High-Low (ft)	Difference Roving-Low (ft)	Δh (ft)
1:55 P.M.	1180	1628	1352	448	172	166
2:15 P.M.	1186	1641	1417	455	231	220
2:20 P.M.	1184	1637	1444	453	260	249

The known difference in elevation between the high base and the low base is $658 - 225 = 433$ ft. This value is represented by the length of the vertical line LH in Fig. 3-6. At the time of the first field reading, the difference between the corrected readings at the high and low bases is 448 ft. This value is represented by the length of the sloped line $L'H$. At the same time, the difference between the corrected readings at the field point and the low base is 172 ft, which is represented by the distance $L'P'$. Then, by proportion, the difference in elevation Δh between the low base and the field point, or the distance LP, is $(172)(433)/448 = 166$ ft. Similarly, the differences in elevation between the low base and the other two points are, respectively, $(231)(433)/455 = 220$ ft and $(260)(433)/453 = 249$ ft.

3-7 Types of Spirit Levels

The instrument most extensively used in leveling is the engineer's level. It consists essentially of a telescope to which a very accurate spirit level is attached longitudinally. The telescope is supported at the ends of a straight bar that is firmly secured at the center to the perpendicular axis on which it revolves. The level is supported on a tripod.

There are three general types of engineer's levels. These are the dumpy level, the tilting level, and the automatic or self-leveling level. The design features will vary in any given type of level, but the operating principles are the same for all.

3-8   Dumpy Level

The engineer's dumpy level is shown in Fig. 3-7. It consists of five basic parts: the telescope a, which is rigidly attached to the cross bar b; the level vial c encased in a housing, which is also attached to the cross bar; the leveling head d; and the tripod e. The vertical axis of rotation is a tapered spindle rigidly attached to the lower side of the cross bar. This tapered spindle fits into a tapered bore through the center of the leveling head. A clamp f either permits (when loose) or prevents (when clamped) rotation of the spindle inside the leveling head. It is used to control the rotation of the telescope in a horizontal plane when sighting. A *slow motion* or tangent screw g is used after the clamp has been clamped to make a fine adjustment to the direction of the telescope when taking a rod reading.

The leveling head contains either three or four leveling screws (the latter shown in Fig. 3-7) that tilt the leveling head against the top of the plate h. This also tilts the vertical spindle to the vertical position, using the level bubble for reference as discussed in Section 3-23. The tripod consists of three legs shod with steel and connected by hinge joints to the metal tripod head. When the level has been unscrewed from the top of the tripod for storage, it should be replaced by a threaded tripod cap to prevent the threads from being damaged. Use of the dumpy level is decreasing in favor of the automatic level (see Section 3-13), although it is still found in use in a great number of surveying firms.

FIGURE 3-7 Dumpy level.

3-9   Telescope

A longitudinal section of a typical telescope used in a surveying instrument is diagrammed in Fig. 3-8. The essential parts are the barrel, the positive objective lens, the negative focusing lens, the reticle, which holds a set of cross hairs, and the eyepiece system. The objective lens is a compound lens composed of both crown and flint glass to eliminate chromatic aberration. Its function is to form an image of the object sighted. The image would be formed ahead of the cross hairs. However, the focusing lens diverges the light rays and brings them into focus in the plane of the cross hairs, as shown in Fig. 3-8. The eyepiece lenses magnify the image together with the cross hairs to give the instrument man the ability to sight accurately and to be able to read the leveling rod graduations accurately. The image formed by the objective and the focusing lens is inverted. Some eyepiece systems erect the image to give a normal view. Other eyepiece systems do not, so the image is seen upside down.

The cross hairs used in some surveying instruments are very fine threads, taken from the cocoon of a brown spider. Many instrument makers use finely drawn platinum wires, some use fine glass threads, and others use a glass diaphragm on which lines have been etched. Levels intended for ordinary work have two cross hairs, one horizontal and one vertical. Those instruments that are intended for precise leveling have two additional horizontal hairs, one above and one below the usual horizontal cross hair. These additional hairs are called *stadia hairs* (see Fig. 14-1 on page 507) and are found in most transits and theodolites.

The cross hairs are attached to a reticle, or cross-hair ring. As shown in Fig. 3-9, this reticle is practically a heavy brass washer that is somewhat smaller in diameter than the inner surface of the telescope barrel. It is thick enough to be tapped and threaded for the capstan-headed screws by which it is held in place in the telescope tube. The holes through the telescope tube are large enough to permit the cross-hair ring to be rotated through a small angle and are therefore covered by beveled washers. The ring can be moved vertically or horizontally by turning the proper capstan screws.

Focusing of the eyepiece is accomplished by changing the distance between it and the cross hairs. On most telescopes this is done by twisting the end of the eyepiece.

FIGURE 3-8 Longitudinal section of telescope.

FIGURE 3-9 Cross-hair reticle.

3-10 Level Tube

The spirit level c in Fig. 3-7 is the part of the instrument on which the accuracy chiefly depends. As indicated in Fig. 3-10, in plan (a) and in side elevation (b), the spirit level consists of a sealed glass tube nearly filled with alcohol or with a mixture of alcohol and ether. The upper surface of the tube, and sometimes also the lower surface, is ground to form a longitudinal circular curve. The sensitiveness of the bubble tube is dependent on the radius of this circular curve, and is usually expressed in terms of the angle through which the tube must be tilted to cause the bubble to move over one division of the scale etched on the glass. This angle may vary from 1″ to 2″, in the case of a precise level, or up to 10″ to 30″ on an engineer's level. The radius to which the tube for an engineer's level is ground is usually between 75 and 150 ft.

In a properly designed level there should be a definite relation between the sensitiveness of the bubble and the magnifying power of the telescope. If, with the telescope focused on a distant object, the leveling screws are turned just enough to cause a barely perceptible movement of the horizontal cross hair, a correspond-

(a)

(b)

FIGURE 3-10 Level tube.

ing displacement of the bubble should be apparent. Conversely, when the bubble is changed very slightly, a movement of the horizontal cross hair should be detected.

The angular value of one division of the level tube can be determined by turning the leveling screws so as to cause the bubble to move over several divisions, and by observing the corresponding vertical distance the horizontal cross hair is raised or lowered on a leveling rod held at a known horizontal distance from the level. Thus, if the bubble is moved over five divisions and the vertical movement of the cross hair is 0.102 ft on a rod 300 ft from the level, the angular value S of one division is given by

$$\tan S = \frac{0.102}{5 \times 300}$$

or

$$S'' = \frac{\tan S}{\tan 1''} = \frac{0.102}{5 \times 300 \times 0.00000485} = 14''$$

The metal case containing the level tube is adjustable vertically at one end.

3-11   Sighting Through the Telescope

When sighting through the telescope of a surveying instrument, whether it's a level, a transit, a theodolite, or an alidade, the observer must first focus the eyepiece system for the observer's eye. This is most easily done by holding an opened field book about six inches in front of the objective lens and on a slant to obscure the view ahead of the telescope and to allow light to enter the objective lens. This is shown in Fig. 3-11. The *eyepiece* is now twisted in or out until the cross hairs are sharp and distinct. Now, with the eyepiece system focused, the telescope is pointed at the object to be sighted, with the observer looking along the top of the telescope barrel (some telescopes are provided with peep sights with which to make this initial alignment). The rotational motion is then clamped. The object to be sighted should now be in the field of view. The tangent screw is then used to bring the line of sight directly on the point.

The *focusing lens* is now focused to bring the image of the point into the plane of the cross hairs. This can be checked by moving the eye up and down or sideways a small amount to detect whether or not the cross hairs appear to move with respect to the object sighted. If they do not appear to move, the telescope is properly focused. If they *do* appear to move, the condition called *parallax* exists between the image and the cross hairs. This condition is shown in Fig. 3-12. If the image formed by the lens lies either in front of or behind the plane of the cross hairs, as shown in Fig. 3-12(a) and (b), and the observer's eye is raised or lowered slightly, the cross hairs seem to move up and down on the face of a leveling rod. Refocusing while observing the cross hairs will bring the focusing lens in the proper position, and the parallax is eliminated, as shown in Fig. 3-12(c).

It is advisable to occasionally check the eyepiece focus by the method shown in Fig. 3-11, particularly when the observer must make a lengthy series of observations.

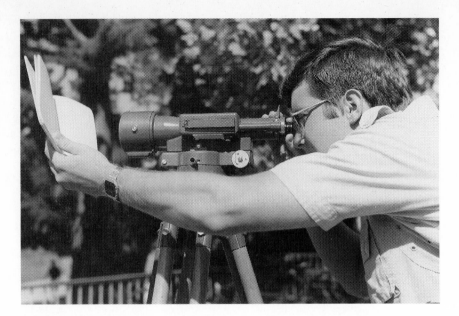

FIGURE 3-11 Focusing the eyepiece.

(a)

(b)

(c)

FIGURE 3-12 Parallax.

3-12 Tilting Level

When a level is used, the level bubble must be centered at the instant of observation. Otherwise serious random errors are introduced. If a conventional dumpy level is being used, the observer must constantly check the level bubble and readjust the leveling screws. The tilting level, shown in Fig. 3-13, is brought to approximate level by means of a circular type or bull's-eye level. When a reading is to be taken, the observer rotates a tilting knob, which moves the telescope through a small vertical angle. With his head in the normal position for viewing the rod through the telescope, he can at the same time look through a window to the left of the eyepiece of the telescope and observe the more sensitive, main-level bubble as two half images of opposite ends of the bubble. These half images are brought into superposition and made visible to the observer by a prismatic arrangement directly over the bubble. The observer then tilts the telescope until the two half images are made to coincide, in which position the bubble is centered. The split bubble before and after coincidence is shown in Fig. 3-14. Note that the apparent discrepancy between the two half images is actually twice the centering error. Because of the fine tilting adjustment and the ability to view both ends of the bubble simultaneously, the observer can center the bubble to within 0.3 mm.

Fig. 3-15 shows a tilting level of more compact design. Its operation is identical with that of the level shown in Fig. 3-13. The eyepiece, however, does not erect

FIGURE 3-13 Tilting level.

FIGURE 3-14 Coincidence bubble.

FIGURE 3-15 Tilting level. Courtesy of Kern Instruments, Inc.

the image of the leveling rod, and this difference accounts for the relatively shorter telescope. In this instrument the image of the coincidence bubble appears directly in the field of view of the telescope rather than through an adjacent eyepiece, and the observer may view the bubble, the cross hairs, and the image of the leveling rod with one eye and from one position.

The tilting level shown in Fig. 3-16 contains a coincidence bubble that is viewed through a separate window, like that of the level shown in Fig. 3-13. Also the level contains a horizontal circle which makes it convenient for layout and construction surveys. The unique feature of this level is the capability of rotating the telescope and the level tube 180° about the telescope's longitudinal axis. This feature permits a direct and reverse reading on a level rod, the mean value of which is free from the error of an inclined line of sight discussed in Section 3-29.

FIGURE 3-16 Tilting level with provision for rotating telescope about longitudinal axis. Courtesy of Leica, Inc.

3-13 Self-Leveling or Automatic Level

The level shown in Fig. 3-17 is said to be self-leveling. When the bull's-eye bubble has been centered, a prism carried on a pendulum supported by two pairs of wires reflects the light rays entering the objective lens on back to the eyepiece end of the telescope. The lengths of the supporting wires and the positions of the points of suspension are so designed that the only rays of light reflected back to the intersection of the cross hairs by the swinging prism are the horizontal rays passing through the optical center of the objective lens. Hence, as long as the prism is free to swing, a horizontal line of sight is maintained, even though the telescope barrel itself is not horizontal. A damping device brings the pendulum to rest quite rapidly, so that the observer does not have to wait until it settles of its own accord. This type of level has the advantage of offering very rapid instrument setups and of eliminating random errors in centering the level bubble.

The self-leveling level shown in Fig. 3-17 is shown in cross section in Fig. 3-18. A small triangular-shaped prism can be seen slightly forward of the eyepiece near the bottom of the telescope. Light on entering the objective lens passes through the focusing lens and then through a fixed prism. It is then reflected off the triangular swinging prism and enters a second fixed prism, from which it is deflected into the eyepiece system.

The automatic level shown in Fig. 3-19 operates by means of a gravity-actuated compensator similar in general principle to that described above. This level contains an optical micrometer, which allows very precise readings of graduated level rods, as described in Section 3-22.

FIGURE 3-17 Self-leveling or automatic level. Courtesy of Carl Zeiss, Inc.

FIGURE 3-18 Section through self-leveling level.

All automatic levels operate generally on the same principle as described in this section.

3-14   Geodetic Level

The geodetic level, used for extreme precision required in first-order leveling conducted by the National Geodetic Survey, is a tilting level made of invar to lessen the effect of temperature. The magnification of the objective-eyepiece combination is about 42 diameters. The geodetic level is equipped with stadia hairs for determining three rod readings at one observation and for determining the lengths of the sights. The tripod on which the geodetic level rests is of such height that the observer may stand erect while observing. The unusually high tripod brings the line of sight somewhat farther from the intervening ground than does the ordinary tripod, helping to lessen differential refraction of the line of sight.

The precise tilting level shown in Fig. 3-20 has a 42-power telescope and contains a very sensitive coincidence-type bubble in addition to an optical micrometer. This level, together with a precise leveling rod, is quite suitable for precise geodetic leveling.

3-15   Electronic Digital Level

The level shown in Fig. 3-21 contains an automatic compensator for maintaining a horizontal line of sight. It is used with a leveling rod, one face of which contains metric graduations (see Section 3-19). The other face contains a bar code, a portion of which is shown in Fig. 3-22. After the circular bubble has been centered, the observer directs the line of sight toward the leveling rod. The visible portion of the light allows him to observe the rod as in the normal operation in leveling. A beam splitter separates the infrared portion of the light entering the telescope and directs it to a sensor. A copy of the bar code is held in the memory of the computer

FIGURE 3-19 Self-leveling level with optical micrometer. Courtesy of Leica, Inc.

contained in the instrument. The computer, in effect, searches through its memory until a match is obtained between the pattern on the rod and that stored in memory. At this point the rod reading and the horizontal distance to the rod are displayed. These data are transferred to a module and are then available for further calculations (see Section 3-25). The data can of course also be entered into a field book or data collector. A check on the rod reading can be made, if desired, by reading the other face of the rod.

FIGURE 3-20 Precise tilting level with optical micrometer used for geodetic leveling. Courtesy of Leica, Inc.

3-16   Hand Level and Clinometer

The hand level, shown in Fig. 3-23, is a brass tube with a small level tube mounted on the top. A 45° mirror on the inside of the main tube enables the user to tell when it is being held horizontally. As the rod viewed through the level is not magnified, the length of sight is limited by the visibility of rod readings with the naked eye.

The hand level is used on reconnaissance surveys where extreme accuracy is unnecessary and in taping to determine when the tape is being held horizontally. It is also used to advantage for estimating how high or how low the engineer's level must be set to be able to read the leveling rod.

The clinometer, shown in Fig. 3-24, can be used in the same manner as the hand level. In addition it can be employed for measuring vertical angles where approximate results are sufficient.

3-17   Leveling Rods

In addition to the bar code rod used with the digital level, there are two other general classes of leveling rods: self-reading and target rods. A self-reading rod has painted graduations that can be read directly from the level. When a target rod is used, the target is set by the rodman as directed by the levelman, and the reading is then made by the rodman. Some types of rods can be used either as self-reading rods or as target rods.

FIGURE 3-21 Wild NA2000 digital level. Courtesy of Leica, Inc.

The graduations on self-reading rods should appear sharp and distinct for the average length of sight. In the United States the rods are ordinarily graduated so as to indicate feet and decimals, the smallest division usually being 0.01 ft. Metric rods are graduated to centimeters, and rod readings are estimated to millimeters.

3-18   Philadelphia Rod

The Philadelphia rod, front and rear views of which are shown in Fig. 3-25, is made in two sections that are held together by the brass sleeves a and b. The rear section slides with respect to the front section, and it can be held in any desired position by means of the clamp screw c on the upper sleeve b. The rod is said to be a *short rod* when the rear portion is not extended, and a *long* or *high rod* when it is extended. The short rod is used for readings up to 7 ft. For readings between 7 and 13 ft, the long rod must be used. When the rod is fully extended, the graduations on the face are continuous.

The rod shown in Fig. 3-25 is graduated to hundredths of a foot by alternate black and white spaces painted on the rod. Each fifth hundredth is indicated by a longer graduation mark, so that an acute angle is formed at one corner of the black

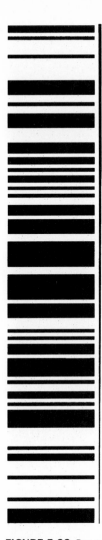

FIGURE 3-22 Bar code on leveling rod used with digital level.

space of which the graduation is a part. The tenths are marked by large black figures, half above and half below the graduation mark, and the feet are shown in a similar manner by red figures (shaded in the illustrations). The graduations can be seen distinctly through the telescope of a level at distances up to about 300 ft, and the rod can thus be used as a self-reading rod.

In case the graduations cannot be read directly from the telescope with sufficient accuracy, the target d is used. The target is a circular or elliptical metal plate divided into quadrants alternating red and white. An opening in the face of the target allows the graduations on the face of the rod to be seen. One side of this opening is beveled to a thin edge, and a scale is marked along this edge so that it is

FIGURE 3-23 Hand level.

FIGURE 3-24 Clinometer.

close to the face of the rod. This scale is used for determining readings between graduation marks. One of its ends is exactly on the horizontal line dividing the colors on the target.

When the clamp *e* is loose, the target can be moved over the face of the rod until the line dividing the colors coincides with the horizontal cross hair of the level. The target can then be fixed in that position by tightening the clamp *e*.

When the target is used on a high rod, it is first set exactly at 7 ft on the extension part of the rod, as shown in Fig. 3-25(a). The extension with the target is then raised until the line dividing the colors coincides with the horizontal cross hair of the telescope, and the rod is held in that position by tightening the clamp *c*. To determine the reading for a high rod, the graduations on the back of the sliding part and a scale on the back of the upper sleeve are used. As shown in view (b), these graduations begin at 7 ft at the top and increase downward to 13 ft.

FIGURE 3-25 Philadelphia rod.

The reading of a high rod is the distance from the base of the rod to the target. Thus, for the position shown in Fig. 3-25(b), the rod reading represents the distance h. When the target is set at 7 ft on the extension part of the rod while the rod is closed, the reading of the high rod is 7 ft, and the 7-ft graduation on the back of the rod is opposite the zero of the scale on the sleeve. As the rod is raised, the 7-ft mark moves upward, whereas the zero mark of the scale remains stationary since it is attached to the lower portion of the rod. The distance h' between these two marks therefore increases as the rod is extended further. Consequently, the distance h' is equal to the amount by which the target is raised above 7 ft, and for any high-rod reading the distance h is equal to 7 ft plus the distance h'. To obviate actual addition, the foot graduations on the back of the rod are numbered downward from 7 to 13. Thus, when the rod is extended 1 ft, the reading on the back is 7 + 1 = 8 ft, and so on. Therefore the numbers must increase downward so the rod readings can become greater as the target is raised.

If the rod has been damaged by allowing the upper portion to slide down with a bang, it is possible that the reading on the back of the rod will be less than 7 ft when the rear section is in the lowered position. In this case the target should be set at the corresponding reading on the face of the rod before extending the rod.

3-19   Precise Leveling Rods

A precise rod is graduated on an invar strip that is independent of the main body of the rod except at the shoe at the bottom of the rod. The graduations are in yards, in feet, or in meters, and the smallest graduations are 0.01 yd, 0.01 ft, and 1 cm, respectively. A yard rod or a meter rod is also graduated in feet on the back of the rod. These foot graduations are painted directly on the main body of the rod. They serve as a check on the more precise readings taken on the invar strip, and help to prevent mistakes in rod readings. The precise rod is equipped with either a bull's-eye level or a pair of level vials at right angles to each other to show when the rod is vertical, and also with a thermometer. Front and back views of a precise rod are shown in Fig. 3-26.

3-20   Reading the Rod Directly

If the target is not used, the reading of the rod is made directly from the telescope. The number of feet is given by the red figure just below the horizontal cross hair when the level has an erecting telescope, or just above the horizontal cross hair when it has an inverting telescope. The number of tenths is shown by the black figure just below or above the hair, the position depending on whether the telescope is erecting or inverting. If the reading is required to the nearest hundredth, the number of hundredths is found by counting the divisions between the last tenth and the graduation mark nearest to the hair. If thousandths of a foot

FIGURE 3-26 Precise leveling rod.

(a)   (b)

FIGURE 3-27 Direct reading of rod.
(a) Rod graduated to 0.01 ft.
(b) Rod graduated to 1 cm.

are required, the number of hundredths is equal to the number of divisions between the last tenth and the graduation mark on the same side of the hair as that tenth, and the number of thousandths is obtained by estimation. If the instrument is equipped with stadia hairs, care must be taken to read the correct (*middle*) horizontal cross hair.

The readings on the rod for the positions x, y, and z in Fig. 3-27(a) are determined as follows: For x, the number of feet below the cross hair is 4, the number of tenths below is 1, and the cross hair coincides with the first graduation above the tenth mark; consequently, the reading is 4.11 ft to the nearest hundredth, or 4.110 ft to the nearest thousandth. For y, the feet and tenths are again 4 and 1, respectively, and the hair is just midway between the graduations indicating 4 and 5 hundredths, therefore the reading to the nearest hundredth can be taken as either 4.14 or 4.15 ft. In determining the hundredths it is convenient to observe that the hair is just below the acute-angle graduation denoting the fifth hundredth, and it is therefore unnecessary to count up from the tenth graduation. If thousandths are required, the number of hundredths is the lower one, or 4; and since the hair is midway between two graduation marks on the rod and the distance between the graduations is 1 hundredth or 10 thousandths of a foot, the number of thousandths in the required reading is $\frac{1}{2} \times 10$, or 5. Hence the reading to the nearest thousandth is 4.145 ft. For z, the reading to the nearest hundredth is 3.96 ft and that to the nearest thousandth is 3.963 ft.

Direct high-rod readings are made with the rod fully extended, as the graduations on the face of the rod then appear continuous.

The metric rod shown in Fig. 3-27(b) is numbered every decimeter and graduated in centimeters. The double dot shown below the decimeter numbers indicates the readings are in the 2-m range. The readings for positions p, q, and r are, respectively, 2.430 m, 2.373 m, and 2.349 m.

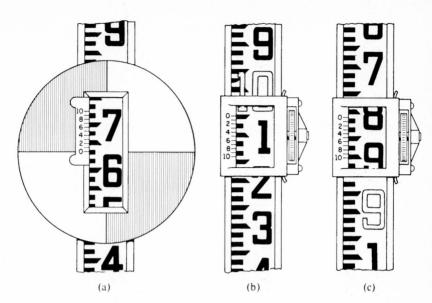

FIGURE 3-28 Verniers on leveling rod.

3-21   Verniers

The vernier is a device for making readings closer than the smallest division of a scale with more certainty than they can be obtained by estimation. The verniers found on a leveling rod are shown in Fig. 3-28. The front view in (a) shows the target and vernier for a short rod, and the rear views in (b) and (c) show the vernier on the back of the rod, which is employed in making high-rod readings when the target is used.

The principle of the vernier is dependent on the fact that the lengths of the divisions on the vernier are slightly less than the lengths of the divisions on the scale. A vernier of the type found on leveling rods is shown in Fig. 3-29. On this vernier a length corresponding to nine of the smallest divisions on the scale, or 0.09 ft, is divided into 10 parts on the vernier. The length of one of these vernier divisions is $\frac{1}{10} \times 0.09$ or 0.009 ft. The difference between the length of one division on the scale and one division on the vernier is equal to 0.01 − 0.009 = 0.001 ft. Hence, if the vernier is shifted from the position shown in Fig. 3-29(a), where the zero of the vernier coincides with the 0.3-ft mark, to the position shown in view (b), where the first division on the vernier coincides with the 0.31-ft mark, the distance the vernier has been moved must be the difference between the length of a main-scale division and the length of a vernier division, or 0.001 ft. In Fig. 3-29(c) the eighth division on the vernier is coinciding with a mark on the main scale, and hence the vernier must have been moved 0.008 ft from the original position in view (a). The number of thousandths of a foot must always be the same numerically as the number of the vernier graduation that coincides with a mark on the main scale. The readings shown in the three positions in Fig. 3-29 are 0.300, 0.301, and 0.308 ft, respectively.

FIGURE 3-29 Verniers.

If s represents the spaces on the main scale and n represents the number of divisions on the vernier, the vernier formula used to determine the least reading l is given as

$$l = \frac{s}{n} \tag{3-6}$$

The least reading is thus determined by dividing the length of the smallest division on the scale by the number of divisions on the vernier between two marks that coincide with marks on the main scale. The least count of the vernier shown in Fig. 3-29 is $0.01/10 = 0.001$ ft.

The readings on the rods shown in Fig. 3-28 are as follows: (a) 0.635 ft, (b) 10.053 ft, (c) 8.807 ft. Mistakes can often be averted by first estimating the number of thousandths and then looking at that part of the vernier where the coincidence must occur.

3-22 Optical Micrometer

The optical micrometer, the principle of which is shown in Fig. 3-30, measures the vertical distance on the rod from the point at which a horizontal line of sight strikes the rod to the next lower graduation mark. It consists of a thick piece of optical glass, the front and rear surfaces of which are ground flat and parallel to one another, together with a micrometer drum graduated so as to indicate the amount by which the plano-parallel element raises or lowers the line of sight. The unit either fits over the barrel of the telescope directly in front of the objective lens or else is an integral part of the instrument. A complete revolution of the drum, or that part of a revolution between the zero graduation and the last graduation, rep-

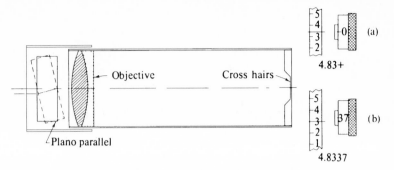

FIGURE 3-30 Optical micrometer principle.

resents a fixed value, such as 0.01 ft, 1 cm, or 5 mm. The rod graduations must therefore correspond to the particular optical micrometer being used.

The micrometer eliminates the necessity for estimating between graduations, and it further subdivides the distance between graduations very precisely into 50 or 100 parts. The position of the horizontal line of sight on the rod shown in Fig. 3-30(a) is between 4.83 and 4.84 ft, and the micrometer drum reads zero. Also the drum is divided into 100 parts. When the micrometer drum is turned and the plano-parallel element is rotated so that the line of sight is lowered to the 4.83-ft rod graduation [Fig. 3-30(b)], the micrometer drum reads 37. This reading means that the line of sight has been lowered $\frac{37}{100}$ of 0.01 ft, or 0.0037 ft. The rod reading is therefore 4.8337 ft.

The optical micrometer seen on the level shown in Fig. 3-31 both raises and lowers the line of sight 0.01 ft from an undeviated position by means of the plano-parallel element. When at the zero or undeviated position, the micrometer drum reads 10; when lowered the maximum amount, it reads 20; and when raised the maximum amount, it reads 0. Three readings using this type of optical micrometer are shown in Fig. 3-32. If the micrometer is set to read 10.0, the cross hair is located about one-fourth of the distance from the 5.24- and the 5.25-ft graduations. This is the undeviated position of the line of sight. If the micrometer is now rotated in order to lower the cross hair to the whole graduation at 5.24, the micrometer drum reading will be, for example, 12.4. This represents 0.0124 ft. The complete reading with this setting is then 5.2400 + 0.0124 = 5.2524 ft. If the micrometer is now rotated to bring the cross hair up to the 5.25-ft graduation, the drum reading will be, say, 2.6, which represents 0.0026 ft. The complete reading with this setting is then 5.2500 + 0.0026 = 5.2526 ft. The mean of the two readings is thus 5.2525 ft.

When using an optical micrometer, the instrument man must constantly check the coincidence of the level bubble, because manipulation of the micrometer knob may cause the telescope to go slightly out of level.

The graduations shown in Fig. 3-32 are characteristic of the Philadelphia rod graduations as well as those of many of the precise rods in use for high-order surveys. Their design requires that a single horizontal cross hair be set exactly on the upper or lower edge of the black bar before the micrometer drum is read. A some-

FIGURE 3-31 Optical micrometer drum.

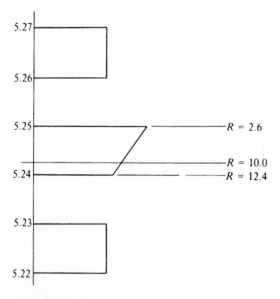

FIGURE 3-32 Micrometer drum reading for different positions of line of sight.

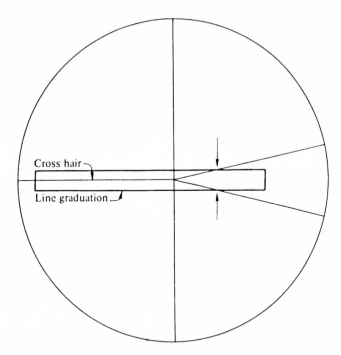

FIGURE 3-33 Wedge type reticle. The thickness of the line graduation is greatly exaggerated. Symmetry is observed at the arrows.

what more satisfactory design for use with the optical micrometer is the line graduation shown in Fig. 3-33. The level used with this type of design contains a horizontal cross hair that splits into a wedge halfway across the field of view. This wedge can be centered on the line graduation with a high degree of accuracy by the principle of symmetry.

3-23   Setting Up the Level

The purpose of direct leveling, as explained in Section 3-4, is to determine the difference of elevation between two points by reading a rod held on the points. These rod readings can be made by the levelman without setting the target, or the target can be set as directed by the levelman and the actual reading made by the rodman. At the instant the readings are made, it is necessary that the line of sight determined by the intersection of the cross hairs and the optical center of the objective be horizontal. In a properly adjusted instrument this line will be horizontal only when the bubble is at the center of the bubble tube.

The first step in setting up the level is to spread the tripod legs so that the tripod head will be approximately horizontal. The legs should be far enough apart to prevent the instrument from being blown over by a gust of wind, and they should be pushed into the ground far enough to make the level stable. Repairs to a damaged instrument are always expensive. For this reason, no instrument should be

(a)   (b)

FIGURE 3-34 Manipulation of four leveling screws.

set up on a pavement or a sidewalk. If such a setup cannot be avoided, additional care should be exercised to protect it from possible mishaps.

If the level contains four leveling screws, the telescope is turned over either pair of opposite leveling screws, as shown in Fig. 3-34(a). The bubble is then brought approximately to the center of the tube by turning the screws in *opposite* directions. The level bubble moves in the direction of the left thumb, a point well worth remembering. No great care should be taken to bring the bubble exactly to the center the first time.

The next step is to turn the telescope over the other pair of screws and to bring the bubble exactly to the center of the tube by means of these screws. This is shown in Fig. 3-34(b). The telescope is now turned over the first pair of screws once more, and this time the bubble is centered accurately. The telescope is then turned over the second pair of screws, and if the bubble has moved away from the center of the tube, it is brought back to the center. When the instrument is finally leveled up, the bubble should be in the center of the tube when the telescope is turned over either pair of screws. If the instrument is in adjustment, the bubble should remain in the center as the telescope is turned in any direction.

The beginner will need considerable practice in leveling up the instrument. It is by practice alone that he is able to tell how much to turn the screws to bring the bubble to the center. The more sensitive the bubble, the more skill is required to center it exactly. For the final centering, when the bubble is to be moved only a part of a division, only one screw need be turned. The screw that has to be tightened should be turned if both are a little loose, and the one that has to be loosened should be turned when they are tight. When the telescope is finally leveled up, all four screws should be bearing firmly but should not be so tight as to put a strain in the leveling head. If the head of the tripod is badly out of horizontal, it may be found that the leveling screws turn very hard. The cause is the binding of the ball-and-socket joint at the bottom of the spindle. The tension may be relieved by loosening both screws of the other pair.

When a three-screw instrument is to be leveled, the level bubble is brought parallel with a line joining any two screws, such as a and b in Fig. 3-35(a). By rotating these two screws in opposite directions, the instrument is tilted about the axis l—l, and the bubble can be brought to the center. The level bubble is now brought perpendicular to the line joining these first two screws, as shown in Fig. 3-35(b). Then only the third screw, c, is rotated to bring the bubble to the center.

FIGURE 3-35 Manipulation of three leveling screws.

This operation tilts the instrument about the axis m—m. The procedure is then repeated to bring the bubble exactly to the center in both directions.

When leveling a three-screw tilting level or an automatic level equipped with a bull's-eye bubble, as shown in Fig. 3-35(c), opposite rotations of screws a and b cause the bubble to move in the direction of the axis m—m. Rotation of screw c only causes the bubble to move in the direction of the axis l—l.

In walking about the instrument, the levelman must be careful not to step near the tripod legs, particularly when the ground is soft. Neither should any part of the level be touched as the readings are being made, because the bubble can be pulled off several divisions by resting the hand on the telescope or on a tripod leg. The bubble will not remain in the center of the tube for any appreciable length of time. The levelman should form the habit of always checking the centering of the bubble just before and just after making a reading. Only in this way can he be sure that the telescope was actually horizontal when the reading was made.

3-24 Signals

In running a line of levels, the levelman and the rodman must be in almost constant communication with each other. As a means of communication, certain convenient signals are employed. It is important that the levelman and the rodman understand these in order to avoid mistakes. When the target is used, it is set by the rodman according to signals given by the levelman. Raising the hand above the shoulder, so that the palm is visible, is the signal for raising the target; lowering the hand below the waist is the signal for lowering the target. The levelman, viewing the rod and the rodman through the telescope, should remember that he can see them much more distinctly than he can be seen by the rodman. Hence his signals should be such that there is no possible chance of misunderstanding. A circle described by the hand is the signal for clamping the target, and a wave of both hands indicates that the target is properly set, or all right. The signal for plumbing the rod is to raise one arm above the head and then to lean the body in the direction in which the rod should be moved.

3-25   Running a Line of Levels

In the preliminary example of direct leveling given in Section 3-4, it was assumed that the difference of elevation between the two points considered could be obtained by a single setting of the level. This will be the case only when the difference in elevation is small and when the points are relatively close together. In Fig. 3-36, rods at points A and K cannot be seen from the same position of the level. If it is required to find the elevation of point K from that of A, it will be necessary to set up the level several times and to establish intermediate points such as C, E, and G. These are the conditions commonly encountered in the field and may serve as an illustration of the general methods of direct leveling.

Let the elevation of the benchmark (abbreviated BM) at A be assumed as 820.00 ft. This is recorded as shown in the leveling notes of Fig. 3-37. The level is set up at B, near the line between A and K, so that a rod held on the BM will be visible through the telescope; the reading on the rod is found to be 8.42 ft. This reading is called a *backsight* reading, or simply a backsight (abbreviated BS), and is recorded as such in the notes. A backsight is the rod reading taken on a point of known elevation to determine the *height of instrument* (abbreviated HI). If the BS of 8.42 ft is added to the elevation of A, the HI is obtained. Thus HI = 820.00 + 8.42 = 828.42 ft. This is shown in the notes.

After the HI has been established, a point C called a *turning point,* is selected that is slightly below the line of sight. This point should be some stable unambiguous object, so that the rod can be removed and put back in the same place as many times as may be necessary. For this purpose, a sharp-pointed solid rock or a well-defined projection on some permanent object is preferable. If no such object is available, a stake or a railroad spike can be driven firmly in the ground and the rod held on top of it. After the turning point at C, designated TP-1,

FIGURE 3-36 Direct leveling.

STA	BS	HI	FS	ELEV			

LEVELING, BMA to BMK
SHATTUCK AVE. SEWER PROJECT AKRON, OHIO   MAY 18, 1981

					LEVEL # 4096	LEVEL J. BROWN	
STA	BS	HI	FS	ELEV	ROD # 18	ROD F. SMITH	
BMA	8.42	828.42		820.00			
TP₁	11.56	838.78	1.20	827.22	BMA is top of iron pipe, S.E. cor.		
TP₂	6.15	843.58	1.35	837.43	Shattuck + Maple Sts.		
TP₃	4.39	837.07	10.90	832.68			
BMK			5.94	831.13	BMK is Bronze disk in sidewalk		
					N.W. Cor. Shattuck + Vincent Sts.		
₰BS	+30.52		₰FS -19.39				
	-19.39				831.13		
	+11.13				-820.00		
					+11.13 checks		

FIGURE 3-37 Level notes.

is set or selected, a reading is taken on the rod held on TP-1. If this reading is 1.20 ft, TP-1 is 1.20 ft below the line of sight, and the elevation of TP-1 is 828.42 − 1.20 = 827.22 ft, as shown in the notes. This rod reading is called a *foresight* (abbreviated FS). An FS is taken on a point of unknown elevation in order to determine its elevation from the height of instrument.

Occasionally, successive foresights and backsights are taken on an overhead point such as on a point in the roof of a tunnel. The foresight taken on such a point is *added* to the HI to obtain the elevation of the point. The backsight taken on the point is *subtracted* from the elevation of the point to determine the HI. Such readings must be carefully noted in the field notes.

While the rodman remains at *C,* the level is moved to *D* and set up as high as possible but not so high that the line of sight will be above the top of the rod when it is again held at *C*. This can be quickly checked by means of a hand level. The reading 11.56 ft is taken as a backsight. Hence, the HI at *D* is 827.22 + 11.56 = 838.78 ft. When this reading is taken, it is important that the rod be held on exactly the same point that was used for a foresight when the level was at *B*.

After the backsight on *C* has been taken, another turning point *E* is chosen, and a foresight of 1.35 ft is obtained. The elevation of *E* is 838.78 − 1.35 = 837.43 ft. The level is then moved to *F* and the backsight of 6.15 ft taken on *E*. The new HI is 837.43 + 6.15 = 843.58 ft. From this position of the level, a foresight of 10.90 ft is taken on *G,* the elevation of which is 843.58 − 10.90 = 832.68 ft. The level is then set up at *H,* from which position a backsight reading of 4.39 ft is taken on *G,* and a foresight reading of 5.94 ft is taken on the new BM at K. The final HI is 832.68 + 4.39 = 837.07 ft, and the elevation of K is 837.07 − 5.94 = 831.13 ft. As the starting elevation was 820.00 ft, the point K is 11.13 ft higher than A.

When using the digital level, the elevation of the beginning benchmark is keyed in. The microprocessor automatically adds backsight readings to obtain successive HIs, and subtracts foresights from preceding HIs to obtain the elevations of successive turning points and new benchmarks. These values, together with point designations keyed in by the instrumentman, are all stored in the memory module.

3-26   Checking Level Notes

To eliminate arithmetical mistakes in the calculation of HIs and elevations, the arithmetic should be checked on each page of notes. Adding backsights gives $\Sigma\,BS$; adding foresights gives $\Sigma\,FS$. Then $\Sigma\,BS - \Sigma\,FS$ should equal the difference in elevation (DE) between the starting point to the last point on the page. This is shown in the notes of Fig. 3-37. $\Sigma\,BS - \Sigma\,FS$ = +11.13 ft, and the calculated DE is also +11.13, which checks the arithmetic. The last point on the page should then be carried to the following page before the BS on that point is recorded in the notes.

3-27   Check Levels

Although the arithmetic in the reduction of the field notes may have been verified, there is no guarantee that the difference of elevation is correct. The difference of elevation is dependent on the accuracy of each rod reading and on the manner in which the field work has been done. If there has been any mistake in reading the rod or in recording a reading, the difference of elevation is incorrect.

The only way in which the difference of elevation can be checked is by carrying the line of levels from the last point back to the original benchmark or to another benchmark whose elevation is known. This is called "closing a level circuit." If the circuit closes on the original benchmark, the last point in the circuit, BMK in Fig. 3-36, must be used as a turning point; that is, after the foresight has been read on the rod at K from the instrument setup at H, the level must be moved and reset before the backsight is taken on K to continue the circuit to closure. Otherwise, if a mistake was made in reading the rod on the foresight to K from the setup at H, this mistake will not be discovered when checking the notes. A plan view of the level circuit between A and K in which the circuit is closed back on A is shown in Fig. 3-38(a). Note that the level has been reset between the FS taken on K and the BS taken on K. In Fig. 3-38(b) the level circuit has been continued to a known benchmark P to close the circuit. BMK is used as a turning point in this instance.

If the line of levels is carried back to BMA in the above example, on return the measured elevation of A should be 820.00 ft. The difference represents the error of closure of the circuit and should be very small. If a large discrepancy exists, the mistake may have been made in adding and subtracting backsight and foresight readings. This will be discovered on checking the notes. Otherwise there was a wrong reading of the rod or a wrong value was entered in the field notes. The adjustment of a level circuit based on the error of closure is discussed in Section 5-11.

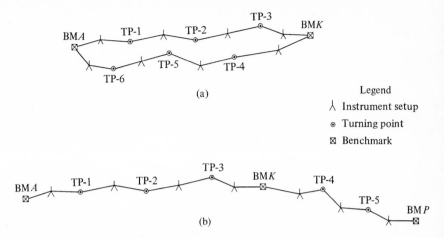

FIGURE 3-38 Closing on known benchmark.

3-28 Sources of Error in Leveling

The principal sources of error in leveling are instrumental defects, faulty manipulation of the level or rod, settling of the level or the rod, errors in sighting, mistakes in reading the rod or in recording or computing, errors due to natural sources, and personal errors.

3-29 Instrumental Errors

The most common instrumental error is caused by the level being out of adjustment. As has been previously stated, the line of sight of the telescope is horizontal when the bubble is in the center of the tube, provided the instrument is in perfect adjustment. When it is not in adjustment, the line of sight will either slope upward or downward when the bubble is brought to the center of the tube. The various tests and adjustments of the level are given in Appendix B.

Instrumental errors can be eliminated or kept at a minimum by testing the level frequently and adjusting it when necessary. Such errors can also be eliminated by keeping the lengths of the sights for the backsight and foresight readings nearly equal at each setting of the level. Since it is never known just when an instrument goes out of adjustment, this latter method is the more certain and should always be used for careful leveling.

In Fig. 3-39 the line of sight with the level at B should be in the horizontal line $EBGK$. If the line of sight slopes upward as shown and a sight is taken on a rod at A, the reading is AF instead of AE. This reading is in error by the amount of $EF = e_1$. When the telescope is directed toward a rod held at C or D, the line of sight will still slope upward through the same vertical angle if it is assumed that the bubble remains in, or is brought to, the center of the tube. The rod reading taken on C will be CH, which is in error by an amount $GH = e_2$. If the horizontal distances BE and BG are equal, the errors e_1 and e_2 will be alike, and the difference between the two

FIGURE 3-39 Errors caused by imperfect adjustment of level.

rod readings *AF* and *CH* will be the true difference of elevation between *A* and *C*. A rod reading taken at *D* will be *DL,* with an error $KL = e_3$. Since *BK* is longer than *BE,* the difference in elevation between *A* and *D* will be in error by an amount equal to the difference between e_3 and e_1. Similar reasoning applies if the line of sight slopes downward instead of upward. This error, which is called the *collimation error,* also exists in an automatic level.

It is not always feasible or even possible to balance a foresight distance with a backsight distance. This situation might occur on account of the terrain or when a series of foresights must be taken from a single instrument setup, as in checking grade over a large area. It is advisable in such cases to check the level and make the necessary adjustments (see Section 3-39 and also Appendix B). An alternative is to employ a level like the one shown in Fig. 3-16 in which the rod can be read with the telescope right-side up and upside down, giving a mean reading free of error from a sloping line of sight. Another alternative is the technique of reciprocal observations discussed in Section 3-37.

The digital level discussed in Section 3-15 has provision for displaying the imbalance between backsight and foresight distances, both for individual setups and for accumulation along the level line between benchmarks (see Section 3-39).

Extremely long sights should also be avoided. The further the rod is from the level, the greater will be the space covered on the rod by the cross hair and the more difficult it will be to determine the reading accurately. For accurate results, sights with the engineer's level should be limited to no more than 150 ft or 50 m.

3-30   Errors Due to Manipulation

As has been previously stated, the careful levelman will form the habit of checking the position of the bubble just before and just after making each rod reading. This is the only way in which he can be certain that he is getting the proper reading.

The amount of the error due to the bubble being off center will depend on the sensitiveness of the bubble. A very convenient way to determine this error for any given bubble and for any given distance is to remember that an error of 1′ in angle causes an error of about $\frac{1}{2}$ in. or 0.045 ft at a distance of 100 ft (15 mm in 50 m). Thus, if a 30″ bubble is off one division at the instant the reading is made, the resulting error will be about 0.02 ft when the rod is 150 ft away (7.5 mm in 50 m). This type of error does not exist in an automatic level.

A common mistake in handling the rod is in not being careful to see that the target is properly set before a high-rod reading is made with the target. Many rods have been damaged by allowing the upper portion to slide down rapidly enough to affect the blocks at the bottom of the lower section and the top of the upper section. If this has been done, it is probable that the reading on the back of a Philadelphia rod will not be exactly 7 ft when set as a low rod. In this case the target should be set at that reading, rather than at exactly 7 ft, for a high-rod reading with the target. When a high rod is being read directly from the level, the rodman should make sure that the rod is properly extended and has not slipped down.

When the target is being used, the levelman should check its position after it has been clamped to make sure that it has not slipped. The beginner will be astonished at the care that must be exercised in making the target coincide exactly with the horizontal cross hair. Readings taken on the same point about 150 ft from the level may vary by several hundredths of a foot if the bubble is not exactly centered and if the target is carelessly set.

3-31   Errors Due to Settlement

If any settlement of the level takes place in the interval between the reading of the backsight and the reading of the foresight, the resulting foresight will be too small, and all elevations beyond that point will be too high by the amount of the settlement. Also, if the turning point should settle while the level is being moved forward after the foresight has been taken, the backsight on the turning point from the new position of the level will be too great and all elevations will be too high by the amount of the settlement.

Errors due to the settlement of the level can be avoided by keeping the level on firm ground. If this is impossible, stakes can be driven in the ground and the tripod legs can be set on these instead of directly on the ground. In precise leveling, two rods and two rodmen are used to make the backsight and foresight readings from a setup more quickly. The error is still further diminished by taking the backsight first at one setup of the level and the foresight first at the next setup.

A proper choice of turning points should eliminate error from their settlement. If soft ground must be crossed, long stakes should be used as turning points.

3-32   Errors in Sighting

If parallax exists between the image formed by the focusing lens and the plane of the cross hairs, an error will be introduced in the rod reading. Parallax is eliminated by the procedure discussed in Section 3-11.

The rod should be plumb when the reading is made. The levelman can tell whether or not the rod is plumb in one direction by noting if it is parallel to the vertical cross hair. He cannot tell, however, if it is leaning toward or away from him. The leveling rods used on precise work are equipped with circular levels, so that the rodman can tell when he is holding the rod vertically. For less accurate work the rodman can balance the rod between his fingers if the wind is not blowing, or he can wave it slowly toward and then away from the level as indicated in

FIGURE 3-40 Waving the rod to obtain lowest reading.

Fig. 3-40. The least reading obtainable is the proper one. If the target is being used, the line dividing the colors should just coincide with the cross hair and then drop away from it. Errors from failure to hold the rod plumb will be much greater on readings near the top of the rod than for those near the bottom. For this reason more care should be exercised when making high-rod readings.

For careful work the lengths of the backsight and the foresight from the same setup should be kept nearly equal (see Section 3-29). If in ascending a steep hill the level is always kept on the straight line between the turning points, the distance to the backsight will be about twice as great as the distance to the foresight, and considerable error may result if the instrument is not in good adjustment. If there are no obstructions, these two distances can be kept nearly equal by setting the level some distance away from the straight line between the turning points. By thus zigzagging with the level, this source of error can be eliminated.

3-33 Mistakes in Reading the Rod, Recording, and Computing

A common mistake in reading the rod is to misread the number of feet or tenths or meters and decimeters. The careful levelman observes the foot and tenth marks both above and below the cross hair. On close sights, no foot mark may appear within the field of the telescope. In this case the reading can be checked by directing the rodman to place his finger on the rod at the cross hair, or if the reading is a high one, by having him slowly raise the rod until a foot mark appears in the telescope. In case of doubt the target can always be used.

Some instruments for precise leveling are equipped with three horizontal cross hairs. All three hairs are read at each sighting. If the hairs are evenly spaced, the difference between the readings of the upper and the middle hairs should equal the difference between the readings of the middle and lower hairs. This comparison is always made before the rodman leaves a turning point.

Where readings to thousandths of a foot are being made with the target, a common mistake in recording is to omit one or more ciphers from such readings as 5.004, and to record instead 5.04 or 5.4. Such mistakes can be avoided by making sure that there are three decimal places for each reading. Thus the second reading, if correct, should be recorded as 5.040, and the third as 5.400. If the values are not so recorded, the inference would be that the levelman was reading only to hundredths of a foot on the second reading and only to tenths on the third.

Other common mistakes of recording are the transposition of figures and the interchanging of backsight and foresight readings. If the levelman will keep the rodman at the point long enough to view the rod again after recording the reading, mistakes of the first type can often be detected. To prevent the interchange of readings, the beginner should remember that ordinarily the first reading taken from each position of the level is the backsight reading and that only one backsight is taken from any position of the level. Any other sights taken are foresights.

Mistakes in computations, as far as they affect the elevations of turning points and benchmarks, can be detected by checking the notes, as described in Section 3-26. This should be done as soon as the bottom of a page is reached, so that incorrect elevations will not be carried forward to a new page.

3-34   Errors Due to Natural Sources

One error due to natural sources is that caused by curvature and refraction, as described in Section 3-2. The error from this source amounts to but 0.0002 ft in a 100-ft sight (0.01 mm/30 m) and to about 0.002 ft in a 300-ft sight (0.7 mm/100 m). So for ordinary leveling it is a negligible quantity. It can be practically eliminated by keeping the backsight and foresight distances from the same setup equal. In precise leveling, if the backsight and foresight distances are not substantially equal, a correction is applied to the computed difference of elevation.

The familiar heat waves seen on a hot day are evidence of refraction, and when they are seen, refraction may be a significant source of error in leveling. When the heat waves are particularly intense, it may be impossible to read the rod unless the sights are much shorter than those usually taken. Refraction of this type is much worse close to the ground. For careful work it may be necessary to discontinue the leveling for 2 or 3 hr during the middle of the day. It may also be possible to keep the error from this source at a low figure by taking shorter sights and by so choosing the turning points that the line of sight will be at least 3 or 4 ft or 1 m above the ground.

Better results will usually be obtained when it is possible to keep the level shaded. If the sun is shining on the instrument, it may cause an unequal expansion of the various parts of the instrument; or if it heats one end of the bubble tube more than the other, the bubble will be drawn to the warmer end of the tube. For precise work the level must be protected from the direct rays of the sun.

To guard against changes in the length of the leveling rod from variations in temperature, the graduations on rods used for precise leveling are placed on strips of invar, which has an extremely small coefficient of expansion. For ordinary leveling, errors from this source are negligible.

3-35   Personal Errors

Some levelmen consistently tend to read the rod too high or too low. This may be caused by defective vision or by an inability to decide when the bubble is properly centered. In general, personal errors tend to be compensating, although some individuals may manipulate the level in such a way as to cause them to be cumulative.

Such a tendency can be discovered by leveling over a line that has been previously checked by several different parties or by running a number of circuits of levels, each of which begins and ends on the same point.

3-36   Limits of Error

If care is used in leveling, most of the errors will tend to be random. For this reason the error in any line can be expected to be proportional to the number of setups. Since the number of setups per mile of levels will be nearly constant, the error will also be proportional to the distance in miles. The precision that is being attained can be determined by comparing the two differences of elevation obtained by running levels in both directions over a line.

Level lines are classified as first order, second order, third order, or fourth order in accordance with the field procedures used and also with the agreement between the results of leveling in both directions over a line. Table 3-2 outlines the standards for all but fourth-order accuracy in vertical control surveys and gives the principal uses to which these levels of accuracy are applied. Trigonometric and barometric leveling are considered as being of fourth-order accuracy or less.

3-37   Reciprocal Leveling

In leveling across a river or a deep valley, it is usually impossible to keep the lengths of the foresight and the backsight nearly equal. In such cases reciprocal leveling is used, except where approximate results are sufficient. The difference of elevation between points on the opposite sides of the river or valley is then obtained from two sets of observations.

The method is illustrated by Fig. 3-41. The level is first set up at L_1 and rod readings are taken on the two points A and B. From these readings a difference of elevation is obtained. The level is then taken across the stream and set at such a position L_2 that $L_2B = L_1A$ and $L_2A = L_1B$. From this second position, readings are again taken on A and B, and a second difference of elevation is obtained. It is probable that these two differences will not agree. Both may be incorrect because of instrumental errors and curvature and refraction. However, the true difference of elevation should be very close to the mean of the two differences thus obtained.

The accuracy of this method will be increased if two leveling rods can be used, so that no appreciable time will elapse between the backsight and the foresight readings. It will be further increased by taking a number of rod readings on the more distant point and using the average of these readings, rather than depending on a single observation. If the distance AB is very great, it is important that the atmospheric conditions be the same for both positions of the level. Otherwise a serious error may be introduced by a changed coefficient of refraction.

◆EXAMPLE 3-2   In Fig. 3-41, with the level set at L_1, a backsight of 5.13 ft is taken on the rod held at A and a foresight of 7.25 ft is taken on the rod held at B. With the level at L_2, a backsight of 4.98 ft is taken on A, and a foresight of 6.82 ft is taken on B. The elevation of A is 298.72 ft. Determine the elevation of B.

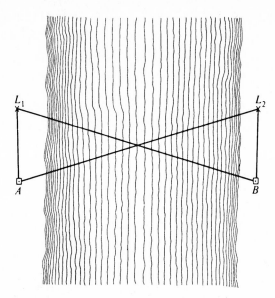

FIGURE 3-41 Reciprocal leveling.

Solution: From the readings taken at L_1, the difference in elevation from A to B is +5.13 − 7.25 = −2.12 ft. From the readings taken at L_2, the difference in elevation from A to B is +4.98 − 6.82 = −1.84 ft. The mean difference in elevation is −1.98 ft. Thus, the elevation of B is 298.72 − 1.98 = 296.74 ft.   ◈

3-38   Three-Wire Leveling

Three-wire leveling, also referred to as precise leveling, is a process of direct leveling wherein three cross hairs, referred to as threads, are read and recorded rather than the single horizontal cross hair. The cross hairs are spaced so as to represent a horizontal distance of about 100 ft for every full foot intercepted on the rod between the upper and lower cross hairs. Any level equipped with three horizontal cross hairs can be used for three-wire leveling.

In the geodetic level discussed in Section 3-14, the cross hairs are spaced so as to represent a horizontal distance of about 350 ft for every foot of rod interval. This spacing has the advantage in that with greater magnification long sights can be made, while at the same time all three hairs can be observed on the rod. Suppose, for example, that the sight length is 400 ft, the reading of the center cross hair is 1.500 ft, and the ratio distance to rod intercept is 100 : 1. The upper cross hair would fall on the 3.500-ft mark, but the lower cross hair would fall 0.500 ft below the bottom of the rod. If the ratio is 350 : 1, the upper cross hair would fall on the 2.071-ft mark, whereas the lower cross hair would fall on the 0.929-ft mark.

The rods used are graduated in feet, yards, or meters. Since three-wire leveling is used for great precision, the precise leveling rods described in Section 3-19 are preferred for use in leveling of this type. Furthermore, the rods are used in pairs.

TABLE 3-2 Classification, Standards of Accuracy, and General Specifications for Vertical Control

Classification	First-Order Class I, Class II	Second-Order Class I	Second-Order Class II	Third-Order
Principal Uses Minimum standards; higher accuracies may be used for special purposes	Basic framework of the National Network and of metropolitan area control; Extension engineering projects; Regional crustal movement investigations; Determining geopotential values	Secondary control of the National Network and of metropolitan area control; Large engineering projects; Local crustal movement and subsidence investigations; Support for lower-order control	Control densification, usually adjusted to the National Network; Local engineering projects; Topographic mapping; Studies of rapid subsidence; Support for local surveys	Miscellaneous local control; may not be adjusted to the National Network; Small engineering projects; Drainage studies and gradient establishment in mountainous areas
Recommended Spacing of Lines National Network	Net A; 100 to 300 km *Class I* Net B; 50 to 100 km *Class II*	Secondary Net; 20 to 50 km	Area Control; 10 to 25 km	As needed
Metropolitan area control; other purposes	2 to 8 km; As needed	0.5 to 1 km; As needed	As needed; As needed	As needed; As needed
Spacing of Marks Along Lines	1 to 3 km	1 to 3 km	Not more than 3 km	Not more than 3 km
Instrument Standards	Automatic or tilting levels with parallel plate micrometers; invar scale rods	Automatic or tilting levels; optical micrometers or three-wire levels; invar scale rods	Geodetic levels and invar scale rods	Geodetic levels and rods

Field Procedures	Double-run; forward and backward, each section	Double-run; forward and backward, each section	Double- or single-run	Double- or single-run
Section length	1 to 2 km	1 to 2 km	1 to 3 km for double-run	1 to 3 km for double-run
Maximum length of sight	50 m *Class I*; 60 m *Class II*	60 m	70 m	90 m
Max. difference in lengths Forward and backward sights per setup	2 m *Class I*; 5 m *Class II*	5 m	10 m	10 m
per section (cumulative)	4 m *Class I*; 10 m *Class II*	10 m	10 m	10 m
Max. length of line between connections	Net A; 300 km Net B; 100 km	50 km	50 km double-run 25 km single-run	25 km double-run 10 km single-run
Maximum Closures[a]				
Section: forward and backward	3 mm $\sqrt{K}$ *Class I:* 4 mm $\sqrt{K}$ *Class II*	6 mm $\sqrt{K}$	8 mm $\sqrt{K}$	10 mm $\sqrt{K}$
Loop or line	4 mm $\sqrt{K}$ *Class I:* 5 mm $\sqrt{K}$ *Class II*	6 mm $\sqrt{K}$	8 mm $\sqrt{K}$	12 mm $\sqrt{K}$

[a]Check between forward and backward runnings where K is the distance in kilometers.

FIGURE 3-42 Direction diagram.

The direction diagram and the form of notes for three-wire leveling are given in Figs. 3-42 and 3-43. The left-hand page is for the backsight readings, and the right-hand page is for the foresight readings. It is to be noted that a station refers to an instrument setup and not to a point on the ground. The procedure is explained with reference to the notes of Fig. 3-43. The levels are carried from BM23 to BM24 in the forward direction. The sun is shining from direction 5, and the wind is blowing from direction 2.

The forward rodman, carrying rod No. 4, selects and turning point, and the rear rodman is at the previous turning point. The level is now set up, and the observer backsights on the back turning point, reading 2487, 2416, and 2345 on the three cross hairs. He must take care that the level bubble is centered at each reading. These readings are made on a meter rod and are actually 2.487, 2.416, and 2.345 m, respectively. The back of the rod is read by using only the middle cross hair, and the reading is 7.93 ft. The upper interval and the lower interval of 71 and 71 are computed and found to agree within 2 mm. So the mean of the three cross-hair readings is computed. This is compared with the equivalent reading in feet as a check against a mistake.

The observer than takes readings on rod No. 4, which is held on the forward turning point. The readings are 0519, 0444, and 0369 from the front of the rod and 1.46 ft from the back of the rod. The half intervals are then computed for agreement, and the mean reading is computed and recorded. The total backsight interval of 142 mm, representing the backsight distance, differs from the total foresight interval of 150 mm, and the sum of the backsight intervals from the point of beginning (BM23) is out of balance with the sum of the foresight intervals.

The back rodman advances to locate the next forward turning point. The next instrument setup is at station 43. At this setup the foresight readings are taken before the backsight readings. Alternating the sequence of readings at each setup tends to eliminate the effect of instrument settlement. Note that rod No. 7 will be sighted on first at each setup. The accumulated sum of the backsight intervals is still out of balance with the sum of the foresight intervals, so the foresight distances must be made shorter. After station 46, the sums of intervals are out of balance by only 1 mm.

As a check on computing the mean readings, all the readings are added, the sum is divided by three, and the result is compared with the sum of the means. This check is shown on the backsight page. A further check is made by adding the middle-thread readings in feet and converting the sum to meters as shown.

3-39   Collimation Correction

In the example of Section 3-38, the sum of the backsight intervals is 725 mm and the sum of the foresight intervals is 724 mm. If the distance-to-intercept ratio is assumed to be 350 : 1, the total of the backsight lengths is $725 \times 350 = 253,750$ mm or about 254 m. The total foresight length is 253 m. This is a very close balance, requiring no measurable correction for the error discussed in Section 3-29 and shown in Fig. 3-39. Usually, however, such a close balance is not possible, and it becomes necessary to apply a collimation correction before the correct difference in elevation can be determined. This correction is commonly referred to as the C-factor correction, in which C represents the inclination of the line of sight when the level bubble is centered.

In Fig. 3-44 it is assumed that the level bubble is centered but the line of sight is inclined by the small angle α. The reading of a rod would be R, instead of R'. The error introduced is $R - R' = D \tan \alpha = D\alpha$, where α is a small angle expressed in radians. The correction to R is then $-D\alpha = DC$, in which C is the value of the C factor. Thus $C = -\alpha$ and it follows that if C is positive, the line of sight is inclined downward.

To correct for an inclined line of sight when precise levels are run, the C factor must be determined from time to time, perhaps as frequently as twice a day. It takes only a few minutes to determine C, and this factor should be evaluated not only for the tilting level but also for the self-leveling level, since the same type of error always exists in this type of instrument.

To compute C, two points at least 200 ft apart are chosen in fairly level terrain, and two independent differences of elevation between the points are determined from two setups. In Fig. 3-45(a) the instrument is set up about 20 ft from A. The distance to A is d_1, and the distance to B is D_1. These distances are measured. The rod reading at A is n_1, and that at B is N_1. The line of sight is shown sloping downward, or C is assumed to be positive. The difference in elevation (abbreviated DE) found from this setup is the corrected backsight at A minus the corrected foresight at B. Thus

$$DE_{AB} = (n_1 + Cd_1) - (N_1 + CD_1) \tag{3-7}$$

The instrument is then set up about 20 ft from B, as indicated in Fig. 3-45(b). Rod readings N_2 and n_2 are taken at A and B, and the distances D_2 and d_2 from the instrument to the two rods are measured. The difference in elevation found from the second setup is

$$DE_{AB} = (N_2 + CD_2) - (n_2 + Cd_2) \tag{3-8}$$

Equating the right-hand sides of Eqs. (3-7) and (3-8) gives

$$(n_1 + Cd_1) - (N_1 + CD_1) = (N_2 + CD_2) - (n_2 + Cd_2)$$

THREE-WIRE LEVELING

DATE : *8-30-59*
Sun : *5*

FORWARD – ~~BACKWARD~~
(Strike out one word)

No. of Station	Thread Reading Back Sight	Mean	Middle Thread Reading Feet	Thread Interval	Sum of Intervals
	2487			71	
42	2416	2416.0	7.93	71	
	2345			142	142
	2800			83	
43	2717	2716.7	8.92	84	
	2633			167	309
	3242			61	
44	3181	3181.0	10.43	61	
	3120			122	431
	2744			72	
45	2672	2672.3	8.77	71	
	2601			143	574
	2602			75	
46	2527	2526.7	8.29	76	
	2451			151	725
	40538	13512.7	44.34		
	−10417	3472.3	11.40		
	3⟌30121	+10040.4	+32.94 ft.	= +10.04	meters
	+10040.3				

FIGURE 3-43 Notes for three-wire levels.

THREE-WIRE LEVELING

From B.M.: 23

Wind: 2

To B.M.: 24

Time: 3⁴⁵

Rod and Temp.	Thread Reading Fore Sight	Mean	Middle Thread Reading Feet	Thread Interval	Sum of Intervals
4	0519			75	
22°C	0444	0444.0	1.46	75	
	0369			150	150
7	0748			82	
	0666	0665.3	2.19	84	
	0582			166	316
4	0663			57	
	0606	0606.3	1.99	56	
	0550			113	429
7	1066			74	
	0992	0991.7	3.25	75	
	0917			149	578
4	0838			73	
24°C	0765	0765.0	2.51	73	
	0692			146	724
	10417	3472.3	11.40		

The result obtained by solving for C is

$$C = \frac{(n_1 + n_2) - (N_1 + N_2)}{(D_1 + D_2) - (d_1 + d_2)} \tag{3-9}$$

Since, as will be shown, corrections for collimation errors are computed from intervals rather than from distances, the two terms in the denominator are expressed in intervals. The quantity C is then expressed in feet per foot of interval or in millimeters per millimeter of interval.

Notes for the determination of C are shown in Fig. 3-46. The leveling rod was graduated in feet and thread readings were recorded in feet. The conditions are represented in Fig. 3-45. When the instrument was set up near A, the readings of the three threads with the rod held at A were 4.600, 4.448, and 4.298, as shown in column 2 of Fig. 3-46. Their mean, which is shown in column 3, is 4.449 ft. This is n_1 in Fig. 3-45. The value of d_1 expressed as an interval is 0.302 ft, as shown in column 4 of Fig. 3-46 and again in column 5. The rod is now taken to B, and the three thread readings are 4.237, 3.515, and 2.790, as shown in column 6. The mean, which is 3.514, is shown in column 7. This is N_1 in Fig. 3-45. The value of D_1 expressed as an interval is 1.447, as shown in columns 8 and 9 of Fig. 3-46.

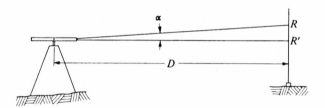

FIGURE 3-44 Inclination of line of sight of level.

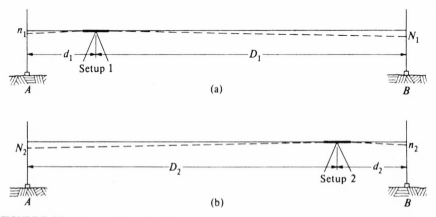

FIGURE 3-45 Setups to determine C factor.

With the instrument set up near B in Fig. 3-45, the rod readings taken on B are recorded on the left-hand page of the notes in Fig. 3-46 in column 2. These rod readings give n_2 and d_2, as shown. In column 5 the quantity $(d_1 + d_2)$, or the sum of the intervals, is shown as 0.692 ft. From this setup, the rod readings taken at A are recorded on the right-hand page in column 6. These readings give N_2 and D_2, as shown. In column 9 the quantity $(D_1 + D_2)$ is shown as 2.845. The sum of two means in column 3 gives $(n_1 + n_2)$ in Eq. (3-9); the sum of the two means in column 7 gives $(N_1 + N_2)$. The remaining computations needed to determine C are shown directly in the notes.

Assume that the results obtained when a line of levels was run between two points A and B in Fig. 3-47 are as follows: sum of backsight means, 31.422 ft; sum of foresight means, 12.556 ft; sum of backsight intervals, 22.464 ft; and sum of

			DETERMINATION OF C-FACTOR						
	BACKSIGHT					FORESIGHT			
No. of Station	Thread Reading Feet	Mean Feet		Thread Interval Feet	Sum of Intervals Feet		Thread Reading Feet	Mean Feet		Thread Interval Feet	Sum of Intervals Feet
(1)	(2)	(3)		(4)	(5)		(6)	(7)		(8)	(9)
	4.600			0.152			4.237			0.722	
1	4.448	4.449=n_1		0.150			3.515	3.514=N_1		0.725	
	4.298		d_1:	0.302	0.302·d_1		2.790		D_1=	1.447	1.447·D_1
	4.400			0.195			5.863			0.698	
2	4.205	4.205·n_2		0.195			5.165	5.164·N_2		0.700	
	4.010		d_2:	0.390	0.692=d_1·d_2		4.465		D_2=	1.398	2.845·D_1·D_2
		8.654 = n_1·n_2						8.678 = N_1·N_2			
		−8.678 = N_1·N_2							(D_1·D_2)−(d_1·d_2) = 2.153		
		−0.024 = (n_1·n_2)−(N_1·N_2)									
		$C = \frac{0.024}{2.153} = -0.0112$ ft /ft of Interval									

FIGURE 3-46 Notes for determining C factor.

FIGURE 3-47 Imbalance of sum of backsight and sum of foresight intervals.

foresight intervals, 27.845 ft. The difference between Σ FS intervals and Σ BS intervals is $27.845 - 22.464 = 5.381$ ft. Therefore the observed value of Σ FS means must be corrected by an amount equal to the C factor times the difference between the intervals, that is, by $-0.0112 \times 5.381 = -0.060$ ft. The corrected value of Σ FS means is thus $12.556 - 0.060 = 12.496$ ft. The corrected difference in elevation is then Σ BS means $- \Sigma$ FS means $= 31.422 - 12.496 = +18.926$ ft.

3-40   Profile Levels

The purpose of profile leveling is to determine the elevations of the ground surface along some definite line. Before a railroad, highway, transmission line, sidewalk, canal, or sewer can be designed, a profile of the existing ground surface is necessary. The route along which the profile is run may be a single straight line, as in the case of a short sidewalk; a broken line, as in the case of a transmission line or sewer; or a series of straight lines connected by curves, as in the case of a railroad, highway, or canal. The data obtained in the field are usually employed in plotting the profile. This plotted profile is a graphical representation of the intersection of a vertical surface or a series of vertical surfaces with the surface of the earth, but it is generally drawn so that the vertical scale is much larger than the horizontal scale in order to accentuate the differences of elevation. This is called vertical exaggeration.

3-41   Stations

The line along which the profile is desired must be marked on the ground in some manner before the levels can be taken. The common practice is to set stakes at some regular interval—which may be 100, 50, or 25 ft, or 30, 20, or 10 m depending on the regularity of the ground surface at each of these points. The beginning point of the survey is designated as station 0. Points at multiples of 100 ft or 100 m from this point are termed *full stations*. Horizontal distances along the line are most conveniently reckoned by the station method. Thus points at distances of 100, 200, 300, and 1000 ft from the starting point of the survey are stations 1, 2, 3, and 10, respectively. Intermediate points are designated as *pluses*. A point that is 842.65 ft from the beginning point of the survey is station 8 + 42.65. If the plus sign is omitted, the resulting figure is the distance, in feet, from station 0.

In the remainder of this chapter, reference will be made to stations of 100 ft. However, 100-m stations are handled in exactly the same way if the leveling is performed in metric units.

When the stationing is carried continuously along a survey, the station of any point on the survey, at a known distance from any station or plus, can be calculated. Thus a point that is 227.94 ft beyond station 8 + 42.65 is $842.65 + 227.94 = 1070.59$ ft from station 0 or at station 10 + 70.59 the distance between station 38 + 66.77 and station 54 + 43.89 is $5443.89 - 3866.77 = 1577.12$ ft.

In the case of a route survey, the stationing is carried continuously along the line to be constructed. Thus, if the survey is for a highway or a railroad, the stationing will be carried around the curves and will not be continuous along the straight lines, which are eventually connected by curves. For the method of stationing that is used in surveys of this sort, see Chapter 13.

3-42   Field Routine of Profile Leveling

The principal difference between differential and profile leveling is in the number of foresights, or −S readings, taken from each setting of the level. In differential leveling only one such reading is taken, whereas in profile leveling any number can be taken. The theory is exactly the same for both types of leveling. A backsight, or +S reading, is taken on a benchmark or point of known elevation to determine the height of the instrument. The rod is then held successively on as many points, whose elevations are desired, as can be seen from that position of the level, and rod readings, called *intermediate foresights* (IFS), are taken. The elevations of these points are calculated by subtracting the corresponding rod readings from the height of the instrument (HI). When profile leveling with the digital level, the instrumentman must indicate via the keyboard that intermediate foresights are being observed from the current instrument setup in order that all IFSs are subtracted from the current HI. When no more stations can be seen, a foresight is taken on a turning point, the level is moved forward, and the process is repeated. Leveling rods up to 25 ft long are used in profile leveling to accommodate low spots along the line.

The method of profile leveling is illustrated in Fig. 3-48. The level having been set up, a sight is taken on a benchmark, not shown in the sketch. Intermediate foresights are then taken on stations 0, 1, 2, 2 + 65, 3, and 4. The sight is taken at station 2 + 65 because there is a decided change in the ground slope at that point. The distance to this point from station 2 is obtained either by pacing or by taping, the better method depending on the precision required. To determine the elevation of the bottom of the brook between stations 4 and 5, the level is moved forward after a foresight reading has been taken on the turning point just beyond station 4. With the level in the new position, a backsight is taken on the turning point, intermediate foresights are taken on stations 4 + 55, 4 + 63, 4 + 75, 5, 5 + 70, 6, 6 + 25, and 7, and lastly a foresight is taken on a turning point near station 7. From the third setup, a backsight is taken on the turning point, intermediate foresights are taken on stations 8, 8 + 75, 9, 10, 10 + 40, and 11, and a foresight is taken on a turning point near station 11. From the final setup shown in the figure, a backsight is taken on this turning point and intermediate foresights are taken on stations 12 and 13. Finally, a foresight is taken on a benchmark not shown in the sketch.

Readings have thus been taken at the regular 100-ft stations and at intermediate points wherever there is a decided change in the slope. The level has not necessarily been set on the line between the stations. In fact, it is usually an advantage to have the level from 30 to 50 ft away from the line, particularly when readings must be taken on intermediate points. More of the rod will then be visible through the telescope and the reading can be made more easily and quickly.

Benchmarks are usually established in the project area by differential leveling prior to running the profile leveling. When running the profile leveling, backsights and foresights on benchmarks and turning points must be taken with the same accuracy as that used to establish the elevations of the project benchmarks, usually to the hundredth of a foot (0.001 m). This is necessary to maintain the overall accuracy of the profile leveling, which minimizes the accumulation of errors. If intermediate foresights along profiles are taken along bare ground, they need only be read to the nearest tenth of a foot (1 cm). However, if the entire profile is a paved surface, it may be required to read the intermediate foresights to the hundredth of a foot, depending on the purpose of the profile. The profile leveling is then adjusted between previously established project benchmarks.

If benchmarks have not been established in advance, they should be established as the work progresses. Benchmarks may be from 10 to 20 stations apart when the differences of elevation are moderate, but the vertical intervals between benchmarks should be about 20 ft where the differences of elevation are considerable. These benchmarks should be so located that they will not be disturbed during any construction that may follow. Their elevations should be verified by running check levels.

The notes for recording the rod readings in profile leveling are the same as those for differential leveling except for the addition of a column for intermediate foresights. The notes and calculations for a portion of the profile leveling shown in Fig. 3-48 are given in Fig. 3-49.

3-43   Plotting the Profile

To facilitate the construction of profiles, paper prepared especially for the purpose is commonly used. This has horizontal and vertical lines in pale green, blue, or orange, so spaced as to represent certain distances to the horizontal and vertical scales. Such paper is called *profile paper.* If a single copy of the profile is sufficient, a heavy grade of paper is used. When reproductions are necessary, either a thin paper or tracing cloth is available. The common form of profile paper is divided into $\frac{1}{4}$-in. squares by fairly heavy lines. The space between each two such horizontal lines is divided into five equal parts by lighter horizontal lines, the distance between these light lines being $\frac{1}{20}$ in. To accentuate the differences of elevation, the space between two horizontal lines can be considered as equivalent to 0.1, 0.2, or 1.0 ft, and the space between two vertical lines as 25, 50, or 100 ft, according to the total difference of elevation, the amount of vertical exaggeration desired, the length of the line, and the requirements of the work.

To aid in estimating distances and elevations, each tenth vertical line and each fiftieth horizontal line are made extra heavy. Profile paper showing the profile for the level notes given in Fig. 3-49 is illustrated in Fig. 3-50. The elevation of some convenient extra-heavy horizontal line is assumed to be 900 ft, and a heavy vertical line is taken as station 0. Each division between horizontal lines represents 1 ft, and each division between vertical lines represents 100 ft, or one station. As the elevation or station of each printed line is known, the points on the ground surface can be plotted easily. When these points are connected with a smooth line, an accurate representation of that ground surface should result.

FIGURE 3-48 Profile leveling.

STA	BS	H.I	FS	IFS	ELEV	LEVEL # 4096	LEVEL J.E. PULLIAM
					PROFILE LEVELING		SEPT. 26, 1986
BM	4.18	902.76			898.58	ROD #18	ROD F. SMITH
0				5.2	897.6		
1				4.6	898.2		
2				3.9	898.9		
2+65				3.8	899.0	ξ BS = 9.06	884.75
3				4.9	897.9	ξ FS =-22.89	- 898.58
4				9.2	893.6	-13.83	-13.83
TP-1	1.21	894.63	9.34		893.42	CHECKS	
4+55				4.4	890.2		
+63				9.9	884.7		
+75				5.3	889.3		
5				5.0	889.6		
5+70				3.9	890.7		
6				4.5	890.1		
6+25				5.4	889.2		
7				2.2	892.4		
TP-2	3.67	897.22	1.08		893.55		
8				2.4	894.8		
8+75				1.1	896.1		
9				1.9	895.3		
10				8.4	888.8		
10+40				11.3	885.9		
11				12.2	885.0		
TP-3			12.47		884.75		
ξ BS=	+9.06	ξ FS =	22.89				

FIGURE 3-49 Profile-level notes.

It is often a convenience to have several related profiles plotted on the same sheet. Thus, in designing a pavement for a city street, three profiles may appear, namely, those of the center line and of the two curb lines. This is practically a necessity when there is any considerable difference in elevation between the two sides of the street. When only a single copy is being made, different colored inks can be used to distinguish one profile from another. When reproductions are to be made, different kinds of lines, such as different combinations of dots and dashes, are used.

As indicated in Section 17-13, it is possible to prepare a profile from a topographic map.

When it is desirable to have both the plan and the profile appear on the same sheet, paper that is half plain and half profile-ruled is used. Plans for highways and sewers often are prepared in this manner, the location plan appearing at the top of the sheet and the profile below it.

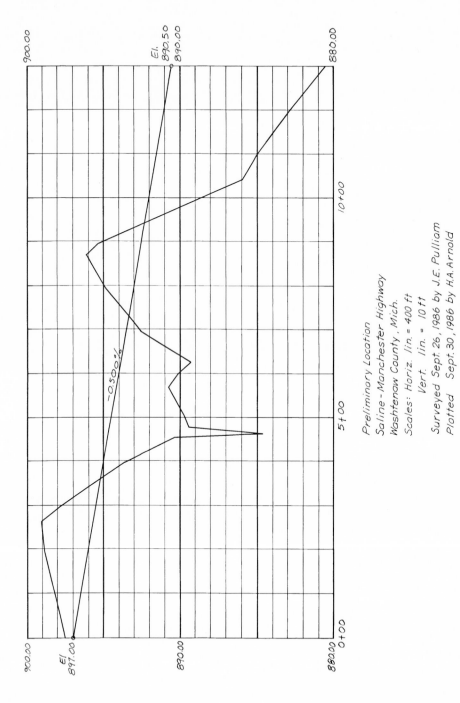

FIGURE 3-50 Plotted profile with grade line superimposed.

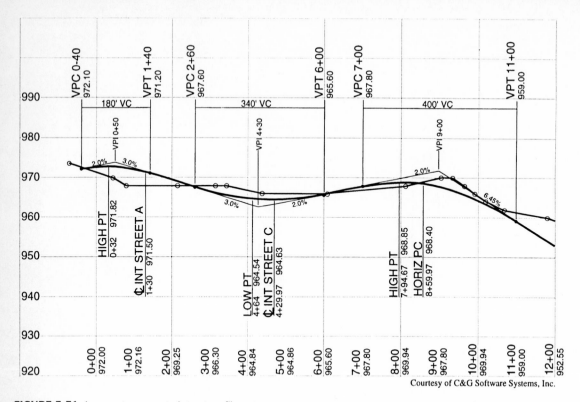

Courtesy of C&G Software Systems, Inc.

FIGURE 3-51 A computer-generated street profile.

Any information that may make the profile more valuable should be added. Thus the names of the streets or streams and the stations at which they are crossed should appear. The locations and elevations of benchmarks may appear as notes on the map. There should be a title giving the following information: what the profile represents, its location, its scales, the date of the survey, and the names of the surveyor and the draftsman.

Figure 3-51 shows the plot of a profile with superimposed grade lines and vertical curves of a street (see Sections 13–15 to 13–20) generated by a computer.

3-44   Grade Lines

The broken line in Fig. 3-50 represents the original ground surface along which the profile has been taken. The line to which this surface is brought by grading operations is called the *grade line.* For railroad planning the grade line represents the proposed position of the base of the rail. For a street or highway the grade line is the finished surface at the center line. For construction purposes it is usually more convenient to let the grade line represent the subgrade, that is, the base of the ballast on a railroad or the bottom of the pavement slab on a paved highway.

The principal purpose for which a profile is constructed is to enable the engineer to establish the grade line. The aim of the engineer is to keep the volume of earthwork at a minimum, to have the grade line straight for considerable distances and its inclination within the allowable limit, and to have the excavation and the embankment balance over reasonably short stretches (see Section 17-17). To determine the best grade line may require considerable study, but the saving of even a few hundred cubic yards of grading will pay for many hours of such study.

Where changes in the inclination of the grade line occur, the straight grades are connected by vertical curves (see Section 13-15).

3-45 Rate of Grade

The inclination of the grade line to the horizontal can be expressed by the ratio of the rise or fall of the line to the corresponding horizontal distance. The amount by which the grade line rises or falls in a unit of horizontal distance is called the *rate of grade* or the *gradient*. The rate of grade is usually expressed as a percentage, that is, as the rise or fall in a horizontal distance of 100 ft. If the grade line rises 2 ft in 100 ft, it has an ascending grade of 2%, which is written +2%. If the grade line falls 1.83 ft in 100 ft, it has a descending grade of 1.83%, which is written –1.83%. The sign + indicates a rising grade line and the sign – indicates a falling grade line.

The rate of grade is written along the grade line on the profile. The elevation of grade is written at the extremities of the line and also at each point where the rate of grade changes. It is common practice to enclose in small circles the points on the profile where the rate of grade changes.

The rate of grade, in percent, is equal to the total rise or fall in any horizontal distance divided by the horizontal distance expressed in stations of 100 ft. The total rise or fall of a grade line in any given horizontal distance is equal to the rate of grade, in percent, multiplied by the horizontal distance in stations. The horizontal distance, in stations of 100 ft, in which a given grade line will rise or fall a certain number of feet, is equal to the amount of the required rise or fall divided by the rate of gradient in percent.

►EXAMPLE 3-3   The grade elevation in a construction project at station 40 + 50 is to be 452.50 ft. The elevation along this same grade line at station 52 + 00 is to be 478.80 ft. What is the percent grade? What is the grade elevation at station 48 + 25? Refer to Fig. 3-52.

FIGURE 3-52 Calculation of grade and elevation.

Solution: The difference in elevation between the two points is 478.80 – 452.50 = +26.30 ft. The difference in stations is 52 – 40.50 = 11.50 stations. The percent grade is thus +26.30/11.50 = +2.287%. Station 48 + 25 lies 7.75 stations beyond station 40 + 50. The rise of the grade line between these two stations is 7.75 × 2.287 = +17.72 ft. The grade elevation at station 48 + 25 is thus 452.50 + 17.72 = 470.22 ft.

◆

PROBLEMS

3-1. A backsight of 3.0455 is taken on a point 60 m from the level. A foresight of 1.1508 m is then taken on a point 220 m from the level. Compute the correct difference in elevation, taking into account the effect of curvature. Neglect the effect of refraction on the line of sight.

3-2. Sighting across a lake 13 miles wide through a pair of binoculars, what is the height of the shortest tree on the opposite shore whose tip the observer can see if his eyes are 5 ft 3 in. above the shore line on which he stands?

3-3. A backsight of 3.865 ft is taken on a point 20 ft from the level. A foresight of 2.680 ft is then taken on a point 220 ft from the level. Compute the correct difference in elevation between the two points, taking into account the effect of curvature. Neglect the effect of refraction.

3-4. A sailor is standing in the crow's nest of a ship. His eyes are 120 ft above the level of the water. Looking through his binoculars, he sees the tip of the mast of an approaching ship. The mast is 150 ft above the water. What is the distance, in miles, between the two ships?

3-5. A vertical angle of –2° 40′ 30″ is read on a target that is 11.0 ft above ground station B. The telescope of the instrument is 5.0 ft above ground at station A. AB = 482.5 m. The elevation of station B is 345.46 m. Compute the elevation of station A in meters.

3-6. A slope distance between two points P and Q is measured using a laser EDM, giving a corrected slope distance of 45,580.50 ft. The elevation of the lower point P is 1542.85 ft. Reciprocal vertical angles are measured at P = +2° 12′ 15″, and at Q = –2° 19′ 15″. Assume the instrument and target are at the same height above ground at the two stations. What is the elevation of point Q?

3-7. In problem 3-6, using only the vertical angle at Q, together with the normal effect of curvature and refraction, compute the elevation of Q.

3-8. A vertical angle of +12° 52′ 25″ is measured to a station at the top of a hill from an instrument set up 2250 feet away from the hilltop station, measured horizontally. The height of the target over the station is 12.28 ft. The telescope of the instrument is 5.20 ft above the lower station, whose elevation is 322.64 ft. Assuming normal refraction conditions, what is the elevation of the hilltop station?

3-9. The slope distance between two points C and D is measured as 12,476.82 m. A vertical angle of +3° 12′ 22″ is measured from C to D. The instrument is 1.63 m above ground at C; the target is 1.63 m above ground at D. Compute the difference in elevation from C to D.

3-10. Three altimeters A, B, and C read as follows when set on a benchmark whose elevation is 457 ft; R_A = 1405, R_B = 1423, R_C = 1413. Altimeter A is kept at the benchmark; altimeter B is taken to a benchmark at an elevation of 1848 ft; altimeter C is used as a field altimeter. The following readings were taken:

Time	Altimeter A	Altimeter B	Altimeter C	Point
2:35	1413	2797	1723	1
3:00	1411	2803	1947	2
3:05	1419	2811	1943	3
3:50	1413	2807	1568	4
4:05	1411	2801	1689	5

Determine the elevations of the five field points.

3-11. If the sensitiveness of a bubble on an engineer's level is $28''$ and the graduations on the tube are 2 mm apart, what is the radius of curvature of the bubble tube in meters?

3-12. A level is set up, and the end of the level bubble is carefully brought to a graduation on the tube by manipulating the leveling screws. A reading of 1.6685 m is taken on a rod held 110 m from the level. The level is then tilted so as to cause the end of the bubble to move over six divisions, and a second rod reading of 1.7466 m is taken. The graduations are 2 mm apart. What is the sensitiveness of the bubble, and what is the radius of curvature of the bubble tube in meters?

3-13. A level is set up 195 ft from a leveling rod. One end of the bubble is carefully brought to a graduation on the tube by means of the leveling screws. A reading of 4.891 ft is taken on the rod. The bubble is moved through five divisions, and a second rod reading is 4.742 ft. The spacing of the graduations is 2 mm. What is the sensitiveness of the bubble, and what is the radius of curvature of the bubble tube, in feet?

3-14. Complete the accompanying set of level notes recorded when running differential levels between two benchmarks. Check the notes to detect arithmetical mistakes.

Station	BS	HI	FS	Elevation (ft)
BM27	4.22			182.64
TP1	6.80		8.91	
TP2	3.16		0,08	
TP3	5.05		4.62	
TP4	12.95		3.16	
BM28	0.16		11.18	
TP5	2.28		7.74	
TP6	6.42		3.02	
TP7	10.98		5.92	
TP8	8.05		6.22	
TP9	8.15		5.80	
BM29	6.62		5.05	
TP10	4.85		5.54	
BM30			10.90	

3-15. The line of sight of a level falls at the rate of 0.122 ft/100 ft when the level bubble is centered. A backsight of 4.782 ft is taken on point A, which is 75 ft from the level. The elevation of A is 425.238 ft. A foresight of 5.162 ft is taken on point B, which is 190 ft from the level. If the bubble was centered for both the backsight and the foresight, what is the elevation of B? Neglect curvature and refraction.

3-16. Assume that an area is to be graded to a level surface, and that the grading is to be controlled by grade stakes. The line of sight of the level rises at the rate of 0.086 ft/100 ft when the bubble is centered. A backsight is taken on a point that defines the grade elevation, and the rod reads 4.28 ft. The backsight point is 32 ft from the level. A series of grade stakes are set at the following distances from the level:

Stake	Distance (ft)	Stake	Distance (ft)	Stake	Distance (ft)
1	20	6	114	11	15
2	68	7	137	12	70
3	36	8	190	13	190
4	102	9	166	14	230
5	128	10	202	15	144

Compute the correct rod readings that will put the foot of the rod at grade at each of these points.

3-17. Assume the graduations of a 13-ft Philadelphia rod to be in the plane of the front of the rod and that the shoe or foot of the rod is $1\frac{7}{8}$ in. wide front to back. What is the effect on a rod reading of 11.000 ft if the rod is held on a flat surface and is waved forward toward the instrument such that the top of the rod moves 10 in.? What is the effect if the rod is waved backward away from the level by the same amount (nearest 0.001 ft)?

3-18. Complete the accompanying set of level notes in running differential levels between two benchmarks. Check the notes for arithmetical mistakes.

Station	BS	HI	FS	Elevation (m)
BM101A	2.087			47.466
TP1	2.110		0.884	
TP2	1.846		1.462	
TP3	0.484		1.598	
TP4	0.256		1.778	
BM102A	1.411		2.324	
TP5	1.798		1.780	
TP6	2.024		1.752	
TP7	3.198		1.046	
TP8	3.148		2.353	
TP9	2.862		3.046	
BM103A	2.140		3.021	
TP10	0.898		2.106	
BM104A			0.989	

3-19. The instrumentman observes a low reading of 11.260 ft and a high reading of 11.285 ft when the rodman waves the rod. How far is the rod out of plumb at the 13-ft mark when the high reading is taken?

3-20. Complete and check the accompanying set of profile level notes.

Station	BS	HI	FS	IFS	Elevation (ft)
BM38	6.94				491.29
22 + 00				6.4	
23 + 00				6.0	
23 + 71				3.2	
24 + 00				2.4	
24 + 18				2.0	
TPl	0.92		8.72		
25 + 00				4.6	
26 + 00				12.9	
26 + 80				9.8	
27 + 00				9.5	
27 + 78				12.4	
28 + 00				12.4	
TP2	5.63		10.10		
28 + 27				3.3	
28 + 47				0.9	
29 + 00				2.8	
30 + 00				6.1	
30 + 60				7.3	
31 + 00				8.6	
32 + 00				10.2	
BM39			9.83		

3-21. Using a horizontal scale of 1″ = one station and a vertical scale of 1″ = 10 ft, plot the profile of the line from the notes of Problem 3-20.

3-22. The grade elevation at station 25 + 00.00 is to be 946.52 ft. What will be the grade elevation at station 34 + 86.20 if the grade line rises at the rate of 1.785% between these two points?

3-23. The grade elevation of station 153 + 50 is 763.60 ft, and the grade elevation of station 161 + 75 is 785.44 ft. What is the grade, in percent, of a slope joining these two stations (nearest 0.001%)?

3-24. Compute the slope of the line of sight of a level from the following observations taken in each instance with the level bubble centered. From the first instrument setup, the reading on a rod at A is 3.729 ft and that on a rod at B is 4.286. The distance from the level to A is 25 ft, and the distance to B is 200 ft. From the second instrument setup, the rod reading at A is 4.368 ft and that at B is 4.908 ft. The distance to A is 185 ft, and that to B is 15 ft.

BIBLIOGRAPHY   1994 Level Survey. *P.O.B.* 20:2.

Balazs, E. I. 1981. The 1978 Houston-Galveston and Texas Gulf Coast vertical control surveys. *Surveying and Mapping* 41:401.

—— 1991. Preliminary test results of the Wild/Lietz NA 2000. *Tech. Papers Ann. Conv. Amer. Cong. Surv. & Mapp.* 1:6.

Berry, R. M. 1969. Experimental techniques for levels of high-precision using the Zeiss Ni2 automatic level. *Proc. 29th Ann. Meet. Amer. Cong. Surv. & Mapp.* p. 309.

—— 1976. History of geodetic leveling in the United States. *Surveying and Mapping* 36:137.

—— 1977. Observational techniques for use with compensator leveling instruments for first-order levels. *Surveying and Mapping* 37:17.

Caddess, H. N., and Cole, G. M. 1983. Precision leveling river crossing techniques using reciprocal vertical angles. *Tech. Papers 43rd Ann. Meet. Amer. Cong. Surv. & Mapp.* p. 549.

Caspary, W. F., and Heister, H. 1988. A new concept towards automation of differential leveling. *Tech. Papers, Ann. Conv. Amer. Cong. Surv. & Mapp.* V. 1, p. 173.

Chapman, W. 1990. Impact of NAVD 88 on the National Mapping Program. *Tech. Papers Fall Conv. Amer. Cong. Surv. & Mapp.* p. 42.

Chen-Yeh Hou, Veress, S. A., and Colcord, J. E. 1972. Refraction in precise leveling. *Surveying and Mapping* 32:231.

Chong, A. K. 1991. Accurate height transfer from benchmarks to awkward sites using GPS. *Tech. Papers Ann. Conv. Amer. Cong. Surv. & Mapp.* 1:38.

Chrzanowski, A. 1983. Economization of vertical control surveys in hilly areas by using modified trigonometric leveling. *Tech. Papers 43rd Ann. Meet. Amer. Cong. Surv. & Mapp.* p. 635.

Holdahl, S. R. 1983. The correction for leveling refraction and its impact on definition of the North American Vertical Datum. *Surveying and Mapping* 43:123.

Hradileh, L. 1972. Refraction in trigonometric and three-dimensional terrestrial networks. *Canadian Surveyor* 26:59.

Huff, L. C. 1987. Rapid precision leveling system. *ACSM Bull. 107, Amer. Cong. Surv. & Mapp.* p. 17.

Hussain, M. 1985. Accuracy evaluation of laser levels. *Tech. Papers Ann. Meet. Amer. Cong. Surv. & Mapp.* p. 20.

Kellie, A. C., and Young, F. R. 1987. Trigonometric leveling using electronic distance measurement. *Tech. Papers Ann. Conv. Amer. Cong. Surv. & Mapp.* p. 79.

Kesler, J. M. 1973. EDM slope reduction and trigonometric leveling. *Surveying and Mapping* 33:61.

Lachapelle, G. 1979. Redefinition of National Geodetic Networks. *Canadian Survey* 33:273.

Lippold, H. R., Jr. 1980. Readjustment of the National Geodetic Vertical Datum. *Surveying and Mapping* 40:155.

Maltby, C. S., and Doskins, E. 1968. Bench mark monumentation. *Surveying and Mapping* 28:45.

McDonnell, P. W. 1980. Motorized leveling at the National Geodetic Survey. *NOAA Technical Memorandum NOS NGS 26.*

—— 1989–1990. P.O.B. 1989 Level Survey. *P.O.B.* 15:2.

O'Connor, D. C. 1968. Some meteorological factors affecting the accuracy of barometric altimetry. *Surveying and Mapping* 28:477.

Richards, J. H., and Young, G. M. 1992. Results of the General Adjustment of the North American Vertical Datum of 1988. *Surveying and Land Information Systems* 52:3.

Roth, A., and Gonzalez, I. 1988. Rotating beam laser leveling systems—Applications for surveyors. *P.O.B.* 14:1.

Rowland, J. B. 1969. Atmospheric refraction in long-line vertical angles. *Surveying and Mapping* 29:59.

Rüeger, J. M. 1996. "Instrument Heights in EDM—Height Traversing." *Surveying and Land Information Systems* 55:4.

Selley, A. D. 1977. A trigonometric level crossing the Strait of Belle Isle. *Canadian Surveyor* 31:249.

Stoughton, H. W. 1992. Some notes on the U.S. coast & geodetic survey's collimation (error) test. *Surveying and Land Information Systems* 52:3.

Vanicek, P. 1991. Vertical datum and NAVD 88. *Surveying and Land Information Systems* 51:2.

Zilkoski, D. B. 1983. Justification of FGCC vertical control specifications and procedures. *Tech. Papers 43rd Ann. Meet. Amer. Cong. Surv. & Mapp.* p. 128.

—— 1986. Impacts and benefits of the North American Datum of 1988. *Tech. Papers 46th Ann. Meet. Amer. Cong. Surv. & Mapp.* p. 128.

—— 1990. NAVD 88 and NGS' responsibility to the surveying and mapping community. *Tech. Papers Fall Conv. Amer. Cong. Surv. & Mapp.* p. 23.

—— and Balazs, E. I. 1987. Status of NAVD 88 datum definition. *Tech. Papers Ann. Conv. Amer. Cong. Surv. & Mapp.* p. 66.

—— and Young, G. M. 1985. New adjustment of the North American vertical datum of 1988 (NAVD 88). *Tech. Papers Fall Conv. Amer. Cong. Surv. & Mapp.* p. 152.

Zilkoski, D. B., Richards, J. H., and Young, G. M. 1992. Results of the general adjustment of the North American vertical datum of 1988. *Surveying and Land Information Systems* 52:3.

4 The Measurement of Angles

4-1 Horizontal Angle

A horizontal angle is the angle formed in a horizontal plane by two intersecting vertical planes. The vertical planes intersect along a vertical line that contains the vertex of the angle. In surveying, an instrument for measuring angles occupies this vertex, and the vertical line formed by the two vertical planes coincides with the vertical axis of the instrument. In Fig. 4-1 two vertical planes, OAZ and OBZ, intersect along the vertical line OZ and form the horizontal angle AOB. Point C lies in the vertical plane containing line OA. Therefore the direction of OC is the same as that of OA. Point D lies in the vertical plane containing line OB, and so the horizontal direction of OD is the same as that of OB. Point O, being the vertex of the

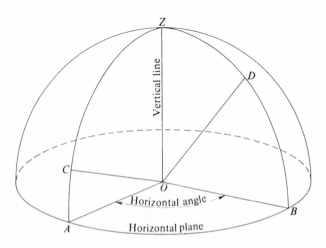

FIGURE 4-1 Horizontal angle.

angle, is the *at* station. The angle at *O* from *A* to *B* is the angle *AOB*, point *A* being the *from* or backsight station and point *B* being the *to* or foresight station. In Fig. 4-1 the horizontal angle at *O* from *C* to *D* is exactly the same as angle *AOB*. The horizontal projection of *OC* is *OA*, and the horizontal projection of *OD* is *OB*. The horizontal angle between two lines therefore is the angle between the projections of the lines onto a horizontal plane.

A horizontal angle in surveying has a direction of sense; that is, it is measured or designated to the right or to the left, or it is considered clockwise or counterclockwise. In Fig. 4-1 the angle at *O* from *A* to *B* is counterclockwise, and the angle at *O* from *D* to *A* is clockwise.

The common methods of measuring horizontal angles are by measuring distances, by transit, by theodolite, or by total station instruments.

4-2   Angles by Measured Distances

It is possible to make simple surveys by using just a tape, the angles being computed from the linear measurements. One method of doing this is to divide the survey area into a series of connected triangles, the sides of which are measured. From the lengths of the resulting triangle sides, the angles at all the vertices are computed by using the formulas of plane trigonometry, and the various separate angles are added to obtain the whole angle at each vertex. If all the sides in Fig. 4-2 are measured, all the angles in the tract can be computed by solving triangles *ABD*, *CDB*, and *EAD*. To obtain the angles at *D* from *E* to *C*, for example, the three angles at *D* from the individual triangles are added.

If the figure is large and if the lines are clear, the distances can be measured using the EDM. The field procedure known as *trilateration* (discussed in Chapter 9) employs the EDM measurements for computing angles in large triangles.

To avoid measuring the interior lines in a tract, as in Fig. 4-2, the angle at each vertex may be determined directly by the chord method or by the tangent method. In the chord method (Fig. 4-3) equal distances are laid off from *B* on the two lines *BA* and *BC*, and points *D* and *E* are set. These points should be lined in carefully.

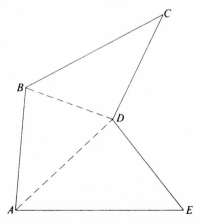

FIGURE 4-2 Survey with a tape.

The chord distance DE, or k, is then measured. The angle is computed from the relation $\sin \frac{1}{2} ABC = k/2BD$. Thus if BD and BE are each 200 ft and k is 72.48 ft, then $\sin \frac{1}{2} ABC = 72.48/(2 \times 200) = 0.1812$, $\frac{1}{2} ABC = 10° 26\frac{1}{2}'$, and $ABC = 20° 53'$.

In the tangent method (Fig. 4-4) a perpendicular to BC is established at P. (See Section 4-3 for a method of laying out a right angle.) Point Q is set on the line BA. The distances BP and PQ are measured with the tape, and the relation $\tan ABC = PQ/BP$ gives the angle at B. This method becomes weaker as the angle increases and it is not satisfactory for angles much over 45° since the line representing the perpendicular may not be a true perpendicular and its intersection with the line, as at Q in Fig. 4-4, will be doubtful. An extremely small angle can be measured quite accurately with the tape by using the tangent method since the perpendicular will be relatively short.

The accuracy of the values obtained by these methods is dependent on the size of the angle, on the care with which the points are set on line, on the accuracy of the measured lengths, and when the tangent method is used, on the care taken in erecting the perpendicular. With reasonable care the value of an angle determined by one of these methods would agree with the value obtained with the transit within 1' or 2'.

4-3 Laying Off Angles with Tape

It is sometimes necessary to stake out an angle more accurately than it can be done with the ordinary transit. Since it is impossible to lay off directly with the transit the angle of 67° 42′ 14″ shown in Fig. 4-5, the angle is laid off as exactly as possible with the instrument, and a point G, 1000 ft from E, is established. The angle DEG is measured by the repetition method described in Section 4-18. A

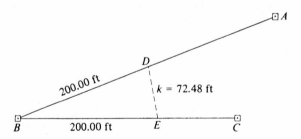

FIGURE 4-3 Measuring angle with tape using chord method.

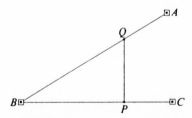

FIGURE 4-4 Measuring angle with tape using tangent method.

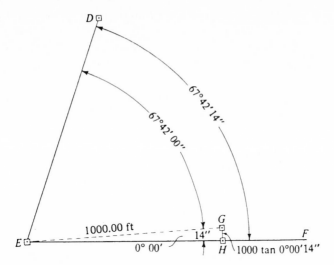

FIGURE 4-5 Laying off small angle with tape.

point H is next located from G at a distance $GH = 1000 \tan (67° \, 42' \, 14'' - DEG)$. The line through E and H will make the required angle with the line ED. This method is used in land surveying, where a line is frequently extended several miles and where a small error in an angle may cause a considerable discrepancy in the location of a point at the end of the line.

A perpendicular to a line can be constructed with a tape by the 3-4-5 method. Since a triangle is right angled if its sides are proportional to the numbers 3, 4, and 5, a right angle can be laid out by constructing a triangle whose sides are in these proportions. Thus in Fig. 4-6 let it be required to erect a perpendicular to the line AD at A. First, a temporary point B is established on line AD and 30 ft or 9 m from A. Then, by using A and B as centers and swinging arcs with radii of 40 and 50 ft, or 12 and 15 m, respectively, point C is located at the intersection of these two arcs. This can be most easily done if two tapes are available. If only a single 100-ft tape is at hand, the zero mark can be held at B and the 100-ft mark held at A, while point C is located by holding the 50- and 60-ft marks together and stretching the

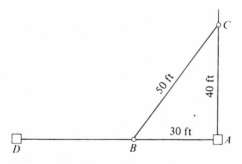

FIGURE 4-6 Constructing right angle with tape.

tape tight. Alternatively, if a 30-m tape is used, the zero mark can be held at *B,* and the 30-m mark is held at *A,* holding the 15- and 18-m marks at *C.* After a stake has been driven at *C,* the exact point on the stake can be found by striking two arcs on the top of the stake. One arc is struck at a time, and the intersection of the two arcs is the required point. Instead of the 30-, 40-, and 50-ft lengths, any other convenient lengths that are in the same proportions can be used.

A perpendicular to a given line through a point off the line can be constructed by first estimating where the perpendicular will intersect the line. To drop a perpendicular from point *P* to line *MN* in Fig. 4-7, first select point *F* on *MN.* The line *FL* is established by the 3-4-5 method, points being set at *A* and at *B.* The offset distance *k* from *FL* to *P* is then measured, and this distance is laid off on line *MN* from *F* to establish point *G.* Thus line *GP* is perpendicular to line *MN.*

4-4   Angles with Transit-Theodolite

The engineer's or surveyor's transit, more commonly referred to as a theodolite (the two terms are used interchangeably), is an instrument of great versatility. By means of a telescope, the line of sight is well defined. As a result long sights and accurate sighting are possible.

The theodolite contains a horizontal circle of either glass or silver. The circle can be read to the nearest 1′ down to an arc as small as 0.2″ using appropriate circular verniers or auxiliary glass scales. The horizontal circle is used when measuring or laying off horizontal angles. The theodolite also contains a vertical circle that is used to measure or lay off vertical angles or zenith angles (see Section 4-19). The standard engineer's transit shown in Fig. 4-8 contains silver circles and circular verniers. It usually contains a compass circle by which magnetic bearings may be observed (see Section 7-11). The optical reading glass-circle theodolites usually do not have an integral compass, but some theodolites have provision for attaching a separate compass needle.

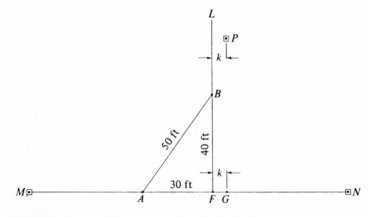

FIGURE 4-7 Perpendicular to a line through a point off the line.

FIGURE 4-8 Engineer's transit.

The reticle containing the cross hairs is equipped with auxiliary stadia hairs which may be used to determine distances by using the optical constants of the telescope. The telescope is provided with a level bubble that is sensitive enough to allow the transit to be used for spirit leveling and to eliminate uncertainties in measuring vertical angles. The telescope can be rotated about its horizontal axis through a complete circle. This last feature gives the transit its name. The word *transit* means to pass over or cross over, and the line of sight of the transit can be made to cross over from one side to the other by rotating the telescope about its horizontal axis.

4-5 Repeating Versus Direction Instruments

Two classes of instruments are used to measure horizontal angles. These are the repeating theodolite and the direction theodolite. The engineer's transit shown in Fig. 4-8 is a repeating instrument in that it is capable of accumulating angles on the horizontal circle by means of two separate rotations, as described in Section 4-8. The measurement of an angle by repetition is described in Section 4-18. A direction theodolite essentially contains only one rotation, and it therefore does not permit angles to accumulate on the horizontal circle. The direction theodolite is discussed in Section 4-22. There is no difference in the vertical circles of the two classes of instruments.

4-6 Parts of the Engineer's Transit

An engineer's transit is made up of three subassemblies. These are shown in Fig. 4-9. The upper part is the *alidade,* which contains a circular cover plate equipped with two level vials at right angles to one another and is rigidly connected to a solid conical shaft called the inner spindle. The vernier for the horizontal circle is contained in the cover plate. The alidade also contains the frames supporting the telescope, which are called the *standards,* the vertical circle and its vernier, the compass box, and finally the telescope and its level vial.

The middle part of a transit contains the *horizontal limb,* which is rigidly connected to a hollow conical shaft called the outer spindle or intermediate center. The inside of the outer spindle is bored with a taper to receive the inner spindle of the alidade. The limb assembly also contains a clamp, called the *upper clamp,* which allows or prevents rotation of the inner spindle within the outer spindle. The graduated horizontal circle is attached to the upper face of the limb.

The lower part of the transit is the *leveling-head assembly.* It contains the leveling screws, a bottom plate that screws onto the tripod, a shifting device that permits the transit to be moved about $\frac{1}{4}$ to $\frac{3}{8}$ in. in any direction while the tripod remains stationary, a half ball that allows the transit to tilt while being leveled, and the four-arm piece or spider, into the center of which fits the outer spindle. The assembly also contains a clamp, called the *lower clamp,* which allows or prevents rotation of the outer spindle, and a hook for attaching a plumb-bob string to the center of the transit. Some of the parts can be seen in Figs. 4-8 and 4-9.

4-7 Transit Telescope

The telescope on the transit shown in Fig. 4-8 is similar to that on an engineer's level, but it is shorter in length. Its main parts are the objective, the internal focusing lens, the focusing wheel, the cross hairs, and the eyepiece. The telescope is attached to the transverse axis, or horizontal axis, which rests on the standards and revolves in bearings at the tops of the standards. The telescope is held at any desired inclination by means of a clamp, called the vertical clamp, and can be rotated slowly in a vertical plane by means of a tangent screw, or slow-motion screw, called the vertical tangent screw, attached to one standard. The telescope in

FIGURE 4-9 Transit showing three subassemblies.

Fig. 4-8 is shown with a sunshade, which shades the objective when the instrument is in use, but this sunshade is replaced by a cap when the transit is not being used.

4-8   Rotation of Spindles

The inner and outer spindles, together with the tapered bore in the four-arm piece, provide for rotation of the alidade and the limb about the vertical axis of the instrument. With the instrument set up on a tripod and with both the lower clamp and the upper clamp in a clamped position, neither the alidade which contains the vernier nor the limb which contains the graduated horizontal circle can move. If the upper clamp remains clamped and the lower clamp is loosened, then the outer spindle is free to rotate in the four-arm piece, but both the alidade and the limb must rotate together about the vertical axis. As a result the vernier will appear opposite the same horizontal circle graduation during rotation since they move together.

If the lower clamp is clamped and the upper clamp is loosened, the horizontal circle is fixed in position, but the alidade is allowed to rotate about the vertical axis and the vernier moves with it. As the vernier moves around the edge of the fixed graduated circle, the reading on the circle changes correspondingly.

To permit a small rotation of the alidade or the inner spindle in the outer spindle, a tangent screw, or slow-motion screw, is provided. Every tangent screw operates against a spring, which reduces play or backlash in the motion. A small rotation of the outer spindle in the four-arm piece is permitted by a tangent screw similar to that controlling the slow motion of the alidade. Neither of these tangent screws will work unless the corresponding clamp is clamped.

Rotation of the inner spindle in the outer spindle is known as the *upper motion* of the transit. Such rotation is controlled by the upper clamp and the upper tangent screw. Rotation of the outer spindle in the four-arm piece or the leveling head is known as the *lower motion* of the transit. This rotation is controlled by the lower clamp and the lower tangent screw. An upper motion of the transit changes the reading on the horizontal circle. A lower motion of the transit does not change the reading on the horizontal circle because when the lower motion is used, the vernier on the alidade moves with the horizontal circle on the limb. In other words either of the motions will rotate the alidade and thus the line of sight about the vertical axis of the instrument, but only the upper motion will change the circle reading.

In practice the terms *upper motion* and *lower motion* are also used to refer to the clamps and tangent screws controlling the motions. For instance, when the lower clamp is tightened, it is said, "the lower motion is clamped"; and when the upper clamp is loosened, it is said, "the upper motion is unclamped." Or when a sight is taken with the upper clamp tightened and the lower clamp loosened, it is said, "a sight is taken by means of the lower motion."

4-9   Horizontal Circle and Vernier

The silver horizontal circle of a transit is located on the periphery of the upper surface of the horizontal limb. It is graduated in various ways, three of which are shown in Fig. 4-10. Each small division on the circle shown in views (a) and (c) is $\frac{1}{2}°$ or 30′. A division in view (b) is $\frac{1}{3}°$ or 20′. One in view (d) is $\frac{1}{6}°$ or 10′. On the circles in views (a), (b), and (c) the graduations for every 10° are numbered both clockwise and counterclockwise from 0° through 360°. This method of numbering is convenient when measuring angles clockwise or counterclockwise. The circle in view (d) is numbered clockwise only. However, it can also be used for measuring counterclockwise angles.

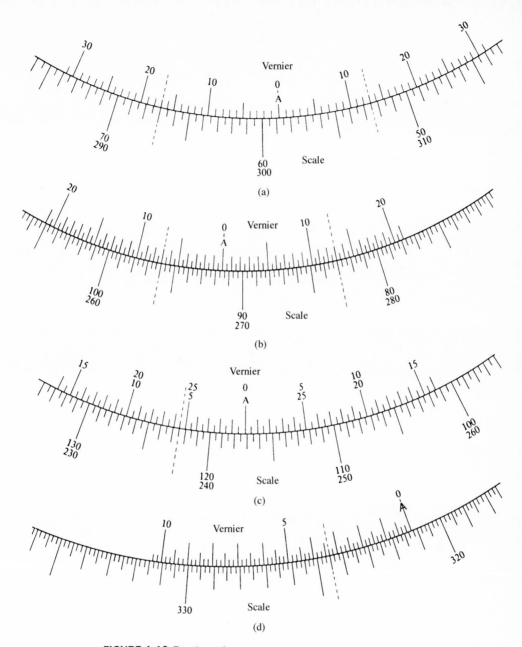

FIGURE 4-10 Transit verniers.

There are two verniers for the horizontal circle of a transit. They are intended to be exactly 180° apart. These are called the A vernier and the B vernier. Each vernier is located in a window in the cover plate and is covered with glass. When the inner spindle is in its proper position in the outer spindle, the outer edge of each vernier comes almost in contact with the inner edge of the graduated circle. Although the scale on the horizontal limb is circular and is graduated in angular units, the principle of a vernier used with this scale is the same as that of a vernier used on a leveling rod and described in Section 3-21.

The vernier shown in Fig. 4-10(a) is one of the two verniers of a 1′ transit. It is actually a double direct vernier, the part to the left of zero being graduated clockwise and the part to the right of zero being graduated counterclockwise. The main scale, or the circle, is graduated to 30′, and there are 30 spaces in each part of the vernier. The least count of the combination is therefore 30′/30 or 1′. The clockwise part of the vernier is used with the numbers on the circle increasing clockwise. The counterclockwise part of the vernier is used with the numbers on the circle increasing counterclockwise. The letter *A* at the zero graduation of the vernier signifies which vernier is being used. The letter *B* would appear in the vernier on the opposite side of the circle.

In Fig. 4-10(b) the circle is divided to 20′ and there are 40 spaces in either the clockwise part of the vernier or the counterclockwise part. Hence the least count of the combination is 20′/40 or $\frac{1}{2}$ ′ or 30″. A transit containing such a circle and vernier is known as a 30″ transit. Except for the least count, the verniers in views (a) and (b) are of the same type.

A folded vernier is represented in Fig. 4-10(c). This type of vernier is used where space for a double vernier is not available. It is to be noted that there are two sets of numbers on the vernier: one set increases clockwise and the other increases counterclockwise. The numbers closest to the circle increase toward the left from 0 at the center up to 15; and from the extreme right, where 15 is repeated, they continue to increase going toward the left to the center mark, which corresponds to 30. The least count of this combination is 30″, since a division on the circle is 30′ and the complete vernier contains 60 spaces. In Fig. 4-10(d) is shown a single vernier with 60 spaces. Since a division on the circle is 10′, the least count is 10′/60 = $\frac{1}{6}$ ′ or 10″.

Any vernier is read by first noting the graduation on the circle beyond which the vernier index lies, and then adding to the value of that graduation the value of the vernier graduation that coincides with a circle graduation. When the circle in Fig. 4-10(a) is read clockwise, the vernier index lies beyond the 58° 30′ graduation, and the vernier shows coincidence at the 17′ mark. Hence the reading is 58° 30′ + 17′, or 58° 47′. The corresponding counterclockwise reading is 301° 13′. For the clockwise reading in Fig. 4-10(b), the vernier index lies beyond the 91° 20′ graduation, and the vernier shows coincidence at the 7′ mark. Hence the reading is 91° 20′ + 7′, or 91° 27′ 00″. The corresponding counterclockwise reading is 268° 33′ 00″. The readings of the other verniers are as follows: in view (c), 117° 05′ 30″ or 242° 54′ 30″, and in view (d), 321° 13′ 20″.

The length of a division on the vernier is slightly smaller than that of a division on the circle. So when an exact coincidence occurs, the two vernier marks on either side of the graduation that coincides with a circle graduation should fall just inside the corresponding circle graduations. If the transit does not have provision for magnification of the graduations by means of an attached microscope, a hand magnifying glass or reading glass must be used. When reading the circle, the observer should look radially along the graduations to eliminate the effect of parallax between the circle and vernier graduations.

4-10 Vertical Circle and Vernier

The vertical circle is rigidly connected to the transverse axis of the telescope and moves as the telescope is raised or depressed. It is clamped in a fixed position by the clamp that controls the vertical motion of the telescope, and after being clamped, it can be moved through a small vertical angle by means of the tangent screw. The vertical circle can be either a full circle or a half circle. The former is graduated from 0° to 90° in both directions and then back to 0; the latter is graduated from 0° to 90° in both directions. Each division on either is $\frac{1}{2}$°, or 30′. On most transits the vernier for reading the vertical circle is mounted on the standard directly beneath the circle and almost comes in contact with the circle. It is a double direct vernier with the graduations numbered from 0 to 30 in both directions. The least count of the vertical circle and its vernier on almost every modern transit is 1′.

When the line of sight, and thus the telescope, of a well-adjusted transit is horizontal, the vernier will read 0° on the vertical circle. If some part of the transit is out of adjustment or if the transit is not carefully leveled, an initial reading other than 0° may be obtained. This is called the *index error*. On a transit with the vernier rigidly attached to the standard, this initial reading must be determined before a vertical angle can be measured correctly (see Section 4-19). On some transits the vernier can be moved slightly on the standard by means of a tangent screw, and its correct position is determined by a level bubble called a *control bubble* or index bubble. With this arrangement the circle will read 0° when the index bubble is centered and the line of sight is horizontal, so that the correct vertical angle can be measured directly.

4-11 Optical Reading Theodolite

The mechanical motions of the optical reading repeating theodolite are essentially the same as those of the engineer's transit, which they are rapidly replacing in practice. That is, they have an upper motion, a lower motion, and a vertical motion together with the appropriate clamps and tangent screws. The telescope contains either an erecting or an inverting eyepiece, but many of the optical reading theodolites do not contain a telescope bubble. Instead, a horizontal line of sight is established by first centering the vertical circle control bubble and then setting off a 90° zenith angle using the vertical clamp and tangent screw. Some theodolites contain a pendulum compensator that eliminates the need for a control bubble. However, the instrument must be leveled up quite closely to allow the compensator to operate properly.

The inner spindle of most theodolites is hollow to provide a line of sight for the *optical plummet,* which takes the place of the plumb bob used to center the theodolite over the point to be occupied. A schematic diagram of the optical plummet is shown in Fig. 4-11. After the instrument has been leveled up over the ground station, the observer loosens the clamp that holds the tribrach to the top of the tripod, and shifts the instrument until the line of sight of the optical plummet, defined by a set of cross hairs as seen through the optical plummet eyepiece, intersects the point marking the ground station. Since the top of the tripod may not be perfectly level, the instrument must be shifted parallel with itself. If it is rotated while being shifted, then the line of sight of the optical plummet will no longer be vertical. Both the leveling and the centering must be checked after this shifting operation.

The glass circles of the optical reading theodolite are much simpler to read than the silver circles of the engineer's transit described in Section 4-9. Figure 4-12 shows a typical repeating theodolite with glass circles. A mirror located on the side of one of the standards directs light through a window into the optical system, thereby illuminating the circles. A light bulb can be attached to the window in lieu of the reflected light in case of poor natural lighting or nighttime operation. The microscope eyepiece located to the side of the telescope eyepiece is used to view both the horizontal and the vertical circles under great magnification through a series of prisms and lenses located inside the housing of the instrument. Figure 4-13 shows the circles of the theodolite shown in Fig. 4-12 as viewed through the microscope eyepiece. The upper part of the field of view is the vertical circle and glass scale showing a zenith angle reading of 57° 53.4′. The lower part is the horizontal circle graduated both clockwise (the upper figures) and counterclockwise (the lower figures) together with the appropriate glass scales containing the minute graduations. The arrows indicate the direction, clockwise or counter clockwise. The clockwise circle reading is 34° 52.6′, and the counterclockwise circle reading is 325° 07.4′.

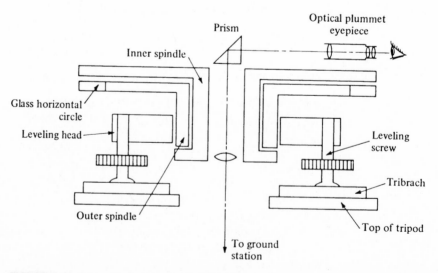

FIGURE 4-11 Schematic diagram of optical plummet.

FIGURE 4-12 Lietz T60D repeating theodolite.

FIGURE 4-13 Circles and scales of Lietz T60D theodolite.

Different methods are employed in various instruments to enable the observer to make the circle reading. The field of view through the reading micrometer of the instrument shown in Fig. 4-14 is shown in Fig. 4-15. The observer sees the horizontal circle marked Hz, the vertical circle marked V, and the optical micrometer, which replaces the glass scale. The micrometer allows an estimation to 3″ of arc. After sighting on the station being observed, the observer turns the micrometer operating screw located on the side of the standard (see Fig. 4-24) and sets the double index lines symmetrical with the degree graduation on either the horizontal or the vertical circle. He then reads the circle, which in this example shows a vertical circle reading of 87° 27′ 09″.

FIGURE 4-14 Leica T1 repeating theodolite.
Courtesy of Leica, Inc.

FIGURE 4-15 Circle reading of T1 theodolite.

4-12   Electronic Theodolites

The most modern theodolites, one of which is shown in Fig. 4-16, contain circular encoders that sense the rotations of the spindles and the telescope, convert these rotations into horizontal and vertical (or zenith) angles electronically, and display the values of the angles on liquid crystal displays (LCDs) or light emitting diode displays (LEDs). These readouts can be recorded in a conventional field book or can be stored in a data collector for future printout or computations. The instrument contains a pendulum compensator or some other provision for indexing the vertical circle readings to an absolute vertical direction. The circles can be set to zero readings by a simple press of a button on the instrument. The horizontal circle readings of some of the electronic theodolites can be preset to any desired angle before a backsight is taken. This is convenient for carrying directions through a traverse (see Chapters 7 and 8) or for determining directions in field mapping (see Chapter 15).

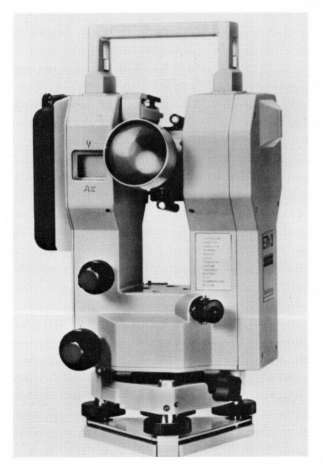

FIGURE 4-16 Zeiss ETh3 electronic theodolite. Courtesy of Carl Zeiss, Inc.

4-13   Instrument Stations

The point over which a transit or theodolite is set up is called an *instrument station.* Such a point should be marked as accurately as possible on some firm object. On many surveys each station is marked by a wooden stake or hub driven flush with the ground, and a tack or small nail marks the exact point on the stake. When the survey is intended to be relatively permanent, each point can be marked by a monument of stone or concrete or of iron pipe set in concrete.

Triangulation stations and *traverse stations* of many of the federal, state, county, and city agencies engaged in survey work are marked by bronze disks set in the tops of firmly planted concrete monuments. A point on rock can be marked by a chiseled cross. For identifying such points, the necessary facts can be written on the stone with keel or lumber crayon. To identify a hub and to indicate its location, a projecting stake, which is called a *guard stake* or *witness or marker stake,* is placed near the hub. On this stake are marked in keel the station number of the hub and any other necessary information, such as the name or purpose of the survey.

4-14   Setting Up an Instrument

In setting up a theodolite it is necessary to have the center of the instrument, which is the point of intersection of the transverse axis and the vertical axis of the instrument, directly over a given point on the ground. Also, the circle and the transverse axis must be horizontal. The position of a transit over a point on the ground may be indicated by a plumb bob suspended from a hook or ring attached to the lower end of the centers. The plumb-bob string should be held by a sliding knot so the height of the bob can be adjusted. The point of the bob should almost touch the mark over which it is desired to set the transit. The optical plummet is used in place of the plumb bob if the instrument contains this feature.

To set up a theodolite, the tripod legs are spread and their points are so placed that the top of the tripod is approximately horizontal and the telescope is at a convenient height for sighting. The instrument should be within a foot of the desired point, but no extra care is taken to set it closely at once. When setting up on rough ground, two legs of the tripod should be set at about the same elevation, and the top of the tripod should be made approximately horizontal by shifting the third leg. If the instrument is more than a few inches from the given point, the tripod is lifted bodily without changing the inclinations of the legs, and the instrument is set as near as possible to the point. By pressing the legs firmly into the ground, the plumb bob or the cross hairs of the optical plummet can usually be brought to within $\frac{1}{4}$ in. of the point. This brings the point within the range of the shifting head of the instrument.

When the tripod head is approximately horizontal and the plumb bob (or cross hairs) is very near the point, the bubbles of the plate levels are brought almost to the centers of the tubes by means of the leveling screws. If there are four screws, the plates are rotated until one level is parallel to a line through each pair of opposite screws. Then each bubble is brought to the center separately by turning the screws of the corresponding pair. When the plates are finally leveled up, all four screws should be bearing firmly but should not be so tight as to put a strain in the leveling head. If the leveling screws turn very hard, the cause may be the binding of the ball-and-socket joint at the bottom of the spindle. The tension may be relieved by loosening both screws of the other pair. If there are three leveling screws, the plates are rotated so that one level is parallel to a line through any pair of screws. Both bubbles are then brought to the center in succession by first turning the screws of that pair in opposite directions and then turning the third screw.

After the bubbles are approximately centered, the instrument is loosened on the tripod plate by loosening two *adjacent* leveling screws, or the retaining clamp, and the plumb bob or optical plummet cross hairs are brought exactly over the point on the ground by means of the shifting head. This arrangement allows the head of the theodolite to be moved laterally without disturbing the tripod. Then the instrument is held securely in position on the tripod plate by tightening the leveling screws that were previously loosened or by clamping the retaining ring. The bubbles of the plate levels are brought exactly to the centers of the tubes, and the position of the instrument over the point is observed. If the instrument has moved off the point in leveling up, it is brought back to the proper position by means of the shifting head, and the instrument is leveled again.

4-15 Sighting with the Theodolite

When an observer sights with the telescope of a theodolite, he must first focus the cross hairs so that they appear sharp to his eye (see Section 3-11). Then, provided the cross hairs are in proper focus, he brings the image of the object sighted on into focus by manipulating the focusing knob that moves the objective lens or the internal focusing lens. The focus should be tested for movement between cross hairs and image by moving the eye back and forth slightly. If parallax exists, it is removed by refocusing the objective or focusing lens. If the cross hairs go out of focus, then both the eyepiece and the objective or focusing lens must be refocused.

When making a pointing with the telescope, it is best to sight directly over the top of the telescope and then clamp the upper or lower motion, whichever is being used. The telescope is then raised or lowered and held in position by means of the vertical clamp. When the object to be sighted appears in the field of view, the vertical cross hair is centered on the object with the upper or lower tangent screw, the final motion of the tangent screw being made *against* the spring of the tangent screw. The vertical tangent screw is then used to bring the object sighted in the center of the field of view, that is, near the center horizontal cross hair. This final movement of the telescope may cause the vertical cross hair to go off the point, in which case it is brought on by the upper or lower tangent screw, whichever is being used.

The sight should be made on as small an object as is feasible or is consistent with the length of the sight. If the station is marked on the ground, the ideal sight is, of course, directly on the marked point. If the station is obscured, a pencil or a taping pin held vertically on the station makes a good target. If these are still

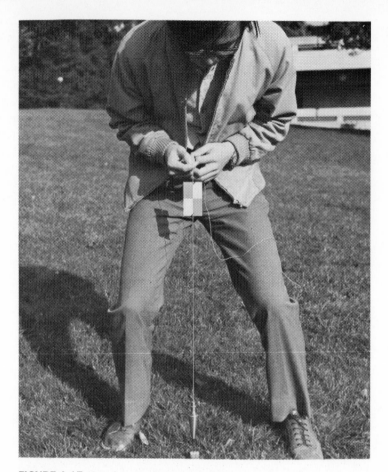

FIGURE 4-17 Use of commercial string target.

obscured, a plumb bob centered over the station, either hand held or suspended from a tripod, is satisfactory. It is often difficult to see a plumb-bob string on a sight more than about 500 ft. To make the string visible, a string target can be used. It may be either a commercial type as shown in Fig. 4-17 or one fashioned on the spot. A sheet of white paper folded twice makes a good string target. The plumb-bob string is held inside and along the second fold, and the paper is held diagonally so that one corner points downward along the freely suspended plumb-bob string. This is shown in Fig. 4-18. A line rod, which is from $\frac{3}{8}$ to $\frac{1}{2}$ in. in diameter, is a good target for sights between 500 and 1000 ft in length. It must be straight and carefully plumbed. The observer should sight as far down the rod as possible. A range pole from $\frac{3}{4}$ to $1\frac{1}{4}$ in. in diameter is satisfactory for sights beyond 500 ft. It, too, must be straight and plumbed over the station. Various sighting targets are used that sit on top of a tripod centered over the point to be observed. One such target is shown in Fig. 4-19.

FIGURE 4-18 Use of string target fashioned on the spot.

4-16 Measuring a Horizontal Angle

With the theodolite set up, centered, and leveled over the *at* station, the A vernier or the glass circle is usually set to read 0° for convenience. This is done by loosening both the upper and lower clamps and rotating the horizontal limb by slight pressure applied with the finger of one hand to its underside while holding the standards in place with the other hand, until the 0° graduation of the circle is brought almost opposite the index of the vernier or glass scale. The upper motion is then clamped and coincidence is obtained between the 0° graduation and the vernier or glass scale index by using the upper tangent screw. If the instrument contains a silver circle, the setting must be made with the aid of the magnifying glass, as this operation constitutes a reading of the circle.

The lower motion being free, the line of sight is directed toward the *from* or backsight station. The lower motion is now clamped, and the line of sight is brought exactly on the backsight station by turning the lower tangent screw. For

FIGURE 4-19 Sighting target.
Courtesy of Kern Instruments, Inc.

this sight the circle reading is observed as 0°00′ and it is recorded as the initial circle reading.

The final reading is obtained by loosening the upper clamp, rotating the alidade about the vertical axis, and directing the line of sight toward the *to* or foresight station. If the alidade has been turned in a clockwise direction, the reading on the clockwise graduations will increase by the angle of rotation. The upper clamp is clamped, and the line of sight is brought exactly on the foresight station by turning the upper tangent screw. This motion of the upper tangent screw will change the circle reading correspondingly. The reading of the circle for this sight is recorded as the final circle reading.

The difference between the initial reading and the final reading is the angle through which the line of sight was turned in going from the backsight station to the foresight station. As stated before, the A vernier or glass scale is usually set to read 0° before the backsight is made. However, it need not be set at any particular reading before the backsight is taken, since it is immaterial what the initial reading is. As an example, suppose it is desired to measure the angle at L from M to P in Fig. 4-20, and when a backsight is taken to station M, the circle reading is 122° 43′. The upper clamp is loosened, the line of sight is rotated so that it is directed to P, and the upper clamp is then clamped. The line of sight is brought exactly on sta-

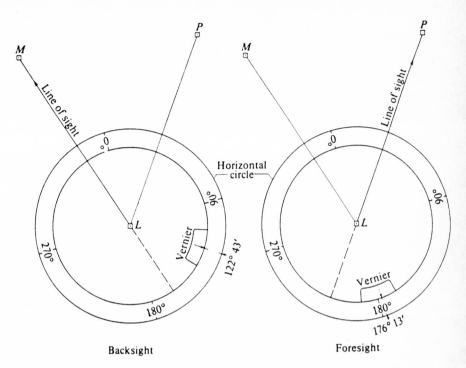

FIGURE 4-20 Horizontal circle readings, backsight and foresight.

tion P by using the upper tangent screw. The circle reading is now 176° 13′. The angle at L from M to P is therefore the difference between the initial reading and the final reading, which is 176° 13′ − 122° 43′ = 53° 30′.

   The foregoing is the basic procedure in measuring a horizontal angle with the theodolite. The principles involved are as follows: (1) The backsight is taken by a lower motion, and the backsight reading is the initial circle reading; (2) the foresight is taken by an upper motion, and the foresight reading is the final circle reading; (3) the horizontal angle is obtained by taking the difference between the initial and final circle readings. The procedure as presented, however, is not sufficient in itself to obtain a reliable measurement of a horizontal angle for the reasons discussed in the following section.

4-17   Double Centering

If the various components of a theodolite are considered as lines and planes, one finds that the graduated horizontal and vertical circles together with their verniers constitute planes. The line of sight, the transverse or horizontal axis of the telescope, the vertical axis, and the axis of any bubble tube are all lines in space. These lines and planes bear definite relations to one another in a well-adjusted instrument. The line of sight is normal to the horizontal axis; the horizontal axis is normal to the vertical axis; the vertical axis is normal to the plane containing the horizontal circle; the line of sight is parallel to the axis of the telescope bubble tube; the axes of the plate levels lie in a plane parallel to the horizontal circle; and the movement of the focusing lens in and out when it is focused is parallel to the line of sight. The inner spindle and the outer spindle must be concentric, and the horizontal circle and the verniers must be concentric. The line joining the indices of the A and B verniers must pass through the center of the horizontal circle. A maladjustment of theodolite in one or more of these respects is to be expected. In fact, a theodolite is never in perfect adjustment.

Two procedures for eliminating almost all of the errors due to maladjustment of the transit are (1) double centering and (2) reading both the A vernier and the B vernier or diametrically opposite sides of the circle (see Section 4-22). Of the two, double centering is the more important. Double centering or double sighting consists of making a measurement of a horizontal or a vertical angle once with the telescope in the direct or erect position and once with the telescope in the *reversed, inverted,* or *plunged* position. The act of turning the telescope upside down, that is, rotating it about the transverse axis, is called *plunging* or *transiting* the telescope. When the telescope is in its plunged position, the telescope bubble is on top of the telescope.

In Section 4-16, the instrument was set up at station L. A backsight was taken on station M with the telescope direct, and the circle reading was 122° 43′. A foresight was then taken on station P with the telescope still direct, and the circle reading was 176° 13′. The value of the angle by the direct measurement is 53° 30′.

To measure the angle with the telescope reversed or plunged, the vertical clamp is loosened and the telescope is plunged about its transverse axis. The circle reading is still 176° 13′. The lower clamp is loosened to allow the circle to move with the vernier or glass scale and thus preserve the reading on the circle. The line of sight is now directed toward station $M,$ the lower clamp is clamped, and the line of sight is brought exactly on the station by using the lower tangent screw. The initial reading for this second measurement is thus 176° 13′, which was the final reading for the first measurement. The upper clamp is loosened, the vernier or glass scale being allowed to move along the circle so as to increase the circle reading, and the line of sight is directed to station $P.$ The upper clamp is clamped and the line of sight is brought exactly on the foresight station with the upper tangent screw. The final circle reading is now observed and recorded. Say this reading is 229° 44′. The second value of the angle at L from M to P is thus 229° 44′ − 176° 413′, or 53° 31′. The mean of the values with the telescope direct and reversed is 53° 30′ 30″, and this mean value is free from errors due to practically all the maladjustments of the instrument.

When an angle is measured with the telescope direct and reversed and the resulting two values disagree by more than the least count of the circle, the whole measurement should be repeated. If double centering is repeated and the discrepancy is always of the same amount and in the same direction, the discrepancy indicates bad adjustment of the theodolite, but the values can be used. A mean of the two readings should be taken as the best value of the angle.

4-18   Angles by Repetition

Study of Section 4-17 indicates that an angle can be measured to the nearest 30″ with a 1′ repeating theodolite by double centering, that an angle can be measured to the nearest 15″ with a 30″ instrument by double centering, and so on. This is true. Moreover, an experienced instrumentman with a good 1′ theodolite is capable of measuring an angle to the nearest 10″ by repeating the angle several times, accumulating the successive readings on the circle, and then dividing by the number of repetitions.

To illustrate the principle of repetition, Table 4-1 gives in the left-hand column the successive values of the angle 25° 10′ 23″ multiplied one, two, three, four, five, and six times. In the middle column the tabulation gives the corresponding values of the angle that would be read on a 1′ transit after one, two, three, four, five, and six repetitions if the initial backsight reading were 0° and if there were no errors in pointing or mistakes in using the wrong motion. The right-hand column gives the values of the angles in the middle column divided by one, two, three, four, five, and six, respectively.

If the repetitions could go on indefinitely, the transit reading on the twelfth repetition divided by 12 would be 25° 10′ 25″, division after the twenty-fourth repetition would give 25° 10′ 22.5″, and so on. The practical limit is reached, however, somewhere between the sixth and twelfth repetition because of errors of graduations, eccentricities of centers, play in the instrument, and inability of the observer to point with sufficient speed and accuracy to preserve precision.

The procedure for repeating an angle six times with a theodolite is to measure it three times with the telescope direct and three times with the telescope reversed, in that order. It is understood that the observer is experienced, that his personal errors will be small, and that he will make no mistakes. Otherwise there is no justification for repeating angles.

The instrument is set up, centered, and leveled. The initial backsight reading is taken on both the A and B verniers with a transit so equipped because it is assumed that a great degree of refinement is necessary. The backsight is taken with a lower motion and with the telescope direct. Then, with the telescope still in the direct position, the upper clamp is loosened and the pointing on the foresight station is made by using the upper clamp and tangent screw. The first foresight reading is made on the circle to determine the approximate value of the angle. Next, with the telescope direct, the lower clamp is loosened and the line of sight is brought back on the backsight station by using the lower clamp and tangent screw. The upper clamp is loosened and the line of sight is swung around to the foresight station. The circle reading is thus increased. A third backsight is taken on the *from* station and the third foresight is taken on the *to* station. Each repetition increases the circle reading. After the third repetition, the telescope is transited to the reversed position, and the angle is turned off three more times with the telescope reversed. The final readings on both the A vernier and the B vernier are taken. The accompanying three sets of notes will illustrate how the final angle is obtained.

TABLE 4-1 Angle by Repetition

Number	Multiple Values of True Angle	Successive Transit Readings	Values Obtained from Transit Readings
1	25° 10′ 23″	25° 10′	25° 10′ 00″
2	50° 20′ 46″	50° 21′	25° 10′ 30″
3	75° 31′ 09″	75° 31′	25° 10′ 20″
4	100° 41′ 32″	100° 42′	25° 10′ 30″
5	125° 51′ 55″	125° 52′	25° 10′ 24″
6	151° 02′ 18″	151° 02′	25° 10′ 20″

In the first set the approximate angle is the difference between the initial backsight and foresight readings and is 28° 10′. When this angle is multiplied by six and the product is added to the initial reading of 0°, the sum is less than 360°. So the vernier or glass scale has not gone completely around the circle.

Readings for Set No. 1

At Station B (30″ Transit)						
From	To	Repetition	Telescope	Circle	A Vernier	B Vernier
A	C	0	D	0° 00′	00″	00″
		1	D	28° 10′	00″	
		6	R	169° 00′	00″	30″

The mean of the initial readings on the A and B verniers is 0° 00′ 00″. The mean of the final readings on the A and B verniers is 169° 00′ 15″. The difference is 169° 00′ 15″. This divided by six is 28° 10′ 02.5″, which is the accepted value.

Readings for Set No. 2

At Station B (30″ Transit)						
From	To	Repetition	Telescope	Circle	A Vernier	B Vernier
A	C	0	D	232° 41′	00″	00″
		1	D	260° 51′		
		6	R	41° 41′	00″	30″

In the second set of notes the difference between the initial backsight and foresight readings is 28° 10′. If the angle is multiplied by six and the product is added to the initial reading of 232° 41′, the sum is greater than 360°. Consequently, the final reading must be increased by 360°. (The number of complete revolutions is determined by inspection.) The mean of the initial readings on the A and B verniers is 232° 41′ 00″, and that of the final readings on the A and B verniers *increased* by 360° is 401° 41′ 15″. The difference is 169° 00′ 15″. This divided by six is 28° 10′ 02.5″, which is the accepted value.

Readings for Set No. 3

			At Station 40 + 12.50 (1′ Transit)			
From	To	Repetition	Telescope	Circle	A Vernier	B Vernier
16 + 22.60	51 + 02.81	0	D	114°	20′	19′
		1	D	274°	52′	
		6	R	357°	33′	33′

In the third set of notes the approximate angle being measured is 274° 52′ – 114° 20′ = 160° 32′. When this angle is multiplied by 6 and the product is added to the initial reading, the sum is 1077° 32′. Obviously, the vernier has passed the 360° mark twice. Consequently, the final reading must be increased by 720°. The mean of the initial readings of the A and B verniers is 114° 19′ 30″; that of the final readings of the A and B verniers *increased* by 720° is 1077° 33′. The difference is 963°13′30″, which is divided by 6. The result, 160° 32′ 15″, is the adopted value.

Many repeating theodolites with glass circles and scales do not have provision for reading diametrically opposite sides of the circle (equivalent to an A- and B-vernier reading), but the principle of repetition applies just as well to these instruments.

4-19 Vertical and Zenith Angles

A vertical angle is an angle measured in a vertical plane from a horizontal line upward or downward to give a positive or negative value, respectively. Positive and negative vertical angles are sometimes referred to as elevation or depression angles, respectively. A vertical angle thus lies between 0° and ±90°.

A zenith angle is an angle measured in a vertical plane downward from an upward directed vertical line through the instrument. It is thus between 0° and 180°.

Some instruments, such as the engineer's transit, measure vertical angles while most optical theodolites measure zenith angles. Referring to the engineer's transit, centering the plate levels brings the vertical axis of the transit truly vertical, provided that the plate levels are in adjustment. Centering the telescope level brings the telescope, and hence the line of sight, into a truly horizontal position, provided that the telescope level and the line of sight are both in adjustment. When the plate-level bubbles are centered and the telescope is brought to a horizontal position by centering the telescope-level bubble, the vernier should read 0° on the vertical circle. If the plate levels are out of adjustment or are not centered at the time a vertical angle is to be measured, then the vertical circle will not read

0° when the telescope is brought horizontal. It will have an initial reading, which may be positive or negative. Furthermore, as the alidade is rotated in azimuth, that is, rotated about its vertical axis, this initial reading will vary. The initial reading is termed the *index error* of the vertical circle. It is determined for a given vertical angle by pointing the telescope in the direction of the desired line, centering the telescope-level bubble, and reading the vertical circle. This method assumes that the line of sight is parallel to the axis of the telescope bubble, and that the vernier is in proper position on the standard.

To measure the vertical angle of the line of sight from one point to another, the transit is set up over the first point, centered, and leveled. The line of sight is brought in the direction of the other point, and the telescope bubble is centered. The reading of the vertical circle is recorded as the initial reading or as the index error. The line of sight is then raised or lowered and is directed accurately to the second point by means of the vertical clamp and tangent screw. The reading of the vertical circle is recorded as the final reading. The correct vertical angle is the difference between the initial and final readings. In recording both the initial and final readings, it is necessary to give each value its correct algebraic sign, that is, the + for an elevation angle and the − for a depression angle. As an example, suppose that when the telescope bubble is centered, the vertical circle reads + 0° 02′. When the line of sight is now raised and directed to the desired point, the vertical circle reads +14° 37′. The correct vertical angle is the difference between the two readings, or +14° 37′ − (+0° 02′) = +14° 35′. As a further example, suppose the initial reading is +0° 02′ and the final reading is −21° 14′. The difference is −21° 14′− (+0° 02′) = −21° 16′, which is the correct vertical angle.

When the initial reading of the vertical circle is recorded as an index error and the final reading is recorded directly as the vertical angle, this angle is corrected by applying the index correction, which is equal to the index error but of opposite sign. Thus, if the index error is +0° 03′ and the recorded vertical angle is −16° 48′, the index correction −0° 03′ is applied to −16° 48′ to obtain −16° 51′, which is the correct vertical angle.

If the plate bubbles are out of adjustment, the transit can be leveled very accurately for measuring vertical angles by using the more sensitive telescope bubble. The procedure is as follows. Level the transit initially using the plate bubbles. Now bring the telescope over a pair of leveling screws and center the telescope bubble by using the vertical clamp and tangent screw. Rotate the telescope 180° in azimuth. If the telescope bubble does not come to the center, bring it halfway to the center by turning the vertical tangent screw, and then center the bubble by using the two leveling screws. Rotate the telescope again 180° in azimuth. The bubble should remain centered. If it does not, repeat the process of bringing it halfway to center with the tangent screw and the rest of the way with the leveling screws. When the bubble remains centered after the telescope is rotated through 180° in azimuth, bring the telescope over the other pair of leveling screws and bring the bubble to center by using the leveling screws only. This operation makes the vertical axis of the transit truly vertical.

If the axis of the transit is truly vertical but the line of sight is not parallel to the axis of the telescope level, or the vertical-circle vernier is displaced, or there is a combination of these maladjustments, an index error will result. This error can be detected in a transit having a full vertical circle by reading the vertical angle with the telescope in the direct position and in the reversed position and comparing the two values. If the two disagree, the index error is one-half the difference between the direct and reversed readings. The mean of the direct and reversed readings is free from index error. Measuring a vertical angle with the telescope direct and reversed will not, however, eliminate the error introduced by not having the vertical axis truly vertical. It is therefore necessary to have the transit leveled carefully when vertical angles are to be measured.

The modern electronic theodolites contain tilt sensing devices that measure the amount by which the vertical axis is out of plumb. Output from the tilt sensor is used internally to compute the correct vertical or zenith angle for display on the screen.

The procedure for measuring a vertical angle with a transit equipped with a movable vertical-circle vernier is to sight directly at the point, center the control-level bubble on the vernier by using the control-level tangent screw, and then read the vertical angle. This is free of index error.

One important difference between the horizontal circle and the vertical circle of the transit is in the placement of the vernier with respect to the graduated circle. On the horizontal circle the vernier is located inside the circle. On the vertical circle the vernier is located outside the circle. When horizontal and vertical angles are measured concurrently, the observer is alternately reading the horizontal-circle and vertical-circle verniers, and he must be on guard against mistakes in reading the circle for the vernier because of the difference in placement of the verniers.

The vertical circle of an instrument that measures zenith angles is graduated from 0° clockwise to 360° (back to 0°). If the circle reading is 0° with the telescope pointed upward, it will read 90° with the telescope in the horizontal *direct* position, 180° when pointed straight down and 270° with the telescope in the horizontal *reversed* position. Then, if Z_D denotes the circle reading with the telescope in the direct position and Z_R denotes the circle reading with the telescope in the reversed position, the zenith angle Z is obtained by

$$Z = Z_D$$
$$Z = 360° - Z_R$$

following which the vertical angle V, if desired, is obtained as $V = 90° - Z$.

If the circle reading is 180° with the telescope pointed upward (as on some theodolites), it will read 90° with the telescope in the horizontal *direct* position, 0° when pointed straight down, and 270° with the telescope in the horizontal *reversed* position. With a vertical circle thus graduated, the zenith angle is

$$Z = 180° - Z_D$$
$$Z = Z_R - 180°$$

following which the vertical angle V is obtained as $V = 90° - Z$.

4-20 Sources of Error in Theodolite Angles

The sources of error in theodolite angles are instrumental, personal, and natural. Most of the following eight instrumental errors have already been discussed.

1. If the line of sight of a telescope is not parallel with the telescope level, an error is introduced in the measurement of vertical or zenith angles. This error is eliminated by reading the vertical or zenith angle with the telescope both direct and reversed.

2. If the vertical-circle vernier is displaced on the standard from its correct position, a corresponding error will result in the measured value of a vertical or zenith angle. This error is eliminated by measuring the angle with the telescope both direct and reversed.

3. If the optical axis of the telescope is not parallel with the movement of the focusing lens when it is being focused, an error will be introduced into the value of a horizontal angle when the backsight and foresight stations are at considerably different distances from the instrument. This error is eliminated by measuring with the telescope direct and reversed. There is no error from this source if the two other stations are at about the same distance from the instrument.

4. If the line of sight is not perpendicular to the horizontal axis of the instrument and the telescope is rotated on that axis, the line of sight generates a cone whose axis is horizontal. In Fig. 4-21, the instrument center is at O, which is the apex of the cone whose base is the dashed half circle HAH'. The line of sight to a point S is an element of that cone. The line OZ is a vertical line through the instrument. The amount by which the line of sight is out of perpendicular with the horizontal axis is designated by the small angle e. The error in the horizontal direction to point S is the angle c. In the spherical triangle ZNS, side NS is perpendicular to side ZN. Angle h is the vertical angle to S. Then by the law of sines for spherical triangles,

$$\frac{\sin c}{\sin N} = \frac{\sin e}{\sin ZS} \qquad (N = 90°)$$

or

$$\sin c = \frac{\sin e}{\sin ZS} = \frac{\sin e}{\cos h} \qquad (h = 90° - ZS) \qquad (4\text{-}1)$$

Since both c and e are small angles, Eq. (4-1) can be expressed as

$$c'' = e'' \sec h \qquad (4\text{-}2)$$

If the line of sight is to the left of the normal to the horizontal axis (as shown in Fig. 4-21), then the clockwise circle reading is too large by the angle c. If a horizontal angle is measured between two stations to which the vertical angles are about the same, there will be no error from this source. If, however, the vertical angles to the backsight and the foresight stations are different from one another, the total error c_t in the horizontal angle is

$$c_t = e(\sec h_F - \sec h_B) \qquad (4\text{-}3)$$

This error is eliminated by double centering.

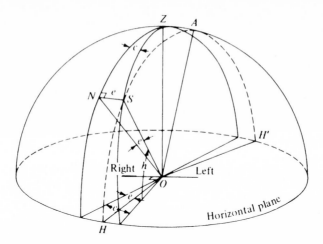

FIGURE 4-21 Line of sight not perpendicular to horizontal axis.

EXAMPLE 4-1   A backsight is taken on the *from* station with the telescope inclined downward at a vertical angle of −12° 22′. A foresight is taken on the *to* station with the telescope inclined upward +42° 50′. The line of sight departs to the left of the normal to the horizontal axis by 25″. The horizontal angle measured clockwise is 76° 15′ 10″. Compute the correct horizontal angle.

Solution: This situation is shown in Fig. 4-21. Thus both the backsight and the foresight readings are too large, and the total error in the measured angle is, by Eq. (4-3),

$$c_t = 25''[\sec 42° 50' - \sec(-12° 22')] = 8.50''$$

The correct angle is thus 76° 15′ 10″ − 8.5″ = 76° 15′ 01.5″.   ◆

5. If the horizontal axis is not perpendicular to the vertical axis and the telescope is rotated on the horizontal axis, the line of sight will generate a plane that is not vertical, as it would be if the instrument were in perfect adjustment. In Fig. 4-22 the right side of the horizontal axis is seen to be inclined upward by small angle e. The line of sight sweeps through the tilted plane $Z'SH'$ instead of the vertical plane ZSH. This produces an error c in the horizontal direction of line OS.

In the spherical triangle $Z'SH$, $c = Z'$. Then by the law of sines for spherical triangles,

$$\frac{\sin Z'}{\sin e} = \frac{\sin HS}{\sin Z'S} = \frac{\sin h}{\sin Z'S} = \frac{\sin h}{\cos(90° - Z'S)} \qquad (4\text{-}4)$$

but since $90° - Z'S = h$ (approximately) and $c = Z'$, Eq. (4-4) becomes

$$\sin c = \sin e \tan h$$

or since c and e are both small angles

$$c'' = e'' \tan h \qquad (4\text{-}5)$$

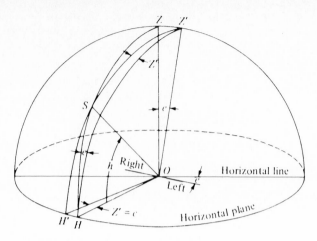

FIGURE 4-22 Inclined horizontal axis.

If the right side of the horizontal axis is inclined upward (as depicted in Fig. 4-22), and if h is positive, then the clockwise circle reading is too large by the angle c. The total error in a measured angle is thus

$$c''_t = e''(\tan h_F - \tan h_B) \qquad (4\text{-}6)$$

Errors from this source are eliminated by double centering.

EXAMPLE 4-2   Given the same data as in Example 4-1, except that the left side of the horizontal axis is inclined upward by 25″, compute the correct horizontal angle.

Solution: Since the left side is high (opposite to that shown in Fig. 4-22), the back-sight reading will be too large since h is negative. The foresight reading will be too small. The total error in the measured angle is thus, by Eq. (4-6),

$$c_t = -25''[\tan 42°\,50' - \tan(-12°\,22')] = -28.7''$$

The correct angle is thus 76° 15′ 10″ + 28.7″ = 76° 15′ 38.7″.

6. If the plate bubbles are out of adjustment, the vertical axis is not truly vertical when the bubbles are centered. This defect causes an error in the horizontal direction of each pointing that varies both with the direction of pointing and with the vertical angle. When the direction of pointing is in the same direction as the inclination of the vertical axis, no error in the horizontal direction will result, no matter what the vertical angle may be. If the direction of pointing is at right angles to the direction of inclination of the vertical axis, the error will be greatest. Furthermore the greater the vertical angle becomes, the greater will be the error of pointing. This error is not of great consequence in ordinary transit work. It *cannot* be eliminated by double centering. It *can*, however, be eliminated by proper use of the plate bubble (or bubbles). If the instrument contains two plate bubbles at right angles to one another, when one bubble is brought over a pair of leveling screws, the other bubble is over the opposite pair (or over the third leveling screw if the instrument contains only three leveling screws). Both bubbles are brought to the center by turning the leveling screws. The instrument is then rotated 180° about its vertical axis. If both bubbles return to the center of the tube, the vertical axis is truly vertical within the sensitivity of the plate levels. If they do not return to the center, each bubble is brought *halfway* back by turning the corresponding leveling screws. This makes the vertical axis truly vertical. The leveling screws should not be moved after this step while angles are being measured. If the instrument contains only one plate level, the above reversal procedure is repeated for directions at right angles to one another. If the theodolite contains a telescope level, the vertical axis can be made truly vertical by the technique described in Section 4-18. Some modern electronic theodolites, as well as the total station instruments described in Section 4-24, contain tilt sensing devices that compute the correct horizontal angle (as well as the zenith angle), thus eliminating this source of error.

7. If the A and B verniers are not set 180° apart, the two readings will be in disagreement. If the inner and outer spindles do not cause the verniers and the horizontal circle to move concentrically with one another, an error will be introduced that varies from zero at one pointing to a maximum at a right angle to the original pointing. These two errors are eliminated by reading both verniers and adopting a mean of the two readings. Since errors due to this source are very small, the A vernier alone is read in most ordinary transit work. Some of the glass circle instruments contain prisms that allow both sides of the circle to be brought into view. The circle reading is designed to eliminate the above error (see Section 4-22).

8. Except in work of high precision, the use of an instrument with very small inequalities in the spacing of the graduations around the horizontal circle or around the verniers will not affect the accuracy of ordinary transit work. If the precision of the work warrants it, the error in an angle at a station can be reduced by measuring the angle several times and distributing the initial backsight reading uniformly around the circle.

It is evident from the foregoing discussion that the error caused by defects in adjustment of the theodolite will be greatest when the survey is being made on hilly terrain with varying vertical angles, and when some points are very close to the instrument and others are far away. Practically all instrumental errors can be eliminated by taking the mean of two angles, one of which is observed with the telescope direct and the other with the telescope inverted.

Personal errors are introduced into theodolite work by the instrumentman and his signalmen. Some of these errors are serious if short sights are encountered, but they are usually negligible on long sights.

Failure to have the instrument centered over the occupied station, bad pointing of the telescope on the target, or failure to have the target directly over the station sighted are all personal errors that affect accuracy in a similar manner.

If the theodolite is not set up exactly over the station, all angles measured at that point will be in error. The error decreases as the length of the sight increases. The diagram in Fig. 4-23 shows the effect of measuring an angle from B' instead of B, the actual station. The correct angle ABC equals the measured angle $AB'C + e_1 + e_2$. The maximum value of e_1 will be obtained when BB' is perpendicular to BC, in which case $\tan e_1 = BB'/BC$. Likewise the value of e_2 will be greatest when BB' is at

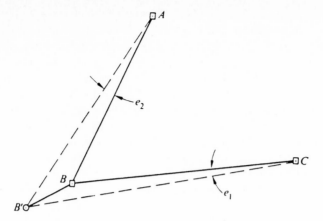

FIGURE 4-23 Instrument not properly centered.

right angles to *BA,* or when $\tan e_2 = BB'/BA$. A convenient relation to remember is that the error in a sight will be about 1′ when the length of the sight is 300 ft and the actual point of setup is about 1 in. off the theoretical line of sight. Thus, if the distances *BA* and *BC* are each about 300 ft and the distance *BB'* is 1 in., both e_1 and e_2 will be less than 1′ and the measured angle will be in error by less than 2′. Consequently, when precision is required and when the lengths of the sights are short, the instrument must be set up more carefully than when the sights are long or when small errors are of no importance.

The error in an angle due to errors of pointing can also be determined from the relationship just given. Thus, when the pointing is made to a point 1 in. to the right or left of a station 300 ft from the transit, the resulting error in the measured angle will be 1′. For a point 100 ft away, this error would be increased to about 3′.

Parallax caused by not having the image of the sighted object focused in the plane of the cross hairs introduces a personal error of varying amounts in the measurement of both horizontal and vertical angles. Its elimination has been discussed in Section 3-11.

Faulty centering of level bubbles is a personal error different from maladjustment of the bubbles. Proper manipulation of the level bubbles has been discussed in previous sections.

If the instrument contains verniers, errors are introduced when the observer reads the verniers. If the observer does not use a reading glass, or if he does not look radially along the graduations when reading the verniers, he may be one or two graduations in either direction from the correct coincidence. Errors in reading the verniers are not to be confused with mistakes in reading the verniers, to be discussed in Section 4-21.

The instrumentman should be very careful in walking around or near the theodolite. The tripod is easily disturbed, particularly when it is set up on soft ground. If a tripod leg is accidentally brushed against, the backsight should be checked at once. Of course, the plates should be releveled first and perhaps a new backsight taken, if necessary.

Some causes of natural errors are gusts of wind, heat from the sun shining on the instrument, settling of the tripod legs, and horizontal and vertical refraction of the line of sight due to atmospheric conditions. These errors are of little consequence in all but extremely precise work. The theodolite can be sheltered from the wind and from the rays of the sun if work under unfavorable weather conditions is necessary and accuracy demands such protections. Stakes may be driven to receive the tripod legs in unstable ground. Boards can be placed around the tripod in the form of a triangle and elevated to reduce the tendency to disturb the tripod. Horizontal refraction is avoided by not allowing the transit lines to pass close to such structures as buildings, smokestacks, and stand pipes, which will radiate a great deal of heat. The effect of vertical refraction can be reduced to a minimum when vertical angles are being measured if the vertical angle can be measured at both ends of the line. Otherwise its effect can be determined by methods discussed in Section 3-3 or Section 9-11.

4-21   Mistakes in Theodolite Angles

All mistakes are of a personal nature, but they are not to be confused with personal errors. They are blunders on the part of the surveyor and are known in the field as "busts." A mistake can have any magnitude and will usually render any observation useless. Some of the more frequent mistakes are forgetting to level the instrument; turning the wrong tangent screw; transposing digits in recording a reading, as writing 291° instead of 219°; dropping a full 20′, 30′, or 40′ from a vernier reading; reading the wrong vernier; reading the wrong circle (clockwise or counterclockwise); reading the wrong side of the vernier; recording a small vertical angle with the wrong algebraic sign; failing to center the control bubble on a vertical circle containing one before reading the circle; sighting on the wrong target; recording the wrong reading on a leveling rod when measuring a vertical angle.

4-22   Direction Theodolite

The direction theodolite, one example of which is shown in Fig. 4-24, does not provide for a lower motion as in a repeating instrument. When it is set up over a point, the horizontal circle is essentially fixed in position. Both the backsight reading and the foresight reading are taken by means of one motion, which is the movement of the reading device with respect to the graduated horizontal circle. The difference between the backsight and the foresight readings is the value of the angle.

The graduations on both the horizontal and the vertical glass circles are further subdivided by means of a circular optical micrometer. This permits directions to be read to the nearest second or smaller. Different methods are employed to enable the observer to make a circle reading.

The direction theodolite shown in Fig. 4-25 has two different kinds of graduations. In the earlier and very widespread model, the view seen through the reading microscope is shown in Fig. 4-26. In the upper part of the field of view can be seen the diametrically opposite graduations of either the horizontal circle or the vertical circle. These are displaced 180° apart and are brought together in the field of view

FIGURE 4-24 Optical-reading theodolite.
(a) Main telescope, (b) Focusing rings,
(c) Reading microscope, (d) Clamp of vertical
circle, (e) Micrometer setting knob, (f) Lighting
mirror, (g) Clamp of horizontal circle,
(h) Horizontal motion tangent screw,
(i) Horizontal level, (j) Optical plummet,
(k) Leveling screw, (l) Circle orienting gear,
(m) Vertical tangent screw.
Courtesy of Kern Instruments, Inc.

FIGURE 4-25 Leica T2 direction theodolite.
Courtesy of Leica, Inc.

Before coincidence

After coincidence

Circle reading.. 265° 40'

Drum reading.. _____7' 23.6"_

265° 47' 23.6"

FIGURE 4-26 Circle reading on earlier T2.

by internal optics. The ultimate reading is thus a mean of two positions on the circle, just like the combined A and B vernier readings of a transit. The lower part of the field of view shows the optical micrometer that can be read directly to 1" and estimated to $\frac{1}{10}$ ". The left part of Fig. 4-26 shows the view the observer sees after he has sighted on his signal using the clamp and tangent screw but before he sets the micrometer operating screw. He now turns the micrometer screw until the diametrically opposite graduations are coincident, as shown in the right part of the figure. After coincidence, it is seen that 265° has its corresponding reading of 85° *lying to the right.* Thus the reading is 265°. The index line also indicates the reading lies somewhere between 265° and 266°. The circle is graduated every 20'. However, if the bottom set of graduations are caused to move 10' to the right, the upper graduations will then move 10' to the left and coincidence will recur. Thus each 20' graduation represents 10' of absolute movement. To determine the number of tens of minutes, the observer counts the number of spaces between 265° and 85°, which in the figure is four. The reading thus far is 265° 40'. On the micrometer drum below, the number that repeats itself is the number of minutes to be added to the multiple of 10', and this is seen to be 7'. The graduations give the number of seconds as 23.6". The reading is therefore 265° 40' + 7' + 23.6" = 265° 47' 23.6".

The most common mistake in reading the micrometer shown in Fig. 4-26 is miscounting the number of spaces between the direct degree number and its corresponding inverted number to get the tens of minutes. This must be carefully guarded against. To eliminate this source of mistake, the newer model of this same instrument employs the technique shown in Fig. 4-27. After the telescope sighting has been made, the instrumentman makes the two sets of three graduations at the top of the field of view coincide using the micrometer screw. The small triangle below the degree number points to the appropriate number of tens of minutes. The optical micrometer drum is then read to obtain the remainder of the reading. The complete reading in the illustration is 94° 12' 44.3".

A switching knob located on the standard opposite the vertical circle of the theodolite shown in Fig. 4-25 is used to present to the observer either the horizon-

Horizontal or vertical circle
94° 12′ 44″

FIGURE 4-27 Circle reading on later T2.

FIGURE 4-28 Leica T3 direction theodolite. Courtesy of Leica, Inc.

Horizontal circle 73° 26′
1′ 56.7″
_____
73° 27′ 56.7″

FIGURE 4-29 Circle reading on T3.

tal circle or the vertical circle. Note that only one or the other appears in the field of view, as indicated in Figs. 4-26 and 4-27.

The view of the circle and optical micrometer of the direction theodolite shown in Fig. 4-28 is presented in Fig. 4-29. The technique for reading is similar to that of Fig. 4-26. Note that the circle is graduated each 4′. Therefore the absolute value of each space between the direct and inverted numbers is 2′. In this example the reading is 73° 26′ plus 1′56.7″, or 73° 27′ 56.7″.

The circle graduations shown in Fig. 4-30 appear in the direction theodolite of Fig. 4-31. This circle is graduated in grads (or gon). The two pairs of graduations opposite either the letter V or the letter H are bisected by the two index lines to give a mean of the positions of diametrically opposite parts of the circle. The tens of minutes are indicated by the small squares shown in the figure. The horizontal circle reading here is $56^g\ 53^c\ 34^{cc}$ or 56.5334^g.

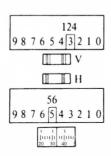

Horizontal circle 400^g
56^g 53^c 34^{cc}

FIGURE 4-30 Circle
reading on Kern DKM 2-A.

FIGURE 4-31 Kern DKM 2–A
direction theodolite. Courtesy of
Kern Instruments, Inc.

The vertical circles of the direction theodolites, which measure zenith angles, are equipped with either an automatic pendulum compensator or a control bubble, which eliminates any index error from the vertical circle reading. The control bubble is usually the coincidence type discussed in Section 3-12 and shown in Fig. 3-14.

In the modern electronic theodolite, the readings of both the horizontal and vertical circles are displayed on the instrument display screen as well as being stored in the electronic field book or data collector.

The direction theodolite contains an optical plummet that allows the instrument to be precisely centered over the instrument station.

An altogether different method for centering the instrument over the ground station is provided by the telescoping plumbing rod, the upper part of which can be seen under the instrument in Fig. 4-31. The bottom of the rod is pointed and is set into the station mark. The top of the rod is moved laterally to bring the rod into a plumb position by means of the tribrach, which moves over the top face of the tripod. The plumb position is indicated by a bull's-eye bubble attached to the rod. The upper end of the rod and, consequently, the tribrach is then locked into position by means of a knurled clamping nut at the upper end of the rod. This method of centering the instrument has the distinct advantage of allowing the theodolite to be removed and replaced by an EDM, a reflector, or a target, without disturbing the tribrach. This technique is referred to as *forced centering*, since each subsequent instrument or target placed on the tripod is automatically centered over the ground station.

Forced centering can be accomplished without the use of the plumbing rod if the tribrach, which is secured to the top of the tripod, is provided with a convenient release of the instrument. Nearly all of the modern theodolites incorporate the forced-centering feature, which allows rapid interchangeability of instruments and targets over the same point.

A cautionary note is needed regarding the optical reading theodolites. Although they are extremely rugged, they should be treated with care in handling. When one of these instruments is being removed from its case to be set up, or is removed from the tripod to be placed back into its case, it should be lifted with one hand by the standard on the side opposite that of the vertical circle and supported underneath by the other hand. This helps to preserve the critical alignment of the internal optical components. Also, as soon as the instrument is set on its tripod or into the tribrach, it should be secured. Otherwise the tripod might inadvertently be moved or even picked up, allowing the instrument to fall to the ground.

4-23   Angles from Measured Directions

Figure 4-32 is an example of recording and reducing the field notes when using the direction instrument. A direction theodolite, reading to 1″, is set up at A to measure the angles between B and C, C and D, and D and E. The telescope in its direct position is pointed at B, the motion is clamped, and the pointing is perfected by means of the tangent screw. The optical micrometer is set, and the circle is read as 132° 20′ 36″ and recorded as shown in Fig. 4-33 opposite D (for *direct*). The telescope is then turned to C, and the reading of 175° 46′ 21″ is recorded as shown. The telescope is next turned to D, and a reading of 192° 06′ 04″ is recorded. Finally, a reading of 232° 15′ 32″ is obtained by a sight to E.

To eliminate instrumental systematic errors, the telescope is reversed, and the line of sight is directed to B. With the telescope in this reverse position, the circle reading is displaced 180° from the direct reading. It is 312° 20′ 42″, which is recorded opposite R (for *reverse*). The discrepancy between the 36″ in D and the 42″ in R reflects sighting errors and instrumental errors. The mean of the two values, or 39″, is suffixed to the direct reading of 132° 20′ to give the mean of the D and R values as shown. This mean is free from instrumental errors but not from

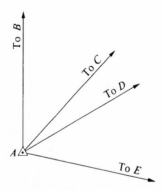

FIGURE 4-32 Angles measured by direction instrument.

STATION	D or R	CIRCLE READING	MEAN D & R
	$\mathcal{I}$	@ STATION A	
B	D	132° 20′ 36″	132° 20′ 39″
	R	312° 20′ 42″	
C	D	175° 46′ 21″	175° 46′ 23″
	R	355° 46′ 25″	
D	D	192° 06′ 04″	192° 06′ 05″
	R	12° 06′ 06″	
E	D	232° 15′ 32″	232° 15′ 35″
	R	52° 15′ 38″	

FIGURE 4-33 Direction instrument notes.

random errors. Sights are then taken to points C, D, and E to obtain the reversed readings, which are recorded as shown in the notes.

The directions to the four points are now used to obtain the angles. The angle at A from B to C is equal to the direction to C minus the direction to B, and so on. Thus

$$\angle BAC = 175°\ 46′\ 23″ - 132°\ 20′\ 39″ = 43°\ 25′\ 44″$$

$$\angle CAD = 192°\ 06′\ 05″ - 175°\ 46′\ 23″ = 16°\ 19′\ 42″$$

$$\angle DAE = 232°\ 15′\ 35″ - 192°\ 06′\ 05″ = 40°\ 09′\ 30″$$

$$\angle BAD = 192°\ 06′\ 05″ - 132°\ 20′\ 39″ = 59°\ 45′\ 26″$$

The direction instrument has provision for advancing the circle for a new set of readings, as just discussed. For the next set the initial reading would then be 132° 20′ 36″ plus a value determined from the number of sets, or *positions,* to be read. This added value is $180°/n$, in which n is the number of positions to be observed. Thus, if three positions are to be observed, the initial reading for the second set would be 132° 20′ 36″ + 60° = 192° 20′ 36″, and the initial reading for the third set would be 252° 20′ 36″. For each position a set of angles can be computed as just shown for the first position. The final angles are then the means obtained from all the sets.

4-24 Total Station Instruments

A total station instrument is essentially an electronic theodolite that contains an integral EDM with coaxial optics, as shown in Fig. 4-34 and Fig. 4-35. Output from the horizontal and vertical circular encoders and from the EDM is stored in a data collector and entered into a built-in microprocessor. The microprocessor converts the measured slope distance (S of Fig. 4-36) to the horizontal distance (H in Fig. 4-36) using the measured vertical or zenith angle (V or Z of Fig. 4-36). The microprocessor also computes the difference in elevation (DE of Fig. 4-36) between the

FIGURE 4-34 Leica TC2000 Theomat total station instrument. Courtesy of Leica, Inc.

FIGURE 4-35 Topcon GT4 total station instrument. Courtesy Topcon Instrument Corporation of America.

instrument center and the reflector target. If the elevation of the instrument center (the HI) and the height of the reflector target above ground are entered, the microprocessor computes the elevation of the target station taking into account the effect of curvature and refraction. For any current horizontal direction, expressed as an azimuth (see Chapter 7) and held in the data collector, the microprocessor computes the resolution of the horizontal distance into a north-south component (N in Fig. 4-36) and an east-west component (E in Fig. 4-36). The N and E components are called latitudes and departures and are discussed in Chapter 8.

In construction layout measurements, the data necessary to establish the direction and distance from a control point to locate a construction point can be entered into the instrument via the keyboard or directly from an office computer. The instrumentman then guides the person holding the reflector along the line of computed direction until the distance to the point to be located agrees with the computed distance. This technique is further discussed in Chapters 13 and 17.

The total station instrument shown in Fig. 4-37 can be used as a "one-person" survey system with the addition of the remote positioning unit (RPU) shown in Fig. 4-38. The RPU contains all of the functions of the instrument itself, including

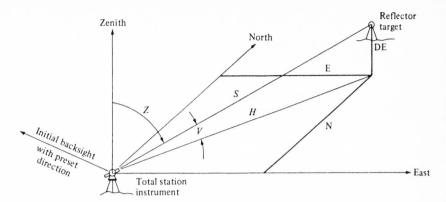

FIGURE 4-36 Resolution of slope distance and horizontal and vertical angles into horizontal and vertical distances and north-south and east-west components.

FIGURE 4-37 Geodimeter 4400 total station instrument. Courtesy of Geotronics of North America, Inc.

FIGURE 4-38 Geodimeter remote positioning unit (RPU). The combination of the Geodimeter 4400 and the RPU is designated Geodimeter System 4000. Courtesy of Geotronics of North America, Inc.

the keyboard and associated microprocessors. The two units interact with one another via a telemetric link. The instrument contains the angle- and distance-measuring instrumentation while the RPU contains the reflector. The instrument can track the reflector as the latter is moved from point to point. All of the data displayed at the instrument are also displayed at the RPU.

After the instrument is set up on a control station and properly oriented, the surveyor then proceeds to a point whose position is to be determined. The instrument then searches for the reflector using the servo-controlled autotracking feature. On command, the instrument measures and records all the data necessary to establish the position of the point occupied by the RPU. These data are also displayed and stored at the RPU. The surveyor then proceeds to the next point, and the instrument tracks the reflector as he moves. The process is then repeated.

The "one-person" combination also allows the surveyor to set out a point at a known distance and direction by observing the readout on the RPU. He moves the RPU until it displays the correct distance and direction. This is observed as the display counts down to zero, indicating that the RPU is located at the desired point.

To attempt to describe all of the different total station instruments in detail would be outside the scope of this text. The student should consult the bibliography at the end of this chapter and obtain descriptive brochures from the instrument manufacturers to gain more insight into the operation of these relatively new instruments. Application of the total station instrument to various survey operations will be discussed in the appropriate sections of the text.

4-25   Accuracy Required in Measuring Angles

The accuracy required in the measurement of horizontal angles is largely determined by the purpose of the survey and by the accuracy with which the linear measurements are made. A balance should be maintained between the accuracy of the angular measurements and the accuracy of the linear measurements.

The location of a point is frequently determined by a horizontal angle from some line of reference and a distance from the vertex of the angle. Thus in Fig. 4-39 a point P can be located by the angle NAP and the distance AP, neither of which can be measured exactly. If the angular error is $\pm e$ and it is assumed that there is no linear error, the position of P would be between P' and P''. On the other hand, if it is assumed that there is no angular error, the effect of linear error would be to locate P somewhere between P_1 and P_2. If the accuracy of the angular measurements is consistent with the accuracy of the linear measurements, the values of $P'P$, P_2P, $P''P$, and P_1P should all be approximately equal, and the located position of P will fall somewhere within an approximate square.

When the allowable linear error in a distance l is assumed to be d, the corresponding allowable angular error e can be computed from the relation $\tan e = d/l$. If the linear error is expressed as a ratio, this ratio will also be $\tan e$. Where the angular error e is assumed, the allowable error in distance will be the length of the line times the tangent of the angle e.

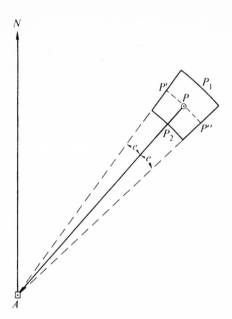

FIGURE 4-39 Relation between angular and linear errors.

In Table 4-2 the allowable errors in angles and the allowable errors in linear measurements are given for various assumed accuracies of linear and angular measurements. From this tabulation it is evident that if the distances are being obtained by the stadia method with the relative accuracy of 1/500, the angles need be measured only to the nearest 5′, whereas if a relative accuracy of 1/10,000 is required, the angles should be known to 20″. If EDMs are used to obtain distances, then the comparable accuracy in angular measurements can only be obtained using an instrument that is accurate to 1″ or 2″. This angular accuracy, however, is not required in all surveys even when EDMs are used. The high accuracy of the EDMs is inherent in the instruments and is obtained with no additional effort.

The accuracy required in the measurement of vertical angles for reducing slope distances and for trigonometric leveling are discussed in Sections 2-11 and 3-3.

4-26   Accuracy of Angles Used in Trigonometric Computations

If the measured angles are to be used in trigonometric computations, the values in Table 4-2 may be somewhat misleading. It appears from the table that with an angular error of 1′ the corresponding linear accuracy should be 1/3440. If the measured angle is used in computing the length of an unknown side, the accuracy of the computed length may be considerably less than this, the exact error depending on the size of the angle and the particular trigonometric function used to compute the distance.

As an example, assume that in Fig. 4-40 the distance AC has been measured with no error and the angle at A is accurate within 1′. The distance BC is $AC \sin A$. With $A = 30° \pm 01′$ and $AC = 500.00$ ft, BC will be 250.00 ± 0.126 ft. The accuracy of this computed value therefore is 0.126/250 or 1/1980. If the base AB of the tri-

TABLE 4-2　Relations Between Linear and Angular Errors

Allowable Angular Error for Given Linear Accuracy		Allowable Linear Error for Given Angular Accuracy						
Accuracy of Linear Measurements	Allowable Angular Error (")	Least Reading in Angular Measurements	Allowable Linear Error in					
			100'	500'	1000'	5000'	Ratio	
1/500	6'53	5'	0.145	0.727	1.454	7.272	1/688	
1/1000	3'26	1'	0.029	0.145	0.291	1.454	1/3440	
1/5000	0'41	30"	0.015	0.073	0.145	0.727	1/6880	
1/10,000	0'21	20"	0.010	0.049	0.097	0.485	1/10,300	
1/50,000	0'04	10"	0.005	0.024	0.049	0.242	1/20,600	
1/100,000	0'02	5"	0.002	0.012	0.024	0.121	1/41,200	
1/1,000,000	0'00.2	2"	0.001	0.005	0.010	0.048	1/103,100	
		1"		0.002	0.005	0.024	1/206,300	

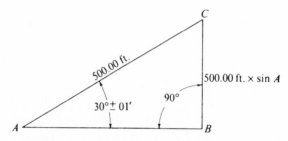

FIGURE 4-40　Accuracy of angles in trigonometric computations.

angle had been measured and BC had been computed from the relation $BC = AB$ tan A, the accuracy of the computed result would be still less than when the sine is used. With $AB = 500.00$ and $A = 30° \pm 01'$, $BC = 288.68 \pm 0.194$ ft and the accuracy of the result is 0.194/288.68 or 1/1490. These results can also be explained from error propagation theory, which is discussed in the following chapter.

　　Table 4-3 shows the accuracy of computed results involving the sines and cosines and the tangents and cotangents of various angles for angular errors of 1', 30", 20", 10", and 5". From this table it is evident that when the sine of an angle of approximately 45° is to be used in a computation and the accuracy of the computed result is to be 1/10,000, the angle must be measured to the nearest 20". When the tangent of 45° is used, the angle must be measured to 10" to maintain the same accuracy. The accuracy of computations involving the sines, tangents, or cotangents of small angles or the cosines, tangents, or cotangents of angles near 90° will be very low, unless these angles are measured much more exactly than is usually the case.

TABLE 4-3   Accuracy of Values Calculated by Trigonometry

Size of Angle and Function		Angular Error				
Sine	Cosine	1′	30″	20″	10″	5″
Sine	Cosine	Accuracy of Computed Value Using Sine or Cosine				
5°	85°	1/300	1/600	1/900	1/1800	1/3600
10°	80°	1/610	1/1210	1/1820	1/3640	1/7280
20°	70°	1/1250	1/2500	1/3750	1/7500	1/15,000
30°	60°	1/1990	1/3970	1/5960	1/11,970	1/23,940
40°	50°	1/2890	1/5770	1/8660	1/17,310	1/34,620
50°	40°	1/4100	1/8190	1/12,290	1/24,580	1/49,160
60°	30°	1/5950	1/11,900	1/17,860	1/35,720	1/71,440
70°	20°	1/9450	1/18,900	1/28,330	1/56,670	1/113,340
80°	10°	1/19,500	1/39,000	1/58,500	1/117,000	1/234,000
	Tan or Cot	Accuracy of Computed Value Using Tan or Cot				
	5°	1/300	1/600	1/900	1/1790	1/3580
	10°	1/590	1/1180	1/1760	1/3530	1/7050
	20°	1/1100	1/2210	1/3310	1/6620	1/13,250
	30°	1/1490	1/2980	1/4470	1/8930	1/17,870
	40°	1/1690	1/3390	1/5080	1/10,160	1/20,320
	45°	1/1720	1/3440	1/5160	1/10,310	1/20,630
	50°	1/1690	1/3390	1/5080	1/10,160	1/20,320
	60°	1/1490	1/2980	1/4470	1/8930	1/17,870
	70°	1/1100	1/2210	1/3310	1/6620	1/13,250
	80°	1/590	1/1180	1/1760	1/3530	1/7050
	85°	1/300	1/600	1/900	1/1790	1/3580

PROBLEMS

4-1. The sides of the triangles shown in Fig. 4-2 were measured with an EDM as follows: $AB = 213.485$ m; $AD = 282.046$ m; $AE = 316.848$ m; $CB = 322.180$ m; $CD = 232.561$ m; $BD = 192.236$ m; $DE = 238.807$ m. Compute the angles in the triangles to the nearest second.

4-2. An angle ABC in Fig. 4-3 is determined by linear measurements in the manner described in Section 4-2. If $BD = BE = 100$ ft and $k = 47.76$ ft, what is the value of the angle (nearest minute)?

4-3. A signal has been displaced 0.18 ft perpendicular to the line of sight of a theodolite. If the signal is 850 ft away from the instrument, what angular error is introduced (nearest second)?

4-4. A line rod is $\frac{1}{2}$ in. in diameter. What angle would be subtended between the two sides at a distance of 180 ft (nearest second)?

4-5. In Problem 4-4, if the instrument is 240 m away, what angle would the line rod subtend (nearest second)?

4-6. An 8-ft range pole is plumbed over a station on which a sight is made with a theodolite. If the sight is made 5 ft up from the ground, and if the top of the pole is $1\frac{1}{2}$ in. out of plumb in a direction normal to the line of sight, what angular error is produced in a 500-ft sight (nearest second)?

4-7. If the diameter of the circle in Fig. 4-10(b) is 7 in., what is the difference in spacings on the circle and those on the vernier (nearest 0.00001 in.)?

4-8. A horizontal angle is measured at E from D to F by the method of double centering. A backsight on D gives a circle reading of $0° 00' 40''$ and a foresight on F gives a circle reading of $146° 40' 20''$. After plunging the telescope, a backsight is made to D and a foresight is made to F. The final circle reading is $293° 19' 40''$. What is the value of the angle?

4-9. A backsight is made at K to J, and the horizontal circle reads $312° 14' 30''$. A foresight is made to L, and the circle reads $14° 42' 00''$. The telescope is reversed, a backsight is again made to J and a foresight to L. The final circle reading is $77° 09' 00''$. What is the value of the angle?

4-10. An angle is measured by repetition with a 1-minute transit. The mean of the A and B verniers on the initial backsight is $359° 59' 30''$. After the first repetition, the circle reads $176° 27' 00''$. After the sixth repetition, the circle reads $338° 44' 00''$. What is the value of the angle?

4-11. An angle is measured by repetition using a 30-second transit. The initial mean backsight reading is $359° 59' 45''$ on the clockwise circle. After the first repetition, the A vernier reads $72° 29' 00''$. After the twelfth repetition, the mean of the A and B verniers is $149° 47' 00''$. What is the value of the angle?

4-12. An angle is measured 6D and 6R using a 5″ theodolite. The initial backsight reading is $0° 00' 15''$. After the first repetition the circle reads $77° 20' 00''$. After the twelfth repetition the circle reads $207° 57' 45''$. What is the value of the angle?

4-13. A vertical angle is measured to a signal on top of a hill. The vertical circle reads $+13° 14'$ with the telescope direct and $+13° 19'$ with the telescope reversed. If the vernier reads zero when the transit is truly leveled and the telescope bubble is centered, what causes the apparent discrepancy? What is the size of the actual discrepancy?

4-14. In laying off a vertical angle, the surveyor sights the transit in the direction of the point to be established. He brings the telescope to a horizontal position and centers the telescope bubble. The vertical circle now reads $–0° 03'$. What reading should be set on the vertical circle to establish a $+15° 20'$ vertical angle?

4-15. A theodolite measures zenith angles in which the circle reads $0°$ with the telescope pointed to the zenith. With the telescope direct, the vertical circle reads $129° 42' 52''$ when the line of sight is directed to a target. With the telescope reversed, the vertical circle reading is $230° 17' 16''$. What is the zenith angle? What is the vertical angle?

4-16. If the instrument of Problem 4-15 is equipped with a control bubble, and the control bubble is centered for both the direct and the reversed readings, what is the rise or fall of the line of sight per 100 m when the circle reads $90° 00' 00''$ when the control bubble is centered?

4-17. The vertical circle of a theodolite reads $180°$ with the telescope pointed to the zenith. If the direct vertical circle reading is $128° 16' 30''$ and the reverse vertical circle reading is $231° 43' 24''$, what is the value of the *vertical* angle?

4-18. The line of sight of a transit rises at the rate of 32 mm per 100 m when the telescope bubble is centered. Assuming that the instrument is in proper adjustment in all other respects, what is the value of the vertical angle to a point 850 m away if the circle reads $–3° 20'$?

4-19. The distance between the upper ends of the standards that provide the points of support for the horizontal axis of the transit is 5.000 in. The left standard is 0.0025 in. high. What error will this produce in the measurement of a clockwise horizontal angle if the backsight vertical angle is −15° 00′ and the foresight vertical angle is +52° 00′? Is the measured horizontal angle too large or too small?

4-20. The right side of the telescope axis of a transit is 0.0012 in. high. The distance between points of support is 6.000 in. On measuring a clockwise horizontal angle, the backsight vertical angle is 0° 00′ and the foresight vertical angle is −42° 20′. What error will this produce in the horizontal angle? Is the measured angle too large or too small?

4-21. The line of sight of a theodolite makes an angle of 89° 57′ with the transverse axis. What error will this produce in the measurement of a horizontal angle if the backsight is level and the foresight is inclined 40° 00′?

4-22. In laying off construction lines with a 1-minute transit, the engineer lays off a line approximately at right angles with a construction baseline, and sets a point 150.00 m away from the instrument. The angle is measured by repetition and found to be 89° 59′ 35″. What offset should be measured perpendicular to the construction line at the set point to establish a true right angle (see Fig. 4-5)?

4-23. If the diameter of the glass circle of Fig. 4-26 is 100.00 mm, what is the spacing of the circle graduations [the upper part of the figure (nearest 0.1 μm)]?

4-24. A direction theodolite is used to measure angles about a point. Four positions of the circle are used. The resulting directions are listed below.

Station Sighted	Mean D and R Position 1	Mean D and R Position 2	Mean D and R Position 3	Mean D and R Position 4
K	27° 16′ 24.8″	72° 15′ 58.3″	117° 17′ 14.6″	162° 15′ 57.2″
L	106° 45′ 32.0″	151° 45′ 09.7″	196° 46′ 26.9″	241° 45′ 08.2″
M	216° 00′ 16.8″	260° 59′ 55.7″	306° 01′ 12.9″	350° 59′ 53.4″
N	262° 24′ 08.0″	307° 23′ 50.7″	352° 25′ 06.2″	37° 23′ 44.2″

Compute the three angles for each position. Compute the mean of each of the three angles from the four positions.

4-25. Convert the mean angles of Problem 4-24 to their grad values.

4-26. The following directions were observed using a direction theodolite that reads in grads. Compute the three angles for each position.

Station Sighted	Mean D and R Position 1	Mean D and R Position 2	Mean D and R Position 3	Mean D and R Position 4
P	175.3460	225.5662	274.8180	324.9050
Q	214.2706	264.4894	313.7420	363.8300
R	10.7828	61.0010	110.2539	160.3426
S	74.7853	125.0031	174.2559	224.3450

4-27. Convert the mean angles of Problem 4-26 to their degree-minute-second values.

4-28. A theodolite is used to measure directions. If it can be assumed that the error in the direction is within a spread of ±5″, how closely must the length of a line 2640 ft long be measured to be consistent with the angular accuracy?

4-29. A 1-minute transit is used to measure an angle by the method of double centering. If the measured angle is in error by no more than 30″, with what accuracy must a line 1320 ft be measured in order that the accuracies be consistent?

4-30. Assume that side BC of Fig. 4-40 is 520.45 ft long with no error. If angle A is measured as 22° 16′ 20″ ± 15″, what will be the accuracy of the computed distance AC?

4-31. Assuming no error in side AC of Fig. 4-40 which is 674.22 m long, angle A is measured as $39.7644^g \pm 20^{cc}$. What will be the accuracy of the computed side BC?

4-32. In Problem 4-31, what will be the accuracy of computed side AB?

BIBLIOGRAPHY   Day, N. B. 1988. Field-to-finish engineering surveying: experience with the wild T-1000/2000 total stations. *Tech. Papers Ann. Conv. Amer. Cong. Surv. & Mapp.* V. 1, p. 60.

Donahue, T. 1986. Shootout at the Metrodome—testing and evaluating total stations and data collectors. *Tech. Papers Ann. Meet. Amer. Cong. Surv. & Mapp.* p. 40.

Knott, D., and Gery, H. 1988. Field procedures for maximum data collection efficiency. *Tech. Papers Ann. Conv. Amer. Cong. Surv. & Mapp.* V. 1, p. 44.

McKelvey, G. R. 1985. Wild T-2000 GRE 3 computation drafting: a complete field to finish system for all types of surveys. *Tech. Papers Ann. Meet. Amer. Cong. Surv. & Mapp.* p. 42.

Moffitt, N. M. 1987. Kern system for positioning and automated coordinate evaluation—a real-time system for industrial measurements. *Tech. Papers Ann. Meet. Amer. Soc. Photogram.* V. 7, p. 57.

Munjy, R., Davies, R., and Hartwig M. 1989. Total Station Survey System (TSSS) software. *Tech. Papers Ann. Conv. Amer. Cong. Surv. & Mapp.* p. 173.

Paiva, J. V. 1985. Evaluating the accuracy of electronic theodolites. *Tech. Papers Fall Conv. Amer. Cong. Surv. & Mapp.* p. 249.

——. 1987. Understanding electronic theodolites and tacheometers. *Tech. Papers Ann. Conv. Amer. Cong. Surv. & Mapp.* V. 3, p. 267.

——. 1989. Improvements in the surveyor-data collector. *Tech. Papers Ann. Conv. Amer. Cong. Surv. & Mapp.* p. 145.

Reilly, J. P. 1993. Surveying with a robotic total station. *P.O.B.* 19:1.

——. 1996. Total Station Survey. *P.O.B.* 21:5.

——. 1992. Transit/Theodolite Survey. *P.O.B.* 18:2.

Roth, A. W. 1990, P.O.B. 1990 Data Collector Survey. *P.O.B.* 15:6.

Stenmark, J. E. 1987. Field to finish: the second generation. *Tech. Papers Ann. Conv. Amer. Cong. Surv. & Mapp.* V. 3, p. 287.

Westerhuis, K. 1988. Total stations and their keyboards and functions. *Geodetical Info.* V. 2, No. 2.

5 Random Errors

5-1 Nature of Random Errors

As described in Chapters 2, 3, and 4, in the process of measuring distance, elevations and angles, several types of errors and mistakes are encountered. Mistakes are eliminated when the observer takes such precautions as measuring a distance in both directions, checking level notes, repeating angles direct and reversed, and using other procedures that will help to avoid or detect the mistakes. Systematic errors can be practically eliminated by using calibrated tapes, by correcting taped distances for temperature, tension, sag, and slope, by determining instrumental constants and making frequency checks in EDM, by balancing backsight and foresight distances in leveling, by alternating the sequence of taking backsight and foresight readings, by having the leveling rod plumb when it is read, and by double centering to measure angles.

Random errors, often called *accidental errors,* are unpredictable in regard to both size and algebraic sign. They are truly accidental and cannot be avoided. The statements in this chapter are based on three important principles of random errors: (1) a plus error will occur as frequently as will a minus error, (2) small errors will occur more frequently than large errors, and (3) very large errors do not occur frequently, in fact rarely.

5-2 Probability of an Error Occurring

If serious thought is given to some of the examples of sources of random errors discussed in Chapters 2, 3, and 4, it will be seen that any one single error can be considered to be in itself the result of an indefinite number of very small elementary errors introduced together at any one time. The size and algebraic sign of the resultant random error are determined by the manner in which these elementary errors build up when combined in a measurement. Each element of error is thought of as having an equal chance of being plus or minus, and each element is further considered as having the same size as each other element. For a given number of these elements, or units, of error acting together, we can determine the possible number of combinations of the units, the size of the resultant error from each combination, and the probability of each such size occurring.

Assume that the size of each unit of error is unity. If one unit acts alone, the size of the resultant error is one and the possible number of combinations is two, since the error may be either +1 or –1. The probability of either combination occurring is one-half, or one out of two. Probability is defined as the number of times a given value should occur divided by the total number of times all values can occur. If two units act together, there are four possible combinations. These are +1 and +1, +1 and –1, –1 and –1, and –1 and +1. The sizes of the resultant errors are, respectively, +2, 0, –2, and 0. The probability of +2 occurring is one-quarter, that of 0 occurring is two-quarters, and that of –2 occurring is one-quarter.

Each time the number of units acting at any one time is increased by one, the number of possible combinations will double. Three units of error acting together give 8 possible combinations: four units give 16 combinations, and so on. Table 5-1 shows the relationship between the number of combinations and the probability that an error of a given size will occur.

Let m designate the number of units acting at any one time. If C is the number of combinations with q specified positive (+) units, then C is given by

$$C = \frac{m(m-1)(m-2)\ldots(m-q+1)}{q!} \tag{5-1}$$

The total number of combinations N that could possibly occur is 2^m. Thus the probability P of q positive units occurring at any one time is

$$P = \frac{C}{N} = \frac{C}{2^m} \tag{5-2}$$

The quantities m, C, N, and P are shown in the headings of Table 5-1.

The results of Table 5-1 are plotted in Fig. 5-1, views (a) through (f). The shaded rectangle represents the following probability. In (a), $\frac{1}{2}$; in (b), $\frac{1}{4}$; in (c), $\frac{1}{8}$; in (d), $\frac{1}{16}$; in (e), $\frac{1}{32}$; and in (f), $\frac{1}{64}$. It is to be noted that the total area in each case is one. In Fig. 5-1(f) the vertical scale has been reduced to one-half of that in the other five views.

If the number of units of error were increased indefinitely, the graphical representation of the size of the resultant errors and the frequency of their occurrence would approach a bell-shaped curve called the *probability curve,* or the *normal error distribution curve.* Such a curve is shown in Fig. 5-2. This curve portrays the three principles of random errors stated in Section 5-1. The equation of this curve is

$$y = ke^{-h^2 x^2} \tag{5-3}$$

in which y is the relative frequency of occurrence of an error of a given size, x is the size of the error, k and h are constants that determine the shape of the curve, and e is the base of natural logarithms. The values of k and h under a given set of circumstances are discussed in Section 5-4. The area under this curve is 1.

5-3 Condition of Least Squares

In Section 1-10 an error is defined as the difference between the true value of a quantity and the measured value of the same quantity. Since the true value of a quantity can never be determined, the errors in the measurements can be determined only to the extent that systematic errors can theoretically be computed and eliminated. Let it be presumed that a quantity has been measured several times

TABLE 5-1 Characteristics of Random Errors

Number of Units Combining at One Time (m)	Units in Combinations (q Specified)	Number of Combinations (C)	Total Number of Combinations (N)	Size of Resultant Error	Probability of Occurrence of Resultant Error (P)
1	+1	1		+1	$\frac{1}{2}$
	−1	1	2	−1	$\frac{1}{2}$
2	+1, +1	1		+2	$\frac{1}{4}$
	+1, −1	2		0	$\frac{2}{4}$
	−1, −1	1	4	−2	$\frac{1}{4}$
3	+1, +1, +1	1		+3	$\frac{1}{8}$
	+1, +1, −1	3		+1	$\frac{3}{8}$
	+1, −1, −1	3		−1	$\frac{3}{8}$
	−1, −1, −1	1	8	−3	$\frac{1}{8}$
4	+1, +1, +1, +1	1		+4	$\frac{1}{16}$
	+1, +1, +1, −1	4		+2	$\frac{4}{16}$
	+1, +1, −1, −1	6		0	$\frac{6}{16}$
	+1, −1, −1, −1	4		−2	$\frac{4}{16}$
	−1, −1, −1, −1	1	16	−4	$\frac{1}{16}$
5	+1, +1, +1, +1, +1	1		+5	$\frac{1}{32}$
	+1, +1, +1, +1, −1	5		+3	$\frac{5}{32}$
	+1, +1, +1, −1, −1	10		+1	$\frac{10}{32}$
	+1, +1, −1, −1, −1	10		−1	$\frac{10}{32}$
	+1, −1, −1, −1, −1	5		−3	$\frac{5}{32}$
	−1, −1, −1, −1, −1	1	32	−5	$\frac{1}{32}$
6	+1, +1, +1, +1, +1, +1	1		+6	$\frac{1}{64}$
	+1, +1, +1, +1, +1, −1	6		+4	$\frac{6}{64}$
	+1, +1, +1, +1, −1, −1	15		+2	$\frac{15}{64}$
	+1, +1, +1, −1, −1, −1	20		0	$\frac{20}{64}$
	+1, +1, −1, −1, −1, −1	15		−2	$\frac{15}{64}$
	+1, −1, −1, −1, −1, −1	6		−4	$\frac{6}{64}$
	−1, −1, −1, −1, −1, −1	1	64	−6	$\frac{1}{64}$

FIGURE 5-1 Plot of size of resultant error versus frequency based on plus and minus unit errors.

and that a slightly different value has been obtained from each measurement. Then, by an extended investigation of Eq. (5-3), the value of the quantity that has the most frequent chance of occurrence, or that has the *maximum probability* of occurrence, is the one that will render the sum of the squares of the errors a minimum.

Suppose, for example, that in a given set of measurements, $M_1, M_2, ..., M_n$ of a quantity, some value is adopted to represent the *best possible value* of the measured quantity obtainable from the set. Let this be denoted as M. The following subtractions are then made:

$$M_1 - M = v_1$$
$$M_2 - M = v_2$$
$$\vdots$$
$$M_n - M = v_n$$

The v's on the right-hand side are termed *residuals*. They are similar to errors (x's), but errors are obtained by subtracting the true value, rather than the best possible value, from the measurements. The equation of the probability curve or probability density function [Eq. (5-3)], can then be written in terms of the residuals as follows:

$$y = ke^{-h^2v^2} \tag{5-4}$$

in which v is the size of the residual and h and k are defined later.

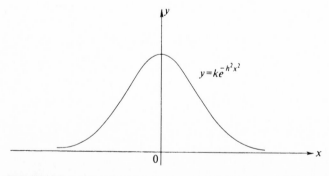

FIGURE 5-2 Probability curve.

The probability of the occurrence of v_1, for example, is the area under the probability curve at this value. This area is obtained by multiplying the ordinate by an arbitrary increment Δv, as shown in Fig. 5-3. If a probability is denoted by P, the probability that $v_1, v_2,..., v_n$ will occur in the set of measurements is as follows:

$$P_{v_1} = y_1 \; \Delta v = ke^{-h^2 v_1^2} \; \Delta v$$

$$P_{v_2} = y_2 \; \Delta v = ke^{-h^2 v_2^2} \; \Delta v$$

$$\vdots$$

$$P_{v_n} = y_n \; \Delta v = ke^{-h^2 v_n^2} \; \Delta v$$

Referring back to Section 5-2, if, for example, four units of error are imagined to be acting at any one time, what is the probability that all four are positive (+) errors at the same time? The probability that any one single error will be positive is obviously $\frac{1}{2}$. The probability that all four will be positive is $\frac{1}{2} \times \frac{1}{2} \times \frac{1}{2} \times \frac{1}{2} = \frac{1}{16}$, as shown in Table 5-1. Thus the probability that a set of events will occur simultaneously is the product of their separate probabilities. Applying this concept to the set of residuals above, the probability that all of the v's will occur simultaneously in the set of measurements (the M_i's) is equal to the product of their separate probabilities. Thus

$$P_{(v_1, v_2..., v_n)} = \left(ke^{-h^2 v_1^2} \; \Delta v\right)\left(ke^{-h^2 v_2^2} \; \Delta v\right) ... \left(ke^{-h^2 v_n^2} \; \Delta v\right)$$

or

$$P_{(v_1, v_2, ..., v_n)} = k^n (\Delta v)^n e^{-h^2(v_1^2 + v_2^2 + ... + v_n^2)} \tag{5-5}$$

The quantity M (the "best" value) should be selected so as to give the maximum probability of the simultaneous occurrence of a certain set of v's. According to Eq. (5-5), that unique set of v's that has the highest probability of occurring is the one that will make $P_{(v_1, v_2, ..., v_n)}$ a maximum. This condition exists when the negative exponent of e in Eq. (5-5) has a minimum value; that is, for maximum probability,

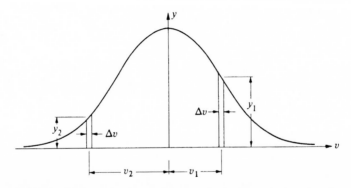

FIGURE 5-3 Probability of occurrence of a residual v_i.

$$v_1^2 + v_2^2 + \cdots + v_n^2 = \text{minimum}$$

or

$$\Sigma v^2 = \text{minimum} \tag{5-6}$$

The expression Σv^2 meaning "sum of the squares of the v's" is also frequently written $[vv]$. The square brackets signify "sum of," and vv is v^2.

Eq. (5-6) expresses the *condition of least squares* that is imposed on a set of measurements when each is made with the same reliability (see Section 5-10) and the resulting set of residuals conforms to the principles relating to random errors stated in Section 5-1.

In matrix notation the condition of least squares is written $\mathbf{V}^T\mathbf{V}$, in which $\mathbf{V}$ is a column matrix of residuals (see Section A-2).

When a quantity is measured directly several times, the best value, or most probable value, of the quantity is the arithmetic mean of the measured values. This can be shown as follows:

$$v_1 = M_1 - M \quad v_1^2 = (M_1 - M)^2$$

$$v_2 = M_2 - M \quad v_2^2 = (M_2 - M)^2$$

$$\vdots$$

$$v_n = M_n - M \quad v_n^2 = (M_n - M)^2$$

$$\Sigma v^2 = (M_1 - M)^2 + (M_2 - M)^2 + \cdots + (M_n - M)^2$$

To make Σv^2 a minimum, in accordance with Eq. (5-6), the standard procedure is to differentiate with respect to M and to set the derivative equal to zero. Thus

$$\frac{d}{dM}(\Sigma v^2) = -2(M_1 - M) - 2(M_2 - M) - \cdots - 2(M_n - M) = 0 \tag{5-7}$$

Then

$$\frac{d^2}{dM^2}(\Sigma v^2) = 2 + 2 + 2 + \cdots + 2 = \text{a positive number}$$

indicating a minimum. Also, setting Eq. (5-7) equal to zero yields

$$M_1 - M + M_2 - M + \cdots + M_n - M = 0$$

or

$$nM = M_1 + M_2 + \cdots + M_n$$

Hence,

$$M = \frac{M_1 + M_2 + \cdots + M_n}{n} = \text{arithmetic mean}$$

5-4   Standard Error

A residual v is treated as a random error in every respect. Suppose that 100 measurements of the length of a line were made and that all systematic errors have been eliminated. The mean of all the measurements is the most probable value. If the mean is now subtracted from each measured value, 100 residuals of varying size would result. If each size of residual is plotted against the frequency of occurrence of that size, the resulting curve will be similar to that of Fig. 5-3. This curve is called the *normal distribution curve* (or *normal density function*) in statistical literature. Its equation is

$$y = \frac{1}{\sigma\sqrt{2\pi}}e^{(-1/2\sigma^2)v^2} \tag{5-8}$$

in which y is the relative frequency of the occurrence of a residual of a given size, σ is a constant to be determined from the measurements, and v is the size of the residual. The value of σ is obtained from the measurements by the formula*

$$\sigma = \sqrt{\frac{\Sigma v^2}{n-1}} \tag{5-9}$$

in which Σv^2 is the sum of the squares of the residuals and n is the number of measurements. The quantity σ is referred to as the *standard error* of the set of measurements. It indicates the precision of the measurements relative to any other set of measurements. The smaller σ becomes, the greater is the precision. The term *standard error* used in this book is also called *standard deviation*.

The square of the standard deviation is called the variance and is a measure of dispersion. The variance is usually derived directly in the statistical literature from the definition $E[M_i - E\,(M_i - E(M_i))^2]$ (using the notation in this chapter). E is a linear operator called the expectation operator.

The quantity $(n-1)$ in the denominator under the radical in Eq. (5-9) represents the number of extra measurements taken to determine a single value. For example, if a line was measured five times, or $n = 5$, there are four extra measurements, since one would have been sufficient although it would not be very reliable. As the number of extra measurements increases, the standard error becomes smaller. In statistics the quantity $(n-1)$ is sometimes referred to as the number of degrees of freedom, and σ as the *estimated* standard deviation. Since a residual is obtained by subtracting the mean value of n measurements from any one of these measurements, the following statement must be true. If the mean of n measurements and any $n-1$ of the measurements are given, the one measurement that was left out can be determined. In other words, it is a dependent measurement.

---

*Some handheld and desk calculators have provision for computing standard errors based on Eq. (5-9) without first having to compute all of the residuals. Only the measurements themselves are entered into the calculator. The student is cautioned, however, that if the numbers involved are large, internal truncation may take place, giving an erroneous value of σ. To avoid this difficulty, if a series of measurements are, for example, 546.55, 546.58, ..., 546.50, they should be entered as 0.55, 0.58, ... , 0.50, and the value 546 added after the mean has been determined in the calculator.

EXAMPLE 5-1   The length of a line is measured five times. Four of the measured lengths are 154.26, 154.29, 154.26, and 154.21 ft. Also, the mean of all the measured lengths is 154.25 ft. What is the value of the dependent measurement?

Solution: The sum of the five measurements is obviously five times the mean, or $5 \times 154.25 = 771.25$ ft. The sum of the four given measurements is 617.02 ft. The dependent measurement is $771.25 - 617.02 = 154.23$ ft. ◆

A comparison of the probability curve given by Eq. (5-3) and the curve representing the distribution of errors derived from an actual set of measurements given by Eq. (5-4) and Eq. (5-8) shows that a residual v is treated as an error x. Also, the two constants k and h are obtained in terms of values derived from actual measurements by the relations

$$k = \frac{1}{\sigma \sqrt{2\pi}}$$

and

$$h^2 = \frac{1}{2\sigma^2}$$

It is seen that as k and h increase, so does the precision of the measurements increase. The quantity h is known as the *precision modulus* of the measurements.

5-5   Histograms

A *histogram* is a visual representation of the distribution of either a set of measurements or a set of residuals. The representations showing the distribution of unit errors in Fig. 5-1 are histograms. A histogram shows by examination whether the measurements or the residuals are symmetrical about some central value, such as the mean value or the zero residual; they show the total spread of values of the measurements or residuals; they show frequencies of the different values; and they show how peaked or how flat the distribution of the values is.

As an example of the development of a histogram, consider a set of 100 measurements of the number of revolutions of a planimeter wheel when measuring the area of a figure. The results of these measurements, arranged in ascending order are shown in Table 5-2. The arithmetic mean of these values, rounded off to the nearest 0.001 revolution (rev), is 1.667. Subtracting the mean from each measured value gives the set of 100 residuals associated with these measurements. The residuals are also arranged in ascending order in Table 5-3, and the unit is the revolution of the wheel.

A histogram can be drawn for each set of tabulated values. In Fig. 5-4, which is the histogram for the first set, the abscissas are the measurements in the range from 1.649 to 1.685 rev plotted at intervals of 0.001 rev. Thus each interval is given a discrete value. The ordinates show the frequency of occurrence of each given abscissa. In Fig. 5-5, which is the histogram for the set of 100 residuals, the abscissas are the sizes of the residuals in the range from −0.018 to +0.018 rev.

TABLE 5-2   Planimeter Measurements
in Revolutions of the Wheel

1.651	1.662	1.665	1.669	1.674
51	63	65	69	74
53	63	66	69	74
53	63	66	69	74
56	63	66	69	74
56	63	66	1,670	75
57	63	66	70	75
57	63	66	70	76
58	63	67	70	76
58	64	67	71	77
58	64	67	71	77
59	64	67	71	78
59	64	67	71	1.680
1.660	64	67	71	80
61	64	67	71	80
61	65	68	71	81
61	65	68	72	83
61	65	68	72	84
61	65	68	73	84
61	65	68	73	85

TABLE 5-3   Residuals in Revolutions of the Wheel

−0.016	−0.005	−0.002	+0.002	+0.007
16	04	02	02	07
14	04	01	02	07
14	04	01	02	07
11	04	01	02	07
11	04	01	03	08
10	04	01	03	08
10	04	01	03	09
09	04	0	03	09
09	03	0	04	+0.010
09	03	0	04	10
08	03	0	04	11
08	03	0	04	13
07	03	0	04	13
06	03	0	04	13
06	02	+0.001	04	14
06	02	01	05	16
06	02	01	05	17
06	02	01	06	17
06	02	01	06	18

An examination of the two histograms—that of the measurements and that of the residuals—shows that they are identical except for the numbering of the abscissa scales. This means that if the values in a set of measurements of a quantity are plotted in the form of a histogram, the residuals can be analyzed visually without also having been plotted as a histogram.

FIGURE 5-4 Histogram of polar planimeter measurements.

The value of σ for the residuals given in Table 5-3 can be computed by Eq. (5-9). The normal distribution curve can then be plotted to some arbitrary scale by applying Eq. (5-8) to compute the ordinates. However, it is usually desirable to plot the normal distribution curve to the same scale as the residual histogram in Fig. 5-5. To do this, Eq. (5-8) must be modified as follows: Let the abscissa interval, which is designated the *class interval*, be denoted by I, and let n denote the number of measurements of residuals. Then Eq. (5-8) takes the form

$$y = \frac{nI}{\sigma\sqrt{2\pi}} e^{(-1/2\sigma^2)v^2} \tag{5-10}$$

After the value of σ has been determined from the residuals, it is possible to compute the following constant quantities for the set:

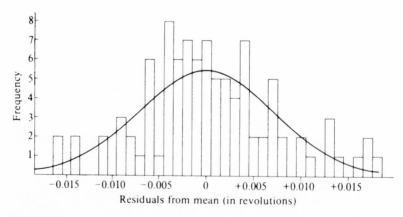

FIGURE 5-5 Histogram and normal distribution curve of residuals.

$$\frac{nI}{\sigma\sqrt{2\pi}} = K_1 \quad \text{and} \quad \frac{1}{2\sigma^2} = K_2$$

When these constants are substituted in Eq. (5-10), that equation becomes

$$y = \frac{K_1}{e^{K_2 v^2}} \tag{5-11}$$

EXAMPLE 5-2 For the normal distribution curve for the set of residuals shown in histogram form in Fig. 5-5, compute the ordinates corresponding to abscissas that are multiples of 0.002 rev. Plot the curve onto the histogram.

Solution: The sum of the squares of the residuals, or Σv^2, is found by computation to be 0.005235. These computations are not shown, but the procedure is similar to that indicated in Example 5-3. The standard error is then

$$\sigma = \sqrt{\frac{0.005235}{100-1}} = \pm 0.0073 \text{ rev}$$

Since $n = 100$ and $I = 0.001$ rev, the value of K_1 in Eq. (5-11) is

$$K_1 = \frac{100 \times 0.001}{0.0073\sqrt{2\pi}} = 5.46$$

Also,

$$K_2 = 9383$$

v	v^2	$K_2 v^2$	$e^{K_2 v^2}$	$y = \dfrac{K_1}{e^{K_2 v^2}}$
0	0	0	1.000	5.46
± 0.002	0.000004	0.0375	1.038	5.26
± 0.004	0.000016	0.1501	1.162	4.70
± 0.006	0.000036	0.338	1.403	3.88
± 0.008	0.000064	0.601	1.82	3.00
± 0.010	0.000100	0.938	2.56	2.13
± 0.012	0.000144	1.351	3.86	1.41
± 0.014	0.000196	1.839	6.30	0.87
± 0.016	0.000256	2.40	11.0	0.50
± 0.018	0.000324	3.04	20.9	0.26

The remainder of the computations are arranged in tabular form, as shown. The computed values of y have been plotted on the histogram in Fig. 5-5, and the normal distribution curve has been drawn as a continuous line through the plotted points. Note that the area under the curve is essentially the same as the area bounded by the histogram, as it should be. Note also that if the histogram of residuals had not been prepared, the normal distribution curve could have been plotted on the histogram of the measurements shown in Fig. 5-4 by renumbering the abscissa scale so that $v = 0$ occurs at the mean measured value of 1.667 rev. ◆

The class interval I of the data given in the above tables and shown in Figs. 5-4 and 5-5 is 0.001 rev. This class interval is somewhat arbitrary and could be taken as some other value, say, 0.005 rev. A limit to the size of the class interval is reached when the plotted histograms no longer reflect the actual distribution of either the measurements or the residuals. To examine the effect of a change of class interval, the planimeter measurements given in Table 5-2 are grouped using a class interval of 0.005 rev in Table 5-4.

The first two columns are plotted in the form of a histogram to give the same area as the histogram of Fig. 5-4. Note that the abscissa scale remains the same but the frequency scale must be reduced by a factor of five. The resulting histogram shown in Fig. 5-6 is considerably more coarse than that of Fig. 5-4, but the picture of the distribution of the measurements is generally the same.

A continuation of the tabulation gives a simplified method for computing the standard error when a great number of measurements are involved. The central value for each class interval is listed in the third column. This value, which represents each measurement in the interval, is multiplied by the corresponding frequency to give the products appearing in column 4. The sum of the products in column 4 is virtually the same as the sum of the 100 individual measurements. The mean, 1.667 rev, is the same in this instance as before. If the mean is now subtracted from the central value of each class interval, the residual for each central

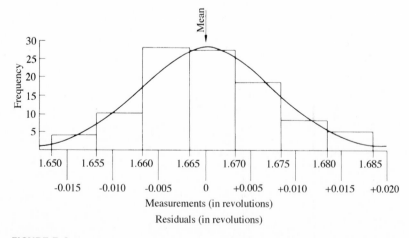

FIGURE 5-6 Histogram of measurements and residuals together with normal distribution curve for I = 0.005 rev.

TABLE 5-4   Planimeter Measurements Grouped Using a Class Interval of 0.005 Rev

(1) Interval	(2) Frequency	(3) Central Value	(4) Frequency × Central Value	(5) Residual at Central Value	(6) v^2	(7) Σv^2
1.651–1.655	4	1.653	6.612	−0.014	0.000196	0.000784
1.656–1.660	10	1.658	16.580	−0.009	0.000081	0.000810
1.661–1.665	28	1.663	46.564	−0.004	0.000016	0.000448
1.666–1.670	27	1.668	45.036	+0.001	0.000001	0.000027
1.671–1.675	18	1.673	30.114	+0.006	0.000036	0.000648
1.676–1.680	8	1.678	13.424	+0.011	0.000121	0.000968
1.681–1.685	5	1.683	8.415	+0.016	0.000256	0.001280
	100		166.745			0.004965

Mean = 1.667 rev

$\sigma = \pm\ 0.0071$ rev

value is obtained and is entered in column 5. This is a representative residual for all those measurements falling in the class interval.

Squaring each residual to give the values in column 6 and then multiplying each squared value by the frequency give the Σv^2 for each class interval. The value is shown in column 7. The values in column 7 are added to give the total Σv^2 = 0.004965. This differs slightly from the value 0.005235 given in Example 5-2. This slight difference is accounted for by the fact that the measurements are not distributed uniformly in each interval.

To plot the normal distribution curve for this distribution, the values of K_1 and K_2 are first computed. Thus

$$K_1 = \frac{100 \times 0.005}{0.0071\sqrt{2\pi}} = 28.1 \quad K_2 = 9919$$

The ordinates to the curve are then computed the same way as shown in Example 5-2.

v	v^2	$K_2 v^2$	$e^{K_2 v^2}$	$y = \dfrac{K_1}{e^{K_2 v^2}}$
0	0	0	1.000	28.1
± 0.005	0.000025	0.2480	1.282	21.9
± 0.010	0.000100	0.9919	2.680	10.5
± 0.015	0.000225	2.232	9.30	3.02
± 0.020	0.000400	3.968	53.2	0.53
± 0.025	0.000625	6.199	490	0.06

The curve is now superimposed on the measurement histogram of Fig. 5-6. Note that the zero residual falls at the 1.667-rev point on the abscissa scale, since the mean value gives zero residual. A close comparison between the normal distribution curves of Figs. 5-5 and 5-6 would show that they are practically identical in shape and amplitude. Thus a reasonable change in the class interval chosen to make an analysis of a set of measurements will have little effect on the results.

5-6 Measures of Precision

Both the standard error σ and the precision modulus h are measures of precision of a set of measurements. If the curve for Eq. (5-8) were integrated between the limits $-\sigma$ and $+\sigma$, the area between these limits would be 0.6826 times the total area under the curve, the total area being unity. In other words, approximately 68% of all the residuals for the set of measurements should be equal to or less than the standard error. In Example 5-2, by actually counting them, 72% of the residuals fall between −0.0073 and +0.0073 rev. For further analysis the curve plotted in Fig. 5-5 is redrawn in Fig. 5-7.

The area under the curve between the limits $-\sigma$ and $+\sigma$ is indicated. Also, as can be seen on this curve and can be shown mathematically, the standard error occurs at the point of inflection on each side of the center.

Consider the problem of finding the size of a residual, plus or minus, within which one-half, or 50%, of the residuals should fall. This size would be the fractional part of σ for which the area under the curve will be 0.50. The limits of this area are the dashed lines in Fig. 5-7. The error of a size embracing 50% of all residuals is called the *probable error*. It is obtained by the formula

$$E = 0.6745\sigma = 0.6745\sqrt{\frac{\Sigma v^2}{n-1}} \tag{5-12}$$

in which E is the probable error of any one of the measurements, and the other symbols have the meanings previously given. Another way of indicating the meaning of probable error is to say that the residual of any one measurement has an equal chance of being either less than E or greater than E. In Example 5-2 the probable error is ±0.0048 rev. It is seen that 55% of the residuals fall in the range between −0.0048 and +0.0048 rev.

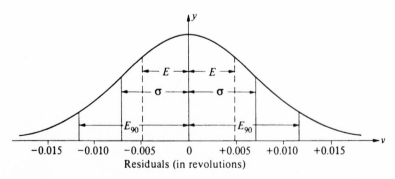

FIGURE 5-7 Measures of precision.

EXAMPLE 5-3   The measurements of the difference in elevation between two points by three-wire leveling are listed in the accompanying table. What are the standard error and the probable error of any one of these measurements?

Measurement Number	Measured Difference in Elevation (ft)	Residual v (ft)	v^2
1	21.232	−0.001	0.000001
2	21.235	+0.002	0.000004
3	21.233	0	0
4	21.225	−0.008	0.000064
5	21.233	0	0
6	21.221	−0.012	0.000144
7	21.239	+0.006	0.000036
8	21.244	+0.011	0.000121
9	21.232	−0.001	0.000001
10	21.234	+0.001	0.000001
11	21.229	−0.004	0.000016
12	21.235	+0.002	0.000004
13	21.237	+0.004	0.000016
14	21.230	−0.003	0.000009
15	21.235	+0.002	0.000004
Sum =	318.494	Σv^2 =	0.000421
Mean =	21.233		

Solution: First find the mean of the measured differences in elevation. Then compute each residual and its square, and get Σv^2. The standard error of any one of these measurements is, by Eq. (5-9),

$$\sigma = \sqrt{\frac{0.000421}{15-1}} = \pm 0.0055 \text{ ft}$$

The probable error of any one of these measurements is, by Eq. (5-12),

$$E = 0.6745(\pm 0.0055) = \pm 0.0037 \text{ ft}$$

Theoretically, one-half of the residuals should lie between +0.0037 and −0.0037 ft. ◆

The probable error E, as well as the quantities h and σ, is a measure of the precision of a given set of measurements and can be computed after the measurements have been made. The quantity E is sometimes referred to as the *50% error*, because 50% of all the residuals are theoretically expected to fall between $+E$ and $-E$. This measure of precision was used extensively in the past for defining the reliability of a set of measurements. However, its use has been almost completely replaced by the concept of standard error.

Another very useful measure of precision is the *90% error,* or the value below which 90% of all residuals of a set of measurements is expected to fall. The 90% error is determined from the standard error by the relation

$$E_{90} = 1.6449\sigma \tag{5-13}$$

This measure of precision is valuable in deciding the worth of a given set of measurements, since it indicates the size of the residual that is exceeded by only 10% of the residuals. In Example 5-2, E_{90} is ±0.0118 rev. Actually, a total of 88% of the residuals fall in the range between –0.0118 and +0.0118 rev. Other common expressions of accuracy include 2σ (95.4%) and 3σ (99.8%).

5-7   Propagation of Random Errors

So far in this chapter, consideration has been given to the measures of precision of directly measured quantities. In Example 5-1 a line was repeatedly measured; in Example 5-2 the area of a figure was measured several times; in Example 5-3 a difference in elevation was measured 15 times. In each of these instances, the measurement was presumed to have been obtained by direct application of the measuring instrument. It could be argued correctly, of course, that a difference in elevation as illustrated by Example 5-3 is the result of the difference between two direct observations of the level rod. The question then arises: If the standard error of each rod reading is known, what then *is* the standard error of the difference in elevation obtained from the rod readings?

The difference in elevation discussed above is what is called a *derived quantity,* or a computed quantity based on directly observed quantities. To investigate how random errors propagate and what the standard error of a derived quantity is, consider a very simple case of the measurement of the line *AC* of Fig. 5-8 in two sections *AB* and *BC*. Letting $X = AB$ and $Y = BC$, then the total length is $U = AC$. But $U = X + Y$, and U is considered to be a derived quantity. Suppose that X and Y are both measured several times and that a mean value for each, designated $\bar{X}$ and $\bar{Y}$, is obtained. Now denoting the difference between the mean and the measured values of X_i as x_i, and the difference between the mean and the measured values of Y_i as y_i, the x's and y's are seen to be residuals. Then

$$
\begin{aligned}
X_1 &= \bar{X} + x_1 & Y_1 &= \bar{Y} + y_1 \\
X_2 &= \bar{X} + x_2 \ \text{ and } \ Y_2 &= \bar{Y} + y_2 \\
&\ \ \vdots & &\ \ \vdots \\
X_n &= \bar{X} + x_n & Y_n &= \bar{Y} + y_n
\end{aligned}
$$

FIGURE 5-8 Sum of two directly measured quantities.

The subscript i above indicates any one of a series of quantities from 1 to n. If $\bar{U}$ denotes the most probable value of the sum of the two distances obtained by adding $\bar{X} + \bar{Y}$, and if u_i denotes the difference between $\bar{U}$ and the value U_i obtained by adding the measurements $X_i + Y_i$, then

$$U_1 = \bar{U} + u_1 = X_1 + Y_1 = \bar{X} + x_1 + \bar{Y} + y_1$$

$$U_2 = \bar{U} + u_2 = X_2 + Y_2 = \bar{X} + x_2 + \bar{Y} + y_2$$

$$\vdots$$

$$U_n = \bar{U} + u_n = X_n + Y_n = \bar{X} + x_n + \bar{Y} + y_n$$

or

$$\bar{U} + u_1 = \bar{X} + x_1 + \bar{Y} + y_1$$
$$\bar{U} + u_2 = \bar{X} + x_2 + \bar{Y} + y_2$$
$$\vdots$$
$$\bar{U} + u_n = \bar{X} + x_n + \bar{Y} + y_n$$

But since

$$\bar{U} = \bar{X} + \bar{Y}$$

then

$$\left.\begin{array}{l} u_1 = x_1 + y_1 \\ u_2 = x_2 + y_2 \\ \quad\vdots \\ u_n = x_n + y_n \end{array}\right\} \qquad \text{(5-14)}$$

Eq. (5-14) states that if AB of Fig. 5-8 is measured with a residual of +0.06 ft, and BC is measured with a residual of +0.03 ft, the sum of AB and BC will have a residual of +0.09 ft; that is, $x_i = +0.06$ ft, $y_i = +0.03$ ft, and $u_i = +0.09$ ft.

Squaring both sides of Eq. (5-14) and adding gives

$$u_1^2 = x_1^2 + 2x_1 y_1 + y_1^2$$

$$u_2^2 = x_2^2 + 2x_2 y_2 + y_2^2$$

$$\vdots$$

$$\dfrac{u_n^2 = x_n^2 + 2x_n y_n + y_n^2}{\Sigma u^2 = \Sigma x^2 + 2\Sigma xy + \Sigma y^2} \qquad \text{(5-15)}$$

Equation (5-15) contains the sum of the products term $2\Sigma xy$, which must be examined. If x_i is a residual, then it can be shown that $\Sigma x = 0$ and likewise $\Sigma y = 0$. Thus the term Σxy tends to be close to zero. Another argument that can be made to show that $\Sigma xy \rightarrow 0$ deals with the correlation between the statistical quantities X_i and Y_i. If the observations X_i and Y_i are uncorrelated, that is, independent, so are the residuals. The term Σxy represents that correlation. In this case the measurements of $X_i = AB$ and $Y_i = BC$ are independent and hence uncorrelated. Therefore, if $\Sigma xy \to 0$ then Eq. (5-15) can be written as

$$\Sigma u^2 = \Sigma x^2 + \Sigma y^2 \qquad (5\text{-}16)$$

Dividing both sides by $n - 1$ gives

$$\frac{\Sigma u^2}{n-1} = \frac{\Sigma x^2}{n-1} + \frac{\Sigma y^2}{n-1}$$

or, by Eq. (5-9),

$$\sigma_U^2 = \sigma_X^2 + \sigma_Y^2$$

and

$$\sigma_U = \sqrt{\sigma_X^2 + \sigma_Y^2} \qquad (5\text{-}17)$$

Thus the standard error of the sum of two quantities is equal to the square root of the sum of the squares of the standard errors of the individual quantities. This concept can be extended to the sum of any number of quantities that are not correlated.

EXAMPLE 5-4   A line is measured in six sections, and the lengths of the sections together with the standard error in the length of each section are given in the accompanying table. What is the standard error of the length of the line?

Section	Length (ft)	σ (ft)
1	961.22	± 0.044
2	433.12	± 0.031
3	1545.90	± 0.060
4	355.40	± 0.021
5	1252.54	± 0.100
6	320.40	± 0.075

Solution: The sum of the lengths is 4868.58 ft, and the standard error of the total length of the line is, by Eq. (5-17),

$$\sigma_s = \sqrt{0.044^2 + 0.031^2 + 0.060^2 + 0.021^2 + 0.100^2 + 0.075^2}$$
$$= \pm 0.150 \text{ ft}$$

◆

EXAMPLE 5-5   A line is laid out using a 200-ft tape. The standard error of each tape length laid out is assumed to be ±0.03 ft. If 12 tape lengths are laid out, giving a total length of 2400 ft, what is the standard error of the length of the line?

Solution: Since each tape length has the same standard error, then Eq. (5-17) can be written as

$$\sigma_s = \sigma\sqrt{n} \qquad\qquad [5\text{-}18]$$

Hence the standard error of the 2400-ft distance is $0.03\sqrt{12} = \pm0.104$ ft . ◆

One measure of precision that is of importance in surveying is the standard error of the mean of a set of like measurements. The standard error of each measurement of the set is given by Eq. (5-9). The standard error of the sum of all the measurements of the set is given by Eq. (5-18) of Example 5-5. Then, since the mean is equal to the sum divided by the number of measurements, the standard error of the mean is given by

$$\sigma_M = \frac{\sigma\sqrt{n}}{n} = \frac{\sigma}{\sqrt{n}} \qquad\qquad (5\text{-}19)$$

Equation (5-19) can be combined with Eq. (5-9) to give a different form of σ_M. Thus

$$\sigma_M = \sqrt{\frac{\Sigma v^2}{n(n-1)}} \qquad\qquad (5\text{-}20)$$

EXAMPLE 5-6   Compute the standard error of the mean of Example 5-3.

Solution: The standard error of the measurements is ±0.0055 ft. Then by Eq. (5-19), the standard error of the mean is $0.0055/\sqrt{15} = \pm0.0014$ ft . ◆

The standard error of the mean indicates what spread would be expected if the set of measurements were to be repeated several times and the means were analyzed. If the set of measurements of Example 5-3 were to be repeated several times under the same field conditions, then, as shown in Example 5-6, about 68% of the means should be expected to fall within ±0.0014 ft of the mean of the means.

The analysis employed to investigate the propagation of standard error through a very simple function $U = X + Y$ can be applied to any function of a set of independent uncorrelated variables whose standard errors are known.

Let U be some function of measured quantities $X, Y, Z, \dots, Q$; that is,

$$U = f(X, Y, Z, \dots, Q)$$

If each independent variable is allowed to change by a small amount $dX, dY, dZ, \dots, dQ$, then the quantity U will change by an amount dU given by the expression from the calculus

$$dU = \frac{\partial U}{\partial X} dX + \frac{\partial U}{\partial Y} dY + \frac{\partial U}{\partial Z} dZ + \dots + \frac{\partial U}{\partial Q} dQ$$

in which $\partial U/\partial X$ denotes the partial derivative of U with respect to X, and likewise for the other variables. Applying this to a set of measurements, and since the residuals are small quantities, let $x_i = dX_i$, $y_i = dY_i$, $z_i = dZ_i$, $\ldots$, $q_i = dQ_i$. Then, letting $u_i = dU_i$,

$$
\left.
\begin{aligned}
u_1 &= \frac{\partial U}{\partial X}x_1 + \frac{\partial U}{\partial Y}y_1 + \frac{\partial U}{\partial Z}z_1 + \cdots + \frac{\partial U}{\partial Q}q_1 \\[2mm]
u_2 &= \frac{\partial U}{\partial X}x_2 + \frac{\partial U}{\partial Y}y_2 + \frac{\partial U}{\partial Z}z_2 + \cdots + \frac{\partial U}{\partial Q}q_2 \\
&\;\;\vdots \\
u_n &= \frac{\partial U}{\partial X}x_n + \frac{\partial U}{\partial Y}y_n + \frac{\partial U}{\partial Z}z_n + \cdots + \frac{\partial U}{\partial Q}q_n
\end{aligned}
\right\}
\tag{5-21}
$$

Squaring both sides of Eq. (5-21) and adding gives

$$
u_1^2 = \left(\frac{\partial U}{\partial X}\right)^2 x_1^2 + 2\left(\frac{\partial U}{\partial X}\right)\left(\frac{\partial U}{\partial Y}\right)x_1 y_1 + \cdots + \left(\frac{\partial U}{\partial Y}\right)^2 y_1^2 + \cdots + \left(\frac{\partial U}{\partial Q}\right)^2 q_1^2
$$

$$
u_2^2 = \left(\frac{\partial U}{\partial X}\right)^2 x_2^2 + 2\left(\frac{\partial U}{\partial X}\right)\left(\frac{\partial U}{\partial Y}\right)x_2 y_2 + \cdots + \left(\frac{\partial U}{\partial Y}\right)^2 y_2^2 + \cdots + \left(\frac{\partial U}{\partial Q}\right)^2 q_2^2
$$

$$
\vdots
$$

$$
u_n^2 = \left(\frac{\partial U}{\partial X}\right)^2 x_n^2 + 2\left(\frac{\partial U}{\partial X}\right)\left(\frac{\partial U}{\partial Y}\right)x_n y_n + \cdots + \left(\frac{\partial U}{\partial Y}\right)^2 y_n^2 + \cdots + \left(\frac{\partial U}{\partial Q}\right)^2 q_n^2
$$

$$
\overline{\Sigma u^2 = \left(\frac{\partial U}{\partial X}\right)^2 \Sigma x^2 + 2\left(\frac{\partial U}{\partial X}\right)\left(\frac{\partial U}{\partial Y}\right)\Sigma xy + \cdots + \left(\frac{\partial U}{\partial Y}\right)^2 \Sigma y^2 + \cdots + \left(\frac{\partial U}{\partial Q}\right)^2 \Sigma q^2}
\tag{5-22}
$$

in which some of the squared and cross-product terms have been omitted for the sake of simplicity. If the measured quantities are independent and are not correlated with one another, the cross products tend toward zero. Dropping these cross products and then dividing both sides of Eq. (5-22) by $n-1$ gives

$$
\frac{\Sigma u^2}{n-1} = \left(\frac{\partial U}{\partial X}\right)^2 \frac{\Sigma x^2}{n-1} + \left(\frac{\partial U}{\partial Y}\right)^2 \frac{\Sigma y^2}{n-1}
$$

$$
+ \left(\frac{\partial U}{\partial Z}\right)^2 \frac{\Sigma z^2}{n-1} + \cdots + \left(\frac{\partial U}{\partial Q}\right)^2 \frac{\Sigma q^2}{n-1}
$$

or

$$
\sigma_U^2 = \left(\frac{\partial U}{\partial X}\right)^2 \sigma_X^2 + \left(\frac{\partial U}{\partial Y}\right)^2 \sigma_Y^2 + \left(\frac{\partial U}{\partial Z}\right)^2 \sigma_Z^2 + \cdots + \left(\frac{\partial U}{\partial Q}\right)^2 \sigma_Q^2
\tag{5-23}
$$

The general expression for propagation of error through any function expressed by Eq. (5-23) can then be applied to any special function.

Let $U = X + Y + Z$, in which X, Y, and Z are three measured quantities. Then from Eq. (5-23),

$$\sigma_U = \sqrt{\sigma_X^2 + \sigma_Y^2 + \sigma_Z^2}$$

which is identical with Eq. (5-17) but for three variables.

Let $U = X - Y$, in which X and Y are two measured quantities. Then, from Eq. (5-23),

$$\sigma_U = \sqrt{\sigma_X^2 + \sigma_Y^2} \qquad (5\text{-}24)$$

Let $U = XY$, in which X and Y are two measured quantities. Then, from Eq. (5-23),

$$\sigma_U = \sqrt{Y^2\sigma_X^2 + X^2\sigma_Y^2} \qquad (5\text{-}25)$$

Let $U = AX$, in which A is a constant and X is a measured quantity. Then, from Eq. (5-23),

$$\sigma_U = A\sigma_X \qquad (5\text{-}26)$$

EXAMPLE 5-7   If the standard error of a backsight reading from a setup of a level is ±0.02 ft, and the standard error of the foresight reading from the same setup is also ±0.02 ft, what is the standard error of the difference in elevation between the two turning points on which the readings are taken?

Solution: The difference in elevation between the two turning points is equal to the backsight minus the foresight. Then by Eq. (5-24) the standard error of the difference in elevation is

$$\sigma = \sqrt{0.02^2 + 0.02^2} = \pm 0.028 \text{ ft}$$

EXAMPLE 5-8   The dimensions of a rectangular field are measured with a steel tape and found to be 550.00 ft with a standard error of ±0.07 ft, and 800.00 ft with a standard error of ±0.12 ft. What are the area of the field and the standard error of the area?

Solution: The area of the field is $550.00 \times 800.00 = 440,000 \text{ ft}^2$. The standard error of the area is, by Eq. (5-25),

$$\sigma = \sqrt{550^2 \times 0.12^2 + 800^2 \times 0.07^2} = \pm 86.6 \text{ ft}^2$$

EXAMPLE 5-9   What is the standard error of the determination of the area in Example 5-8, in acres?

Solution: Since 1 acre equals 43,560 ft^2, the standard error of the area in acres is, by Eq. (5-26),

$$\sigma = \left(\frac{1}{43,560}\right)(\pm 86.6) = \pm 0.00199 \text{ acre}$$

EXAMPLE 5-10   The diameter D of the base of a cone is measured as 3.002 in., with a standard error of ±0.0005 in. The height h of the cone is measured as 5.45 in. with a standard error of ±0.01 in. What are the volume of the cone and the standard error of the volume?

Solution: The volume of a cone is

$$V = \frac{\pi D^2 h}{12} = \frac{\pi \times 3.002^2 \times 5.45}{12} = 12.86 \text{ in.}^3$$

The standard error of the volume can be expressed, according to Eq. (5-23), as follows:

$$\sigma_V = \sqrt{\left(\frac{\partial V}{\partial D}\sigma_D\right)^2 + \left(\frac{\partial V}{\partial h}\sigma_h\right)^2}$$

Then

$$\frac{\partial V}{\partial D} = \frac{\pi D h}{6} = \frac{\pi \times 3.002 \times 5.45}{6} = 8.567 \text{ in.}^3/\text{in.}$$

and

$$\frac{\partial V}{\partial h} = \frac{\pi D^2}{12} = \frac{\pi \times 3.002^2}{12} = 2.359 \text{ in.}^3/\text{in.}$$

Hence the standard error of the volume is

$$\sigma_V = \sqrt{(8.567 \times 0.0005)^2 + (2.359 \times 0.01)^2} = \pm 0.024 \text{ in.}^3 \qquad \blacklozenge$$

EXAMPLE 5-11   Two sides and the included angle of a triangle were measured with the following results: $a = 472.58$ ft and $\sigma_a = \pm 0.09$ ft; $b = 214.55$ ft and $\sigma_b = \pm 0.06$ ft; $C = 37°15'$ and $\sigma_C = \pm 30''$. Compute the area of the triangle, in square feet, and compute the standard error of the area.

Solution: The area A of the triangle is given by the relationship

$$A = \tfrac{1}{2} ab \sin C = \tfrac{1}{2} \times 472.58 \times 214.55 \sin 37°15' = 30{,}686 \text{ ft}^2$$

The standard error of the area is, by Eq. (5-23),

$$\sigma_A = \sqrt{\left(\frac{\partial A}{\partial a}\sigma_a\right)^2 + \left(\frac{\partial A}{\partial b}\sigma_b\right)^2 + \left(\frac{\partial A}{\partial C}\sigma_C\right)^2}$$

Then

$$\frac{\partial A}{\partial a} = \frac{1}{2} b \sin C = \frac{1}{2} \times 214.55 \sin 37°15' = 64.93 \text{ ft}^2/\text{ft}$$

$$\frac{\partial A}{\partial b} = \frac{1}{2} a \sin C = \frac{1}{2} \times 472.58 \sin 37°15' = 143.02 \text{ ft}^2/\text{ft}$$

and

$$\frac{\partial A}{\partial C} = \frac{1}{2}ab \, \cos \, C = \frac{1}{2} \times 472.58 \times 214.55 \, \cos \, 37°15' = 40{,}354 \text{ ft}^2/\text{rad}$$

Since σ_C expressed in radians (rad) is $30 \times 0.00000485 = 0.0001455$ rad, the standard error of the computed area is

$$\sigma_V = \sqrt{(64.93 \times 0.09)^2 + (143.02 \times 0.06)^2 + (40{,}354 \times 0.0001455)^2}$$
$$= \pm 12 \text{ ft}^2 \qquad \blacklozenge$$

5-8   Weighted Measurements

The *weight* of a measurement can be thought of as the value or the worth of that measurement relative to any other measurement. To investigate the relationship between the weight of a given observation and the measures of precision σ and h, consider the length of a line to have been measured 15 times by using the same measuring technique, the same degree of refinement, the same experienced personnel, and so on. Suppose that the 15 values are as shown in the accompanying table.

Measurement Number	Value (ft)	Measurement Number	Value (ft)	Measurement Number	Value (ft)
1	432.33	6	432.33	11	432.36
2	432.33	7	432.36	12	432.33
3	432.36	8	432.31	13	432.33
4	432.33	9	432.33	14	432.36
5	432.31	10	432.36	15	432.33

Each of the listed values has equal weight because all conditions of measurement were identical. The most probable value of the length of the line may therefore be obtained by adding all the values and dividing by 15, that is, by finding the mean value. Since the value 432.33 appears eight times, the value 432.36 appears five times, and the value 432.31 appears twice, the mean can be computed as follows:

$$8 \times 432.33 = 3458.64$$
$$5 \times 432.36 = 2161.80$$
$$\underline{2 \times 432.31 = 864.62}$$

$$15 \qquad\qquad\qquad 6485.06/15 = 432.337 \text{ ft}$$

The 15 measurements can be considered equivalent to 3 measurements with the relative worths, or weights, or 8, 5, and 2. By analysis of the arithmetic in the assumed example, an expression for a *weighted mean* can be formed as follows:

$$\overline{M} = \frac{p_1 M_1 + p_2 M_2 + \ldots + p_n M_n}{p_1 + p_2 + \ldots + p_n} = \frac{\Sigma(pM)}{\Sigma p} \qquad (5\text{-}27)$$

in which $\bar{M}$ is the weighted mean of several measurements of a quantity; M_1, M_2, and so on are the measurements; and p_1, p_2, and so on are the relative weights of the measurements.

Now suppose that the standard error of each of the 15 measurements of the line in the assumed example is ±0.02 ft. The mean of the eight values of 432.33 ft has a standard error of ±0.02/$\sqrt{8}$ ft by Eq. (5-19); the mean of the five values of 432.36 ft has a standard error of ±0.02/$\sqrt{5}$ ft, and the mean of the two values of 432.31 ft has a standard error of ±0.02/$\sqrt{2}$ ft. For convenient reference the values of the three means, their weights, their standard errors, and the squares of the standard errors are given in the accompanying table.

Value (ft)	Weight p	Standard Error (ft)	σ^2 (ft^2)
432.33	8	±0.02/$\sqrt{8}$	0.0004/8
432.36	5	±0.02/$\sqrt{5}$	0.0004/5
432.31	2	±0.02/$\sqrt{2}$	0.0004/2

The example demonstrates the following very important relationship between relative weights and measures of precision: The weight of a measured value is inversely proportional to the square of the standard error of the measurement, and directly proportional to the square of the precision modulus, that is,

$$p \propto \frac{1}{\sigma^2} \propto h^2 \tag{5-28}$$

In the preceding discussion the value ±0.02 is assumed to be the standard error of each individual measurement of the set, and each measurement has the same weight. This quantity is sometimes referred to as the standard error of unit weight, and is designated σ_0.

The standard error of unit weight of a set of n measurements of the same quantity, each having different weights, is given by the expression

$$\sigma_0 = \sqrt{\frac{\Sigma(pv^2)}{n-1}} \tag{5-29}$$

in which the v's are the differences between the measurement and the weighted mean; that is, $v_i = M_i - \bar{M}$. The standard error of each weighted value can then be expressed in terms of its weight and the standard error of unit weight, as follows:

$$\sigma_p = \sigma_0 \sqrt{\frac{1}{p}} \tag{5-30}$$

in which σ_p is the standard error of a weighted value and p is the weight. Of course, if the weight of a measurement is unity, then $\sigma = \sigma_0$. The quantity $1/p$, which is the reciprocal of the weight, is referred to as a *weight number* or *cofactor*.

EXAMPLE 5-12   Three lines of levels were run between BM20 and BM21 by different routes, but each setup and each rod reading were made with the same degree of care. On the first line, 17 setups were required and the measured difference in elevation was 29.492 ft; on the second line, 9 setups were required and the measured difference in elevation was 29.440 ft; on the third line, 10 setups were required and the measured difference in elevation was 29.480 ft. What is the weighted mean of the measured differences in elevation between BM20 and BM21?

Solution: According to Eq. (5-18) in Example 5-5, the standard error of the determination of the difference in elevation between two points is proportional to the square root of the number of setups required. Therefore the standard errors of lines 1, 2, and 3 are in proportion to the quantities $\sqrt{17}$, $\sqrt{9}$, and $\sqrt{10}$. The relative weights of the values of the difference in elevation obtained from lines 1, 2, and 3, are, by Eq. (5-28), $\frac{1}{17}$, $\frac{1}{9}$, and $\frac{1}{10}$, respectively. Then by Eq. (5-27) the weighted mean of the measured differences in elevation is

$$\overline{M} = \frac{(\frac{1}{17})(29.492) + (\frac{1}{9})(29.440) + (\frac{1}{10})(29.480)}{\frac{1}{17} + \frac{1}{9} + \frac{1}{10}} = 29.466 \text{ ft} \qquad \blacklozenge$$

If the weighted mean is computed by Eq. (5-27), the standard error of the weighted mean is given by the expression

$$\overline{\sigma} = \sqrt{\frac{\Sigma(pv^2)}{\Sigma p(n-1)}} \qquad [5\text{-}31]$$

The similarity with Eq. (5-20) can be observed.

EXAMPLE 5-13   Determine the standard error of the weighted mean difference in elevation in Example 5-12.

Solution: The calculations may be arranged as follows:

$$v_1 = 29.492 - 29.466 = +0.026 \quad v_1^2 = 0.000676$$

$$v_2 = 29.440 - 29.466 = -0.026 \quad v_2^2 = 0.000676$$

$$v_3 = 29.480 - 29.466 = +0.014 \quad v_3^2 = 0.000196$$

$$p_1 v_1^2 = \frac{1}{17} \times 0.000676 = 0.0000398 \quad \Sigma p = 0.2699$$

$$p_2 v_2^2 = \frac{1}{9} \times 0.000676 = 0.0000751 \quad \overline{\sigma} = \sqrt{\frac{0.0001345}{0.2699(3-1)}} = \pm 0.0158 \text{ ft}$$

$$p_3 v_3^2 = \frac{1}{10} \times 0.000196 = 0.0000196$$

$$\overline{\Sigma(pv^2) = 0.0001345} \qquad \blacklozenge$$

EXAMPLE 5-14   What is the standard error of each of the measured differences in elevation in Example 5-12?

Solution: By Eq. (5-29),

$$\sigma_0 = \sqrt{\frac{0.0001345}{2}} = \pm 0.0082 \text{ ft}$$

Then by Eq. (5-30),

$$\sigma_1 = 0.0082\sqrt{17} = \pm 0.0338 \text{ ft} \qquad \sigma_2 = 0.0082\sqrt{9} = \pm 0.0246 \text{ ft}$$

$$\sigma_3 = 0.0082\sqrt{10} = \pm 0.0259 \text{ ft}$$

5-9   Simple Adjustment of Measurements

If the three angles in a triangle were to be measured by methods described in Chapter 4 and each angle was measured with the same degree of precision, then each of the three measured angles would carry the same weight. If the sum of the three measured angles is not 180°, then it is expected that each measured angle would be corrected by the same amount. The reasonableness of this adjustment is proved by an extended investigation of the principle of least squares developed in Section 5-3.

If three angles are measured about a point, as the angles A, B, and C in Fig. 5-9, and the total angle, as E, is also measured and if all measurements are made with the same precision, $A + B + C$ should equal E. If the sum of the measured values of A, B, and C does not equal the measured value of E, then each of the measured angles must receive the same amount of correction because their weights are identical.

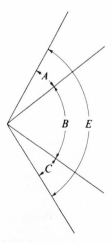

FIGURE 5-9 Angles measured about a point.

EXAMPLE 5-15   In Fig. 5-9 the measured angles are $A = 22°\ 10'\ 10''$, $B = 71°\ 05'\ 50''$, $C = 24°\ 16'\ 00''$, and $E = 117°\ 31'\ 00''$. What are the corrected angles?

Solution: The computations may be arranged as in the accompanying table.

Angle	Measured Value	Correction	Adjusted Value
A	22° 10′ 10″	−15″	22° 09′ 55″
B	71° 05′ 50″	−15″	71° 05′ 35″
C	24° 16′ 00″	−15″	24° 15′ 45″
Sum =	117° 32′ 00″		117° 31′ 15″
E	117° 31′ 00″	+15″	117° 31′ 15″
Error =	1′ 00″		

If a series of angles closing the horizon about a point are measured, their total should be 360°. If the total is not 360° and all angles were measured with the same degree of refinement, each should be corrected by an equal amount. Obviously this amount is found by taking the difference between 360° and the measured total and dividing that difference by the number of angles.

5-10   Adjustment of Weighted Measurements

The corrections to be applied to uncorrelated weighted measurements are inversely proportional to their weights. Equation (5-6) is the statement of the condition of least squares that applies to measurements of equal or unit weight. Assume now a set of measurements $M_1, M_2, \dots, M_n$ with varying weights $p_1, p_2, \dots, p_n$ and the corresponding residuals $v_1, v_2, \dots, v_n$. By Eq. (5-28), p is directly proportional to h^2. The probability P that $v_1, v_2, \dots, v_n$ will occur in the set is as follows:

$$P_{v_1} = y_1\,\Delta v = k_1 e^{-h_1^2 v_1^2}\,\Delta v$$

$$P_{v_2} = y_2\,\Delta v = k_2 e^{-h_2^2 v_2^2}\,\Delta v$$

$$\vdots$$

$$P_{v_n} = y_n\,\Delta v = k_n e^{-h_n^2 v_n^2}\,\Delta v$$

Also, since the probability that residuals will occur simultaneously in the set is equal to the product of their separate probabilities,

$$P_{(v_1, v_2, \dots, v_n)} = \left(k_1 e^{-h_1^2 v_1^2}\Delta v\right)\left(k_2 e^{-h_2^2 v_2^2}\Delta v\right)\cdots\left(k_n e^{-h_n^2 v_n^2}\Delta v\right)$$

or

$$P_{(v_1, v_2, \ldots, v_n)} = (k_1 k_2 \ldots k_n) \Delta v^n e^{-(h_1^2 v_1^2 + h_2^2 v_2^2 \cdots + h_n^2 v_n^2)}$$

Obviously, if P is to be a maximum in order to give the most probable value from the measurements, the negative exponent of e must be made a minimum; that is,

$$h_1^2 v_1^2 + h_2^2 v_2^2 + \cdots + h_n^2 v_n^2 = \text{a minimum}$$

Since p is proportional to h^2, then

$$p_1 v_1^2 + p_2 v_2^2 + \cdots + p_n v_n^2 = \text{a minimum}$$

or

$$\Sigma (pv^2) = \text{minimum} \tag{5-32}$$

or

$$[pvv] = \text{minimum}$$

In matrix notation, the sum of the weighted squared residuals is written as $\mathbf{V}^T \mathbf{P} \mathbf{V}$ in which $\mathbf{V}$ is a column vector of residuals and $\mathbf{P}$ is a square matrix of weights (see Section A-2).

The residual v for any value is defined as the measured value minus the most probable value, or $v_i = M_i - M$, in which M_i is the ith measured value and M is the most probable value. Thus the most probable value, or the adjusted value, is $M = M_i - v_i$. It can be seen that if the quantity $-v$ is added to the measured value, the adjusted value is obtained. But, $-v$ must be the correction to the measured value. Thus

$$c = -v \tag{5-33}$$

in which c is the correction.

Note that c^2 has both the same magnitude and the same algebraic sign as v^2. Then, if Eq. (5-32) is to be satisfied, the corrections must be inversely proportional to the weights. For example, if one measurement is, say, five times as good as another like measurement, then the correction to the first measurement should be only one-fifth of the correction to the second measurement. As an illustration of the adjustment of weighted measurements, consider the three angles of the triangle in Fig. 5-10. Assume that the mean value of six measurements of angle A is 32° 40′ 15″, the mean value of three measurements of angle B is 84° 33′ 15″, and a single measurement of angle C is 62° 46′ 00″. Also assume that each measurement of an angle is made with the same degree of refinement, and that the standard error of each single measurement is ±1′. Then, by Eq. (5-19), the standard error of the mean of the six measurements of A is ±1″/$\sqrt{6}$; the standard error of the mean of the three measurements of B is ±1″/$\sqrt{3}$; and the standard error of the measurement of C is ±1″/$\sqrt{1}$. So by Eq. (5-28) the relative weights are $p_A = 6$, $p_B = 3$, and $p_c = 1$. In general the number of repetitions of an angle indicates its relative weight.

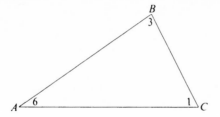

FIGURE 5-10 Angles in a triangle.

The computations for determining the adjusted angles may be arranged as shown in the accompanying table. Since the sum of the observed angles is 179° 59′ 30″, the total of the three corrections must be 30″. Also, the correction to the mean value of angle A is only one-half that of the correction to the mean value of angle B and is only one-sixth that of the correction to the mean value of angle C. The relative corrections to the three angles A, B, and C are therefore $\frac{1}{6}$, $\frac{1}{3}$, and $\frac{1}{1}$ or $\frac{1}{6}$, $\frac{2}{6}$, and $\frac{6}{6}$. Now the sum of these relative corrections must be 30″, and the correction to angle C can be found by solving the equation

$$\frac{6x}{6} + \frac{2x}{6} + \frac{x}{6} = 30''$$

hence $x = 20''$. Renormalizing the ratios to $\frac{1}{9}$, $\frac{2}{9}$, and $\frac{6}{9}$ as shown in the accompanying table yields the corrections to angles A, B, and C.

Angles	Number of Measure- ments	Mean Value of Angle	Relative Corrections		Corrections	Adjusted Value
A	6	32° 40′15″	$\frac{1}{6}$ or $\frac{1}{9}$		$\frac{1}{9} \times 30'' = 3.3''$	32° 40′18.3″
B	3	84° 33′15″	$\frac{2}{6}$ or $\frac{2}{9}$		$\frac{2}{9} \times 30'' = 6.7''$	84° 33′21.7″
C	1	62° 46′00″	$\frac{6}{6}$ or $\frac{6}{9}$		$\frac{6}{9} \times 30'' = 20.0''$	62° 46′20.0″
		179° 59′30″	$\frac{9}{6}$	$\frac{9}{9}$		180° 00′00.0″

5-11 Adjustment of a Level Circuit

The adjustment of a level circuit is a special case of adjusting weighted observations. When several lines of levels between two points are compared, the weight of each line is inversely proportional to the number of setups, as shown in Example 5-12. It was also pointed out that the standard error of a line is proportional to the square root of the length of the line. Then, by Eq. (5-28), the weight of a line of levels is inversely proportional to the distance along the line. This fact is the basis of the following principle for adjusting a line or circuit of levels: Corrections to measured elevations of points are proportional to the distances from the beginning of the line or circuit to the points, or are proportional to the number of setups between the beginning of the line or circuit and the points.

EXAMPLE 5-16 A line of levels is run from BM30 to BM33, the elevations of which have been previously established. As shown in Fig. 5-11, two intermediate benchmarks, BM31 and BM32 are established. The fixed elevation of BM30 is 453.52 ft, and the fixed elevation of BM33 is 482.10 ft. As a result of the field work the elevation of BM31 is 440.98 ft; that of BM32 is 464.25 ft; and that of BM33 is 482.23 ft. The error of closure is therefore +0.13 ft. There are a total of 20 setups distributed as shown in Fig. 5-11. The relative corrections for BMs 31, 32, and 33 are therefore $\frac{5}{20}$, $\frac{17}{20}$, and $\frac{20}{20}$, respectively. The correction to the elevation of BM31 is $\frac{5}{20} \times 0.13 =$ 0.03 ft; that to the elevation of BM32 is $\frac{17}{20} \times 0.13 = 0.11$ ft; and that to the elevation of BM33 is $\frac{20}{20} \times 0.13 = 0.13$ ft. The adjusted elevations of BMs 32, 32, and 33 are, respectively, 440.95, 464.14, and 482.10 ft. This solution should be reconciled with the previous discussion in Sections (5-8) and (5-10). ◆

EXAMPLE 5-17 A circuit of levels is shown in Fig. 5-12. It starts and ends at BM8 and passes through BM10, BM11, and BM14. The distances between the several benchmarks, in thousands of feet, are given on the diagram. Also, the known elevation of BM8 is 445.108 ft, and the measured elevations of the benchmarks are as shown in the accompanying table. What are the adjusted elevations of BMs 10, 11, and 14?

FIGURE 5-11 Lines of levels between two points.

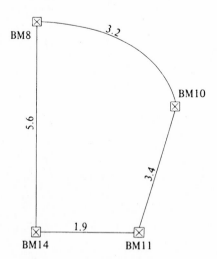

FIGURE 5-12 Circuit of level lines.

Benchmark	Measured Elevation (ft)	Distance from BM8 (1000) (ft)	Correction (ft)	Adjusted Elevation (ft)
8	445.108	0	0	445.108 (fixed)
10	430.166	3.2	3.2/14.1 x 0.074 = 0.017	430.183
11	433.421	6.6	6.6/14.1 x 0.074 = 0.035	433.456
14	440.380	8.5	8.5/14.1 x 0.074 = 0.045	440.425
8	445.034	14.1	14.1/14.1 x 0.074 = 0.074	445.108

Solution: To the closing error on BM8 is 445.108 − 445.034 = 0.074 ft. The corrections to the measured elevations of the benchmarks are computed as shown in the table. Since the measured elevation of BM8 is too small, the corrections are added to the measured elevations of the other benchmarks. ◆

5-12   Other Adjustments

The necessity for applying corrections to observations that are subject to random errors enters into all types of surveying measurements. The adjustment of the various types of measurements will be discussed in the separate chapters where the need arises. Appendix A shows the adjustment of elementary surveying measurements by the principle of least squares.

PROBLEMS

5-1. If 10 coins are tossed in the air at random, what is the probability that only 1 coin will come up heads?

5-2. In Problem 5-1, what is the probability that exactly three coins will come up heads?

5-3. If six coins are tossed in the air at random, what is the probability that exactly one-half will come up heads?

5-4. In Problem 5-3, what is the probability that at least four coins will come up heads?

5-5. Plot a histogram similar to those shown in Fig. 5-1 for the coins of Problem 5-1.

5-6. Five dice are rolled at random. Determine the probability that the sum of the marks on the upper face of the dice will be **(a)** five, **(b)** six, and **(c)** seven.

5-7. Plot a histogram to show the theoretical frequency distribution of the sum of the marks on three dice if they are rolled 500 times.

5-8. The difference in elevation between two points was measured for a total of 12 times with the following results:

Measurement Number	Difference in Elevation (m)	Measurement Number	Difference in Elevation (m)
1	9.1044	7	9.1054
2	9.1032	8	9.1037
3	9.1048	9	9.1057
4	9.1046	10	9.1052
5	9.1055	11	9.1066
6	9.1063	12	9.1054

Compute the mean, the probable error, standard error, and 90% error of this set of measurements.

5-9. Compute the standard error of the mean of Problem 5-8.

5-10. As an experiment in reaction time, five periods of a pendulum were timed by a stopwatch. This measurement was repeated 100 times with the following results, given in seconds of time:

8.08	8.08	8.02	7.90	7.92
8.00	8.05	8.10	8.01	7.94
7.86	8.18	8.19	7.98	8.07
7.79	7.94	7.98	8.23	7.81
8.04	8.00	8.02	8.15	8.12
7.87	7.95	7.75	7.93	7.89
7.93	8.00	8.05	7.99	8.00
8.12	8.00	8.00	7.98	8.01
8.08	8.15	8.02	7.98	8.05
7.92	7.98	8.02	7.86	8.00
7.91	7.90	8.00	8.15	8.10
7.95	8.00	7.91	7.82	7.80
8.00	8.08	7.98	8.10	7.98
7.93	7.93	8.11	8.09	8.00
8.01	7.98	8.00	8.04	7.98
8.15	8.14	8.15	8.05	7.90
8.05	8.15	7.97	8.25	8.08
8.01	8.01	7.99	8.02	7.95
8.02	7.89	7.88	8.02	8.05
7.85	8.07	8.10	7.95	8.05

(a) Plot these observations in the form of a histogram with a class interval of 0.05^s, beginning with 7.75^s. **(b)** Compute the mean of the observed values, the standard error of the set of observations, the standard error of the mean, and the probable (50%) error of the mean. **(c)** Using the standard error of the set of observations, compute ordinates to the normal distribution curve representing the set. Plot the curve on the histogram in (a). **(d)** Locate the position of $+\sigma$ and $-\sigma$ on the curve in (c).

5-11. What percent of the measurements are within $\pm\sigma$ from the mean of the measurements of Problem 5-10?

5-12. To determine the precision of the setting of the optical micrometer of a 1-second direction theodolite, the following independent readings, in seconds, were made:

30.9	33.9	32.7	34.8	32.5
29.0	35.2	33.5	30.5	38.0
32.1	35.1	37.9	37.9	32.1
35.4	30.3	34.5	37.8	35.7
34.7	35.4	36.4	33.2	30.6
29.8	32.2	38.0	31.2	36.1
28.7	36.3	33.8	30.4	32.9
29.2	35.5	35.1	34.4	36.2
36.0	32.0	30.4	34.2	33.5
34.1	33.2	37.2	36.4	31.8
34.2	39.2	33.2	33.0	36.8
33.1	36.5	36.7	34.8	31.2
33.1	33.1	36.9	34.3	35.0
34.4	37.2	30.8	36.6	31.8
34.9	32.0	36.8	35.0	36.1
34.3	35.7	36.5	29.1	34.5
30.2	35.4	39.0	36.9	32.1
36.1	32.2	36.2	34.5	30.9
34.8	31.2	33.0	37.5	34.0
34.9	35.0	33.0	33.0	37.2

(a) Plot these readings in the form of a histogram with a class interval of 1″ beginning with 28″. (b) Compute the mean of the readings, the standard error of the set of readings, the standard error of the mean, and the probable (50%) error of the mean. (c) Using the standard error of the set of readings, compute ordinates to the normal distribution curve representing the set. Plot the curve on the histogram in (a). (d) Locate the position of $+\sigma$ and $-\sigma$ on the curve in (c).

5-13. What percent of the measurements are within $\pm\sigma$ from the mean of Problem 5-12?

5-14. Assume that the standard error of a rod reading in leveling due to multiple sources of random errors is ±0.003 ft. The distance between a fixed benchmark and one whose elevation is to be determined is 2465 ft. If the average length of sight is 85 ft, what is the standard error of the measured elevation of the new benchmark?

5-15. A line was measured to be 965.56 m using a 30-m tape. If the standard error of each tape length is ±0.0025 m, what is the standard error of the measured length?

5-16. The line JN is broken into sections for measurement with a steel tape. The results are $JK = 285.00$ ft ±0.035 ft, $KL = 1240.00$ ft ± 0.055 ft, $LM = 685.00$ ft ± 0.045 ft, and $MN = 802.55$ ft ± 0.048 ft. Compute the length of the line and the standard error of the length.

5-17. A rectangular field has measured dimensions of 325.62 m ± 0.025 m (1 σ) and 580.22 m ± 0.033 m (1 σ). What is the area of the field, in m² and what is the standard error of the area, in m²?

5-18. What is the area and standard error of the area of Problem 5-17 in hectares?

5-19. The volume of a pyramid is $\frac{1}{3}\,bh$. The base of the pyramid is a rectangle with measured sides 18.4 ± 0.4 in. ($1\,\sigma$) and 27.2 ± 0.4 in. ($1\,\sigma$). The height of the pyramid is measured as 10.0 ± 0.02 in. ($1\,\sigma$). Compute the volume of the pyramid and the standard error of the computed volume.

5-20. Three sides of a triangle were measured with the following results: $a = 742.66$ m $\pm$ 0.015 m, $b = 425.16$ m ± 0.025 m, $c = 1050.27$ m ± 0.030 m. Compute the angles in the triangle together with the standard errors of the angles.

5-21. What is the standard error of the area of the triangle in Problem 5-20: **(a)** in m^2, **(b)** in hectares, and **(c)** in acres.

5-22. Two sides and the included angle of a triangle are measured as follows: $a = 478.24$ ft $\pm$ 0.38 ft, $b = 686.20$ ft ± 0.06 ft, and $C = 81°14' \pm 25''$. Compute the area of the triangle in ft^2 and in acres. Compute the standard error of the area in ft^2 and in acres.

5-23. Compute side c of Problem 5-22 and its standard error.

5-24. Compute angles A and B of Problem 5-22 and the standard error of these angles.

5-25. One side and two adjacent angles of a triangle are measured to compute the lengths of the other two sides and the angle opposite the measured side. Side c measures 372.624 m ± 0.020 m, angle A measures $72° 26' 10'' \pm 10''$, and angle B measures $60° 16' 30'' \pm 15''$. Compute angle C, side a, and side b. Compute the standard error of each quantity.

5-26. If the Line JN of Problem 5-16 is measured with an EDM as 3012.46 ft with a standard error of ± 0.018 ft, what is the most probable length of the line?

5-27. The difference in elevation between two benchmarks was measured by each of three field parties using different kinds of leveling instruments. The results are Party 1, $DE = 12.46 \pm 0.04$ ft; Party 2, $DE = 12.43 \pm 0.03$ ft; Party 3, $DE = 12.50$ ft ± 0.06 ft. What is the weighted difference in elevation between the two benchmarks (nearest 0.001 ft)?

5-28. What is the standard error of the weighted difference in elevation of Problem 5-27?

5-29. A line of levels is carried from BMA, whose elevation is 264.252 m, to a new BMK, requiring ten setups. The measured difference in elevation is $+4.246$ m. A line is carried from BMB, whose elevation is 273.936 m, to BMK, requiring four setups. The measured difference in elevation is -5.446 m. A line is carried from BMC, whose elevation is 266.364 m, to BMK, requiring seven setups. The measured difference in elevation is $+2.128$ m. Compute the weighted elevation of BMK and the standard error of this elevation.

5-30. What is the standard error of each of the three measured differences in elevation of Problem 5-29?

5-31. The weight of an angle is assumed to be proportional to the number of times it has been repeated (see Section 4-18). Five angles in a five-sided figure are measured with the following results:

Angle	Observed Value	Number of Repetitions
1	86° 15′20″	2
2	134° 44′30″	8
3	75° 48′55″	6
4	167° 02′15″	4
5	76° 08′40″	6

Compute the adjusted values of the angles (nearest second).

5-32. Three contiguous angles were measured separately at 0, and then the total angle was measured. The results are as follows:

From Station	To Station	Observed Value	Number of Repetitions
P	Q	12° 28′44″	6
Q	R	71° 02′23″	8
R	S	65° 52′18″	4
P	S	149° 23′08″	12

Compute the adjusted angles (nearest tenth of a second).

5-33. A line of levels was run between BM 50 at an elevation of 742.65 ft and BM 56, the fixed elevation of which is 576.80 ft. Five intermediate benchmarks—BMs 51, 52, 53, 54, and 55—were set prior to running the levels. The results of the field work are shown in the table. What are the adjusted elevations of the benchmarks to the nearest 0.01 ft?

Benchmark	Observed Elevation (ft)	Number of Setups Between Benchmarks
50	742.65 (fixed)	
51	718.90	4
52	702.86	6
53	646.40	11
54	619.18	8
55	572.06	6
56	576.67	10

BIBLIOGRAPHY

Barry, B. A. 1978. *Errors in Practical Measurement in Science, Engineering, and Technology.* New York: Wiley.

Benjamin, J. R., and Cornell, C. A. 1970. *Probability, Statistics and Decision for Civil Engineers.* New York: McGraw-Hill.

Brownlee, K. A. 1965. *Statistical Theory and Methodology in Science and Engineering.* New York: Wiley.

Buckner, R. B. 1983. *Surveying Measurements and Their Analysis.* Rancho Cordova, CA: Landmark Enterprises.

Crandall, K. C., and Seabloom, R. W. 1970. *Engineering Fundamentals in Measurements Probability Statistics and Dimensions.* New York: McGraw-Hill.

Meyer, P. L. 1965. *Introductory Probability and Statistical Applications.* Reading, MA: Addison-Wesley.

Mikhail, E. M. 1976. *Observations and Least Squares.* New York: Harper & Row.

———, and Gracie, G. 1981. *Analysis and Adjustment of Surveying Measurements.* New York: Van Nostrand Reinhold.

Parzen, E. 1960. *Modern Probability Theory and Its Applications.* New York: Wiley.

Uotila, U. A. 1973. Useful statistics for land surveyors. *Surveying and Mapping* 33:67.

6 Field Operations with the Transit or Theodolite

6-1 Introduction

Chapter 4 dealt with the construction of the transit, the theodolite, and the total station instrument and the methods by which horizontal and vertical or zenith angles are measured with them. Also the various errors introduced into the measurement of angles when using the transit were discussed. This chapter describes the various operations for which the theodolite is used in the field, without going into the fundamentals of angle measurement or into instrumental errors except to point out instances in which they are eliminated by certain field techniques.

6-2 Measuring Interior Angles

The angles formed within a closed figure between the adjacent sides are known as *interior angles*. In Fig. 6-1 there are five interior angles to be measured. They are the angles at L, M, N, O, and P. If the theodolite is set up at each station in turn in the order just given, all the interior angles would be measured in a counterclockwise direction. Thus the interior angle at L is measured from P to M, that at M from L to N, that at N from M to O, that at O from N to P, and that at P from O to L. On the other hand, if the instrument is set up in turn at L, P, O, N, and M, then all the interior angles would be measured clockwise. In the first instance the graduation numbers of the horizontal circle and vernier or scale that increase in a counterclockwise direction would be used; in the second instance those that increase in a clockwise direction would be used.

The form of notes used to record the interior angles is important since the notes are the record of the measurements. Fig. 6-2 shows one method of recording interior angles measured at stations M and N of Fig. 6-1; each angle is measured once with the telescope direct and once with it reversed.

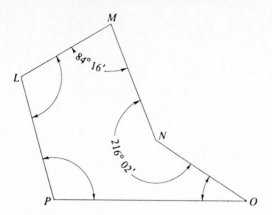

FIGURE 6-1 Interior angles.

The occupied station appears under the column headed *Sta.* The backsight station and the foresight station appear above and below, respectively, under the column headed $\frac{From}{To}$ *Sta.* The three readings under the column headed *Circle* are the initial backsight reading, the initial foresight reading, and the second foresight reading. The first value of the interior angle at station M from L to N, measured with the telescope direct, is obtained by subtracting 0° 00′ from 84° 16′ and is recorded under the column headed *Interior Angle* as 84° 16′. The second value, measured with the telescope reversed, is obtained by subtracting 84° 16′ from 168° 32′, giving 84° 16′. This is entered under the *Interior Angle* column. The first value of the interior angle at N from M to O, measured with the telescope direct, is obtained by subtracting 0° 00′ from 216° 02′ and is recorded under the column headed *Interior Angle* as 216° 02′. The second value, measured with the telescope reversed, is obtained by subtracting 216° 02′ from (72° 05′+ 360°) and is recorded as 216° 03′. The mean of the two angles is recorded as 216° 02′ 30″. At both M and N the counterclockwise circle was read because the angles are counterclockwise interior angles.

The sum of the interior angles in a closed plane figure must equal $(n - 2) \times$ 180°, in which n is the number of sides in the figure. This relation furnishes a check on the accuracy of the field measurements of the angles and a basis for distributing the errors in measurement, as discussed in Sections 5-9 and 5-10.

6-3   Measuring Angles to the Right

Most surveys performed to establish horizontal control by traversing (see Section 8-1) between fixed control points are made by measuring all the angles from the backsight to the foresight station in a clockwise direction. Angles to the right are shown in Fig. 6-3, in which a traverse is run from station Park to station Church through stations 1, 2, 3, and 4. The form for recording the notes is the same as those for recording readings for interior angles.

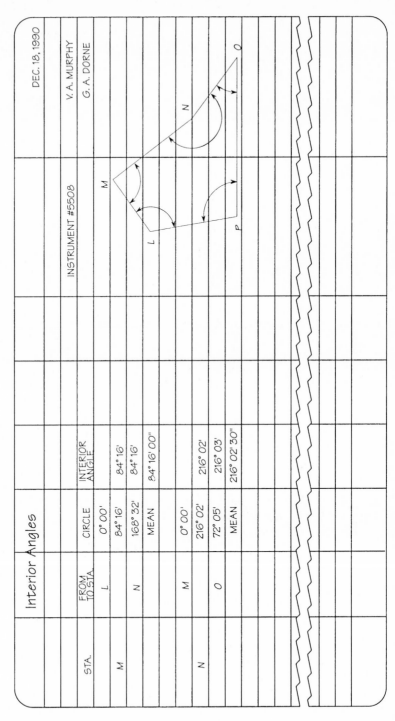

STA.	FROM TO STA.	CIRCLE	INTERIOR ANGLE		INSTRUMENT #5508	DEC. 18, 1990
						V. A. MURPHY
						G. A. DORNE
M	L	0° 00'	84° 16'			
	N	168° 32'	84° 16'			
		MEAN	84° 16' 00"			
N	M	0° 00'	216° 02'			
	O	72° 05'	216° 03'			
		MEAN	216° 02' 30"			

Interior Angles

FIGURE 6-2 Notes for recording interior angles.

FIGURE 6-3 Angles to the right.

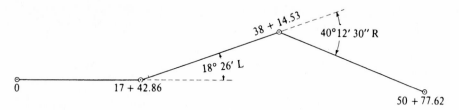

FIGURE 6-4 Deflection angle.

The check on the angles measured to the right is made by comparing the final azimuth, as computed from the angles, with a fixed azimuth (see Section 7-6). This check is fully discussed in Section 8-6.

6-4　Measuring Deflection Angles

A deflection angle, as shown in Fig. 6-4, is the angle made by the foresight with the prolongation of the backsight. Thus at station 17 + 42.86 a deflection to the *left* of 18° 26′ is shown; at station 38 + 14.53 the deflection is 40° 12′ 30″ to the *right*.

To measure the deflection angle with the transit occupying station 17 + 42.86, a backsight is taken on station 0 with the telescope direct. The circle is read and the value is recorded as the initial circle reading. Say the reading is 0° 00′. The telescope is plunged to the reversed position so that the line of sight is now directed along the prolongation of the line from station 0 to station 17 + 42.86. The circle still reads 0° 00′ since no angle has been turned.

The upper clamp is loosened, and the line of sight is turned to the left and directed to station 38 + 14.53. When the line of sight is brought exactly on the point with the upper tangent screw, the circle is read. The counterclockwise circle is read since the telescope was turned to the left, or counterclockwise. The reading has increased from 0° to 18° 26′.

The angle is now measured a second time. With the telescope *still in the reversed position,* the lower clamp is loosened, and the line of sight is directed toward the backsight station. The line of sight is brought exactly on station 0 by means of the lower tangent screw. The counterclockwise circle still reads 18° 26′. The telescope is plunged again (it is thus brought back to its direct position) so that the line of sight is now directed along the prolongation of the backsight. The counterclockwise circle still reads 18° 26′.

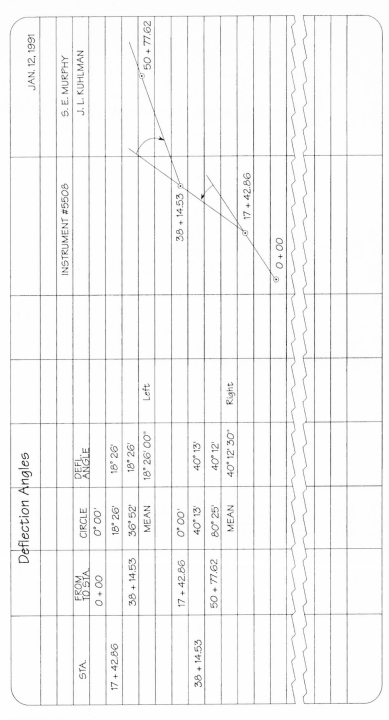

FIGURE 6-5 Deflection-angle notes.

The upper clamp is loosened, and the line of sight is turned to the left and brought on station 38 + 14.53. The circle is read to obtain the final reading. The reading will have increased from 18° 26′ to 36° 52′ on the counterclockwise circle, the difference being 18° 26′. The direction of the deflection angle, in this case *left*, is as important as the size of the angle itself. The readings are recorded as shown in Fig. 6-5.

With the instrument at station 38 + 14.53, a backsight is made on station 17 + 42.86 with the circle set to 0° 00′. The telescope is plunged and then turned to the right, using the upper motion, to make the sight on station 50 + 77.62. The clockwise circle reading has increased from 0° 00′ to 40°13′. On the second measurement, the clockwise circle reading has increased from 40° 13′ to 80° 25′. The readings are recorded as shown in Fig. 6-5.

When a deflection angle is measured only once, the more serious instrumental errors are introduced into the measurement. If a deflection angle is measured by double centering, most of the instrumental errors are eliminated, and the second value of the angle affords a check on the first value.

Deflection angles are used frequently in surveys for highways, railroads, transmission lines, and similar routes that extend in the same general direction, although they are also employed by many surveyors and engineers in running traverse lines. The principal advantages of using deflection angles are (1) azimuths are easily computed from deflection angles (see Section 8-5), (2) the deflection angles are used in the computation of circular curves for highway and railroad work, and (3) deflection angles are easily plotted. One distinct disadvantage is the likelihood of the notekeeper making a mistake in recording the *direction* of the deflection angle.

In a closed figure in which the sides do not cross one another, as in Fig. 6-6, the difference between the sum of the right deflection angles and the sum of the left deflection angles should equal 360°. In a closed survey in which the lines cross once, as in Fig. 6-7, the sum of the right deflection angles should equal the sum of the left deflection angles. If the lines do not cross or cross any *even* number of times, the difference between the sums should equal 360°. If the lines cross any *odd* number of times, the sums should equal one another.

FIGURE 6-6 Deflection-angle traverse in which no lines cross.

FIGURE 6-7 Deflection-angle traverse in which lines cross once.

6-5 Laying Off Angles

In all types of construction work when laying out highways, bridges, buildings, and other structures, it becomes necessary to establish lines in given directions and thus to lay off both horizontal and vertical angles. A line called a *baseline* is established in the vicinity of the work. The baseline may or may not be a portion of a control traverse, its position depending on circumstances. Nevertheless, it acts as the control for the construction. Stations along the baseline are occupied by the theodolite, and backsights are made to other baseline stations. The angles called for on construction plans and drawings are laid off with respect to the baseline. It should be noted here that construction plans often contain point numbers with rectangular coordinates. In this instance, the coordinates of the occupied point and those of the point to be set are used to compute the direction and distance to each point. If the total station instrument is used, the coordinates are loaded into the data collector, and the directions and distances are all computed in the microprocessor on command. If the total station instrument is not available, these computations, which are discussed in Chapter 8, are performed and the listing of directions (or angles off the baseline) and distances furnished to the survey party.

The principle of double centering is employed in turning angles from the baseline. If construction lines are quite long, the principle of repetition may be employed. For example, to lay off an angle of 55° 12′ 30″ by double centering with a 1′ transit, a backsight is taken on a baseline station with the circle reading 0° 00′. The upper clamp is loosened and the telescope is turned until the circle reading is approximately 55° 12′. The upper clamp is tightened and the vernier is made to read 55° 12′ as closely as possible by turning the upper tangent screw. A point is set on the line of sight, preferably being marked by a tack in a stake or a hub. The telescope is inverted and a second backsight is taken on the baseline station by a lower motion. When the backsight has been made, the upper clamp is loosened, and the telescope is turned until approximately 2 × 55° 12′ 30″, or 110° 25′, is read. The upper clamp is tightened, and the exact reading of 110° 25′ is made by turning the upper tangent screw. A second point is set on the line of sight alongside the first point. The second angle is 55° 13′. A third point halfway between the other two points is the correct point, and the line joining this point and the transit station makes the desired angle with the baseline. To lay off an angle by repetition, see Sections 4-3 and 4-18.

If a direction theodolite is used to lay off a horizontal angle, the backsight is made along the baseline and the circle is read. The angle to be laid out is added to or subtracted from the backsight circle reading (depending on whether the angle is to be laid off clockwise or counterclockwise) to obtain the circle reading for directing the line of sight to the point to be set. The optical micrometer is then set to the correct number of minutes and seconds, and the clamp and tangent screw are used to make the remainder of the circle reading. The set point should be checked by measuring the angle with the telescope in the reversed position.

To lay off a vertical angle with a theodolite having a full vertical circle, carefully level the instrument by using either the plate levels as described in Section 4-20 or the telescope level as described in Section 4-19. Rotate the telescope in the direction of the point to be set and incline the telescope until the vertical circle reads approximately the angle to be laid off. Clamp the vertical clamp and set off the exact reading by using the vertical tangent screw. The point is marked by reference to the center horizontal cross hair. Invert the telescope and rotate the transit on its vertical axis until the line of sight is in the desired direction. Sight the first point and clamp the vertical clamp. If the vertical circle reads the desired angle when the horizontal cross hair is brought on the point previously set, it may be concluded that the index error was zero and the point has been correctly set. If the vertical circle does not read the desired angle, the vertical tangent screw is turned until it does. A second point is set above or below the first point. A third point halfway between the two will be in very nearly the correct position.

To lay off a vertical angle with a transit having a half vertical circle, level the instrument as before and rotate the telescope until the line of sight is in the direction of the desired point. Bring the telescope horizontal and center the telescope bubble by using the vertical tangent screw. Read the vertical circle to obtain the index error. Add the index error algebraically to the desired vertical angle to obtain the correct circle reading. Set off the correct circle reading by using the vertical clamp and tangent screw. The line of sight is now directed along the desired vertical angle and the point is set on line.

If the instrument incorporates the tilt compensator, errors due to slight misleveling are eliminated. However, the vertical angle must still be double centered to eliminate any possible collimation error or index error.

If the theodolite reads zenith angles, then any vertical angle to be laid off must be converted to the equivalent zenith angle beforehand.

6-6 Straight Line Extension by Double Centering

Whenever a straight line is to be accurately extended for any considerable distance, the method of double centering should be used. If in Fig. 6-8 the line *AB* is to be extended to the right, the instrument is set up at *B*, and a backsight is taken on *A* with the telescope in the normal or direct position. The telescope is then plunged. If the theodolite is in adjustment, the line of sight will then be the line

FIGURE 6-8 Straight line extension by double centering.

BC. If the theodolite is not in adjustment, it may be either the line *BC'* or the line *BC''*. A stake is driven on the line of sight, and the exact line is marked by a point on the stake. The backsight is first checked to see that the telescope has not moved off line. Then either the upper clamp or the lower clamp is loosened, and a second backsight is taken on *A,* this time with the telescope in the inverted position. The telescope is again plunged and the line of sight is again directed toward *C.* If the instrument is in adjustment, the line of sight will strike the point previously set. If it is out of adjustment, the line of sight will miss that point, and a second point should be set on this line. A third point, midway between the two points already set, will be on the line *AB* extended. Thus if *C'* is the first point set, *C''* will be the second one and the correct line will be marked by *C,* which is midway between *C'* and *C''.*

The line can be extended further by moving the theodolite to *C* and repeating the procedure, a fourth point *D* being located midway between *D'* and *D''.* The first backsight might, of course, be made with the telescope in the inverted position. If this is done, the second backsight is taken with it normal. In Fig. 6-8 the scale is very much distorted. With an instrument in reasonably good adjustment, the three points *C', C'',* and *C* should all fall on the same 2-in. stake at a distance of 1200 to 1500 ft from the instrument. If they do not, the instrument should be adjusted. Otherwise, when this kind of work is being done, unnecessary time will be spent in driving stakes.

Establishing a straight line by the method of double centering is nothing more than laying off a zero deflection angle twice, the first backsight being taken with the telescope direct and the second one with it reversed.

6-7 Establishing Points on a Straight Line

The simplest case of establishing points on a straight line is when the two ends of the line are marked and the entire line is visible from one end or the other. Thus if it is necessary to establish additional points between *A* or *B* in Fig. 6-8, the transit can be set up at either *A* or *B* and a sight taken to the other end of the line. The rodman is then directed by motions to the right or the left to the point that is on the line. The rodman may use a line rod if the lengths are considerable. When they are short and the ground is clear, he may have the instrumentman line in the stake or other object on which the point is to be marked. The instrumentman should make sure that the stake or other marker is itself started vertically. A point on the top of the stake may be obtained by lining in the line rod again or a pencil or a plumb-bob string.

If the ground surface is so shaped that the telescope does not have to be raised or lowered appreciably after taking the sight on the opposite end of the line and if the focus of the telescope is likewise unchanged, this work can be done with a theodolite that is not in perfect adjustment. If the inclination of the telescope must be changed considerably and the focus altered, important points should be checked by taking two sights, one with the telescope normal and another with it inverted.

If points are to be established between B and C, the distant point C should first be established by the method of double centering and a foresight should be taken on C before line is given for the intermediate points. Where points are being set at regular intervals, as in a route survey, the instrument is moved only when the foresight distance becomes so great that it is impossible to give line accurately or when the ground conditions prevent further accurate sighting. The forward station should be set by the method of double centering. The careful instrumentman will use this procedure as a check, even when he is reasonably certain that his instrument is in good adjustment.

6-8 Balancing In

It often happens that the two ends of a line are not intervisible, but some intermediate point can be found from which the ends can be seen. Thus in Fig. 6-9 both A and B can be seen from the top of an embankment. The method of finding, by trial, the location of a point C that will be on the line AB is called *balancing in*. The transit is first set up at C', which is as nearly on the line as can be estimated. A backsight is taken on A and the telescope is plunged. From the sketch it is evident that the line of sight strikes considerably to the left of B. The amount by which it fails to strike B is measured or estimated, the transit is moved a proportionate amount (in the sketch about one-third) of this distance, and the procedure is repeated. After a few trials the line of sight should strike B. The final movement of the instrument can often be made with the shifting head of the transit.

To check the position of the transit, backsight on A with the telescope in the inverted position, plunge the telescope, and check to see whether or not the line of sight is directed on B. If it is, then the transit is on line. If it is to one side of B, measure the distance by which it misses B and move the transit in the opposite direction a proportionate amount of this distance. A final check is made by double

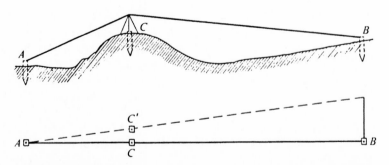

FIGURE 6-9 Balancing in.

centering as described in Section 6-6. Point B should fall at the midpoint of the two points B' and B'' (analogous to C' and C'' of Fig. 6-8).

After the correct position of the transit at C has been established, any additional points between C and A or between C and B can be set by the method described in the preceding section.

6-9   Random Lines

In Fig. 6-10 the distance from A to B may be very great, and the ground may be so thickly wooded or so hilly that it is impossible to see directly from A to B or to find an intermediate point from which both A and B can be seen. Under any of these conditions, a trial straight line, or *random line,* is run as near as can be estimated to the true line. This line, *ACDEF,* is run by the method of double centering to a point from which B can be seen. The length AF of the random line, the distance FB, and the angle at F from A to B are then measured. In the triangle ABF two sides and the included angle are known and the angle at A can be computed. The instrument is set up again at A, the computed angle is turned off from the line AF by using the method described in Section 4-3 or Section 6-5, and the line AB is run. If this new line fails to strike B exactly, points along the line can be shifted slightly to bring them on line, the amounts they are to be moved being obtained by proportion.

A second method of locating points on the required line is to set up the instrument at points on the random line, turn off angles equal to the measured angle AFB, and measure distances obtained by proportion. Thus in Fig. 6-10, if the angle at E is made equal to the angle at F, then $EG = FB \times (AE/AF)$. Points adjacent to C and D may be set in a similar manner.

Another method of locating points along the line AB is discussed in Chapter 8. In this method a closed random traverse is run instead of a straight line, and points on the traverse are located with respect to the line AB by computation. From these traverse points computed angles and distances are laid off to determine additional points on AB. Furthermore, these additional points may be set on AB at any desired distances from A or B.

6-10   Intersection of Two Straight Lines

Two examples of the intersection of straight lines are shown in Fig. 6-11. To find the intersection of lines AB and CD in view (a), the transit is set up at either A or B and a sight is taken along the line AB. Stakes, known as *straddle stakes,* are then set at E' and E''', being so located that the point of intersection will fall somewhere between them. Three stakes should be set so close that a string can be

FIGURE 6-10 Random line.

stretched between them. If it is difficult to estimate where the intersection is to be, more than two such stakes can be set.

The transit is next moved to C or D and a sight is taken along the line CD. The point E is then found at the intersection of the line of sight and the string stretched between E' and E''. If two transits are available, one can be set up on each line and E can be found at the intersection of the two lines of sight.

In Fig. 6-11(b) the intersection of the lines AB and CD is on the prolongations of both lines. If the lines are short, the stakes at E' and E'' can be set by placing the transit at A, sighting on B, and extending this line to E' and E''. When the lines are long, the transit is set up at B and points on the straddle stakes are set by double centering. The intersection is determined by moving the transit to either C or D and finding where the new line of sight strikes the string stretched between E' and E''. If the transit is at C, the method of double centering should be used.

6-11   Obstacles on a Line

When a boundary line is being extended by the method developed in Section 6-6 or when the center line for a highway or railroad is being staked out in preparation for construction, the problem of extending the line through heavily wooded or congested areas is invariably faced. Where timber cannot be cut, as in a

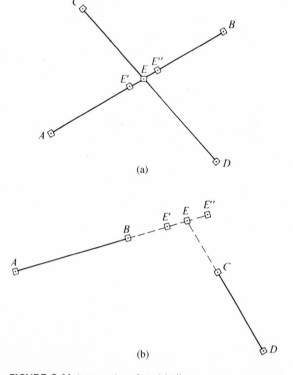

FIGURE 6-11 Intersection of straight lines.

boundary survey, or where buildings are not removed until immediately before construction begins and an obstacle is on the line, it becomes necessary to leave the line in order to pass the obstacle. Two methods for passing an obstacle are given here.

The first method, illustrated in Fig. 6-12, consists of establishing two equal perpendicular offsets from the desired line AF, as BB' and CC'. To provide a check, an equal offset from A to A' is also laid off. The line $A'B'C'$ is parallel to the line ABC and is extended beyond the obstacle to points D', E', and F' by the methods described in Section 6-6 and 6-7, the point F' being set as a check. Perpendicular offsets equal in length to BB' and CC' are established at these points to locate points D, E, and F. If measurements are to be carried along the line being run, then the offsets at C and D' should be made parallel by using the theodolite, so that the unmeasurable distance CD will be equal to $C'D'$, which can be measured.

The second method, illustrated in Fig. 6-13, is used where there are many obstacles on the desired line and where running an offset parallel line as just described is not feasible. This method consists of laying off a small deflection angle α from the desired line at a convenient point as A, setting point E at a measured distance from A, then laying off a deflection angle 2α from line AE in the opposite direction, and setting point B on the new line of sight so that $EB = AE$. If the deflection angle α is now turned from line EB at B, the line of sight will be directed along the desired line. If there is an obstacle between B and C, the line EB may be extended to a point at F. The angle 2α is then laid off at F, and the distance FC is laid off equal to BF. The procedure is repeated until the obstructed area has been traversed. Finally, the direction of the desired line beyond C can be established by laying off the angle α at C.

This method is convenient in that isosceles triangles are laid off, and the angles α and 2α can have the same values throughout. A random traverse discussed in Chapter 8 will accomplish the same results with perhaps more speed in the field, but more computations will be needed. Using a random traverse also requires computed distances and angles to be laid out, necessitating additional field work.

FIGURE 6-12 Prolonging line past obstacle.

FIGURE 6-13 Passing obstacles on line.

6-12   Parallel Lines

The principles of plane geometry are used in the establishment of parallel lines. In Fig. 6-12 the lines AF and $A'F'$ are made parallel by making the distance between them the same at B, C, D, and E. Equal offsets at two points are sufficient to establish the position of the parallel line, but the field work can be checked if offsets are made at three or more points.

In Fig. 6-14 the line CD is made parallel to the line AB by making the opposite interior angles at B and C equal. In the field the angle ABC would be measured and the direction of CD determined by laying off the angle BCD equal to the measured angle ABC. If the length of CD is more than a very few feet, these angles should be measured by repetition.

6-13   Location of a Point

The field location of a point involves the measurement of enough angles and distances to enable a person in the office to compute the position of the point or to plot the point in its correct position, either with respect to two fixed points or with respect to a fixed point and a fixed line through that point. As the two fixed points and the required point are the vertexes of some triangle, the position of the required point will also be fixed when the triangle is determined.

A triangle is determined by (1) two sides and the included angle, (2) a side and the two adjacent angles, (3) three sides, (4) two sides and the angle opposite one of these sides. The field applications of these methods of locating a point are shown in Fig. 6-15. In view (a) the point P is located from A by the measurement of an angle from the known line AB and a distance. In view (b) two angles are measured at the known points A and B, and in view (c) two distances from A and B are measured. In view (d) an angle at the known point A and the side opposite from the known point B are measured. In view (e) the measurements are the perpendicular distance CP from P to line AB and the distance BC or AC. In view (f) point P is at the intersection of the known lines AB and CD. The location of P in view (g) is determined by measuring the angles at P between the lines to three fixed points A, B, and C.

Although each of these methods has its particular uses, the first one shown—the measurement of an angle and a distance—is used much more frequently than the others. The shape of the figure may sometimes preclude the use of some particular method. Thus in view (b) where two angles are measured, the

FIGURE 6-14 Parallel lines.

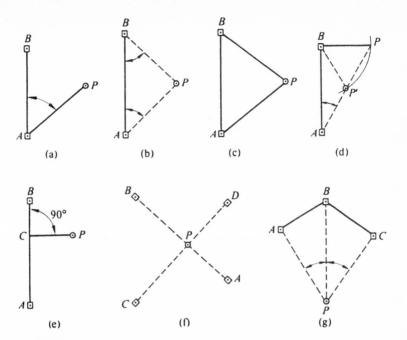

FIGURE 6-15 Locating a point.

location will be very uncertain if the angles at A and B are so small as to give an indefinite intersection when the lines AP and BP are, for example, drawn on a map. This method is used when the measurement of distances is difficult or impossible. It is also the basis of triangulation where, with one side and all the angles of a triangle known, the unknown sides are computed.

The method in view (c) can be used when the survey is being made with only a tape or an EDM. The location of P will be best when angle BPA is near 90° and will be uncertain when this angle approaches 0° or 180°. In such cases the method shown in view (d) will give a better location. This method can also be used when the measurement of the side AP is impossible. When this method is adopted, the approximate shape of the triangle must be known as there generally will be two solutions of a triangle when two sides and the angle opposite one of them are known. In every case it must be known whether the located point is to the right or left of the line of reference.

The method represented in view (e) is frequently used in route surveying to locate objects with respect to the center line. It is most advantageously used when extreme precision is unnecessary, as considerable field work is involved in locating the point C exactly. On construction work, important points that are liable to be disturbed during the process of the work can be replaced if their location is determined by the intersection of lines, as shown in view (f). The location will be more certain if the distances from P to A, B, C, and D are also measured.

View (g) shows a *three-point resection* to determine the position of the point P occupied by an instrument when this position is not known before the instrument is set up. The positions of points *A, B,* and *C* are known, and the angles at *P* from *A* to *B* and from *B* to *C* are measured. The computations necessary to determine the position of *P* are described in Section 9-14.

In Fig. 6-15(c), if the rectangular coordinates of *A* and *B* are known, the position of point *P* can be determined by occupying this point using a total station instrument as discussed in Section 9-14.

PROBLEMS

6-1. The interior angles in a seven-sided figure are measured as follows: **(a)** 83° 22′ 00′, **(b)** 193° 04′ 30″, **(c)** 88° 58′ 00″, **(d)** 149° 16′ 00′, **(e)** 116° 41′ 30″, **(f)** 133° 40′ 00″, **(g)** 134° 54′ 30″. All angles have equal weights. Compute the adjusted interior angles.

6-2. The interior angles in a six-sided figure are measured as follows: **(a)** 126° 10′, **(b)** 43° 32′, **(c)** 222° 58′, **(d)** 91° 04′, **(e)** 116° 17′, **(f)** 119° 56′. All angles have equal weight. Compute the adjusted interior angles.

6-3. The interior angles in a six-sided figure are measured as follows: **(a)** 90° 00′ 15″, **(b)** 89° 58′ 00″, **(c)** 216° 30′ 30″, **(d)** 55° 14′ 45″, **(e)** 147° 00′ 30″, **(f)** 121° 16′ 30″. All angles have equal weight. Compute the adjusted interior angles.

6-4. The following equal-weight deflection angles were measured in a five-sided traverse that closes on itself. *A,* 88° 14′L; *B,* 70° 52′L; *C,* 121° 30′L; *D,* 36° 03′R; *E,* 115° 37′L. Compute the adjusted deflection angles.

6-5. The following equal-weight deflection angles were measured in a seven-sided traverse that closes on itself: *P,* 55° 14′ 15″R; *Q,* 52° 10′ 00″R; *R,* 93° 38′ 30″R; *S,* 72° 53′ 30″L; *T,* 107° 40′ 00″L; *U,* 111° 14′ 30″L; *V,* 90° 47′ 00″R. Compute the adjusted deflection angles.

6-6. The line of sight of a transit makes an angle of 89° 57′10″ with the transverse axis. What would be the angular error in the line *CD* in Fig. 6-8 if each extension of the line *AB* was made with only one plunging and each backsight was made with the telescope in the direct position?

6-7. If each of the distances *AB, BC,* and *CD* in Problem 6-6 is 350 ft, how far off the true prolongation of *AB* will point *D* be set?

6-8. In Problems 6-6 and 6-7, if the backsight at *B* were made with the telescope direct and the backsight at *C* were made with the telescope inverted, and if only one plunging was performed at each point, what will be the resulting angular error in the direction *CD*? What will be the offset error of point *D*?

6-9. In Fig. 6-10 *AC* = 580.00 ft, *CD* = 445.00 ft, *DE* = 430.00 ft, *EF* = 305.00 ft, and *FB* = 6.97 ft. The angle at *F* from *B* to *E* is measured as 93° 53′. What is the angle at *A* from *F* to *B*? What is the distance *EG,* if *EG* is parallel to *FB*?

6-10. Compute the length of line *AB* of Problem 6-9.

6-11. In Fig. 6-10 *AC* = 137.00 m, *CD* = 130.00 m, *DE* = 125.00 m, *EF* = 225.00 m, and *FB* = 3.352 m. The angle at *F* from *B* to *E* is measured as 115° 32′. Compute the angle at *A* from *F* to *B*. What is the distance *EG,* if *EG* is parallel to *FB*?

6-12. Compute the length of line *AB* of Problem 6-11.

6-13. In Fig. 6-10 *AC* = 295.50 ft, *CD* = 280.00 ft, *DE* = 280.00 ft, *EF* = 250.00 ft, and *FB* = 8.02 ft. The angle at *F* from *B* to *E* is 100° 22′. Compute the angle at *A* from *F* to *B*.

6-14. What is the distance *EG* in Problem 6-13?

6-15. What is the length of the line *AB* in Problem 6-13?

6-16. In Fig. 6-13, $AE = EB = 180.00$ ft, $BF = FC = 425.00$ ft, and angle $\alpha = 1°\ 10'$. What is the length of line AC?

6-17. In Fig. 6-13 $AE = EB = 150.00$ m; $BF = FC = 315.00$ m, and angle α is $2°\ 25'$. What is the length of AC?

6-18. In Fig. 6-15(c), $AB = 175.252$ m. The measured distances to point P are: $AP = 105.915$ m, and $BP = 172.644$ m. What is the perpendicular distance from P to line AB? How far is the foot of the perpendicular from point A?

7 Direction of a Line

7-1 Astronomical Meridian

A plane passing through a point on the surface of the earth and containing the earth's axis of rotation defines the *astronomical meridian* at the point. The direction of this plane may be established by observing the position of the sun or a star, as described in Chapter 12. By popular usage the intersection of this meridian plane with the surface of the earth is known as the *true meridian*.

7-2 Magnetic Meridian

The earth acts very much like a bar magnet with a north magnetic pole located considerably south of the north pole defined by the earth's rotational axis. The magnetic pole is not fixed in position but changes its position continually. A magnetized needle freely suspended on a pivot will come to rest in a position parallel to the magnetic lines of force acting in the vicinity of the needle. Generally, the greatest component of the magnetic force at a point is that created by the earth's magnetic field, but other components may be created by other magnetic fields such as those around electric power lines, reinforcing bars in roads and structures, and iron deposits. The direction of the magnetized needle defines the *magnetic meridian* at the point at a specific time. Unlike the true meridian, whose direction is fixed, the magnetic meridian varies in direction.

A gradual shift in the earth's magnetic poles back and forth over a great many years causes a secular change, or variation, which amounts to several degrees in a cycle. An annual variation of negligible magnitude is experienced by the earth's magnetic field. A daily variation causes the needle to swing back and forth through an angle of not much more than one-tenth of a degree each day.

Local attraction, the term applied to the magnetic attractions other than that of the earth's magnetic field, may change at a given location. This change may be caused by a variation in the voltage carried by a power line or by a gradual increase or decrease of a magnetic field in reinforcing bars, wire fences, underground utility pipes, or other metal parts.

7-3 Assumed Meridian

For convenience in a survey of limited extent, any line of the survey may be assumed to be a meridian or a line of reference. An assumed meridian is usually taken to be in the general direction of the true meridian.

7-4 Convergence of Meridians

True meridians on the surface of the earth are lines of geographic longitude, and they converge toward each other as the distance from the equator toward either of the poles increases. The amount of convergence between two meridians in a given vicinity depends on (1) its distance north or south of the equator and (2) the difference between the longitudes of the two meridians. Magnetic meridians tend to converge at the magnetic poles, but the convergence is not regular and is not readily obtainable.

7-5 Grid Meridian

Another assumption is convenient in a survey of limited extent: When a line through one point of the survey has been adopted as a reference meridian, whether true or assumed north, all the other meridians in the area are considered to be parallel to the reference meridian. This assumption eliminates the necessity for determining convergence. The methods of plane surveying assume that all measurements are projected to a horizontal plane and that all meridians are parallel straight lines. These are known as *grid meridians*.

Two basic systems of grids are used in the United States to allow plane surveying to be carried statewide without any appreciable loss of accuracy. In each separate grid one true meridian is selected. This is called the *central meridian*. All other north-south lines in that grid are parallel to this line. The two systems, which are known as the *Lambert conformal projection* and the *transverse Mercator projection,* are the subject of Chapter 11.

7-6 Azimuth of a Line

The azimuth of a line on the ground is the horizontal angle measured from the plane of the meridian to the vertical plane containing the line. Azimuth gives the direction of the line with respect to the meridian. It is usually measured in a clockwise direction with respect to either the north meridian or the south meridian. In astronomical and geodetic work, azimuths are measured from the south meridian. In plane surveying, azimuths are generally measured from north.

A line may have an azimuth between 0° and 360°. In Fig. 7-1 the line *NS* represents the meridian passing through point *O,* with *N* toward the north. The azimuth

FIGURE 7-1 Azimuths.

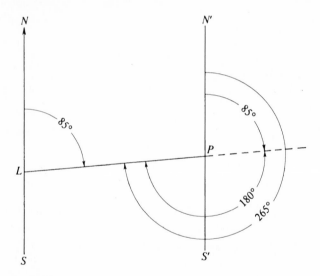

FIGURE 7-2 Relationship between azimuth and back azimuth.

of the line OA measured from the north is 70°; that of OB is 145°; that of OC is 235°; and that of OD is 330°. Azimuths are called *true* azimuths when measured from the true meridian, *magnetic* azimuths when measured from the magnetic meridian, *assumed* azimuths when referred to an arbitrary north-south line, and *grid* azimuths when referred to the central meridian in a grid system.

In the field of geodesy and geodetic surveying, the azimuth of a line is termed *geodetic* azimuth. It differs slightly from the true azimuth because a line at a point normal to the reference spheroid (see Sections 1-2 and 9-9) does not exactly follow the direction of gravity at that point, whereas true azimuth is related to the local vertical line.

7-7 Back Azimuth

When the azimuth of a line is stated, it is understood to be that of the line directed from an original point to a terminal point. Thus a line *LP* has its origin at *L* and its terminus at *P*. If the azimuth of *LP* is stated as 85°, then the azimuth of the line *PL* must have some other value since *LP* and *PL* do not have the same direction. One direction is the reverse of the other. For the purpose of discussing back azimuth, consider the line *LP* in Fig. 7-2. In the diagram *NS* is the meridian through *L*, and *N'S'* is the meridian through *P*. According to the assumption in plane surveying, the two meridians are parallel to each other. If the azimuth of *LP* is 85°, then the azimuth of *PL* is 85° + 180°, or 265°. Thus the back azimuth of *LP* is the same as the azimuth of *PL*. In Fig. 7-3 the azimuth of *CD* is 320° and the back azimuth of *CD*, which is the azimuth of *DC*, is 320° – 180°, or 140°.

From the preceding explanation, it is seen that the back azimuth of a line can be found from its forward azimuth as follows: If the azimuth of the line is less than 180°, add 180° to find the back azimuth. When the azimuth of the line is greater than 180°, subtract 180° to obtain the back azimuth.

7-8 Bearing of a Line

The bearing of a line also gives the direction of the line with respect to the reference meridian. Unlike an azimuth, which is always an angle measured in a definite direction from a definite half of the meridian, a bearing angle is never greater than 90°. The bearing states whether the angle is measured from the north or the south and also whether the angle is measured toward the east or toward the west. For

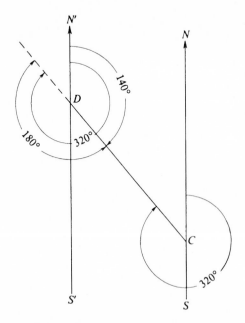

FIGURE 7-3 Azimuth and back azimuth.

example, if a line has a bearing of S 35° E (called south 35° east), the bearing angle 35° is measured from the south meridian eastward. Figure 7-4 shows four lines, the origin in each case being point O. Each line lies in a different quadrant. The bearing of OA is N 70° E; that of OB is S 35° E; that of OC is S 55° W; and that of OD is N 30° W.

It is apparent from Fig. 7-4 that the bearing angle of a line must be between 0° and 90°. A stated bearing is a *true* bearing, a *magnetic* bearing, an *assumed* bearing, or a *grid* bearing, according to whether the reference meridian is true, magnetic, assumed, or grid. In land surveying and in the conveyance of title to property, references are made to maps, notes, and plats of previous surveys that are recorded with the various counties throughout the United States. When reference is made to a bearing in such a previous survey, the term *record bearing* is used. If reference is made to a property deed, the term *deed bearing* is used. The terms *deed bearing* and *record bearing* are commonly used interchangeably.

7-9   Back Bearing

The bearing of a line in the direction in which a survey containing several lines is progressing is called the *forward* bearing, whereas the bearing of the line in the direction opposite to that of progress is the *back* bearing. The back bearing can be obtained from the forward bearing by simply changing the letter N to S or S to N and also changing E to W or W to E. In Fig. 7-5 the bearing of the line AB is N 68° E, and the bearing of BA, which is in the reverse direction is S 68° W.

7-10   Relation Between Azimuths and Bearings

To simplify computations based on survey data, bearings may be converted to azimuths and azimuths may be converted to bearings. The instances when such conversion is convenient will become apparent in later chapters. The conversion itself is quite simple.

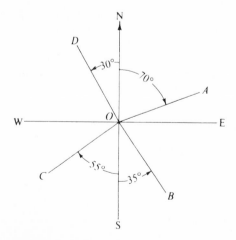

FIGURE 7-4 Bearings.

An inspection of Fig. 7-1 will show that the line *OA,* whose azimuth from north is 70°, lies in the northeast quadrant since the angle eastward from the meridian is less than 90°. Furthermore it is apparent that the bearing angle and the azimuth are identical. Therefore the bearing of *OA* is N 70° E. The line *OB* is 145° from the north meridian. It lies south of a due-east line and is therefore in the southeast quadrant. The problem in this case is to determine the angle from the south meridian. Since the north meridian and the south meridian are 180° apart, the problem is solved by subtracting the azimuth, or 145°, from 180° to arrive at the bearing angle, which is 35°. Therefore the bearing of the line *OB* is S 35° E. The line *OC* is 235° from the north meridian in a clockwise direction and is beyond the south meridian in a westerly direction by an angle of 235° – 180°, or 55°. Therefore if the azimuth is 235°, the bearing is S 55° W. For the line *OD* the angle from the north meridian is 330° in a clockwise direction. The angle from the north meridian in a counterclockwise or westerly direction is 360° – 330°, or 30°. The line *OD* lies in the northwest quadrant, and its bearing is N 30° W.

The rules to observe in converting from azimuths to bearings are very quickly established in a person's mind after the rules have been put to practice a few times. They are as follows: (1) If an azimuth from north is between 0° and 90°, the line is in the northeast quadrant, and the bearing angle is equal to the azimuth; (2) if an azimuth from north is between 90° and 180°, the line is in the southeast quadrant, and the bearing angle is 180° minus the azimuth; (3) if the azimuth from north is between 180° and 270°, the line is in the southwest quadrant, and the bearing angle is the azimuth minus 180°; (4) if the azimuth from north is between 270° and 360°, the line is in the northwest quadrant, and the bearing angle is 360° minus the azimuth.

To convert from bearings to azimuths, it is only necessary to reverse the foregoing rules, and the computations should be made with the same mental ease. The azimuth of a line in the northeast quadrant is equal to the bearing angle; that of a line in the southeast quadrant is 180° minus the bearing angle; that of a line in the southwest quadrant is 180° plus the bearing angle; that of a line in the northwest quadrant is 360° minus the bearing angle. As an example, the azimuth of a line whose bearing is N 77° 43′ 16″ W is 282° 16′ 44″.

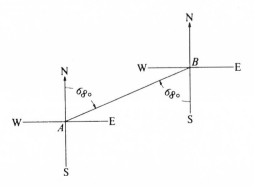

FIGURE 7-5 Bearing and back bearing.

7-11   Magnetic Compass

Since a freely suspended magnetized needle will lie in the magnetic meridian, the direction of a line can be determined with respect to the needle, and thus to the magnetic meridian, by measuring the angle between the line and the needle. The magnetic compass is constructed so as to allow a needle to swing freely on a pivot when in use, and to allow a line of sight to be directed from the occupied point to a terminal point. As shown in Fig. 7-6, a graduated circle is rotated as the line of sight is rotated. The north-seeking end of the compass needle is read against the circle to obtain the angle between the magnetic meridian and the line of sight.

The circle and the needle are encased in a metal compass box and are covered with a glass plate. The line of sight normally is fixed in line with the zero mark or the north graduation on the circle. Thus, if a line of sight is directed along the north

FIGURE 7-6 Surveyor's compass.

magnetic meridian, the needle will point to the zero mark or to the north graduation. As the line of sight is turned clockwise from magnetic north, the needle remains in the magnetic meridian, but the graduated circle is turned clockwise through the corresponding angle. When the line of sight is turned exactly 90° east of north, then the letter E is brought opposite the north end of the compass needle. The circle thus indicates that the magnetic bearing is due east, as it should. If the line of sight is turned exactly 180° from north, the letter S is brought opposite the north end of the needle, and the magnetic bearing of the line of sight is shown to be due south.

The surveyor's compass (Fig. 7-6) is the instrument that was used in the past to run surveys of reasonable accuracy. Its main part is the compass box containing the graduated circle between 4 and 6 in. in diameter, the magnetic needle, one or two level bubbles to allow the circle to be brought horizontal, long sight vanes to allow fairly accurate pointings along rather steeply inclined lines, and a screw for lifting the needle off the pivot and against the glass cover plate. The compass box is fitted with a spindle that rotates in a socket. The socket fits in a leveling head consisting of a ball-and-socket joint, and the whole compass assembly is leveled about the ball-and-socket against friction applied by a clamping nut. The leveling head is fastened to a tripod or to a staff and is used in the field in this fashion. The circle is usually graduated in half degrees. The sight vanes are aligned so that the line of sight passes through the north graduation on the circle.

The transit compass is similar to the surveyor's compass in all respects save for the method of leveling and for the line of sight. The compass box sets in the center of the horizontal limb of the transit, astride of which are the standards that hold the telescope (see Fig. 4-8). The method of leveling the transit, and thus the compass box, was described in Section 4-14. The line of sight for the transit compass is the collimation line of the telescope defined by the center of the objective lens and the intersection of the cross hairs. The telescope of the transit is capable of being raised or depressed for sighting along inclined lines. The letter N on the compass circle is normally under the objective end of the telescope when the telescope is in the direct position (see Section 4-17). The letter S is under the eyepiece. Therefore this arrangement keeps the north graduation directed along the line of sight just as in the case of the surveyor's compass.

7-12 Determining Directions with a Magnetic Compass

When using a surveyor's compass or a transit, the observer occupies one end of the line, centers the instrument over the point, and levels the instrument. He releases the clamp holding the needle, allowing it to rest on the pivot. He then directs the line of sight toward a point at the other end of the line and brings it on that point. When the needle comes to rest, he reads the circle at the north end of the needle to obtain the bearing of the line from the occupied point to the point sighted on. He then clamps the needle before disturbing the instrument.

To obtain the back magnetic bearing as a check, he occupies the second point and sights back to the first point, reading the north end of the needle as before. The bearing angles should show reasonable agreement, and the letters N and S as well as E and W should be reversed on the back bearing. This check should be made before leaving the point.

7-13   Magnetic Declination

The magnetic poles do not coincide with the poles defined by the earth's rotational axis, and certain irregularities in the earth's magnetic field cause local and regional variations in the direction of the needle. Therefore, except in a very narrow band around the earth, the magnetic needle does not point in the direction of true north. It some areas the needle points east of true north; in other areas the needle points west of true north. The amount and direction by which the magnetic needle is off the true meridian is called the *magnetic declination* (formerly called *variation*). Declination is positive or plus when the needle points east of true north, and it is negative or minus when the needle points west of true north. The declination varies from +23° in the state of Washington to –22° in the state of Maine; the total range over the United States is 45°.

The declination in a given vicinity can be obtained by observing the magnetic bearing of a line of known true azimuth or bearing.

◀**EXAMPLE 7-1**   The magnetic bearing of a line of known true azimuth is observed as S55° 40′E. The true azimuth of this line is 109° 05′. What is the magnetic declination at the point?

Solution: The magnetic azimuth according to Section 7-10 is 124° 20′. As shown in Fig. 7-7, the difference is 15° 15′, and the declination is 15° 15′W.   ◆

7-14   Horizontal Angles from Compass Bearings

An angle at a point can be obtained, within the limits of accuracy of compass readings, by observing the bearings to the backsight station and the foresight station. The resulting angle is free from the effects of the declination and of any local attraction affecting the compass needle since both sights are referred to the compass needle.

FIGURE 7-7 Magnetic declination.

EXAMPLE 7-2   The following bearings were observed at the points in the four-sided traverse shown in Fig. 7-8. Compute and adjust the angles.

Line	Observed Mag Bearing	Unadjusted Angles	Adjusted Angles
AD	S 77° 00′ E		
		$(180° - 77° 00′) + 12° = 115° 00′$	114° 45′
AB	N 12° 00′ W		
BA	S 12° 00′ E		
		$82° 30′ - 12° 00′ = 70° 30′$	70° 15′
BC	S 82° 30′ E		
CB	N 82° 00′ W		
		$180° - (82° 00′ + 20° 00′) = 78° 00′$	77° 45′
CD	S 20° 00′ W		
DC	N 20° 00′ E		
		$20° + 77° 30′ = 97° 30′$	97° 15′
DA	N 77° 30′ W		$\overline{361° 00′}$ 360° 00′

Solution: The calculations are shown in the accompanying table. The student should refer to Fig. 7-8 to follow these calculations. Since the angles do not total $(n-2)180°$, they are adjusted equally by subtracting 15′ from each angle.   ◆

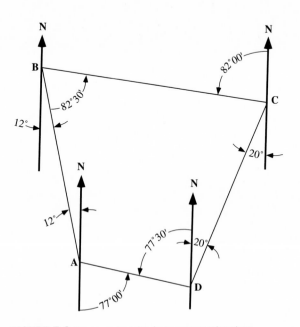

FIGURE 7-8 Horizontal angles from compass bearings.

PROBLEMS

7-1. Convert the following bearings to azimuths: **(a)** S 15° 25′ 20″ E, **(b)** N 18° 16′ 30″ E, **(c)** N 18° 16′ 30″ W, **(d)** S 27° 20′ 15″ W, **(e)** S 88° 16′ 40″ E.

7-2. Convert the following bearings to azimuths: **(a)** N 45° 27′ 45″ W, **(b)** N 27° 20′ 35″ E, **(c)** S 27° 20′ 35″ E, **(d)** S 77° 20′ 15″ W, **(e)** N 89° 10′ 15″ W.

7-3. Convert the following azimuths to bearings: **(a)** 106° 22′ 10″, **(b)** 57° 18′ 50″, **(c)** 327° 14′ 20″, **(d)** 267° 53′ 15″, **(e)** 178° 14′ 20″.

7-4. Convert the following azimuths to bearings: **(a)** 20° 12′ 30″, **(b)** 90° 04′ 50″, **(c)** 166° 14′ 40″, **(d)** 270° 14′ 15″, **(e)** 354° 26′ 15″.

7-5. The true azimuth of a line is 187° 10′. The observed magnetic bearing of the line is S 2° 15′ E. What is the magnetic declination of the point of observation?

7-6. In a five-sided traverse, ABCDE, the following magnetic bearings were observed:

Line	Observed Magnetic Bearing
AE	S 7° 00′ W
AB	N 75° 30′ E
BA	S 75° 30′ W
BC	S 45° 30′ W
CB	N 46° 00′ E
CD	S 15° 00′ E
DC	N 14° 30′ W
DE	N 82° 30′ W
ED	S 82° 30′ E
EA	N 6° 30′ E

Compute the interior angles in the traverse and adjust them equally.

8 Traverse Surveys and Computations

8-1 Traverse

A traverse is a series of connected lines of known length related to one another by known angles. The lengths of the lines are determined by direct measurement of horizontal distances, by slope measurement, or by indirect measurement based on the methods of tacheometry. These methods are discussed in Chapters 2 and 14. The angles at the traverse stations between the lines of the traverse are measured with the instruments discussed in Chapter 4. They can be interior angles, deflection angles, or angles to the right, which are discussed in Chapter 6.

The results of field measurements related to a traverse will be a series of connected lines whose lengths and azimuths, or whose lengths and bearings, are known. The lengths are horizontal distances; the azimuths or bearings are either true, magnetic, assumed, or grid.

A completely different method of traversing can be performed using an inertial system. This system incorporates a gyro-stabilized platform containing three accelerometers oriented at right angles to one another, a precise clock, and an onboard computer. The accelerometers sense accelerations in the north–south, east–west, and vertical directions from which displacements in these three directions are computed.

The satellite constellation of the Global Positioning System (GPS) is used in traversing by the method referred to as *kinematic GPS surveying* and is described in detail in Chapter 10.

In general, traverses are of two classes. One of the first class is an open traverse. It originates either at a point of known horizontal position with respect to a horizontal datum or at an assumed horizontal position, and terminates at an unknown horizontal position. A traverse of the second class is a closed traverse, which can be described in any one of the following three ways: (1) It originates at an assumed horizontal position and terminates at that same point; (2) it originates at a known horizontal position with respect to a horizontal datum and terminates at that same point; (3) it originates at a known horizontal position and terminates at another known horizontal position. A known horizontal position is defined by its geographic latitude and longitude, by its Y and X coordinates on a grid system, or by its location on or in relation to a fixed boundary.

Traverse surveys are made for many purposes and types of projects, some of which follow:

- To determine the positions of existing boundary markers
- To establish the positions of boundary lines
- To determine the area encompassed within the confines of a boundary
- To determine the positions of arbitrary points from which data may be obtained for preparing various types of maps, that is, to establish *control* for mapping
- To establish ground control for photogrammetric mapping
- To establish control for gathering data regarding earthwork quantities in railroad, highway, utility, and other construction work
- To establish control for locating railroads, highways, and other construction work

8-2   Open Traverse

An open traverse is usually run for exploratory purposes. There are no arithmetical checks on the field measurements. Since the figure formed by the surveyed lines does not close, the angles cannot be summed to a known quantity, for example, the angles in a plane triangle sum to 180°. None of the positions of the traverse stations can be verified, since no known or assumed position is included except that of the starting station. To strengthen an open traverse, that is, to render it more reliable, several techniques may be employed. Each distance can be measured in both directions and can be roughly checked by using the stadia hairs of the theodolite (see Chapter 14). The measurements of the angles at the stations can be repeated by using the methods of Chapters 4 and 6 and checked approximately by observing magnetic bearings. The directions of the lines can be checked by observing the sun or the stars to determine true azimuths or bearings of selected lines in the traverse. An open traverse should not be run for any permanent project or for any of the projects indicated in Section 8-1 because it does not reveal mistakes or errors and the results are always open to doubt.

8-3   Closed Traverse

A traverse that closes on itself immediately affords a check on the internal accuracy of the measured angles, provided that the angle at each station has been measured. As will be discussed later, a traverse that closes on itself gives an indication of the consistency of measuring distances as well as angles by affording a check on the position closure of the traverse. Unless astronomical observations for azimuths (see Chapter 12) have been made at selected stations of a traverse that closes on itself, the only provision for verifying the directions of the lines is that afforded by the angular closure. In this type of traverse, there is no check on the systematic errors introduced into the measurement of lengths. Therefore, when this type of traverse is executed for a major project, the measuring apparatus must be carefully calibrated to determine the systematic errors and to eliminate them.

A traverse that originates at a known position and closes on another known position is by far the most reliable, because a check on the position of the final point checks both the linear and angular measurements of the traverse. When a point of known position is referred to, it is understood that such a point has been located by procedures as precise as, or more precise than, those used in the traverse being executed. These procedures are the methods of either traversing, triangulation, trilateration, or GPS surveying. Triangulation and trilateration are the subjects of Chapter 9.

8-4   Interior-Angle Traverse

An interior-angle traverse is shown in Fig. 8-1. The azimuth or bearing of the line *AP* is known. The lengths of the traverse lines are measured to determine the horizontal distances. With the transit or theodolite at *A*, the angle at *A* from *P* to *B* is measured to determine the azimuth of line *AB*. The angle at *A* from *E* to *B* is also measured, as this is one of the interior angles in the figure. The instrument is then set up at *B*, *C*, *D*, and *E* in succession, and the indicated angles are measured. The notes for the angles are recorded as indicated in Chapter 6. Had the azimuth of one of the traverse sides been known, there would have been no necessity for measuring the angle at *A* from *P* to *B*.

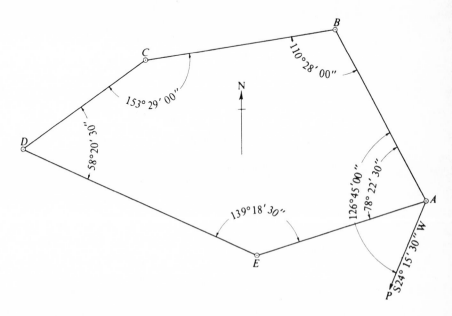

FIGURE 8-1 Interior-angle traverse.

TABLE 8-1   Adjustment of Angles
in an Interior-Angle Traverse

Station	Measured Angle	Correction	Adjusted Angle
A	78° 22′ 30″	+18″	78° 22′ 48″
B	110° 28′ 00″	+18″	110° 28′ 18″
C	153° 29′ 00″	+18″	153° 29′ 18″
D	58° 20′ 30″	+18″	58° 20′ 48″
E	139° 18′ 30″	+18″	139° 18′ 48″
	539° 58′ 30″		540° 00′ 00″
	−540° 00′ 00″		
	closure = −1′30″		

To test the internal angular closure, the interior angles are added and their sum compared to $(n - 2)180°$, which in this example is $(5 - 2)180° = 540°$. The total angular error, or the closure, is $1′ 30″$. In the tabulation in Table 8-1 the measured angles are adjusted by assuming that the angular error is of the same amount at each station. This assumption may not be valid, because an error in angular measurement, all other things being equal, will increase as the lengths of the adjacent sides decrease. In the traverse of Fig. 8-1 the interior angles are presumed to have been measured to the nearest 30″. The manner of adjusting the $1′ 30″$ closure can be somewhat arbitrary in this instance. For example, instead of applying an 18″ correction to each angle, the discrepancy could be distributed by correcting the angles at A, B, and C by 30″ each, or by correcting the angles at A, C, and D by 30″ each.

After the adjusted angles are computed, they should always be added to see whether their sum is, in fact, the proper amount. A mistake in arithmetic either in adding the measured angles or in applying the corrections will become apparent.

The azimuth of the line AP in Fig. 8-1 is known, and the azimuths of all the traverse sides can be determined by using the measured angle at A from P to B and the adjusted interior angles in the closed figure. Since the bearing of AP is S 24° 15′ 30″ W, the azimuth of AP reckoned from north is 204° 15′ 30″. As indicated in Fig. 8-2, the azimuth of AB is equal to the azimuth of AP plus the angle at A from P to B, or 204° 15′ 30″ + 126° 45′ 00″ = 331° 00′ 30″. The azimuths of the other lines are computed systematically by applying the adjusted angle at each station to the azimuth of each backsight line in turn. To determine the azimuth of BC, compute the azimuth of BA by subtracting 180° from the azimuth of AB, and then add the adjusted angle at B from A to C. A similar procedure is adopted at each station (see Table 8-2).

The azimuths of the lines can also be computed by applying the angles at A, E, D, C, and B in that order. Then each interior angle would be *subtracted* from the azimuth of the proper back line. This is apparent from a study of Fig. 8-1.

The bearing of each line, if desired, is determined from its azimuth. Since the conversion from azimuths to bearings is easily made, it is unwise to attempt to compute the bearing of each line directly by applying the proper angle in the traverse to the bearing of the adjacent line.

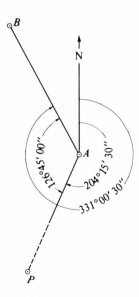

FIGURE 8-2 Determining azimuth of line *AB*.

TABLE 8-2 Computation of Azimuths and Bearings Using Adjusted Interior Angles

Line	Azimuth	Bearing
AP	204° 15′ 30″	S 24° 15′ 30″ W
	(+ ∠*A*) +126° 45′ 00″	
AB	331° 00′ 30″	N 28° 59′ 30″ W
BA	151° 00′ 30″	
	(+ ∠*B*) +110° 28′ 18″	
BC	261° 28′ 48″	S 81° 28′ 48″ W
CB	81° 28′ 48″	
	(+ ∠*C*) +153° 29′ 18″	
CD	234° 58′ 06″	S 54° 58′ 06″ W
DC	54° 58′ 06″	
	(+ ∠*D*) + 58° 20′ 48″	
DE	113° 18′ 54″	S 66° 41′ 06″ E
ED	293° 18′ 54″	
	(+ ∠*E*) +139° 18′ 48″	
EA	432° 37′ 42″	
EA	72° 37′ 42″	N 72° 37′ 42″ E
AE	252° 37′ 42″	
	(+ ∠*A*) + 78° 22′ 48″	
AB	331° 00′ 30″ check	N 28° 59′ 30″ W

In a traverse of n sides or stations, the closure of the measured angles should not exceed the least count of the vernier or scale of the instrument times the square root of n. With care in sighting, in centering the instrument, and in reading the verniers or scales, the angular closure can be expected to be half this allowable amount. Of course, with few traverse stations, this precision may not be obtainable since the opportunities for random errors to compensate are few.

One very important point to observe in the preceding example of an interior-angle traverse is that there is no check on the angle at A from P to B. If this angle is in error, then the error affects the azimuth of each line in the traverse. To avoid the possibility of making a mistake in measuring this angle, the clockwise angle at A from B to P should also be measured. This measurement immediately affords a check at station A since the angle at A from P to B and the angle at A from B to P should total 360°. If the azimuth of one of the sides of the traverse were known, then this uncertainty would have been avoided, because the test of the accuracy in this case is simply that the sum of the interior angles should equal $(n-2)180°$.

8-5   Deflection-Angle Traverse

A deflection-angle traverse that originates at station $0 + 00$ and closes on station $22 + 20$ is shown in Fig. 8-3. The azimuth of the line FG is fixed as 196° 35′ and the azimuth of EM is fixed as 306° 43′, these directions having been established by previous surveys. The traverse is made to fit between these two fixed azimuths. After the deflection angles and the lengths of the several courses have been measured, the check on angular closure is made by carrying azimuths through the traverse by applying the deflection angles. In Fig. 8-4 the azimuth of GF is 16° 35′, obtained by

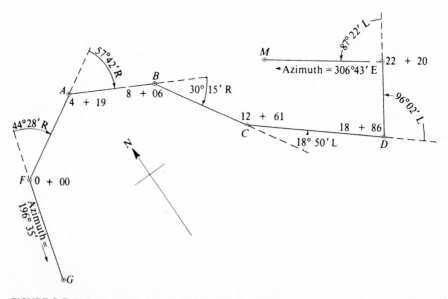

FIGURE 8-3 Deflection-angle traverse between fixed azimuths.

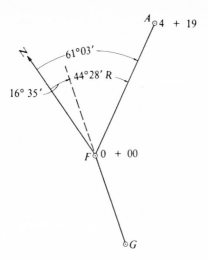

FIGURE 8-4 Determining azimuth of line *FA.*

subtracting 180° from the azimuth of *FG.* So the azimuth of the line *FA* is 16° 35′ plus 44° 28′, which is the right deflection angle at *F* from *G* to *A.* The addition gives 61° 03′ as the azimuth of *FA.* Now to determine the azimuth of the next line *AB,* the right deflection angle at *A* from *F* to *B* is added to the azimuth of *FA* prolonged. This sum is 61° 03′ + 57° 42′ = 118° 45′. Observe that no back azimuths need be computed as was the case for the interior-angle traverse. To compute an azimuth by using a deflection angle, simply add a right deflection angle to the *forward* azimuth of the backsight to obtain the *forward* azimuth of the foresight; or subtract a left deflection angle from the *forward* azimuth of the backsight to obtain the *forward* azimuth of the foresight. A computation of azimuths in a deflection-angle traverse is given in Table 8-3.

In the example in Fig. 8-3, the computed azimuth of *EM* failed to check by 3′. Since six deflection angles were measured, the correction to each angle is 30″. Instead of applying this correction to each angle and recomputing the azimuths, the azimuths themselves are adjusted. The azimuth of *FA* receives a 30″ correction, since this azimuth was obtained by considering only one measured angle; the azimuth of *AB* receives a 1′ 00″ correction, since this azimuth was obtained by using two angles, and so on. The correction to the last azimuth is 6 × 30″ = 3′ 00″, since this azimuth was obtained by using all six deflection angles.

Figure 8-5 illustrates deflection-angle traverse that closes on the point of origin at *L.* The bearing of the line *OP* is known to be N 66° 02′ W. Before the bearings of the remaining sides are computed, the angles must be adjusted so that the difference between the sum of the right deflection angles and the sum of the left deflection angles is 360°. It is found that the sum of the right deflection angles must be reduced and the sum of the left deflection angles must be increased. The correction in this case is distributed equally among all of the angles. Had the angular closure been, for example, only 01′, then 30″ could have logically been added to the angle at *N* and subtracted from the angle at *Q,* since these two angles are

TABLE 8-3 Computation of Azimuths and Bearings Using Deflection Angles

Line	Azimuth	Correction	Adjusted Azimuth	Adjusted Bearing
GF	16° 35″	fixed		N16° 35′ 00″ E
	(+ ∠F) + 44° 28′			
FA	61° 03′	−0′ 30″	61° 02′ 30″	N 61° 02′ 30″ E
	(+ ∠A) + 57° 42′			
AB	118° 45′	−1′ 00″	118° 44′ 00″	S 61° 16′ 00″ E
	(+ ∠B) + 30° 15′			
BC	149° 00′	−1′ 30″	148° 58′ 30″	S 31° 01′ 30″ E
	(−∠C) − 18° 50′			
CD	130° 10′	−2′ 00″	130° 08′ 00″	S 49° 52′ 00″ E
	(−∠D) − 96° 02′			
DE	34° 08′	−2′ 30″	34° 05′ 30″	N 34° 05′ 30″ E
DE	394° 08′			
	(−∠E) − 87° 22′			
EM	306° 46′	−3′ 00″	306° 43′ 00″	N 53° 17′ 00″ W
	306° 43′	fixed		
	closure = + 03′			

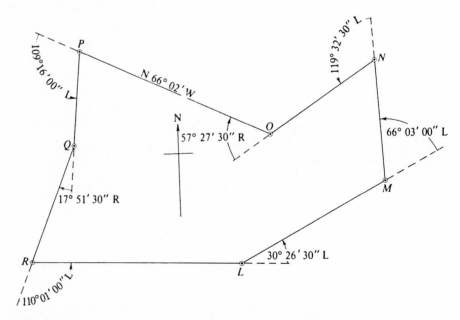

FIGURE 8-5 Deflection-angle traverse that closes on point of origin (adjusted angles are shown).

TABLE 8-4   Adjustment of Deflection Angles

Station	Deflection Angle	Correction	Adjusted Deflection Angle
L	30° 26′ 00″ L	+ 30″	30° 26′ 30″ L
M	66° 02′ 30″ L	+ 30″	66° 03′ 00″ L
N	119° 32′ 00″ L	+ 30″	119° 32′ 30″ L
O	57° 28′ 00″ R	− 30″	57° 27′ 30″ R
P	109° 15′ 30″ L	+ 30″	109° 16′ 00″ L
Q	17° 52′ 00″ R	− 30″	17° 51′ 30″ R
R	110° 00′ 30″ L	+ 30″	110° 01′ 00″ L

Σ right   75° 20′ 00″   Σ right   75° 19′ 00″

Σ left   435° 16′ 30″   Σ left   435° 19′ 00″

difference   359° 56′ 30″   360° 00′ 00″ check

closure   3′ 30″

formed by the shortest lines of sight. The tabulation of Table 8-4 shows this adjustment of deflection angles.

Bearings of the sides of the traverse are computed by first converting the bearing of *OP* to an azimuth, then applying the adjusted deflection angles to obtain azimuths, and finally converting the azimuths to bearings as shown in Table 8-5.

The angular closure of a deflection-angle traverse should be no more than the least count of the vernier of the instrument times the square root of the number of angles in the traverse. A deflection angle should never be measured without double

TABLE 8-5   Computation of Azimuths and Bearings Using Adjusted Deflection Angles

Line	Azimuth	Bearing
OP	293° 58′ 00″	N 66° 02′ 00″ W
	(− ∠P) − 109° 16′ 00″	
PQ	184° 42′ 00″	S 4° 42′ 00″ W
	(+ ∠Q) + 17° 51′ 30″	
QR	202° 33′ 30″	S 22° 33′ 30″ W
	(− ∠R) − 110° 01′ 00″	
RL	92° 32′ 30″	S 87° 27′ 30″ E
	(− ∠L) − 30° 26′ 30″	
LM	62° 06′ 00″	N 62° 06′ 00″ E
LM	422° 06′ 00″	
	(− ∠M) − 66° 03′ 00″	
MN	356° 03′ 00″	N 3° 57′ 00″ W
	(− ∠N) − 119° 32′ 30″	
NO	236° 30′ 30″	S 56° 30′ 30″ W
	(+ ∠O) + 57° 27′ 30″	
OP	293° 58′ 00″ check	N 66° 02′ 00″ W

centering the instrument, because the error caused by the line of sight not being normal to the horizontal axis of the instrument may be too large to be tolerated.

8-6   Angle-to-the-Right Traverse

Either an open traverse or a closed traverse can be executed by measuring angles to the right. The method of measuring the angles is described in Section 6-3. The method of computing azimuths from a given fixed azimuth is similar to that employed in an interior-angle traverse. A forward azimuth is always obtained by *adding* the angle to the right to the azimuth of the backsight. Angles to the right are always employed when a traverse is executed with a direction instrument.

Figure 8-6 is an illustration of an angle-to-the-right traverse executed by use of a direction instrument. Station P is occupied, a backsight is taken on station T, and the circle is read. A foresight is taken on station Q, and the circle is read. This procedure is repeated at each station along the traverse, each backsight and foresight being observed with the telescope both direct and reversed. Notes for a 10″ direction instrument are given in Table 8-6.

Assume that the azimuth of PT is known to be 88° 20′ 10″. The azimuth of the line PQ is the azimuth of PT plus the angle to the right at P. The angle to the right at P from T to Q can be obtained by subtracting the mean circle reading to T from the mean circle reading to Q. This angle, as computed from notes, is 92° 16′ 00″ − 51° 54′ 50″ = 40° 21′ 10″. Then the azimuth of PQ is 88° 20′ 10″ + 40° 21′ 10″ = 128° 41′ 20″. This computation can be simplified if the mean circle reading to Q is added to the azimuth of PT, and then the mean circle reading to T is subtracted from the sum. In this manner the circle reading for the foresight is always added and the circle reading for the backsight is always subtracted.

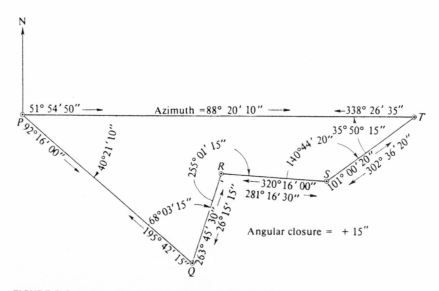

FIGURE 8-6 Angles-to-the-right from direction instrument readings.

TABLE 8-6   Notes for Measuring Angles to the Right
When Using a Direction Instrument

Station		D or R	Reading	Mean
At P	T	D	51° 54′ 50″	51° 54′ 50″
		R	231° 54′ 50″	
	Q	D	92° 15′ 50″	92° 16′ 00″
		R	272° 16′ 10″	
At Q	P	D	195° 42′ 10″	195° 42′ 15″
		R	15° 42′ 20″	
	R	D	263° 45′ 20″	263° 45′ 30″
		R	83° 45′ 40″	
At R	Q	D	26° 15′ 10″	26° 15′ 15″
		R	206° 15′ 20″	
	S	D	281° 16′ 30″	281° 16′ 30″
		R	101° 16′ 30″	
At S	R	D	320° 16′ 00″	320° 16′ 00″
		R	140° 16′ 00″	
	T	D	101° 00′ 10″	101° 00′ 20″
		R	281° 00′ 30″	
At T	S	D	302° 36′ 10″	302° 36′ 20″
		R	122° 36′ 30″	
	P	D	338° 26′ 30″	338° 26′ 35″
		R	158° 26′ 40″	

The computation is shown in Table 8-7 as it would be made without the aid of a calculating machine designed for adding and subtracting angles on the sexagesimal system. If such a calculating machine is used, only the forward azimuth of each line of the traverse need be recorded on the computation sheet. The resulting azimuths are then adjusted to the fixed azimuth, and bearings may be obtained from the adjusted azimuths.

8-7 Traverse-by-Azimuth Method

In running a traverse for the purpose of establishing a lower order of control for mapping and for locating the positions of ground objects with respect to such a supplementary traverse, the transit or theodolite can be handled so that the clockwise circle reading at all times indicates the azimuth of the line of sight. This procedure eliminates the need for computing azimuths from interior angles, deflection angles, or angles to the right.

Figure 8-7 shows a portion of a supplementary traverse in which the line *DE* has an azimuth of 96° 22′. This may be a true, magnetic, assumed, or grid azimuth. The instrument is set up at station *E,* and the clockwise circle is set to read the azimuth of the line *ED,* which is 276° 22′. A lower motion is used to take a backsight along the line *ED.* The theodolite and the horizontal circle are now oriented. In other words, when any sight is taken by an upper motion, the clockwise circle reading will always indicate the azimuth of the line of sight. If a pointing is made

TABLE 8-7 Computations of Azimuths
Using Angle-to-the Right Notes

Line	Azimuth	Correction	Adjusted Azimuth
PT	88° 20' 10" fixed		
	+ 92° 16' 00"		
	180° 36' 10"		
	− 51° 54' 50"		
PQ	128° 41' 20"	−3"	128° 41' 17"
QP	308° 41' 20"		
	+ 263° 45' 30"		
	572° 26' 50"		
	− 195° 42' 15"		
QR	376° 44' 35"		
QR	16° 44' 35"	−6"	16° 44' 29"
RQ	196° 44' 35"		
	+ 281° 16' 30"		
	478° 01' 05"		
	− 26° 15' 15"		
RS	451° 45' 50"		
RS	91° 45' 50"	−9"	91° 45' 41"
SR	271° 45' 50"		
	+ 101° 00' 20"		
	372° 46' 10"		
	− 320° 16' 00"		
ST	52° 30' 10"	−12"	52° 29' 58"
TS	232° 30' 10"		
	+ 338° 26' 35"		
	570° 56' 45"		
	− 302° 36' 20"		
TP	268° 20' 25"	−15"	268° 20' 10"
PT	88° 20' 25" fixed		
	88° 20' 10"		
	closure + 15"		

on station F and the clockwise circle reads 68° 02', this is the azimuth of the line EF and is recorded as such.

When the transit is set up at F, the circle is oriented by setting the azimuth of FE, which is 248° 02', on the clockwise circle and backsighting along the line FE by a lower motion. A sight taken on point P by an upper motion will give the azimuth of the line FP directly on the clockwise circle. The readings to P, Q, and G are shown to be, respectively, 57° 25', 192° 52', and 113° 10'.

The advantages of executing a traverse by observing azimuths are that the transit or theodolite is allowed to do the work of adding and subtracting angles, and that the field notes are easily reduced to map form by having all lines related to the same meridian directly in the notes.

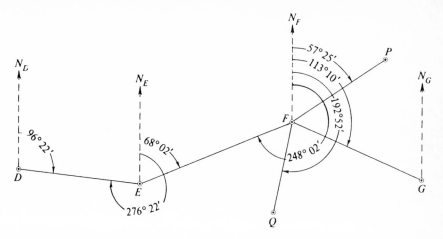

FIGURE 8-7 Traverse by azimuth method.

The disadvantage of such a procedure is in not realizing the benefit of double centering, which eliminates instrumental errors and which makes mistakes in reading the circle quite obvious. The purpose of the traverse, however, is usually such that small errors are of little consequence (see Section 14-8).

If a traverse is to be run on magnetic azimuths and the directions of the lines are to be consistent with one another, the initial setup determines the specific magnetic meridian to which all other lines are referred. To orient on the magnetic meridian, set the clockwise circle to read zero, unclamp the compass needle, loosen the lower clamp, and rotate the transit until the compass needle points to the north point of the compass circle. Tighten the lower clamp, and make an exact setting by using the tangent screw. The transit is then oriented for measuring magnetic azimuths. The procedure for carrying azimuths through the remaining lines in the traverse is the same as that previously described.

8-8 Azimuth Traverse

An azimuth traverse is a continuous series of lines of sight related to one another by measured angles only. The distances between the instrument stations are not measured. An azimuth traverse serves one of two purposes. The first purpose is to permit the determination of directions far removed from a beginning azimuth without the necessity of measuring distances. As an example, consider a pair of intervisible stations, such as Ridge 1 and Ridge 2 in Fig. 8-8, situated high on a ridge, and assume that the azimuth of the line joining these stations is known. A traverse is to be run in an adjacent valley, and the basis of azimuths of this traverse is to be the same as that for the line along the ridge.

To carry the azimuth down into the valley, one end of the known line, Ridge 1, is occupied and the other end of the line, Ridge 2, is used as a backsight for measuring an angle to the right, a deflection angle, or, if the traverse is to close by occupying Ridge 2, an interior angle. All the other points are occupied by the

FIGURE 8-8 Azimuth traverse.

instrument, and angles are measured. That portion of the survey from Ridge 1 to A and the portion from G to Ridge 2 are azimuth traverses. The lengths of the lines from AB through FG are measured along with the angles. The angular closure can be computed, and the traverse from A to G will be on the same basis of azimuths as is the straight line from Ridge 1 to Ridge 2.

The second purpose of an azimuth traverse is to avoid carrying azimuths through extremely short traverse sides. Since the angular error will increase as the lines of sight become shorter, the desirable traverse is one with long sights. Such sights may not be practical, however, for several reasons. The ground over which the traverse must follow may be rough and the sides may have to be short to alleviate difficult taping; the traverse may run through a city where every street corner or every point at a break in the street center line is a transit station; the traverse may be run along a curving right of way where intervening brush would interfere with the measurement of the lengths of long traverse sides. In these situations the angular errors accumulating from short lines of sight can be isolated by employing an azimuth traverse of one or more stations through which the azimuth is computed. This use of an azimuth traverse is illustrated in Fig. 8-9. The azimuth is carried through lines MN, NO, OP, PQ, QR, and RS, although the azimuths of the lines forming the loops that have been cut off must also be determined. The lengths of the cutoff lines OP and PQ would not be measured. All the remaining distances would be measured.

The angular closure along the route $MNOPQRS$ is adjusted by methods discussed in the previous sections. This adjustment fixes the directions of the lines OP and PQ. The loop O, O-1, O-2, O-3, O-4, P is then adjusted to the azimuth of OP. The loop P, P-1, P-2, P-3, P-4, Q is then adjusted to the azimuth of PQ.

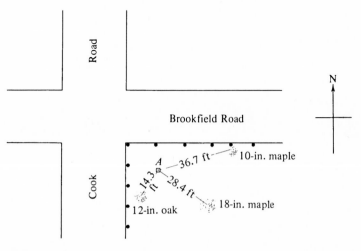

FIGURE 8-9 Azimuth traverse by cutoff lines.

8-9   Referencing a Traverse Station

As an aid in relocating a point that may become hidden by vegetation or buried beneath the surface of the ground, or as a means of replacing a point that may have been destroyed, measurements are made to nearby permanent or semipermanent objects. This process is known as *referencing* or *witnessing* the point. Property corners and, on important surveys, all instrument stations are usually referenced.

Figure 8-10 shows the locations of the witness points with respect to station *A*. If the stake at *A* cannot be found at a later time, its approximate position can be determined by locating the intersection of arcs struck with the trees as centers. The point would, of course, be determined by two arcs, but a third measurement is taken to serve as a check. If it is likely that any of the witness points will be destroyed, additional witnesses are located.

The method of recording the witnesses is the same as that used in notes for all United States land surveys (see Chapter 18). First the object is described. If it is a tree, its diameter is given. Next the bearing from the station to the witness point is given. Last the distance is recorded. If the measurement is made to any definite

FIGURE 8-10 Referencing a point.

point, such as a nail driven in the root or side of a tree, that fact should be stated. To be of the most value in replacing a missing station, the witnesses should be less than 100 ft from the station and, if possible, the arcs should intersect approximately at right angles.

On many surveys no permanent objects may be available as witnesses. In such cases additional stakes can be driven. The method illustrated in Fig. 6-15(f) is the most satisfactory way of referencing a point so that it can be replaced in its original position. In this figure point P is at the intersection of lines AB and CD. If the distances from P to the four stakes are measured carefully, the point can be replaced if any two of the stakes remain, and the relocation can be checked if three of them can be found. This method is commonly used in referencing transit stations on route surveys, where it is known that all centerline stakes will be destroyed as soon as grading operations are begun.

In addition to the specific witnesses, a general description of the location of the traverse station should be given, so that a person searching for it will have a fairly good idea of where to begin searching.

8-10 Traverse Computations

The result of the field work in executing a traverse of any kind is a series of connected lines whose directions and lengths are known. The angular closure is distributed to give a series of preliminary adjusted azimuths or bearings. Errors in the measured lengths of the traverse sides, however, will tend to alter the shape of the traverse. The steps involved in adjusting a traverse whose preliminary adjusted azimuths have been determined are as follows:

1. Determine the distance that each line of the traverse extends in a north or south direction, and the distance that each line extends in an east or west direction. These distances are called, respectively, *latitudes* and *departures*.

2. Determine the algebraic sum of the latitudes and the algebraic sum of the departures, and compare them with the fixed latitude and departure of a straight line from the origin to the closing point. This presumes a closed traverse.

3. Adjust the discrepancy found in step 2 by apportioning the closure in latitudes and the closure in departures on a reasonable and logical basis.

4. Determine the adjusted position of each traverse station with respect to some origin. This position is defined by its Y coordinate and its X coordinate with respect to a plane rectangular coordinate system, the origin being the intersection of the Y axis and the X axis with the Y axis being in the direction of the meridian.

The above sequence of computations can be altered to one in which the plane rectangular coordinates are computed after step 1, and then the coordinates of the last point of the traverse are compared with the fixed coordinates of this point. This comparison establishes the closure error that is used to adjust the computed coordinates. Both of these sequences are presented in the sections that follow.

When the computations have been performed, the position of each traverse station is known with respect to any other traverse station. Furthermore each station is related in position to any other point that is defined on the same coordinate system, even though the point is not included in the traverse.

Surveying computations are performed by handheld or desk calculators, or by electronic computers for which a variety of programs are readily available. The built-in microprocessor of a total station instrument and some field data collectors are capable of solving many surveying computations (see Section 8-20).

8-11 Latitudes and Departures

The latitude of a line is the distance the line extends in a north or south direction. A line running in a northerly direction has a plus latitude; one running in a southerly direction has a minus latitude.

The departure of a line is the distance the line extends in an east or west direction. A departure to the east is considered plus; a departure to the west is minus.

In Fig. 8-11 the bearing of the line MP is N 23° 34′ 20″ E and its length is 791.28 ft. The bearing of PR is S 51° 05′ 10″ E and its length is 604.30 ft. The line MP has a latitude of +725.25 ft and a departure of +316.44 ft. The line PR has a latitude of −379.60 ft and a departure of +470.20 ft. From Fig. 8-11 it is seen that the latitude of each line is the length of the line times the cosine of the bearing angle, and that the departure of each line is the length of the line times the sine of the bearing angle. If D represents the length of the line and B is the bearing angle, then

$$\text{latitude} = D \cos B \tag{8-1}$$

$$\text{departure} = D \sin B \tag{8-2}$$

FIGURE 8-11 Latitudes and departures.

If the direction of the line is given in terms of its azimuth from north and A represents that azimuth, then

$$\text{latitude} = D \cos A \tag{8-3}$$

$$\text{departure} = D \sin A \tag{8-4}$$

When bearings are used, the sine and cosine of the bearing angle are always considered positive, and the algebraic signs of the latitude and departure are obtained from the quadrant. The signs of the sine and cosine of an azimuth in each of the four quadrants (northeast, southeast, southwest, and northwest) are shown in Fig. 8-12, in which Roman numerals are used for the quadrants. These signs agree, respectively, with the algebraic signs of the functions of angles lying in the four quadrants defined in trigonometry. Thus, the correct algebraic signs are automatically generated in the calculator or computer.

The calculation of latitudes and departures for the traverse of Fig. 8-13 is shown in Table 8-8 as it would be set up for using a desk or handheld calculating machine. The bearings given in the tabulation are those computed after an adjustment was made for the angular closure of the traverse. Several variations of the form of computation will be found in practice. The form can be extended to include the calculation of coordinates. If the computer generates the sine and cosine function internally, the trigonometric functions shown in Table 8-8 are not necessary.

Many handheld calculators such as the Hewlett Packard contain internal programs to convert from polar to rectangular coordinates. In computing latitudes and departures, the azimuth clockwise from north and the length of the line (the polar coordinates) are first entered. The polar-to-rectangular coordinate program is then executed. The first display gives the Y coordinate which is interpreted as the latitude of the line with its correct algebraic sign. The second display gives the departure of the line with its correct algebraic sign.

8-12 Closure in Latitudes and Departures

In a traverse that closes on the point of origin, the algebraic sum of the latitudes should be zero and the algebraic sum of the departures should be zero. In the traverse of Fig. 8-13, the sum of the plus latitudes is 948.92 ft and the sum of the minus latitudes is 948.48 ft. The closure in latitude is therefore +0.44 ft. The sum of the plus departures is 972.68 ft and the sum of the minus departures is 973.05 ft. Thus the closure in departure is −0.37 ft. These closures result from errors in

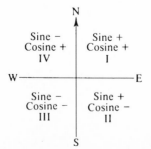

FIGURE 8-12 Algebraic signs of azimuth-angle functions.

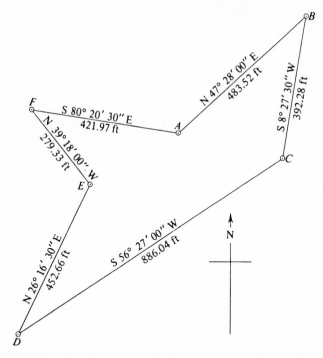

FIGURE 8-13 Traverse showing lengths and bearings.

TABLE 8-8 Calculation of Latitudes and Departures by Handheld Calculator or Computer

Station	Bearing	Length	Cosine	Sine	Latitude +	Latitude −	Departure +	Departure −
A								
	N 47° 28′ 00″ E	483.52	0.676019	0.736884	326.87		356.30	
B								
	S 8° 27′ 30″ W	392.28	0.989123	0.147090		388.01		57.70
C								
	S 56° 27′ 00″ W	886.04	0.552665	0.833404		489.68		738.43
D								
	N 26° 16′ 30″ E	452.66	0.896680	0.442680	405.89		200.39	
E								
	N 39° 18′ 00″ W	279.33	0.773840	0.633381	216.16			176.92
F								
	S 80° 20′ 30″ E	421.97	0.167773	0.985826		70.79	415.99	
A								
		2915.80			948.92	948.48	972.68	973.05
					−948.48		−973.05	
					+0.44		−0.37	

measuring angles and distances when executing the traverse. Even though an adjustment has been made to eliminate the angular closure, each angle has not necessarily received the correct amount of adjustment.

In a traverse that originates at one known position and closes on another known position, the algebraic sum of the latitudes and the algebraic sum of the departures must equal, respectively, the latitude and departure of the line joining the origin and the closing point. The latitude and departure of this line are usually given as a difference in Y coordinates and a difference in X coordinates, respectively, between the point of beginning and the closing point. An example of a traverse computation between two known positions will be given in Section 8-19 following a discussion of coordinates.

If the error in latitudes or the error in departures is large, a mistake is indicated. The first place to search for the mistake is in the computations. Sources of mistakes are, of course, very numerous. If no mistakes can be found in the computations, then a large discrepancy in latitudes or departures, or both, indicates a mistake in the field work of measuring either an angle or a distance. If the length of one line contains a mistake, the line itself may be selected from the remainder of the traverse by analyzing the traverse closure.

8-13   Traverse Closure

The closure of a traverse is the line that will exactly close the traverse. In Fig. 8-14 the line $A'A$ is the closure of a traverse that theoretically closes on itself. The angle B that this line makes with the meridian is determined by the relation

$$\tan B = \frac{\text{closure in departure}}{\text{closure in latitude}} \tag{8-5}$$

The length D of the line is given by the equation

$$D = \sqrt{(\text{closure in latitude})^2 + (\text{closure in departure})^2} \tag{8-6}$$

In the traverse computed in Section 8-11 and shown in Fig. 8-13, the bearing angle of the traverse closure is $\tan^{-1}(0.37/0.44) = 40°$.

Since the algebraic sum of the latitudes is positive, the latitude of the closure must be negative; and since the algebraic sum of the departures is negative, the departure of the closure must be positive. So the bearing of the closure must be S 40° E. The length of the closure is $\sqrt{0.37^2 + 0.44^2} = 0.57$ ft.

A *closure precision* for a given traverse may be obtained by dividing the length of the traverse closure by the total length of the traverse. It is to be expected that the length of the traverse closure will be greater in long surveys than in short ones if the same precautions are taken in both cases. Therefore the lengths of the traverse closures of two surveys of different lengths do not indicate the relative degrees of accuracy of the surveys. The closure precisions, however, do indicate the relative degrees of precision of the two surveys. The length of the closure of the traverse computed in Section 8-11 is 0.57 ft, and the total length of the traverse is 2915.80 ft. The closure precision is therefore 0.57/2915.80 or about 1/5120. If a traverse that was twice as long had the same length of closure, its clo-

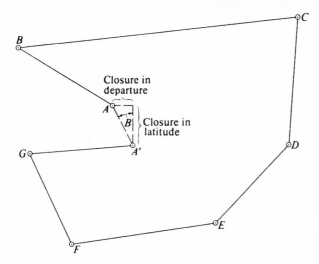

FIGURE 8-14 Traverse closure.

sure precision would be about 1/10,240. Traverse work is often rated as to order of accuracy on the basis of the closure precision.

It should be noted that if the traverse begins and closes on the same point, the traverse closure is the result of random errors rather than of systematic errors. Thus if a traverse of this kind is run with a tape of incorrect length or with an EDM containing a frequency error, the actual error in the length of each line in the traverse can be considerable, even though the traverse closure and the closure precision are very small. Also, if the beginning azimuth or bearing is wrong and if the azimuths or bearings of the remaining sides have been determined by using the adjusted angles in the figure, then the azimuths or bearings of all the sides are wrong by approximately the same amount. If a traverse begins on one known position and closes on another known position, then the amounts by which it fails to close on the basis of the computations reflects both systematic and random errors in the field work. The amount by which it reflects the systematic errors depends on the shape of the traverse. If the traverse is run in a direct line between the two known points, then all the systematic errors are represented. If the two points lie on an east-west line, for example, and if the traverse is run north for a considerable distance, then east, and then south for a considerable distance to the point of closure, the systematic errors accumulated in traveling north will be canceled by those accumulated in traveling south, insofar as they affect the traverse closure.

8-14   Balancing a Traverse

Before the results of a traverse are usable for determining areas or coordinates, for publishing the data, or for computing lines to be located from the traverse stations, the traverse must be mathematically consistent; that is, the closures in latitudes and departures must be adjusted out. Applying corrections to the individual latitudes and departures so that they will sum up to a given condition is called *balancing a traverse.*

At this point it will be well to examine Table 4-2 to appreciate the relationship between the accuracy in angular measurements and the accuracy in linear measurements. For example, according to Table 4-2, an accuracy of 1 part in 5000 in linear measurements requires that the error in angular measurements must not be greater than about $0'\ 41''$. This accuracy is obtained by using a 1-minute transit if the angles are turned once direct and once reversed. On the other hand, if angles are measured within $1'\ 00''$, the distances should be measured to an accuracy of one part in 3440, which is about $1\frac{1}{2}$ ft/mile. In each of these examples, the two types of measurements will be consistent with one another.

Only three basic conditions can exist: (1) The angular accuracy is higher than the linear accuracy; (2) the angular accuracy is the same as the linear accuracy; (3) the angular accuracy is lower than the linear accuracy. An extreme example of the first condition is a traverse, executed in rugged terrain, in which angles are measured directly to $1''$ and distances are determined by holding the tape horizontally throughout the measurements. The angular accuracy is expected to be high, whereas the linear accuracy is expected to be low because it is necessary to break tape. The second condition is exemplified by a traverse, executed in fairly level terrain, in which angles are measured with a 30-second or 1-minute transit and the tape is handled by experienced personnel; or a similar traverse in which angles are measured with a 1-second instrument and the tape is supported on posts or taping bucks and is handled by experienced personnel; or else the distances are measured with an EDM. The third condition is represented by a traverse in which directions are obtained by compass bearings and distances are measured with an EDM. There are, of course, various degrees of these three basic conditions.

To balance a traverse properly, the conditions should be known because the conditions themselves govern the selection of the method employed. In general, traverses are executed so as to satisfy the second condition as a matter of policy, economy, and common sense. The method of balancing based on this condition is known as the *compass rule*. In using the compass rule to balance or adjust a traverse, the errors, and consequently the corrections, are presumed to be in direct proportion to distance. Note that this is the basis for the adjustment of a level circuit.

8-15   Balancing by the Compass Rule

According to the compass rule, the correction to the latitude of a side is to the length of that side as the closure in latitude of the traverse is to the total length of the traverse; and the correction to the departure of a side is to the length of that side as the closure in departure of the traverse is to the total length of the traverse. For example, in the traverse computed in Section 8-11 the compass rule states that the correction c_L to the latitude of the line CD may be found from the proportion $c_L/886.04 = 0.44/2915.80$; and the proportion for finding the correction c_D to the departure of the line CD is $c_D/886.04 = 0.37/2915.80$. Both c_L and c_D have algebraic signs opposite to those of the closure in latitude of the traverse and the closure in departure of the traverse. Thus for the line CD, $c_L = -0.13$ ft and $c_D = +0.11$ ft. The corrections to the latitudes and departures for each line in the traverse are shown in Table 8-9.

TABLE 8-9 Corrections to Latitudes and Departures

Station	Length	Latitudes +	Latitudes −	Departures +	Departures −	c_L	c_D	Balanced Latitudes +	Balanced Latitudes −	Balanced Departures +	Balanced Departures −
A											
B	483.52	326.87		356.30		−0.07	+0.06	326.80		356.36	
C	392.28		388.01		57.70	−0.06	+0.05		388.07		57.65
D	886.04		489.68		738.43	−0.13	+0.11		489.81		738.32
E	452.66	405.89		200.39		−0.07	+0.06	405.82		200.45	
F	279.33	216.16			176.92	−0.04	+0.04	216.12			176.88
A	421.97		70.79	415.99		$\overset{7}{−0.06}$	+0.05		70.86	416.04	
	2915.80	948.92	948.48	972.68	973.05	$\begin{array}{c}−0.43\\−0.44\end{array}$	+0.37	948.74	948.74	972.85	972.85
		−948.48		−973.05							
		+0.44		−0.37							

After the corrections have been computed, they should be added to check whether, in fact, their sums equal the closures in latitude and departure. In the example, the sum of the latitude corrections as first computed was less than the closure in latitude by 0.01 ft. This is a rounding-off error and is corrected by adding the 0.01 ft to that correction which lies nearest to the next higher 0.01 ft. Thus the correction to the latitude of the line *FA* has been changed from −0.06 to −0.07.

After the balanced latitudes and departures have been computed, they should be added to provide a check on the computations.

8-16   Least-Squares Adjustment

The least-squares method of adjustment of a traverse applies corrections to the latitudes and departures so that the weighted sum of the squares of the adjustments to the measured angles and the measured lengths is a minimum according to the principles developed in Chapter 5. If the assumption can be made that the measured lengths and the measured angles are equally reliable, that is, their weights are the same, then an adjustment of the traverse by the method of least squares will give essentially the same results as the compass rule. This is according to the second condition discussed in Section 8-14. If the weights are not equal, however, then the method of least squares can accommodate to this fact and give theoretically better results.

The vast majority of traverses executed in the practice of surveying can be adjusted by the compass rule. When traverses of great extent and very high accuracy are run, and when the lines in the traverse join one another at junction points, the overall adjustment should be made by the least-squares method. A method of adjusting a traverse net of crossed lines and junction points by least squares is given in Appendix A.

The traverse described in Section 8-15 was adjusted using the least-squares technique described in Appendix A, and the results are shown in Table 8-10. The weights for the angles were assumed to be $1/\sigma_a^2$, where $\sigma_a = \pm 30''$, and the weights for the distances were $1/\sigma_d^2$, where $\sigma_d = \pm .01/100$ or one hundredth of a foot per 100 ft. Therefore the weighting between angles and distances is slightly unequal. It is important to note that the balanced angles were used as observations for the least-squares adjustment in an attempt to compare compass rule procedures with least-squares procedures. This is not a proper procedure—in a least-squares adjustment, the observed angles should be used as the starting values. Nonetheless, the results are interesting, showing that the largest difference between compass rule and least squares is 0.12 ft, given a total closure of ± 0.57 ft. As time goes on, more and more handheld calculators will be capable of performing these adjustments.

The development of the rigorous methods of least squares for the adjustment of traverses and traverse nets is outside the scope of this book. The interested reader is referred to the readings cited at the end of this chapter.

TABLE 8-10   Least-Squares Adjustment of Traverse

Station	Adjusted Bearing	Adjusted Length	Least-Square Latitude +	Least-Square Latitude −	Least-Square Departure +	Least-Square Departure −
A						
	N 47° 28′ 40″ E	483.52	326.80		356.37	
B						
	S 8° 27′ 10″ W	392.28		388.02		57.65
C						
	S 56° 25′ 50″ W	886.02		489.93		738.24
D						
	N 26° 17′ 00″ E	452.65	405.85		200.44	
E						
	N 39° 18′ 00″ W	279.32	216.14			176.92
F						
	S 80° 20′ 10″ E	421.98		70.84	416.00	
A			948.79	948.79	972.81	972.81

8-17   Remarks on Adjustments

A traverse is balanced to make the figure formed by the traverse geometrically consistent. If the traverse is run to establish control points for other traverses of equal or lower order, the adjusted positions of the points furnish checks on the work to follow. If the traverse is run to determine areas, the consistency in the figure affords a check on the calculations. In many surveys, however, no balancing will be required, particularly when the latitudes and departures are to be used only in plotting the positions of the stations on a map and when the closure is too small to be scaled on the map.

8-18   Rectangular Coordinates

The rectangular coordinates of a point are the distances measured to the point from a pair of mutually perpendicular axes. As in analytic geometry the distance from the X axis is the Y coordinate of the point and the distance from the Y axis is the X coordinate of the point. In plane surveying conducted in the United States, the north part of the meridian on which the survey is based is the positive Y axis, and a line perpendicular to this meridian through an assumed origin is the X axis. The X axis is positive in an easterly direction. Also, the Y and X coordinates are expressed in feet or meters. In Fig. 8-15 the Y and X coordinates of M are, respectively, + 200.00 ft and + 200.00 ft; those of P are + 925.25 ft and + 516.44 ft; and those of R are + 545.65 ft and + 986.64 ft.

In general the origin of coordinates is situated far enough south and west of the area to make the coordinates of all points in the survey positive quantities. This practice reduces the chances of making mistakes in computations involving the coordinates. Furthermore there is no need to assign an algebraic sign to the coordinates of a point, since it is understood that all coordinates are positive. In subsequent examples in this chapter the origin is selected so as to render all the coordinates positive.

Before the coordinates of the stations of a traverse can be computed, the coordinates of at least one station must be known or else they must be assumed. In Fig. 8-15 the coordinates of station M were assumed to be 200.00 and 200.00. The Y coordinate of station P is determined by adding the latitude of the line MP to the Y coordinate of station M; the X coordinate of P is determined by adding the departure of MP to the X coordinate of M. If the subscript 1 is used for the first station and the subscript 2 is used for the second station of the line 1-2, then

$$Y_2 = Y_1 + \text{latitude of 1-2} \tag{8-7}$$

$$X_2 = X_1 + \text{departure of 1-2} \tag{8-8}$$

These equations can be applied to obtain Y and X coordinates all the way around a traverse. If a traverse originates and closes on the same point, then the final Y and X coordinates of the point of beginning computed by using the latitudes and departures of the traverse sides should be the same as the beginning Y and X coordinates. This will be true provided the traverse has been balanced. An addition to the table of computations for the traverse of Section 8-11 is given in Table 8-11 to show the calculation of coordinates of the points in the traverse. The columns headed Balanced Latitudes and Balanced Departures from Table 8-9 are repeated for continuity. The coordinates of station A are known to be $Y = 4166.20$ and $X = 6154.22$.

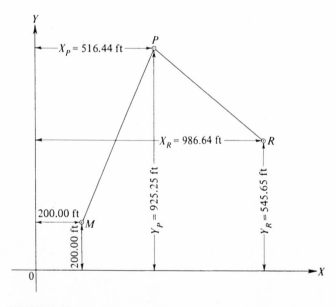

FIGURE 8-15 Rectangular coordinates.

TABLE 8-11 Computation of Coordinates with Calculator

Station	Balanced Latitudes +	Balanced Latitudes −	Balanced Departures +	Balanced Departures −	Y Coordinates	X Coordinates
A					4166.20	6154.22
	326.80		356.36			
B					4493.00	6510.58
		388.07		57.65		
C					4104.93	6452.93
		489.81		738.32		
D					3615.12	5714.61
	405.82		200.45			
E					4020.94	5915.06
	216.12			176.88		
F					4237.06	5738.18
		70.86	416.04			
A					4166.20	6154.22

A check on the arithmetic is provided by computing the coordinates of station A from those of station F to see whether or not they are the same as the given coordinates of A.

8-19 Adjustment of Traverse by Coordinate Adjustment

The coordinates of traverse stations may be computed directly from the measured lengths of the sides of the traverse, the preliminary azimuths or bearings of the sides, and the coordinates of the point of beginning. This work is done conveniently with a desk or handheld calculator by first putting the Y coordinate of the point of beginning into the calculator, adding the product of the distance and the cosine of the bearing angle of the first side, and using the result as the Y coordinate of the second point. This result is simply the sum of the latitude of the first line and the Y coordinate of the point of beginning, as called for by Eq. (8-7). The sum is combined with the product $D \cos B$ of the second line, and the procedure is repeated for successive lines. Due regard must be paid to algebraic signs. Each intermediate result is the Y coordinate of a point in the traverse. In a closed traverse the computed Y coordinate of the point on which the traverse closes will differ from its known Y coordinate by the closure in latitude.

Next, the X coordinate of the point of beginning is put into the calculator, and to this is added the product $D \sin B$ for the first line. The result is the X coordinate of the second point, as the operation is simply the application of Eq. (8-8), or the addition of the departure of the first line to the X coordinate of the point of beginning. Similarly, the product $D \sin B$ for each successive line is used, due regard being paid to algebraic signs, and each intermediate result is taken as the X coordinate of a point in the traverse. The computed X coordinate of the point on which the traverse closes will differ from the known X coordinate by the closure in departure.

If the directions of the lines are expressed as azimuths clockwise from north, the polar-to-rectangular coordinate conversion of a handheld calculator is convenient for computing successive preliminary coordinates.

After the preliminary coordinates have been computed, the correction to be applied to each Y coordinate is found by multiplying the closure in latitude by the ratio of the distance of the point from the point of beginning to the total length of the traverse. An adjustment is then made to each X coordinate in the same manner; that is, the correction to a Y coordinate is to the discrepancy in Y coordinates (closure in latitude) as the distance from the point of beginning is to the total length of the traverse. Similarly, the correction to an X coordinate is to the discrepancy in X coordinates (closure in departure) as the distance from the point of beginning is to the total length of the traverse. Since corrections are proportionate to length, this is to be recognized as a compass-rule adjustment. If a discrepancy is plus, the corrections to all coordinates are minus; if a discrepancy is minus, the corrections to all coordinates are plus.

Tables 8-12 and 8-13 show the computation and the adjustment of coordinates of traverse stations. The traverse originates on station Richmond, whose coordinates are $Y = 28,221.34$ and $X = 20,370.66$, and closes on station Kenney, whose coordinates are $Y = 18,972.77$ and $X = 17,146.82$. The azimuths are prelimi-

TABLE 8-12   Computation of Coordinates with Handheld Calculator or Computer

Station	Distance	Total Distance	Azimuth	Cosine	Sine	Y	X
Richmond						28,221.34	20,370.66
	2872.45		110° 10′ 20″	−0.3448432	+0.9386603		
A		2,872				27,230.80	23,066.91
	1977.14		184° 22′ 15″	−0.9970916	−0.0762115		
B		4,850				25,259.41	22,916.23
	1440.40		177° 49′ 54″	−0.9992841	+0.0378355		
C		6,290				23,820.04	22,970.73
	1635.97		167° 02′ 11″	−0.9745127	+0.2243322		
D		7,926				22,225.77	23,337.73
	3230.84		212° 18′ 28″	−0.8451893	−0.5344671		
E		11,157				19,495.10	21,610.95
	1491.48		284° 50′ 23″	+0.2561160	−0.9666461		
F		12,648				19,877.09	20,169.22
	2231.98		270° 42′ 25″	+0.0123382	−0.9999239		
G		14,880				19,904.63	17,937.41
	1220.96		220° 20′ 12″	−0.7622542	−0.6472777		
Kenney		16,101				18,973.95	17,147.11
					fixed	18,972.77	17,146.82
					closure	+1.18	+0.29

$$\text{closure precision} = \frac{\sqrt{1.18^2 + 0.29^2}}{16,101} = \frac{1}{13,200}$$

TABLE 8-13   Adjustment of Coordinates

Station	Total Distance	Y Correction	X Correction	Adjusted Y	Adjusted X
Richmond				28,221.34	20,370.66
A	2,872	−0.21	−0.05	27,230.59	23,066.86
B	4,850	−0.35	−0.09	25,259.06	22,916.14
C	6,290	−0.46	−0.11	23,819.58	22,970.62
D	7,926	−0.58	−0.14	22,225.19	23,337.59
E	11,157	−0.82	−0.20	19,494.28	21,610.75
F	12,648	−0.93	−0.23	19,876.16	20,168.99
G	14,880	−1.09	−0.27	19,903.54	17,937.14
Kenney	16,101	−1.18	−0.29	18,972.77	17,146.82

nary azimuths obtained after an adjustment of the measured angles in the traverse. The column headed Total Distance is the distance from the point of beginning to the station in question. It is needed only to the nearest foot, because it is used to compute the corrections that are, or should be, relatively small.

The computed Y coordinate of station Kenney is too large by 1.18 ft. This is comparable to a closure in latitude of +1.18 ft. The computed X coordinate of station Kenney is too large by 0.29 ft. Thus it is apparent that the Y coordinate of each point must be reduced by an amount proportionate to the distance from the beginning point to the point under consideration, and that the X coordinate of each point must be reduced in the same proportion.

The Y coordinate of station A is reduced by an amount equal to $(1.18)(2872)/16,101 = 0.21$ ft; the Y coordinate of B is reduced by $(1.18)(4850)/16,101 = 0.35$ ft; and so on. The X coordinate of A is reduced by $(0.29)(2872)/16,101 = 0.05$ ft; the X coordinate of B is reduced by $(0.29)(4850)/16,101 = 0.09$ ft; and so on.

The corrections to all the coordinates, together with the adjusted coordinates, are shown in Table 8-13. Actually these results can be combined with the previous computations of Table 8-12 in one tabulation, which would show the unadjusted and adjusted coordinates. Each adjusted coordinate may be entered directly below the corresponding unadjusted coordinate or alongside it in a separate column; or the last few digits of the unadjusted coordinate may be struck out, and the adjusted figures entered above them. The two separate tabulations are shown here for the sake of clarity.

The traverse computations and adjustments discussed in Sections 8-10 through 8-19 can be performed quite readily using certain handheld calculators as well as personal computers by using various coordinate geometry (COGO) programs available to the surveyor. These same programs can be used also in solving problems discussed in Section 8-21.

8-20   Traversing Using Total Station Instruments

As discussed in Section 4-24, the microprocessor contained in the total station instrument computes the horizontal distance from the instrument station to the reflector target based on the slope distance measured with the EDM and the slope angle measured electronically by means of the vertical circular encoder. The microprocessor also computes the latitude (N of Fig. 4-36) and the departure (E of Fig. 4-36) based on the horizontal distance and the azimuth of the line. These calculations are based on Eqs. (8-3) and (8-4).

Assume that the rectangular coordinates of the initial traverse station as well as a starting azimuth are known. After the instrument has been set up on the initial station, the starting coordinates are entered into the computer. The starting azimuth is then set on the horizontal circle display. A backsight is taken on the azimuth mark using the "lower motion" of the instrument. A foresight is then taken on the next traverse station using the "upper motion" of the instrument. The horizontal circle display will now show the forward azimuth stored in memory. The EDM measures the slope distance, which is reduced to the horizontal distance. The microprocessor computes the latitude and departure of the forward line and adds these components to the Y and X coordinates of the initial station to give the Y and X coordinates of the forward station based on Eqs. (8-7) and (8-8). These new coordinates are stored in memory. On the next setup, the back azimuth is computed (or 180° is added to or subtracted from the stored azimuth) in the microprocessor and displayed. The updated coordinates are recalled from memory, and the entire procedure is repeated at the next station and so on throughout the traverse.

When a known position is reached, the fixed Y and X coordinates are entered into the computer and compared to the coordinates obtained by traversing from the initial station. The error of closure is computed, and an adjustment is made to the coordinates of the intermediate traverse stations. These values are stored for later printing or plotting through the data collector.

The above procedure is seen to be identical to that discussed in the previous sections of this chapter. The method of coding points, entering data, executing calculations, and storing the values varies from one type of instrument to the next. The necessary documentation is given in the user's manual that comes with the instrument. Thus no particular procedure can be fully described in this textbook.

8-21   Use of Rectangular Coordinates

The rectangular coordinates of a point uniquely define its position with respect to a known horizontal datum. Therefore the use of coordinates is the most convenient method of publishing the horizontal positions of points that have been located in the field. The coordinates of a point give its position with respect to any other point located on the same coordinate system or horizontal datum. Coordinates are used to plot the positions of survey points on a map sheet as explained in Section 8-33.

If the coordinates of two stations 1 and 2 are known, the latitude and departure of the line from station 1 to station 2 are obtained by transposing terms of Eqs. (8-7) and (8-8) to give the following relations:

$$\text{latitude of 1-2} = Y_2 - Y_1 \tag{8-9}$$

$$\text{departure of 1-2} = X_2 - X_1 \tag{8-10}$$

Referring to Fig. 8-16, the tangent of the bearing angle B of the line joining the two stations is given by the equation

$$\tan B = \frac{X_2 - X_1}{Y_2 - Y_1} \tag{8-11}$$

The distance D between the two points is given by the following equation:

$$D = \sqrt{(Y_2 - Y_1)^2 + (X_2 - X_1)^2} \tag{8-12}$$

If the numerator of Eq. (8-11) is positive, the line from 1 to 2 bears east, and if the denominator is positive, the line bears north.

The computation made to determine the bearing and length of a line from the coordinates of the two points defining the line is called *inversing* the line; this is one of the most frequent computations made in surveying.

The azimuth of A_N of a line 1-2 measured from the north meridian is given by the equation

$$\tan A_N = \frac{X_2 - X_1}{Y_2 - Y_1} \tag{8-13}$$

The azimuth A_S of the same line measured from the south meridian is given by the equation

$$\tan A_S = \frac{X_1 - X_2}{Y_1 - Y_2} \tag{8-14}$$

In Eqs. (8-9), (8-10), (8-11), (8-13), and (8-14), due consideration must be given to the algebraic signs of the coordinates and the differences in coordinates.

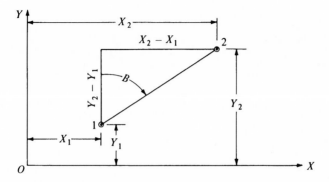

FIGURE 8-16 Calculation of bearing angle.

If a handheld calculator such as the Hewlett Packard contains the program for conversion from rectangular to polar coordinates, inversing a line is easily performed. The X coordinate of the end of the line is entered, and the X coordinate of the beginning of the line is subtracted. The Y coordinate of the end of the line is entered, and the Y coordinate of the beginning of the line is subtracted. The rectangular-to-polar coordinate program is then executed. The first display gives the length of the line according to Eq. (8-12). The second display gives the azimuth of the line according to Eq. (8-13). If the azimuth display is between 0 and 180°, the display is the azimuth measured clockwise from north. If the azimuth display is between 0 and $-180°$, the azimuth measured clockwise from north is the displayed value plus 360°. Since different calculators use different methods for the rectangular-to-polar coordinate conversion, the student is cautioned to inverse a line of known end coordinates and known length and azimuth to establish the proper sequence of entry and the displayed values.

The equations of straight lines and second-degree curves given by the methods of analytic geometry are equally applicable to problems in surveying when coordinates are used. The slope of a straight line in analytic geometry is given as $m = \tan \alpha$, in which α is the angle measured from the positive X axis. As seen in Fig. 8-17, the slope of the line OP is $m = \tan \alpha$. The angle α is the complement of the bearing angle B of the line or of its azimuth from north. Therefore, in surveying, the slope of the line is given as $\cot B$ or $\cot A$, the choice depending on whether bearings or azimuths are used. When bearings are used, $\cot B$ is positive in the NE and SW quadrants, and is negative in the NW and SE quadrants. The following equations from analytic geometry, applicable to surveying, are given here for convenience.

1. Equation of a straight line: When the coordinates (X_1, Y_1) and (X_2, Y_2) of two points are given,

$$\frac{Y - Y_1}{Y_2 - Y_1} = \frac{X - X_1}{X_2 - X_1} \tag{8-15}$$

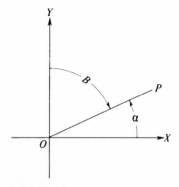

FIGURE 8-17 Slope of a line.

When the coordinates (X_1, Y_1) of one point and the bearing angle B of the line are given,

$$Y - Y_1 = \cot B(X - X_1) \tag{8-16}$$

When the line passes through the origin and its bearing angle is given,

$$Y = X \cot B \tag{8-17}$$

Equations (8-15) and (8-16) can be reduced to the general form

$$aX + bY + c = 0 \tag{8-18}$$

Then, referring to Fig. 8-18, the azimuth of the line is given by

$$\tan A = -\frac{b}{a} \tag{8-19}$$

The intercepts of the line on the X and Y axes are

$$X \text{ intercept} = -\frac{c}{a} \tag{8-20}$$

$$Y \text{ intercept} = -\frac{c}{b} \tag{8-21}$$

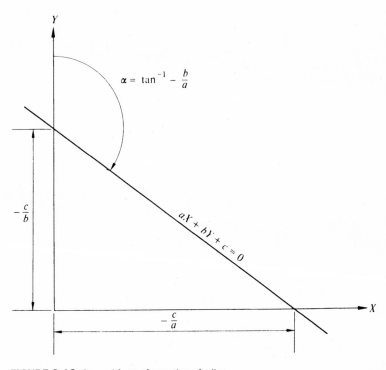

FIGURE 8-18 General form of equation of a line.

One of the most convenient forms of the equation of a line in solving for the distance from the line to a point off the line is the normal form given in terms of the coefficients a, b, and c of Eq. (8-18). If these are known, the normal form is then given by

$$\frac{a}{\pm\sqrt{a^2+b^2}}X + \frac{b}{\pm\sqrt{a^2+b^2}}Y + \frac{c}{\pm\sqrt{a^2+b^2}} = 0 \tag{8-22}$$

The algebraic sign of the radical in the denominator is chosen to be the same as that of b. To determine the distance d shown in Fig. 8-19 between the line and a point P off the line, the coordinates of P are substituted into Eq. (8-22) the left side of which becomes the desired distance.

2. Equation of a circle of radius R: When its center is at the origin,

$$X^2 + Y^2 = R^2 \tag{8-23}$$

When coordinates of its center are $X_c = H$ and $Y_c = K$,

$$(X-H)^2 + (Y-K)^2 = R^2 \tag{8-24}$$

3. Area of polygon:

$$2A = X_1(Y_2 - Y_n) + X_2(Y_3 - Y_1) + \cdots$$
$$+ X_{n-1}(Y_n - Y_{n-2}) + X_n(Y_1 - Y_{n-1}) \tag{8-25}$$

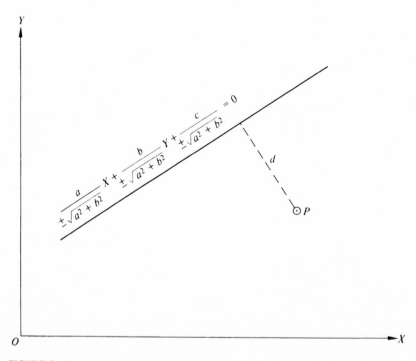

FIGURE 8-19 Normal form of equation of a line.

or

$$2A = Y_1(X_2 - X_n) + Y_2(X_3 - X_1) + \cdots$$
$$+ Y_{n-1}(X_n - X_{n-2)} + Y_n(X_1 - X_{n-1)} \tag{8-26}$$

4. Dividing a line into two parts in the proportion R_1/R_2:

$$X_0 = \frac{X_1 R_2 + X_2 R_1}{R_1 + R_2} \qquad Y_0 = \frac{Y_1 R_2 + Y_2 R_1}{R_1 + R_2} \tag{8-27}$$

in which the coordinates of the points at the ends of the line are (X_1, Y_1) and (X_2, Y_2) and the coordinates of the division points are (X_0, Y_0).

The foregoing equations are but a few of many that can be taken from analytic geometry and applied to surveying computations. Engineers and surveyors in the past have been prone to apply the more cumbersome methods of trigonometry to the solution of surveying problems, but when a programmable handheld calculator or computer is available, the methods of analytic geometry are much more efficient than are trigonometric methods. These equations are the basis for many of the solutions used in COGO programs. The following examples are given to show how easily some of the methods of analytic geometry are applied.

EXAMPLE 8-1   For the traverse computed in Section 8-11, determine the Y and X coordinates of the intersection of the line joining points E and C with the line joining points D and A.

Solution: The equation of each of these lines may be written from the coordinates computed in Section 8-18 by using Eq. (8-15). The equation of the line EC is

$$\frac{Y - 4020.94}{4104.93 - 4020.94} = \frac{X - 5915.06}{6452.93 - 5915.06} \tag{1a}$$

and the equation of the line DA is

$$\frac{Y - 3615.12}{4166.20 - 3615.12} = \frac{X - 5714.61}{6154.22 - 5714.61} \tag{2a}$$

Because the actual coordinates are relatively large numbers that give terms with many significant figures in the calculations, the values of the coordinates can be reduced by subtracting a constant for each Y and X coordinate. In this problem each Y coordinate is reduced by 3000 ft, and each X coordinate is reduced by 5000 ft for computational efficiency. The two line equations then become

$$\frac{Y' - 1020.94}{1104.93 - 1020.94} = \frac{X' - 915.06}{1452.93 - 915.06} \tag{1b}$$

$$\frac{Y' - 615.12}{1166.20 - 615.12} = \frac{X' - 714.61}{1154.22 - 714.61} \tag{2b}$$

Rearranging and collecting terms, gives

$$83.99X' - 537.87Y' = -472,277.11 \tag{1c}$$

$$551.08X' - 439.61Y' = +123,394.38 \tag{2c}$$

These two equations are solved simultaneously to find the coordinates of the intersection. If Eq. (2c) is multiplied by 537.87/439.61, and Eq. (1c) is subtracted from the new equation, the work is as follows:

$$\begin{array}{rl}
674.26X' - 537.87Y' &= +150{,}975.04 \\
83.99X' - 537.87Y' &= -472{,}277.11 \\
\hline
590.27X' \qquad\quad &= +623{,}252.15 \\
X' \qquad\quad &= +1055.876
\end{array}$$

Substituting the value of X' in Eqs. (1c) and (2c) to provide a check gives Y' as 1042.93. The coordinates of the point of intersection therefore are

$$Y = 1042.93 + 3000 = 4042.93$$

$$X = 1055.88 + 5000 = 6055.88 \qquad \blacklozenge$$

◀EXAMPLE 8-2   For the traverse computed in Section 8-11, determine the coordinates of the intersection of a line passing through point B and parallel to line CD with the line AF.

Solution: The equations of these two lines are written and then solved simultaneously to obtain the coordinates of the point of intersection. The bearing of the line passing through B is the same as that of CD. Therefore $\tan B_{CD} = (5714.61 - 6452.93)/(3615.12 - 4104.93)$ or $\cot B = (3615.12 - 4104.93)/(5714.61 - 6452.93) = +0.663412$, and the equation of the line through B is, by Eq. (8-16),

$$Y - 4493.00 = +0.663412(X - 6510.58) \tag{1a}$$

The equation of the line AF is, by Eq. (8-15),

$$\frac{Y - 4166.20}{4237.06 - 4166.20} = \frac{X - 6154.22}{5738.18 - 6154.22} \tag{2a}$$

To keep the number of significant figures in the calculations as small as possible, each Y coordinate is reduced by 4000 ft and each X coordinate by 5000 ft. The two line equations then become

$$Y' - 493.00 = +0.663412(X' - 1510.58) \tag{1b}$$

$$\frac{Y' - 166.20}{237.06 - 166.20} = \frac{X' - 1154.22}{738.18 - 1154.22} \tag{2b}$$

Rearranging and collecting terms gives

$$0.663412X' - Y' = +509.14 \tag{1c}$$

$$70.86X' + 416.04Y' = +150{,}933.88 \tag{2c}$$

These equations are solved simultaneously to give the coordinates of the intersection. When Eq. (1c) is multiplied by 70.86/0.663412 and the new equation is subtracted from Eq. (2c), the work is as follows:

$$\begin{array}{rl}
70.86X' - 106.81Y' &= +54{,}381.98 \\
70.86X' + 416.04Y' &= +150{,}933.88 \\
\hline
+522.85Y' &= +96{,}551.90 \\
Y' &= +184.665
\end{array}$$

From Eqs. (1c) and (2c) it is found that $X' = 1045.81$ and $X' = 1045.81$. The final coordinates therefore are

$$Y = \ \ 184.66 + 4000 = 4184.66$$
$$X = 1045.81 + 5000 = 6045.81$$

◆

EXAMPLE 8-3   Where does the line CD of Example 8-2 intersect the X and Y axes?

Solution: See Fig. 8-20. The equation of CD is written in the form of Eq. (8-15),

$$\frac{Y - 4104.93}{3615.12 - 4104.93} = \frac{X - 6452.93}{5714.61 - 6452.93}$$

which reduces to the general form given by Eq. (8-18); thus

$$489.81X - 738.32Y - 129{,}957.73 = 0$$

Then by Eqs. (8-20) and (8-21),

$$X \text{ intercept} = -\frac{-129{,}957.73}{489.81} = +265.32 \text{ ft}$$

$$Y \text{ intercept} = -\frac{-129{,}957.73}{-738.32} = -176.02 \text{ ft}$$

◆

EXAMPLE 8-4   Compute the distance PQ from the line CD of Example 8-3 to a point P whose coordinates are $X_P = 6802.52$ and $Y_P = 2654.88$.

Solution: Point P is shown in relation to the line CD in Fig. 8-20. Either of two methods can be used to solve this problem. The first is to substitute the coordinates

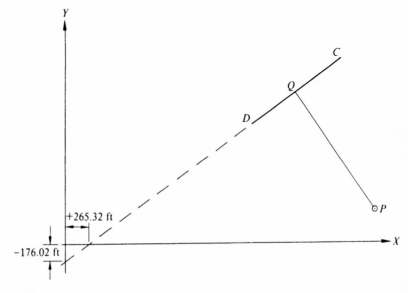

FIGURE 8-20 Distance from a point to a line.

of P into the normal form of the equation of CD given by Eq. (8-22) and solve for the distance. The general form of the equation of CD is already given in Example 8-3. Then, since b is negative,

$$-\sqrt{a^2 + b^2} = -\sqrt{489.81^2 + 738.32^2} = -886.02$$

giving

$$\frac{489.81}{-886.02}X + \frac{-738.32}{-886.02}Y + \frac{-129,957.73}{-886.02} = 0$$

or

$$-0.552820X + 0.833299Y + 146.68 = 0$$

which is the normal form of the equation of CD. The distance from CD to P is then obtained by substituting the coordinates of P into the normal form. Thus

$$PQ = (-0.552820)(6802.52) + (0.833299)(2654.88) + 146.68 = -1401.58$$

The negative sign is not significant because the desired quantity is the absolute distance between the line and the point.

The second method for solving this problem is to keep the line CD in the two-point form given by Eq. (8-15) and then write the equation of a perpendicular line through P in the point-slope form. The slope of this line is 90° different from that of CD. These two equations are solved simultaneously to give a point of intersection Q on line CD. PQ is then solved by Eq. (8-12). ◆

EXAMPLE 8-5 The coordinates of the center of a circular curve defining the center line of a highway are $Y = 2655.10$ and $X = 17,255.35$. The radius of the curve is 750.00 ft. Determine the coordinates of the intersection of this center line with a line passing through a point whose coordinates are $Y = 3844.20$ and $X = 17,000.00$ and having a bearing S 30° E (see Fig. 8-21).

Solution: The equation of the circular curve is, by Eq. (8-24),

$$(X - 17,255.35)^2 + (Y - 2655.10)^2 = 750.00^2 \tag{1a}$$

The equation of the straight line is, by Eq. (8-16),

$$Y - 3844.20 = -\cot 30° \ (X - 17,000.00) \tag{2a}$$

The cotangent is minus because the bearing is in the southeast quadrant. The calculations may be simplified by subtracting 2000 ft from each Y coordinate and 17,000 ft from each X coordinate before the equations are solved, and then adding these amounts to the computed coordinates. Therefore

$$(X' - 255.35)^2 + (Y' - 655.10)^2 = 750.00^2 \tag{1b}$$

$$Y' - 1844.20 = -1.732051 \ (X' - 0) \tag{2b}$$

To solve these simultaneous equations, first the value of Y' obtained from Eq. (2b) is substituted in Eq. (1b). Thus

$$Y' = -1.732051X' + 1844.20 \qquad\qquad (2c)$$

$$(X' - 255.35)^2 + (-1.732051X' + 1844.20 - 655.10)^2 = 750.00^2 \qquad (1c)$$

Expanding, rearranging, and collecting terms gives

$$4X'^2 - 4629.86X' + 916{,}662.43 = 0$$

This is a quadratic equation of the form $ax^2 + bx + c = 0$ whose solution is $x = (-b \pm \sqrt{b^2 - 4ac})/2a$, and its solution will give two values of X', as follows:

$$X' = \frac{4629.86 \pm \sqrt{21{,}435{,}603.6196 - 14{,}666{,}598.8800}}{8}$$

$$X' = 903.95 \text{ or } X' = 253.52$$

Substituting the values X' in Eq. (2c) gives

$$Y' = 278.51 \text{ or } Y' = 1405.09$$

FIGURE 8-21 Intersection of circular curve and straight line.

As a check on the computation, both values of X' and Y' may be substituted in Eq. (1b) to see whether both values satisfy the equation. An inspection of Fig. 8-21 shows that the values obtained by using the smaller value of X' are the coordinates of the desired point. The final coordinates of the point are obtained by adding 2000 ft to Y' and 17,000 ft to X'. The results are $Y = 3405.09$ ft and $X = 17,253.52$ ft.

8-22 Coordinates of Unoccupied Points

To determine the coordinates of points that cannot be included in the traverse because it is not possible to set the instrument up over such points, spur lines or tie lines must be run from the traverse to these points. For example, in Fig. 8-22 the boundaries of a parcel of land are defined by an iron pipe, a nail in a tree, and two fence posts. The only corner that can be occupied is the one marked by the iron pipe. The problem is to determine the lengths and bearings of the property lines.

It will be assumed that the coordinates of the iron pipe at point A and the bearing of the line AB are known. The procedure is then as follows:

FIGURE 8-22 Traverse to determine boundary.

1. Measure the lengths of the traverse sides *AB*, *BC*, *CD*, and *DA*, and also the lengths of the three tie lines *BP*, *CR*, and *DS*.
2. Measure the angles in the traverse *ABCD*, and also the angles between traverse lines and the tie lines to *P*, *R*, and *S*, as shown in Fig. 8-22.
3. Adjust the angles in the traverse *ABCD*, and compute the bearings of lines *BC*, *CD*, and *DA* based on the known bearing of the line *AB*.
4. Balance the traverse *ABCD*.
5. Compute the coordinates of points *B*, *C*, and *D*.
6. Determine the bearings of the lines *BC*, *CD*, and *DA* based on the coordinates computed in step 5.
7. Compute the bearings of the tie lines from the bearings computed in step 5 and the measured angles.
8. Compute the latitudes and departures of the tie lines.
9. Compute the coordinates of the unoccupied stations from the coordinates of traverse stations *B*, *C*, and *D* and the latitudes and departures of the lines *BP*, *CR*, and *DS*, respectively.

When the coordinates of *P*, *R*, and *S* have been computed, the bearings and lengths of the boundary lines may be computed by Eqs. (8-11) and (8-12).

EXAMPLE 8-6   The following deflection angles were measured at stations *A*, *B*, *C*, and *D* of the traverse in Fig. 8-22.

Station	From	To	Deflection Angle	Adjusted Deflection Angle
A	*D*	*B*	89° 23′ 30″ L*	89° 24′ 00″ L
B	*A*	*C*	109° 52′ 00″ L*	109° 52′ 30″ L
B	*A*	*P*	69° 08′ 00″ L	—
C	*B*	*D*	66° 58′ 00″ L*	66° 58′ 30″ L
C	*B*	*R*	128° 02′ 00″ R	—
D	*C*	*A*	93° 44′ 30″ L*	93° 45′ 00″ L
D	*C*	*S*	124° 30′ 00″ L	—
			check	360° 00′ 00″

The lengths as determined in the field are as follows:

Line	Length (ft)
AB	280.20
BC	294.54
BP	40.15
CD	183.60
CR	64.24
DA	286.92
DS	29.80

Determine the lengths and bearings of the boundary lines *AP*, *PR*, *RS*, and *SA*.

Solution: The sum of the four angles in the closed traverse shown by an asterisk (*) is 359° 58′. These angles are adjusted to total 360°, as shown in the last column. The bearing of the line *AB* is known to be N 3° 42′ E. The bearings of the other traverse sides (not including the tie lines) are computed by subtracting each left deflection angle from the forward azimuth of each backsight, as outlined in Section 8-5. When these azimuths are converted to bearings, the results are as follows:

Line	Adjusted Bearing
AB	N 3° 42′ 00″ E
BC	S 73° 49′ 30″ W
CD	S 6° 51′ 00″ W
DA	S 86° 54′ 00″ E

From the adjusted bearings and the measured lengths of the traverse sides, the latitudes and departures are computed and balanced. From the balanced latitudes and departures, the coordinates of *B*, *C*, and *D* are determined, a check being made by computing the coordinates of *A* from point *D*. For the sake of brevity Table 8-14 shows only unbalanced and balanced latitudes and departures of the traverse sides, together with the computed coordinates of the traverse stations.

From the coordinates of the traverse stations, the azimuths of the lines *AB*, *BC*, and *CD* are computed, since these azimuths are needed to determine the azimuths of the tie lines. Then for line *AB* by Eq. (8-13),

$$\tan A_N = \frac{+18.13}{+279.68} \qquad A = 3° 42′ 30″$$

TABLE 8-14   Computation of Coordinates of Traverse Stations

Station	Latitude +	Latitude −	Departure +	Departure −	Balanced Latitude +	Balanced Latitude −	Balanced Departure +	Balanced Departure −	Y Coordinate	X Coordinate
A									1000.00	1000.00
	279.62		18.08		279.68		18.13			
B									1279.68	1018.13
		82.05		282.88		81.98		282.82		
C									1197.70	735.31
		182.29		21.90		182.25		21.86		
D									1015.45	713.45
		15.52	286.50			15.45	286.55			
A									1000.00	1000.00
	279.62	279.86	304.58	304.78	279.68	279.68	304.68	304.68		

For line *BC*,

$$\tan A_N = \frac{-282.82}{-81.98} \qquad A = 253° \; 50' \; 00''$$

For line *CD*,

$$\tan A_N = \frac{-21.86}{-182.25} \qquad A = 186° \; 50' \; 30''$$

The azimuths and bearings of the tie lines are determined as shown in Table 8-15.

From the bearings and lengths of the tie lines, the latitudes and departures and then the coordinates are computed, as shown in Table 8-16.

Since the coordinates of each boundary corner are known, the bearings and lengths of the boundary lines may now be computed by Eqs. (8-11) and (8-12).

For line *AP*,

$$\tan B = \frac{981.62 - 1000.00}{1296.38 - 1000.00} \qquad B = \text{N } 3° \; 33' \; 00'' \text{ W}$$

$$AP = \sqrt{(981.62 - 1000.00)^2 + (1296.38 - 1000.00)^2} = 296.95 \text{ ft}$$

For line *PR*,

$$\tan B = \frac{759.24 - 981.62}{1257.32 - 1296.38} \qquad B = \text{S } 80° \; 02' \; 20'' \text{ W}$$

$$PR = \sqrt{(759.24 - 981.62)^2 + (1257.32 - 1296.38)^2} = 225.78 \text{ ft}$$

For line *RS*,

$$\tan B = \frac{739.84 - 759.24}{1029.28 - 1257.32} \qquad B = \text{S } 4° \; 51' \; 50'' \text{ W}$$

$$RS = \sqrt{(739.84 - 759.24)^2 + (1029.28 - 1257.32)^2} = 228.86 \text{ ft}$$

For line *SA*,

$$\tan B = \frac{1000.00 - 739.84}{1000.00 - 1029.28} \qquad B = \text{S } 83° \; 34' \; 40'' \text{ E}$$

$$SA = \sqrt{(1000.00 - 739.84)^2 + (1000.00 - 1029.28)^2} = 261.80 \text{ ft}$$

The above calculation of lengths and bearings from coordinates can be done quite conveniently on most hand calculators by means of the rectangular-to-polar coordinate conversion function. ◆

It is to be noted that the adjusted azimuth of line *AB* is 30″ greater than the given azimuth (3° 42′ 30″ − 3° 42′ 00″). If the problem is to determine only the lengths and bearings of the boundary lines, then the computed azimuths of *AP*, *PR*, *RS*, and *SA* should all be reduced by 30″. However, if the coordinates of points *P, R,* and *S,* based on a fixed azimuth of *AB*, are required, a transformation must be performed as discussed in Section 8-23.

TABLE 8-15   Computation of Azimuths of Tie Lines

Line	Azimuth	Line	Azimuth	Line	Azimuth
AB	3° 42′ 30″	BC	253° 50′ 00″	CD	186° 50′ 30″
AB	363° 42′ 30″	(+ ∠C)	+128° 02′ 00″	(−∠D)	−124° 30′ 00″
(−∠B)	−69° 08′ 00′	CR	381° 52′ 00″	DS	62° 20′ 30″
BP	294° 34′ 30″	CR	21° 52′ 00″		
BP	N 65° 25′ 30″ W	CR	N 21° 52′ 00″ E	DS	N 62° 20′ 30″ E

TABLE 8-16   Computation of Coordinates of Boundary Corners

Station	Length	Bearing	Latitude +	Latitude −	Departure +	Departure −	Y Coordinate	X Coordinate
B							1279.68	1018.13
	40.15	N 65°25′30″ W	16.70			36.51		
P							1296.38	981.62
C							1197.70	735.31
	64.24	N 21°52′00″ E	59.62		23.93			
R							1257.32	759.24
D							1015.45	713.45
	29.80	N 62°20′30″ E	13.83		26.39			
S							1029.28	739.84

8-23   Coordinate Transformations

It frequently becomes necessary in surveying to transform rectangular coordinates in one coordinate system to their positions in another coordinate system. This can be done if the coordinates of at least two points are known in both systems. In Fig. 8-23(a), point A has coordinates X_A and Y_A in the XY coordinate system. The azimuth AB is designated α, obtained by inversing AB. In Fig. 8-23(b) point A has coordinates X'_A and Y'_A in the $X'Y'$ coordinate system. The azimuth of AB in this second system is α'. If the $X'Y'$ axes are rotated through angle $\theta = (\alpha' - \alpha)$, a new set of coordinates (X'_A) and (Y'_A) relative to the dashed (X'), (Y') coordinate axes are obtained. These coordinates are computed by the rotation equations from analytical geometry. Thus

$$(X'_A) = X'_A \, \cos\,\theta - Y'_A \, \sin\,\theta$$

$$(Y'_A) = X'_A \, \sin\,\theta + Y'_A \, \cos\,\theta \qquad (8\text{-}28)$$

If the dashed axes are now translated in the X direction by $X_A - (X'_A) = c$ and in the Y direction by $Y_A - (Y'_A) = d$, X_A and Y_A are obtained. Thus

$$X_A = X'_A \, \cos\,\theta - Y'_A \, \sin\,\theta + c$$

$$Y_A = X'_A \, \sin\,\theta + Y'_A \, \cos\,\theta + d \qquad (8\text{-}29)$$

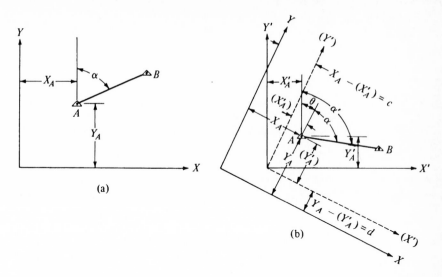

FIGURE 8-23 Coordinate transformation.

The value of θ is obtained by inversing the line in each system. Then Eq. (8-28) is solved to determine (X'_A) and (Y'_A). The c and d terms are then obtained by

$$c = X_A - (X'_A)$$
$$d = Y_A - (Y'_A) \tag{8-30}$$

The coordinates of any other point such as point C in the $X'Y'$ system can be determined by Eq. (8-29).

If the scale of the two coordinate systems are different, then the correct coordinates would be obtained by multiplying the terms on the right side of Eq. (8-28) by some scale s, giving

$$(X'_A) = (s \cos \theta)X'_A - (s \sin \theta)Y'_A$$
$$(Y'_A) = (s \sin \theta)X'_A + (s \cos \theta)Y'_A \tag{8-31}$$

Eq. (8-29) becomes

$$X_A = (s \cos \theta)X'_A - (s \sin \theta)Y'_A + c$$
$$Y_A = (s \sin \theta)X'_A + (s \cos \theta)Y'_A + d \tag{8-32}$$

Let $a = s \cos \theta$ and $b = s \sin \theta$. Then, for any point, Eq. (8-32) becomes

$$X = aX' - bY' + c$$
$$Y = bX' - aY' + d \tag{8-33}$$

Eq. (8-33) is called a similarity transformation or a four-parameter transformation, the four parameters being a, b, c, and d. The parameters are solved by substituting the known coordinates of two points in both systems into Eq. (8-33), as shown by the following example.

EXAMPLE 8-7   The coordinates of A and B are given for two systems. The coordinates of two additional points M and N in the second system are also given.

	First System		Second System	
Point	X	Y	X'	Y'
A	2120.00	1180.50	1145.25	1345.20
B	3570.80	1925.70	2595.35	599.82
M			2095.18	1278.48
N			2502.55	1296.80

Compute the coordinates X_M, Y_M, X_N, Y_N in the first system.

Solution: Substituting the known coordinates of A and B in the first system in the left side of Eq. (8-33) and their coordinates in the second system in the right side gives four equations in the four unknowns a, b, c, and d. These are solved simultaneously.

$$2120.00 = 1145.25a - 1345.20b + c$$
$$1180.50 = 1345.20a + 1145.25b + d$$
$$3570.80 = 2595.35a - 599.82b + c$$
$$1925.70 = 599.82a + 2595.35b + d$$

The solution of these equations gives $a = +0.582440$; $b = +0.813281$; $c = +2546.99$; $d = -534.41$. Substituting the coordinates of M and N in the second system into the right side of Eq. (8-33) gives

$$X_M = 0.582440 \times 2095.18 - 0.813281 \times 1278.48 + 2546.99 = 2727.54$$
$$Y_M = 0.813281 \times 2095.18 + 0.582440 \times 1278.48 - 534.41 = 1914.20$$
$$X_N = 0.582440 \times 2502.55 - 0.813281 \times 1296.80 + 2546.99 = 2949.91$$
$$Y_N = 0.813281 \times 2502.55 + 0.582440 \times 1296.80 - 534.41 = 2256.17 \quad \blacklozenge$$

If the scale is identical between the two systems, then $a = \cos\theta$ and $b = \sin\theta$, and $\sqrt{a^2 + b^2} = 1$. In this example, $\sqrt{a^2 + b^2} = 1.000331$, indicating that the scale of the first system is larger than that of the second system.

If there are more than two points known in both systems, the parameters a, b, c, and d can be solved by the method of least squares, using the method of observation equations given in Sections A-3(a) and (b) of Appendix A.

The rotation formulae given by Eq. (8-28) can be used to compute the correct coordinates of points P, R, and S of Example 8-6, in which the azimuth of AB is to be held fixed as 3° 42′ 00″. The amount of rotation θ is 3° 42′ 30″ − 3° 42′ 00″ = + 30″. The rotation takes place about point A as the origin (see Fig. 8-23b). Thus the coordinates of each of the points P, R, and S must first be reduced to an origin at A by subtracting the coordinates of A as shown in Table 8-17.

TABLE 8-17 Rotation of Coordinates

Point	Coordinates Reduced to Origin at A		Coordinates After Rotation, Origin at A		Final Coordinates	
	Y′	X′	(Y′)	(X′)	Y	X
A	0	0	0	0	1000.00	1000.00
P	296.38	−18.38	296.38	−18.42	1296.38	981.58
R	257.32	−240.76	257.28	−240.80	1257.28	759.20
S	29.28	−260.16	29.24	−260.16	1029.24	739.84

The reduced coordinates are then rotated using Eq. (8-28). Finally the coordinates of A are added to obtain the final coordinates of P, R, and S, as shown in Table 8-17.

8-24 Location of a Line Based on Computations

The problem of locating a desired line in the field by measurement from stations in a control traverse is constantly faced in location surveys, construction surveys, and boundary surveys. In Sections 6-9 and 6-11, two methods were discussed for laying out a straight line in the field by means of a straight random traverse and by a traverse, such as that illustrated in Fig. 6-13. Either of these problems can be solved by running a control random traverse between the two ends of the line. Latitudes and departures can be computed for the lines in the random traverse, and coordinates can be computed for the traverse stations. From the coordinates of the two ends of the desired line, its length and bearing can be computed. Furthermore, the coordinates of any point on the desired line can be computed.

If the control random traverse is on the same system of coordinates as that of the two ends of the line along which points are to be set, the calculated closure error at the far end of the line is used as the basis of a compass rule adjustment of the intermediate points on the random traverse.

If the control random traverse begins at one end of the line and some assumed bearing of the starting line is used to compute the azimuths of the lines in the control traverse, then the two coordinate systems will be different, and a coordinate transformation is made. This transformation determines the coordinates of the random traverse stations in the system of the given line on which points are to be set.

By use of the computed coordinates at selected points on the desired line, the lengths and bearings of tie lines from the control-traverse stations can be computed. The angle to be laid off from a control-traverse line to run a tie line is computed from the known bearing of the tie line and that of the traverse line. In this manner all the necessary angles are computed, and the points are set on the desired line by reoccupying the traverse stations and running the tie lines on the computed bearings and for the computed distances.

◆**EXAMPLE 8-8**   The coordinates of the two ends of line AP of Fig. 8-24 are $X_A = 1452.60$ m, $Y_A = 2086.12$ m; $X_P = 2533.14$ m, $Y_P = 2461.85$ m. Points are to be set on this line at 1, 2,... every 200 m beginning at A. A random traverse $ABCDEP$ is run by deflection angles and distances on an assumed meridian with the following results:

Station	Deflection Angle	Line	Measured Distance (m)
B	21°06′10″ R	AB	271.40
C	26°42′30″ L	BC	295.35
D	36°02′10″ R	CD	239.50
E	49°36′10″ R	DE	307.65
		EP	150.80

No backsight was available at A, and thus no deflection angle was measured. The approximate direction of AB was determined by observing a compass bearing at B and applying the magnetic declination. The approximate bearing of AB is N 50° 15′ E. Compute the necessary distances and azimuths to be run from B to points 1 and 2, from C to point 3, from D to point 4, and from E to point 5.

Solution: The approximate directions of BC, CD, DE, and EP (which are consistent with one another to within the accuracy of the measured angles) are computed as discussed in Section 8-5. The bearings are shown in Table 8-18. The calculations to determine the tentative coordinates are also shown.

The coordinates of the points given in Table 8-18 must now be transformed into the system in which line AP lies by means of Eq. (8-33). The values of a, b, c, and d are found using points A and P. The fixed coordinates of A and P are placed on the left side of Eq. (8-33), and the tentative coordinates of A and P from Table 8-18 are placed on the right side giving

$$1452.60 = 1452.60a - 2086.12b + c$$

$$2086.12 = 2086.12a + 1452.60b + d$$

$$2533.14 = 2528.04a - 2476.82b + c$$

$$2461.85 = 2476.82a + 2528.04b + d$$

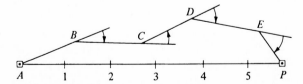

FIGURE 8-24 Random traverse run to establish points on line AP.

Solving gives $a = +\,0.999\,7225$; $b = -\,0.013\,8195$; $c = -\,28.43$; and $d = +\,20.66$. The value of $\sqrt{a^2 + b^2}$ is $0.999\,8180$ which gives an indication of the closure precision of the random traverse, assuming that the fixed coordinates of A and P are without error. Using the values of the tentative coordinates of the random traverse points B, C, D, and E from Table 8-18 in the right side of Eq. (8-33) gives their coordinates in the system of the fixed line AP.

$$X_B = 0.999\,7225 \times 1661.26 + 0.013\,8195 \times 2259.66 - 28.43 = 1663.60$$
$$Y_B = 0.999\,7225 \times 2259.66 - 0.013\,8195 \times 1661.26 + 20.66 = 2256.74$$
$$X_C = 0.999\,7225 \times 1941.11 + 0.013\,8195 \times 2354.10 - 28.43 = 1944.67$$
$$Y_C = 0.999\,7225 \times 2354.10 - 0.013\,8195 \times 1941.11 + 20.66 = 2347.28$$
$$X_D = 0.999\,7225 \times 2109.41 + 0.013\,8195 \times 2524.50 - 28.43 = 2115.28$$
$$Y_D = 0.999\,7225 \times 2524.50 - 0.013\,8195 \times 2109.41 + 20.66 = 2515.31$$
$$X_E = 0.999\,7225 \times 2413.00 + 0.013\,8195 \times 2574.32 - 28.43 = 2419.48$$
$$Y_E = 0.999\,7225 \times 2574.32 - 0.013\,8195 \times 2413.00 + 20.66 = 2560.92$$

The transformed coordinates of the points in the random traverse are now inversed to obtain the correct azimuths of the lines. This is shown in Table 8-19. These azimuths are needed to compute the necessary angles to turn when running the tie lines.

The coordinates of points 1 through 5 of Fig. 8-24 are now computed as shown in Table 8-20. The bearing of AP is found by inversing to give N 70° 49′ 30″ E.

The tie lines can now be inversed to determine their azimuths and lengths. These inverse computations are shown in Table 8-21.

The angles necessary to lay off in the field from the random traverse to the points on the desired line are obtained from the azimuths involved. For example in Fig. 8-25, with the theodolite at B, the right deflection angle from A to point 2 is 103° 14′ 20″ − 51° 02′ 20″ = 52° 12′ 00″ R. This angle should be laid off by double centering. The calculated distance of 171.36 m is measured off to locate the desired point. ◆

TABLE 8-18 Computation of Tentative Coordinates

Station	Bearing	Latitude	Departure	Y'	X'
A				2086.12	1452.60
	N 50°15′00″ E	+173.54	+208.66		
B				2259.66	1661.26
	N 71°21′10″ E	+94.44	+279.85		
C				2354.10	1941.11
	N 44°38′40″ E	+170.40	+168.30		
D				2524.50	2109.41
	N 80°40′50″ E	+49.82	+303.59		
E				2574.32	2413.00
	S 49°43′00″ E	−97.50	+115.04		
P				2476.82	2528.04

TABLE 8-19   Inversing to Obtain Azimuths of Lines in Random Traverse

Line	ΔX	ΔY	Azimuth
AB	+211.00	+170.62	51° 02′ 20″
BC	+281.07	+90.54	72° 08′ 40″
CD	+170.61	+168.03	45° 26′ 10″
DE	+304.20	+45.61	81° 28′ 20″
EP	+113.66	−99.07	131° 04′ 40″

TABLE 8-20   Calculation of Coordinates of Points on Line *AP*

Station	Length	Latitude[a]	Departure[a]	Y	X
A				2086.12	1452.60
	200.00	65.691	188.904		
1				2151.81	1641.50
	200.00	65.691	188.904		
2				2217.50	1830.41
	200.00	65.691	188.904		
3				2283.19	2019.31
	200.00	65.691	188.904		
4				2348.88	2208.22
	200.00	65.691	188.904		
5				2414.58	2397.12

[a]Carried an extra place to avoid accumulated rounding-off errors.

TABLE 8-21   Inversing Tie Lines

Line	ΔX	ΔY	Azimuth	Distance
B-1	− 22.10	−104.93	191° 53′ 40″	107.23
B-2	+166.81	−39.24	103° 14′ 20″	171.36
C-3	+74.64	−64.09	130° 39′ 00″	98.38
D-4	+92.94	−166.43	150° 49′ 10″	190.62
E-5	−22.36	−146.34	188° 41′ 10″	148.04

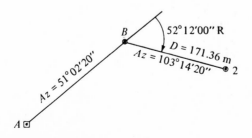

FIGURE 8-25   Deflection angle to tie line.

8-25   Radiation Traverse

A radiation traverse is one that is executed at a single station. It is applicable in an open area with no obstructed line of sight to the other points in the survey. It can be used either to determine the positions of existing points or to lay out points as in a construction project. Assume that the position of some point in the area, defined by its plane coordinates, is known and also that the direction to some other point is known. For example, the position of point A in Fig. 8-26 and the azimuth of the line AP are both known. Point A is occupied with the transit or theodolite, and a backsight is made to point P. The angles between all the lines AB through AH are measured with respect to line AP. The theodolite is then replaced by the EDM (unless both features are incorporated into the same instrument). The distances to points B through H are then measured and reduced to horizontal distances. The angles are used to determine the azimuths of all the lines with reference to the known azimuth of AP. The azimuths (or bearings) together with the horizontal distances are used to compute latitudes and departures of the lines AB through AH. These values are then added to the coordinates of A to get the coordinates of B through H by Eqs. (8-7) and (8-8).

To provide a check on the measurements and discover any possible blunders, a second point A' is carefully set on the line AP. This can be done at the time the initial backsight is made on P. The distance AA', which can be fairly short, say 10 to 12 ft, is then measured. The coordinates of A' can be computed later by applying the latitude and departure of AA' to the coordinates of A using Eqs. (8-7) and (8-8). After the measurements are made at A, the instruments are moved to A', a backsight is made along the line $A'P$, and the angles between this line and the lines $A'B$ through $A'H$, shown by dashed lines in Fig. 8-26, are measured. The EDM is then used to measure the lengths of the lines. These lengths, reduced to horizontal distances, are then used together with the azimuths of $A'B$ through $A'H$ to obtain the latitudes and departures of all the lines. The coordinates of the points computed from A' are then compared with those computed from A. If they are in reasonable agreement, the mean of each pair is adopted. Otherwise the mistake must be found in the field.

The total station instrument is employed to good advantage when executing a radiation traverse. After the instrument is set up, the coordinates of the occupied station (point A of Fig. 8-26) are entered into the microprocessor memory. The azimuth of the reference line is then entered and displayed. A backsight is taken on the reference line, and the reflector target is taken successively to the perimeter stations. In this case the microprocessor is directed to add the successive latitudes and departures to the coordinates of the occupied station to compute the coordinates of the perimeter stations. The process is then repeated at the second point (point A' of Fig. 8-26) to provide a check on the perimeter points.

The main drawback to the application of the radiation traverse is the weak angular relationship between the lines joining the perimeter points such as B through H of Fig. 8-26. For example, if the line EF is relatively short, then a small uncertainty in the length of AE or AF, or both, will have a relatively large effect on

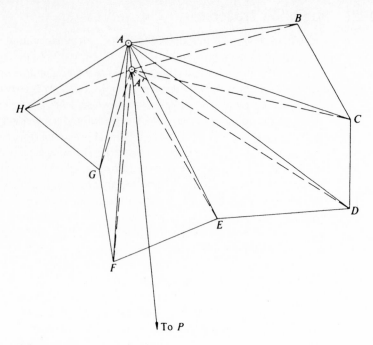

FIGURE 8-26 Radiation traverse.

the direction of *EF.* Consequently, the lines on the perimeter cannot be used as starting lines in subsequent surveys in the area.

When using the radiation traverse to set out points, it is presumed that the coordinates of the points are known, having been either derived from the geometry of the construction lines or scaled from the construction plans. Each radiating line in the layout scheme can then be inversed to determine its length and bearing, using Eqs. (8-11) and (8-12). The inversed values of the bearings are used to compute the angles or circle readings to be set off in the field with respect to the reference line, such as *AP* in the discussion above. The distances along the radiating lines are then laid out according to methods discussed in Section 2-30.

When a total station instrument is used for layout, the coordinates of all of the points to be laid out are loaded into the instrument's memory. As each point is called up, the microprocessor computes the distance and azimuth to be laid out to set that point. The person on the reflector is then guided by the instrumentman until the difference in azimuth (ΔA) and the difference in distance (ΔH) are both reduced to zero as seen on the instrument display screen.

After all the points have been set, their coordinates can be checked from an auxiliary station such as A' as outlined above. This check precludes the possibility of setting out a point in the wrong place. It also verifies the inverse calculations made to determine distances, angles, and bearings.

8-26 Area from Rectangular Coordinates

The area of a figure for which coordinates have been computed is determined by either Eq. (8-25) or Eq. (8-26) of Section 8-21. In evaluating each term of this equation, some of the terms will be positive and some negative, giving the possibility of a net positive area or a net negative area. This is of no consequence unless the result is subject to further analytical treatment (see Section 8-30). The application of Eq. (8-25) to the area enclosed within the *boundary* of Fig. 8-22, the coordinates of which were computed in Section 8-22, is shown in Table 8-22.

8-27 Areas from Maps

When a map of a required area is available, the area can be obtained, within the limits of the accuracy of the map, by dividing the figure into geometrical shapes (triangles, trapezoids, and rectangles), and computing the area of these figures from the dimensions scaled from the map. This method is limited to figures that are bounded by approximately straight lines. When the three sides a, b, and c of a triangle have been scaled from the map, the area may be found by the formula

$$\text{area} = \sqrt{s(s-a)(s-b)(s-c)} \qquad (8\text{-}34)$$

where $s = \frac{1}{2}(a + b + c)$. These lengths are sometimes the result of direct field measurements.

When two sides a and b and the included angle C of a triangle are known, the formula for the area is

$$\text{area} = \tfrac{1}{2}ab \sin C \qquad (8\text{-}35)$$

If the figure is very irregular in shape, as is generally the case when the boundaries are bodies of water, its area can be obtained most accurately by the use of the planimeter. The planimeter can be used also to check an area that has been computed from the field measurements.

TABLE 8-22 Computation of Area Within a Boundary by Coordinates

	+	−
$X_1(Y_2 - Y_4) = 1000.00(1296.38 - 1029.28) =$	267,100	
$X_2(Y_3 - Y_1) = 981.62(1257.32 - 1000.00) =$	252,590	
$X_3(Y_4 - Y_2) = 759.24(1029.28 - 1296.38) =$		202,793
$X_4(Y_1 - Y_3) = 739.84(1000.00 - 1257.32) =$		190,376
	519,690	393,169
	− 393,169	
	2)126,521	
area =	63,260 ft^2	

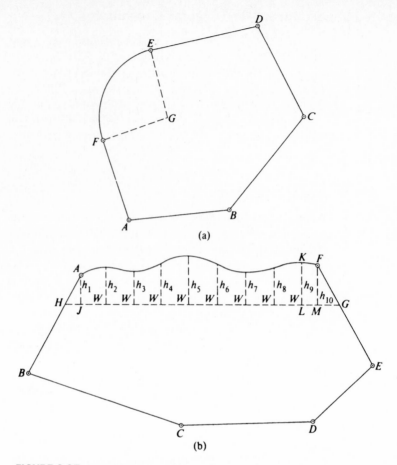

FIGURE 8-27 Area with one curved boundary.

The larger planimeters having adjustable arms are generally more carefully constructed and hence are more accurate than the smaller ones. The precision of the resulting area is dependent on the scale to which the map is drawn, and on the skill of the operator of the planimeter. With care, areas that are accurate within 1% can be obtained by means of the planimeter.

8-28 Area with One Curved Boundary

Two examples of areas with one curved boundary are shown in Fig. 8-27. In view (a) one side of the figure is a circular arc with radius R and central angle FGE. This area can be calculated in two parts, namely the sector EFG and the part $ABCDEGFA$ bounded by straight lines. The area of the sector is $\pi R^2 \times$ angle $FGE/360°$.

In Fig. 8-27(b) the boundary AF is an irregular curve. The area of this figure can be divided into two parts by the straight line GH, and computations for the portion bounded by straight lines can be made by any of the ordinary methods. If perpendicular offsets from the straight line GH to the curved boundary are measured, either on the ground or on a map, the irregular portion can be considered as a series of triangles and trapezoids. These offsets should be measured at all breaks in the curved boundary. The area of each triangle and each trapezoid can be computed separately. When the boundary is sufficiently regular, the offsets can be taken at equal intervals along the line GH, and the computations can thus be simplified considerably. The area of the regular portion $AJLK$ in Fig. 8-27(b) can be obtained by a single multiplication by applying the trapezoidal rule, which is

$$\text{area} = W\left(\frac{h_1 + h_n}{2} + h_2 + h_3 + \cdots + h_{n-1}\right) \qquad (8\text{-}36)$$

in which W is the common spacing of the offsets and n is the number of offsets.

EXAMPLE 8-9   Determine the area $AFGH$ in Fig. 8-27(b) by using the trapezoidal rule for the regular portion. The distances are as follows:

$h_1 = 22.6$ ft	$h_6 = 36.9$ ft	$HJ = 14$ ft
$h_2 = 28.0$ ft	$h_7 = 30.0$ ft	$LM = 13.5$ ft
$h_3 = 27.1$ ft	$h_8 = 31.5$ ft	$MG = 21$ ft
$h_4 = 30.6$ ft	$h_9 = 34.8$ ft	$W = 25$ ft = common spacing
$h_5 = 38.5$ ft	$h_{10} = 35.0$ ft	

Solution: The computations follow:

$$\text{area } AJH = \frac{(14.0)(22.6)}{2} = 158 \text{ ft}^2$$

$$\text{area } KFML = \frac{(34.8 + 35.0)(13.5)}{2} = 471 \text{ ft}^2$$

$$\text{area } FGM = \frac{(21.0)(35.0)}{2} = 368 \text{ ft}^2$$

$$\text{area } AJLK = 25\left(\frac{22.6 + 34.8}{2} + 28.0 + 27.1 + 30.6 + 38.5 + 36.9 + 30.0 + 31.5\right)$$

$$= 6283 \text{ ft}^2$$

The total area $AFGH$ is therefore 7280 ft^2.   ◆

The assumption made in using the trapezoidal rule is that the curved boundary is composed of chords connecting the ends of the offsets. When the offsets are taken closely enough together and when the curves are flat, no considerable error is introduced by this assumption.

A more accurate value for the area between a straight line and an irregular boundary may be obtained by taking offsets at regular intervals and applying Simpson's one-third rule to that portion lying between an *odd* number of offsets. Simpson's one-third rule assumes that the curve through each successive three points is a portion of a parabola. In Fig. 8-28 the curve through points A, B, and C is

FIGURE 8-28 Area by Simpson's one-third rule.

assumed to be a segment of a parabola cut off by the chord AC. The area between the chord and the parabola, found by the calculus, is equal to two-thirds the area of the parallelogram $AFGC$. The area of $ABCDE$ is therefore

$$\text{area} = 2W\left(\frac{h_1 + h_3}{2}\right) + \frac{2}{3} \times 2W\left(h_2 - \frac{h_1 + h_3}{2}\right)$$

Expanding and collecting terms gives

$$\text{area} = \frac{W}{3}(h_1 + 4h_2 + h_3)$$

By the same reasoning, the area for the next three offsets would be

$$\text{area} = \frac{W}{3}(h_3 + 4h_4 + h_5)$$

Extending this reasoning to any *odd* number of offsets and summing up the results gives the following relation:

$$\text{area} = \frac{W}{3}[h_1 + 2(h_3 + h_5 + \cdots + h_{n-2}) + 4(h_2 + h_4 + \cdots + h_{n-1}) + h_n]$$

in which h_1 is the first offset; h_n is the last odd-numbered offset; W is the common interval at which the offsets are taken.

In its condensed form Simpson's one-third rule may be expressed as follows:

$$\text{area} = \frac{W}{3}(h_1 + h_n + 2\Sigma h_{\text{odd}} + 4\Sigma h_{\text{even}}) \tag{8-37}$$

◄EXAMPLE 8-10 Determine the area $AFGH$ in Fig. 8-27(b) by using Simpson's one-third rule where it applies.

Solution: Obviously Simpson's one-third rule applies only between the first and the ninth offsets. The area of the remaining portion is computed by the trapezoidal rule. Thus

$$\text{area } AJLK = \tfrac{25}{3} [22.6 + 34.8 + 2(27.1 + 38.5 + 30.0) + 4(28.0 + 30.6 + 36.9 + 31.5)]$$

$$= 6305 \text{ ft}^2$$

The remaining area, as computed in the preceding example, is 997 ft². Therefore the total area *AFGH* is 7302 ft², which is 22 ft² more than that obtained by the trapezoidal rule. This difference is due to the fact that the irregular boundary is generally concave toward the line *HG*. If the irregular boundary is concave away from the straight line from which offsets have been measured, Simpson's one-third rule will give a smaller area than will the trapezoidal rule. ◆

8-29 Problems in Omitted Measurements

The common examples of omitted measurements are shown in Fig. 8-29. In view (a) the length, the azimuth, or both of side *EA* may be unknown. The latitudes and departures of the known sides are computed. Coordinates are then computed, and *EA* is inversed to obtain its bearing and length.

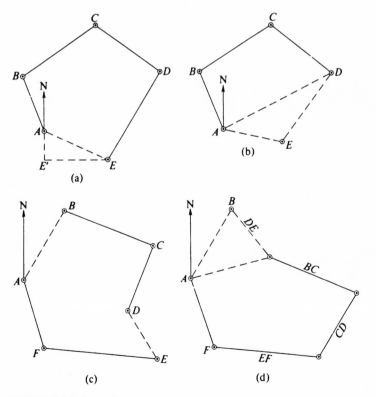

FIGURE 8-29 Omitted measurements.

In Fig. 8-29(b) two adjacent sides are involved in the missing measurements. Both lengths, both azimuths, or the azimuth of one line and the length of the other may be unknown. From the calculated coordinates of A and D, the azimuth and length of the closing line DA are computed. The triangle ADE can now be solved. When the lengths of DE and EA are missing, the triangle is solved from the computed length DA and the three angles, which can be calculated from the known azimuths. When the azimuths of DE and EA are unknown, the three sides of the triangle are known and the angles are computed. When the azimuth of one of the sides and the length of the other are unknown, the triangle is solved from the two known sides and the angle opposite one of the known sides. Frequently there will be two solutions in this last case, and the approximate shape of the figure must then be known. The trigonometric formulas for the solution of oblique triangles are given in Table C-3 in Appendix C.

In Fig. 8-29(c) the omitted measurements occur on sides that are not adjacent. Since the latitudes and departures of equal parallel lines are equal, this problem can be solved by shifting the line DE until it is adjacent to AB, so as to form the closed figure shown in view (d). The solution of this problem is the same as that outlined for view (b).

8-30 Parting Off Land

One of the common problems of land surveying is the division of an irregular polygon into two or more parts with known areas. The division may be made by a line of known direction, or it may be made by a line from a given starting point on the perimeter of the polygon.

In Fig. 8-30 let the problem be the division of the polygon *ABCDEF* into two equal parts by a line parallel to *AF.* The azimuths and lengths of the sides being known, the first step is the calculation of the entire area. If a rough sketch is made

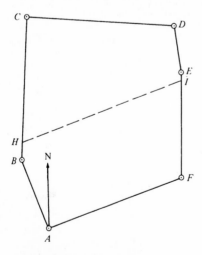

FIGURE 8-30 Parting off land.

from the field measurements or from calculated coordinates, the approximate position of the line *HI,* which is to divide the area into two equal parts, can be determined. From the figure it is evident that *H* will be somewhere on the line *BC,* and *I* will be somewhere on the line *EF.*

The solution to this problem is obtained by the methods of analytic geometry. The rectangular coordinates of the corners are determined from the latitudes and departures of the sides. For convenience the *Y* axis should be chosen through *B* and the *X* axis through *A.* The coordinates are as shown in Table 8-23 along with the determination of the area of the entire field found by Eq. (8-25).

This area is –539,870 ft². The algebraic sign is significant in this solution, since one-half this area, or –269,935 ft², is a term in one of the equations to be used in solving the problem analytically.

From Fig. 8-30 and from the given conditions, the following four statements are apparent: (1) Point *H* lies on the line *BC;* (2) point *I* lies on the line *EF;* (3) the area of *ABHIF* is one-half the total area, or –269,935 ft²; and (4) the line *HI* has the same bearing as does the line *AF.* If the coordinates of the unknown point *H* are denoted by Y_H and X_H and the coordinates of the unknown point *I* by Y_I and X_I, the following four equations may be written as follows: by Eq. (8-15),

$$\frac{Y_H - Y_B}{Y_C - Y_B} = \frac{X_H - X_B}{X_C - X_B}$$

by Eq. (8-15),

$$\frac{Y_I - Y_E}{Y_F - Y_E} = \frac{X_I - X_E}{X_F - X_E}$$

by Eq. (8-25),

$$X_A(Y_B - Y_F) + X_B(Y_H - Y_A) + X_H(Y_I - Y_B) + X_I(Y_F - Y_H) + X_F(Y_A - Y_I) = 2(-269{,}935)$$

and by Eq. (8-11),

$$\frac{X_F - X_A}{Y_F - Y_A} = \tan B_{AF} = \tan B_{HI} = \frac{X_I - X_H}{Y_I - Y_H}$$

TABLE 8-23 Computation of Area by Coordinates

Station	Y	X		+	−
A(= 1)	0	108.5	108.5(301.5 − 197.0) =	11,338	
B(= 2)	301.5	0	0 (917.1 − 0) =		
C(= 3)	917.1	60.7	60.7(843.8 − 301.5) =	32,918	
D(= 4)	843.8	710.9	710.9(644.1 − 917.1) =		194,076
E(= 5)	644.1	729.9	729.9(197.0 − 843.8) =		472,099
F(= 6)	197.0	710.8	710.8(0 − 644.1) =		457,826
				+44,256	− 1,124,001
					+44,256
				2)	− 1,079,745
				area =	−539,873

Substituting the known values of the coordinates in these equations gives the following four equations that may be solved simultaneously to obtain the values of the four unknowns X_H, Y_H, X_I, and Y_I:

$$\frac{Y_H - 301.5}{615.6} = \frac{X_H - 0}{60.7} \tag{1a}$$

$$\frac{Y_I - 644.1}{-447.1} = \frac{X_I - 729.9}{-19.1} \tag{2a}$$

$$(108.5)(104.5) + 0(Y_H - 0) + X_H(Y_I - 301.5)$$
$$+ X_I(197.0 - Y_H) + 710.8(0 - Y_I) = -539,870 \tag{3a}$$

$$\frac{X_I - X_H}{Y_I - Y_H} = \frac{602.3}{197.0} \tag{4a}$$

The results obtained by cross multiplying and rearranging and collecting terms are

$$615.6X_H - 60.7Y_H + 18,301.05 = 0 \tag{1b}$$

$$447.1X_I - 19.1Y_I - 314,035.98 = 0 \tag{2b}$$

$$301.5X_H - 197.0X_I - X_HY_I + X_IY_H + 710.8Y_I - 551,208 = 0 \tag{3b}$$

$$X_H - X_I - 3.05736Y_H + 3.05736Y_I = 0 \tag{4b}$$

Expressing Y_H in terms of X_H from Eq. (1b) gives

$$Y_H = 10.14168X_H + 301.5 \tag{1c}$$

Expressing Y_I in terms of X_I from Eq. (2b) gives

$$Y_I = 23.40838X_I - 16,441.7 \tag{2c}$$

When these values of Y_H and Y_I are substituted in Eqs. (3b) and (4b), the results are

$$16,743.2X_H + 16,743.2X_I - 13.26670X_HX_I - 12,237,968 = 0 \tag{3c}$$

$$30.00677X_H - 70.56784X_I + 51,190 = 0 \tag{4c}$$

or

$$X_I = 0.42522X_H + 725.40$$

Substituting the value of X_I from Eq. (4c) in Eq. (3c) gives

$$5.64127X_H^2 - 14,239X_H + 92,451 = 0 \tag{3d}$$

From this quadratic equation, $X_H = 6.5096$. By Eq. (4c), $X_I = 728.17$. By Eq. (1c), $Y_H = 367.52$. By Eq. (2c), $Y_I = 603.58$. By Eq. (8-12), $BH = 66.34$ ft; $EI = 40.56$ ft; $HI = 759.2$ ft.

In Fig. 8-31 let the problem be the division of the polygon $ABCDEF$ of Fig. 8-30 into two equal parts by a line GH from the middle point of the side AF. This problem is solved by analytic geometry by first computing coordinates of the traverse stations and then computing the area from the coordinates. The coordinates of point G are $Y = 98.5$ and $X = 409.6$, which are equal to the means of the coordinates of A and F since G is the middle point of the line AF. From Fig. 8-31 and from the conditions in the problem, two statements are apparent: (1) Point H lies on

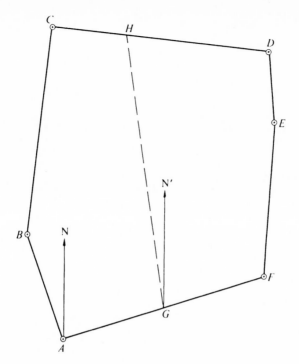

FIGURE 8-31 Parting off land.

line CD and (2) the area of $ABCHG$ is one-half the total area. Two equations suffice to compute the coordinates X_H and Y_H of the unknown point H. The equations are as follows: by Eq. (8-15),

$$\frac{Y_H - Y_C}{Y_D - Y_C} = \frac{X_H - X_C}{X_D - X_C}$$

and by Eq. (8-25),

$$X_A(Y_B - Y_G) + Y_B(Y_C - Y_A) + X_C(Y_H - Y_B) + X_H(Y_G - Y_C) + X_G(Y_A - Y_H) = -2(269{,}935)$$

It is only necessary to substitute the values of the known coordinates in these equations and solve the simultaneous equations thus obtained. The work follows:

$73.3X_H + 650.2Y_H -$	$600{,}747.7 \quad = 0$	(5a)
$818.6X_H + 348.9Y_H -$	$543{,}594.4 \quad = 0$	(6a)
$73.3X_H + 650.2Y_H -$	$600{,}747.7 \quad = 0$	(5b)
$1525.5X_H + 650.2Y_H -$	$1{,}013{,}026.9 \quad = 0$	(6b)

$$1452.2X_H \qquad\qquad\qquad -412{,}279.2 \quad = 0$$

$$X_H = 283.9 \qquad\qquad Y_H = 891.9$$

By Eq. (8-12), $CH = 224.6$ ft and $GH = 803.3$ ft. By Eq. (8-11), the bearing of GH is N 9° 00′ W.

8-31   Methods of Plotting a Survey

Traverses as well as calculated figures such as parcels of land can be plotted to suitable scale by using the computed coordinates of the traverse stations or the points in the figures, or by laying off distances and bearings. The latter method is used to advantage when the traverse has not been computed and balanced. The method of coordinates is the more reliable of the two methods, however, because most of the work in plotting has been arrived at numerically.

8-32   Plotting by Distance and Bearings

The most convenient approach to plotting traverse lines by their lengths and bearings is to reference all the work on the map sheet to a very carefully constructed square that measures for convenience 10 in. on a side. In Fig. 8-32 a 10-in. square has been constructed on the map sheet. This will be used to plot the traverse from Section 8-11 and shown in Fig. 8-13. The lines, their lengths and bearings, and either the tangent or the cotangent of the bearing angle are listed in Table 8-24. If the bearing angle is less than 45°, the tangent is listed; if it is greater than 45°, the cotangent is listed.

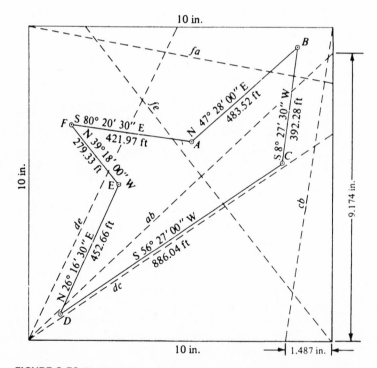

FIGURE 8-32 Plotting traverse by tangents or cotangents of bearing angles.

TABLE 8-24 Data for Plotting a Traverse by Lengths and Bearings

Line	Length	Bearing	Tangent B	Cotangent B
AB	483.52	N 47° 28′ 00″ E		0.91740
BC	392.28	S 8° 27′ 30″ W	0.14871	
CD	886.04	S 56° 27′ 00″ W		0.66314
DE	452.66	N 26° 16′ 30″ E	0.49369	
EF	279.33	N 39° 18′ 00″ W	0.81849	
FA	421.97	S 80° 20′ 30″ E		0.17018

The first traverse station is plotted at some tentative point on the map sheet, the position of which depends on the size of the sheet, the scale of the plotting, and the shape of the traverse. This tentative position may need repositioning if the traverse begins to run off the limits of the map sheet. In the present instance station A is plotted in the position shown in Fig. 8-32.

Next the direction of the line AB is established. Since its bearing is N 47° 28′ 00″ E, if we imagine commencing at the southwest corner of the square, going eastward 10 in. and then northward 10 cot 47° 28′ 00″ = 9.174″ along the right edge of the square, a point is plotted 9.174″ up from the lower right southeast corner of the square. A line joining this point with the southwest corner of the square defines the dashed line ab, which bears N 47° 28′ 00″ E. This direction is transferred to station A, and a line through A of indefinite length is drawn. The length AB is now laid off along this line according to the plotting scale. This defines the position of B.

Next the direction of BC is established. The bearing of BC is S 8° 27′ 30″ W. Thus commencing at the northeast corner of the square, we go south 10″ along the right side of the square and then west 10 tan 8° 27′ 30″ = 1.487″ along the lower edge of the square. When joined with the northeast corner, a point plotted at this position gives the dashed line cb, which is in the proper direction. A line is drawn through B parallel with cb, and the length BC is laid off to scale in order to locate point C.

The above procedure is used with each succeeding line to lay off the correct direction and distance. The end of the last line FA should fall on the initial point A. If the survey is of reasonable accuracy, even though it has not been balanced, it should close on paper with no detectable error. Any discrepancy between the beginning and ending positions of the plotted, closed traverse therefore is due to either errors or mistakes in plotting or else a combination of plotting errors and mistakes.

A graphical application of the compass rule is used to adjust a misclosure. The amount of closure should first be scaled to examine whether a unit such as 10 or 100 ft has been misplotted. The direction of closure can be an indication of which line might be plotted with the wrong length, because the closure line and the faulty line should be parallel with one another. If the closure is relatively small, then each plotted point can be shifted by an amount proportionate to its distance from the point of beginning and in the direction required to move the last plotted point into correct position.

8-33 Plotting by Rectangular Coordinates

The method of plotting a traverse or a figure by rectangular coordinates is more accurate than plotting by lengths and bearings and has advantages not possessed by the latter method. It is known in advance that the traverse closes. The erroneous plotting of any point does not affect the location of any of the succeeding points, as the plotting of each point is independent of the rest. Each point can be checked as soon as plotted by comparing the scaled distance to the preceding station with the length measured in the field. The proper size of sheet for the map may be determined by an inspection of the coordinates.

A series of *grid* lines is carefully constructed. The lines of the two sets are perpendicular to one another, and those of each set are spaced at intervals of 50, 100, 500, or 1000 ft, or at some other regular interval, the distance depending on the scale to which the traverse or map is to be plotted. All these grid lines should be plotted in as brief a period of time as possible, so that the measurements are consistent with one another under the given ambient conditions of temperature and humidity. Each grid line is labeled with its proper Y coordinate or X coordinate and, finally, the several points are plotted, according to their coordinates, from the nearest grid lines. Plotting by coordinates is shown in Fig. 8-33, where the stations of the traverse of Section 8-11 are plotted by using a 200-ft grid spacing. When all the points have been plotted, their positions are checked by scaling the distances between successive points and comparing them with the known distances.

FIGURE 8-33 Plotting by rectangular coordinates.

Grid lines and coordinates can be plotted mechanically by means of an instrument called a *coordinatograph.* One is shown in Fig. 8-34. A girder travels in the X direction, and its motion is capable of being read or set by means of a scale and rotating dial to an accuracy of approximately ± 0.003 in. or less than 0.1 mm. A pencil, pen, or needle holder travels along the girder in the Y direction, and its motion also can be read or set to the same accuracy. Grid lines are ruled in both directions by appropriate settings of the scales and dials, which are determined by the choice of the plotting scale. Each point is then plotted, according to its coordinates, by making appropriate settings of the scales and dials.

The plotter shown in Fig. 8-35 contains a control unit that has been preprogrammed to perform several operations that a draftsman would ordinarily do. It contains two or more pens that can be instructed to plot and annotate points according to their rectangular coordinates, draw grid lines, draw straight lines between points, and draw circles and curves. The input to the plotter can be obtained through the keyboard of a computer, from a stereoscopic plotting instrument (see Section 16-9), or directly from the output of a data collector discussed in Sections 4-12, 4-24, and 8-20.

The value of a map may be increased considerably by retaining the grid lines on the finished map, either as complete lines or as tick marks at the edges of the sheet and cross marks at the grid intersections in the interior portion. The map user can then detect any change in scale in any or all portions of the maps by measuring between grid lines or between grid ticks. Where accurate lengths are to be determined between well-defined points, the coordinates of points defining the ends of the lines are scaled from the nearest grid lines. The scaled distances are then either increased or decreased to allow for the scale change caused by temperature and humidity changes, or by the reproduction process. The lengths may then be computed by Eq. (8-12). In Fig. 8-36 the grid was originally plotted with 5-in.

FIGURE 8-34 Coordinatograph. Courtesy of Unitech Corp.

FIGURE 8-35 Plotting table with control unit. Courtesy of Kern Instruments, Inc.

spacings representing 1000 ft. The scaled distances to point P from the grid lines representing $Y = 6000$ ft and $X = 10,000$ ft are 420 ft and 775 ft, respectively. The actual distance between the grid lines in the Y direction is 5.012 in. and the distance in the X direction is 4.980 in. The Y coordinate of the point P is therefore 6000 + (420)(5)/5.012 = 6419 ft and the X coordinate is 10,000 + (775)(5)/4.980 = 10,778 ft.

8-34   Scales

The topographic maps prepared by the U.S. Geological Survey are plotted to so-called natural scales. These scales are expressed as ratios. On a map drawn to a scale of 1/20,000, one unit on the map represents 20,000 units on the ground. Thus 1 in. on the paper represents 20,000 in. on the ground, 1 ft represents 20,000 ft, and 1 m represents 20,000 m on the ground. The scales commonly used on maps intended for engineering purposes are those found on the edges of an engineer's scale. Thus 1 in. on the paper represents some multiple of 10, 20, 30, 40, 50, or 60 ft on the ground. Maps are generally classed as large scale when the scale is greater than 1 in. = 100 ft,[†] as intermediate scale when the scale is between 1 in. = 100 ft and 1 in. = 1000 ft, and as small scale when the scale is less than 1 in. = 1000 ft.

[†]Obviously 1″ cannot equal 100′; however, this notation is common practice among engineers and surveyors.

FIGURE 8-36 Scaling coordinates.

The large scales are used when a large amount of detail is to be shown, or where the area covered by the map is small. In general, the scale should be as small as possible and still represent the detail with sufficient precision. Maps intended for the design of engineering projects are commonly plotted to scales between 1 in. = 20 ft and 1 in. = 800 ft, the exact scale depending on the detail to be shown and the area covered.

Surveys made for architects are often plotted to the scales used by architects, namely 1 in. = 4 ft, 1 in. = 8 ft, and 1 in. = 16 ft.

By the decimal nature of the metric system, ratios or *representative fractions* (RF) are used exclusively to express scale. Some approximate equivalents between engineer's scales and representative fractions used in the metric system in addition to the actual ratios are given in Table 8-25.

As previously stated, the scale of a map may change slightly because of the shrinkage or stretching of the drawing paper. The scale is changed also when the map is reproduced. The change may be slight in blueprinted reproductions, or it may be considerable when photographic reductions or enlargements are made. Since the original scale no longer applies to the duplicates, a graphical scale, similar to one of those shown in Fig. 8-37, should appear on the map.

8-35   Meridian Arrow

A meridian arrow should appear on the map. It should be simple in design and, although it should be of sufficient length to permit the reasonably accurate scaling of directions on any part of the map, it should not be so conspicuous as to convey the impression that the map is a drawing of an arrow. The proportions of an arrow

TABLE 8-25 Approximate Equivalents Between Engineer's Scale and Metric Representative Fractions

Engineer's Scale	Approximate RF for Metric System		Actual Ratio
1 in = 20 ft	$\frac{1}{200}$ or $\frac{1}{250}$	1:200 or 1:250	1:240
1 in = 40 ft	$\frac{1}{500}$	1:500	1:480
1 in = 50 ft	$\frac{1}{500}$	1:500	1:600
1 in = 100 ft	$\frac{1}{1000}$	1:1000	1:1200
1 in = 200 ft	$\frac{1}{2000}$ or $\frac{1}{2500}$	1:2000 or 1:2500	1:2400
1 in = 250 ft	$\frac{1}{2000}$ or $\frac{1}{2500}$	1:2000 or 1:2500	1:3000
1 in = 1000 ft	$\frac{1}{10000}$	1:10000	1:12,000
1 in = 2000 ft	$\frac{1}{20000}$ or $\frac{1}{25000}$	1:20000 or 1:25000	1:24,000

FIGURE 8-37 Graphical scales.

that can be constructed easily with drawing instruments are shown in Fig. 8-38. The dotted circle does not appear on the finished drawing. When both true and magnetic directions are known, two arrows are combined as in Fig. 8-38—a full-headed arrow usually indicating true north and a half-headed arrow indicating magnetic north. The magnetic declination is shown in figures. If the map contains grid lines, or grid ticks, then grid north is indicated by the grid lines running in the *Y* direction.

8-36 Titles

On an engineer drawing the title appears in the lower-right corner of the sheet. This same convention is followed as far as possible on maps, although on many maps the shape of the area covered may be such that the title is placed elsewhere on the sheet to give the drawing a more balanced appearance.

FIGURE 8-38 Meridian arrow.

As maps may pass through many hands and may be intended as permanent records, more time is spent on the lettering than would be justified on the shop drawings for an engineering project. The very ornate lettering found on older maps is no longer used, carefully constructed Roman or Gothic capitals being used instead. Legends and notes, separated from the main title, may be shown in single-stroke Reinhardt letters.

Various automatic drafting tables, such as that shown in Fig. 8-35, can be programmed to generate titles with all the necessary information lettered in various sizes and fonts.

The title should be sufficiently complete to permit the identification of the map. It should include the purpose of the survey, the name of the owner or the organization for whom the survey is made, the location (city or township, county, and state), the date of the survey, the name of the engineer or surveyor responsible for the work, and the scale. The name of the draftsman may also appear.

8-37 Property Survey Plats

Figure 8-39 shows a typical plan plat produced by P & L Systems of Columbus, Ohio. Note the scale, north arrow, and tie to section corner 15-14-22-23 (T5N, R17W), which was the point of beginning (P.O.B.) of the survey. Section corners are discussed in Chapter 18.

FIGURE 8-39 Plat plan. Courtesy of P and L Systems, LTD.

PROBLEMS

8-1. The adjusted *clockwise* interior angles in a traverse that originates and closes on point *A* are as follows:

Station	From	To	Adjusted Clockwise Interior Angles
A	F	B	27°43′ 45″
B	A	C	152°36′ 55″
C	B	D	67°20′ 30″
D	C	E	200°38′ 45″
E	D	F	128°27′ 30″
F	E	A	143°12′ 35″

The bearing of *AB* is S 2° 22′ 15″ E. Compute the azimuths and bearings of the remaining sides of the traverse.

8-2. Given the same *clockwise* angles as in Problem 8-1 and the bearing of *DE* as S 15° 25′ 00″ W, compute the azimuths and bearings of the remaining sides of traverse.

8-3. Given the same *clockwise* angles as in Problem 8-1 and the bearing of *CD* as due east, compute the azimuths and bearings of the remaining sides of the traverse.

8-4. Given the same *clockwise* angles as in Problem 8-1 and the bearing of *EF* as N 10° 22′ 15″ W, compute the azimuths and bearings of the remaining sides of the traverse.

8-5. The following deflection angles were measured in a traverse that originates at station *A* with a backsight on station *R* and closes on station *G* with a foresight on station *S*.

Station	From	To	Measured Deflection Angle
A	R	B	88° 16′ R
B	A	C	26° 50′ R
C	B	D	46° 32′ L
D	C	E	61° 25′ R
E	D	F	8° 16′ R
F	E	G	61° 18′ L
G	F	S	62° 48′ L

The fixed azimuth of *AR* is 97° 15′ 30″, and the fixed azimuth of *GS* is 291 °21′ 00″. What are the adjusted azimuths and bearings of the intermediate traverse sides?

8-6. Solve Problem 8-5 if the fixed azimuth of *AR* is 282° 50′ 00″ and that of *GS* is 117° 06′ 00″.

8-7. Solve Problem 8-5 if the fixed azimuth of *AR* is 186° 18′ 20″ and that of *GS* is 20° 26′ 10″.

8-8. Solve Problem 8-5 if the fixed azimuth of *AR* is 8° 57′ 30″ and that of *GS* is 203° 10′ 00″.

8-9. The following angles to the right were measured with a 1-second theodolite in a traverse that originates at point *K* with a backsight on point *J*, and closes on point *O* with a foresight on point *P*.

Station	From	To	Measured Angle to Right
K	J	L	212° 21′ 16.2″
L	K	M	161° 14′ 58.6″
M	L	N	215° 30′ 08.4″
N	M	O	90° 50′ 01.4″
O	N	P	148° 47′ 23.0″

The fixed azimuth of JK is 179° 52′ 55.5″, and the fixed azimuth of OP is 108° 36′ 53.6″. Compute the adjusted azimuths of the intermediate traverse sides.

8-10. Solve Problem 8-9 if the fixed azimuth of JK is 17° 15′ 52.5″ and the fixed azimuth of OP is 305° 59′ 50.1″.

8-11. Solve Problem 8-9 if the fixed azimuth of JK is 246° 14′ 18.3″ and the fixed azimuth of OP is 174° 58′ 04.4″.

8-12. Solve Problem 8-9 if the fixed azimuth of JK is 350° 40′ 18.8″ and the fixed azimuth of OP is 279° 24′ 09.9″.

8-13. The measured lengths and adjusted *counterclockwise* angles in an eight-sided traverse are given as follows:

Station	Adjusted Counter-clockwise Interior Angles	Line	Measured Length (ft)
A	148° 04′ 30″	AB	195.402
B	146° 32′ 00″	BC	351.005
C	97° 58′ 10″	CD	384.863
D	161° 04′ 50″	DE	398.474
E	101° 18′ 40″	EF	338.508
F	98° 44′ 40″	FG	296.162
G	239° 15′ 30″	GH	252.862
H	87° 01′ 40″	HA	309.337

The coordinates of A are X_A = 1000.000; Y_A = 1000.000. The azimuth of AB is 58° 25′ 50″. Compute the azimuths of the traverse sides and show a check. Compute the latitudes and departures of the traverse lines.

8-14. Balance the traverse of Problem 8-13 by the compass rule. Then compute the coordinates of the stations.

8-15. Using the results of Problem 8-14, compute the area bounded by the traverse in Problem 8-13 by Eq. (8-25).

8-16. Given the data of Problem 8-13 except that the azimuth of AB is 58° 39′ 10″, compute the azimuths of the traverse sides and show a check. Compute the latitudes and departures of the traverse sides.

8-17. Balance the traverse of Problem 8-16 by the compass rule. Then compute the coordinates of the stations.

8-18. Using the results of Problem 8-17, compute the area bounded by the traverse in Problem 8-16 by Eq. (8-25).

8-19. Given the data of Problem 8-13 except that the azimuth of *AB* is 58° 52′ 30″, compute the azimuths of the traverse sides and show a check. Compute the latitudes and departures of the traverse sides.

8-20. Balance the traverse of Problem 8-19 by the compass rule. Then compute the coordinates of the stations.

8-21. Using the results of Problem 8-20, compute the area bounded by the traverse in Problem 8-19 by Eq. (8-25).

8-22. Given the data of Problem 8-13 except that the azimuth of *AB* is 59° 05′ 50″, compute the azimuths of the traverse sides and show a check. Compute the latitudes and departures of the traverse sides.

8-23. Balance the traverse of Problem 8-22 by the compass rule. Then compute the coordinates of the stations.

8-24. Using the results of Problem 8-23, compute the area bounded by the traverse in Problem 8-22 by Eq. (8-25).

8-25. Compute the length and bearing of the line *AD* in Fig. 8-13 by using the adjusted coordinates of the traverse given in Section 8-18.

8-26. Compute the length and bearing of the line *AE* in Fig. 8-13 by using the adjusted coordinates of the traverse given in Section 8-18.

8-27. Compute the length and bearing of the line *BD* in Fig. 8-13 by using the adjusted coordinates of the traverse given in Section 8-18.

8-28. Compute the length and bearing of the line *BE* in Fig. 8-13 by using the adjusted coordinates of the traverse given in Section 8-18.

8-29. A line through point *A* of Fig. 8-13 bears S 25° 40′ W. At what distance from point *D* does this line intersect line *CD*?

8-30. A line through point *A* of Fig. 8-13 bears N 72° 12′ E. At what distance from point *C* does this line intersect line *BC*?

8-31. A line through point *E* of Fig. 8-13 bears N 3° 27′ E. At what distance from point *A* does this line intersect line *FA*?

8-32. A line through point *B* of Fig. 8-13 bears S 21° 15′ W. At what distance from point *C* does this line intersect line *CD*?

8-33. Compute the coordinates of the intersection of lines *BC* and *FA* of Fig. 8-13.

8-34. Compute the coordinates of the intersection of lines *CD* and *EF* of Fig. 8-13.

8-35. Compute the coordinates of the intersection of lines *AB* and *DE* of Fig. 8-13.

8-36. What is the perpendicular distance from point *A* to line *CD* of Fig. 8-13?

8-37. What is the perpendicular distance from point *E* to line *CD* of Fig. 8-13?

8-38. What is the perpendicular distance from point *E* to line *FA* of Fig. 8-13?

8-39. Given the following coordinates of points in two coordinate systems:

	System 1		System 2	
Point	X	Y	X	Y
A	14,385.26	8,790.20	28,601.84	44,952.62
B	15,588.22	8,305.98	28,836.43	46,228.27
C			27,638.80	45,515.42
D			29,402.27	45,597.40
E			28,166.64	46,207.15

Transform the coordinates of points C, D, and E into System 1.

8-40. Compute the scale of System 1 relative to System 2 of Problem 8-39.

8-41. Compute the rotation angle of Problem 8-39.

8-42. In Fig. 8-26 the azimuth of line AP is 176° 10′ 22″. The following directions and horizontal distances were measured at A using a total station instrument:

Point	Direction	Distance (ft)
B	120° 44′ 16″	462.88
C	145° 33′ 08″	634.10
D	163° 17′ 20″	742.25
E	190° 10′ 04″	521.72
P	211° 59′ 51″	
F	222° 08′ 11″	579.90
G	232° 15′ 43″	342.20
H	277° 18′ 05″	331.28

The coordinates of A are $X_A = 25{,}452.84$, $Y_A = 16{,}780.92$. Compute the coordinates of the other points in the closed figure.

8-43. Compute the area of the closed figure of Problem 8-42 using Eq. (8-25).

8-44. Compute the area of the closed figure of Problem 8-42 using Eq. 29 in Table C-3.

8-45. The following perpendicular offsets in feet are measured from a straight line to an irregular boundary at regular intervals of 20 ft.

$$h_1 = 24.8 \qquad h_6 = 40.8 \qquad h_{11} = 60.2$$
$$h_2 = 41.4 \qquad h_7 = 45.6 \qquad h_{12} = 47.7$$
$$h_3 = 36.6 \qquad h_8 = 50.3 \qquad h_{13} = 36.6$$
$$h_4 = 32.6 \qquad h_9 = 44.7 \qquad h_{14} = 36.0$$
$$h_5 = 36.7 \qquad h_{10} = 51.9$$

Compute the area lying between the straight line and the irregular boundary by the trapezoid rule.

8-46. Solve Problem 8-45 using Eq. (8-25).

8-47. Compute the area of Problem 8-45 by Simpson's one-third rule: (**a**) using h_1 as the first offset, and (**b**) using h_{14} as the first offset.

8-48. The following perpendicular offsets in meters are measured from a straight line to an irregular boundary at regular intervals of 10 meters:

$$h_1 = 42.23 \qquad h_6 = 75.82 \qquad h_{11} = 92.88$$
$$h_2 = 41.04 \qquad h_7 = 69.07 \qquad h_{12} = 64.04$$
$$h_3 = 33.26 \qquad h_8 = 95.47 \qquad h_{13} = 43.36$$
$$h_4 = 36.83 \qquad h_9 = 93.00 \qquad h_{14} = 28.51$$
$$h_5 = 53.30 \qquad h_{10} = 93.31 \qquad h_{15} = 17.84$$

Compute the area in hectares lying between the straight line and the irregular boundary by the trapezoid rule.

8-49. Compute the area in Problem 8-48 by Simpson's one-third rule.

8-50. Compute the area in Problem 8-48 by Eq. (8-25).

8-51. In the traverse of Fig. 8-29(b) the following lengths and bearings were measured:

Side	Length (ft)	Bearing
AB	300.11	N 27° 53′ W
BC	412.54	N 56° 13′ E
CD	369.12	S 51° 45′ E
DE	405.27	—
EA	—	S 85° 49′ W

Compute the missing length and bearing.

8-52. In the traverse of Fig. 8-29(b) the following lengths and bearings were measured:

Side	Length (ft)	Bearing
AB	424.36	N 32° 58′ W
BC	583.33	N 51° 08′ E
CD	521.93	S 56° 50′ E
DE	573.05	—
EA	—	S 80° 44′ W

Compute the missing length and bearing.

8-53. In the traverse of Fig. 8-29(b) the following lengths and bearings were measured:

Side	Length (m)	Bearing
AB	133.25	N 38° 42′ W
BC	183.17	N 45° 24′ E
CD	163.89	S 62° 34′ E
DE	179.94	—
EA	—	S 75° 00′ W

Compute the missing length and bearing.

8-54. In the traverse of Fig. 8-29(c) the following lengths and bearings were measured:

Side	Length (ft)	Bearing	Side	Length (ft)	Bearing
AB	—	N 23° 06′ E	DE	—	S 29° 52′ E
BC	317.95	N 89° 30′ E	EF	356.35	S 83° 26′ W
CD	420.19	S 6° 28′ W	FA	261.64	N 28° 26′ W

Compute the missing sides.

8-55. In the traverse of Fig. 8-29(c) the following lengths and bearings were measured:

Side	Length (ft)	Bearing	Side	Length (ft)	Bearing
AB	—	N 28° 18′ E	DE	—	S 24° 40′ E
BC	529.70	S 85° 18′ E	EF	593.68	S 88° 38′ W
CD	700.04	S 11° 40′ W	FA	435.89	N 23° 14′ W

Compute the missing sides.

8-56. In the traverse of Fig. 8-29(c) the following lengths and bearings were measured:

Side	Length (m)	Bearing	Side	Length (m)	Bearing
AB	—	N 33° 20′ E	DE	—	S 19° 38′ E
BC	141.172	S 80° 16′ E	EF	158.224	N 86° 20′ W
CD	186.563	S 16° 42′ W	FA	116.176	N 18° 12′ W

Compute the missing sides.

8-57. Part off the southwesterly one-quarter portion of the area enclosed by the traverse of Fig. 8-13 by a line passing through point *E*. Give the position of the partition point along line *CD* with respect to point *C*.

8-58. Part off the easterly one-third of the area enclosed by the traverse of Fig. 8-13 by a line parallel to *BC*. Give the positions of the partition points with respect to points *A* and *D*.

8-59. Plot the traverse *ABCD* of Section 8-22 by the method of distance and bearings outlined in Section 8-32. Use a plotting scale 1 in. = 50 ft.

8-60. Plot the boundary *APRS* of Section 8-22 by coordinates using a scale 1 in. = 50 ft with 100-ft grid spacings.

BIBLIOGRAPHY

Burkholder, E. F. 1986. Coordinates, calculators, and intersections. *Surveying and Mapping* 46:29.

Graff, D. R. 1973. Survey coordinate systems. *Proc. Fall Conv. Amer. Cong. Surv. & Mapp.* p. 635.

Harris, L. W. 1970. A practical system for adjusting a network of survey traverses. *Proc. 30th Ann. Meet. Amer. Cong. Surv. & Mapp.* p. 635.

Hillman, H. F. 1978. Further uses and refinements of analytic geometry in surveying computations. *Proc. Ann. Meet. Amer. Cong. Surv. & Mapp.* p. 322.

McKelvey, G. R. 1975. Wild T-2000 GRE 3 computation drafting: a complete field to finish system for all types of surveys. *Tech. Papers Ann. Meet. Amer. Cong. Surv. & Mapp.* p. 42.

McReynolds, J. E. 1991. Open ended electronic survey traversing. *Tech. Papers Ann. Conv. Amer. Cong. Surv. & Mapp.* 1:180.

Mikhail, E. M., and Gracie, G. 1981. *Analysis and Adjustment of Survey Measurements.* New York: Van Nostrand Reinhold.

Richardus, P. 1966. *Project Surveying.* New York: Wiley.

Smith, R. J., and Toler, J. P. 1978. Mapping control with simultaneous double traverse system using radial side shots. *Proc. Fall Conv. Amer. Cong. Surv. & Mapp.* p. 345.

Stoughton, H. W. 1975. Computing missing elements in a polygon. *Surveying and Mapping* 35:217.

9 Horizontal Control Networks

9-1 Introduction

The methods of measuring distances presented in Chapter 2 and those of measuring angles presented in Chapter 4 were combined in the form of traverse discussed in Chapter 8. The objective of traverse is to establish the horizontal positions of points in a survey, these positions being expressed as X and Y coordinates on a plane coordinate system. The positions are determined by measurement, computations, and adjustments.

The horizontal positions of points in a network developed to provide accurate control for subsidiary surveys can be obtained in a number of different ways in addition to traversing. These methods, called *triangulation, trilateration, intersection, resection,* and *satellite positioning,* are generally used.

The method of surveying called *triangulation* is based on the trigonometric proposition that if one side and the three angles of a triangle are known, the remaining sides can be computed. Furthermore, if the direction of one side is known, the directions of the remaining sides can be determined. In Fig. 9-1(a) the side BA of triangle CBA is known, having been measured directly with a tape or EDM, or else computed from a previous triangle side. The three angles have been measured in the field. The triangle can be solved by the law of sines to give the lengths of AC and CB.

If the azimuth of $AB,$ and thus of $BA,$ is known, then the azimuths of AC and BC can be determined from the measured angles. Also, since this is a closed figure and since the three measured angles should add up to 180° plus any spherical excess (see Section 9-5), the internal consistency of the angle measurements can be determined. The triangle is now like a closed traverse because the directions and lengths of all the lines are known. If the coordinates of either A or B are known, the coordinates of the other two points can be computed by application of principles developed in Chapter 8.

The methods of triangulation are quite demanding. They require a considerable amount of precise angle measurement with a minimum amount of distance measurement. The triangles are developed into a net of interconnected figures, and certain lines, called *baselines,* must be measured to compute the lengths of the other lines in the net. The baselines must be measured with extreme precision, since errors propagated through the system originate with these measured lines.

To eliminate the effects of random errors as much as possible, triangulation systems always include more than the minimum number of measurements necessary to fix the positions of the points in the triangulation net. These extra, or redundant, measurements provide the data necessary for the adjustment of the net, usually by the method of least squares.

Since the advent of long-range EDMs, a method of surveying, called *trilateration,* was adopted to combine with triangulation in order to extend horizontal control. Trilateration is based on the trigonometric proposition that if the three sides (laterals) of a triangle are known, the three angles can be computed. In Fig. 9-1(b) all three sides of the triangle *FED* have been measured with EDM. Thus the three angles can be computed by formulas given in Table C-3 of Appendix C. This accomplishes the same thing as triangulation, namely the determination of all the sides and all the angles in the triangle. Directions of the lines and positions of the points in the triangle can then be computed as in traverse and triangulation.

Frequently in surveying an occasion arises in which a point, whose position is to be determined, is not occupied—for example, a prominent church spire. Point J in Fig. 9-1(c) is such a point. This point can be determined by measuring the two angles G and H at the two ends of a known line and determining the third angle at J by computation since $G + H + J = 180°$ plus spherical excess. The angle J is called a *concluded angle.* The method is referred to as *intersection* because J lies at the intersection of the two lines from G and H.

The intersection shown in Fig. 9-1(c) has no geometric check because angle J is concluded. This is the major difference between the triangulation solution shown in view (a) and the intersection solution shown in view (c). As a practical matter, an intersection station such as J is observed from three or more regular tri-

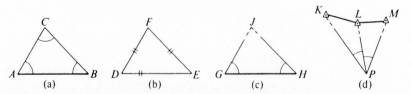

FIGURE 9-1 Basic horizontal control configurations. (a) Trangulation. (b) Trilateration. (c) Intersection. (d) Resection.

angulation stations in the net. This provides the necessary checks on the accuracy and reliability of the intersection station.

The method of locating a point P in Fig. 9-1(d) from at least three known points K, L, and M without having occupied the known points is called *resection*. In this system only the point to be determined is occupied, and the angles between the known stations are measured. A minimum of three known points is required to determine the position of an unknown point by the method of resection. The situation shown in Fig. 9-1(d) thus has no check because it just meets the minimum conditions. To provide the check necessary to test the reliability of the computed position of the unknown point, four or more known stations in the net are observed.

The modern total station instrument has the capability of measuring both horizontal distances and angles. Furthermore, the on-board microprocessor can perform COGO calculations in virtually real time. This instrument can be used to perform a *two-point resection*, as discussed in Section 9-14, in order to compute the position of the occupied point.

In satellite positioning, a ground receiver at the point to be located receives signals that are continuously transmitted from passing satellites. The output from the receiver is used to compute the X, Y, and Z coordinates of the occupied point. Today the dominant satellite system is the Global Positioning System (GPS), which is explained in Chapter 10.

9-2   Control Stations

The system of control points covering the United States is now called the National Spatial Reference System (NSRS) and includes horizontal control points (latitude and longitude or plane coordinates), vertical control (orthometric or ellipsoidal heights or elevations), gravity values and, in rare cases, crustal motion values (the velocity of points). Any given point may have one or more of these values attached to it. There are about 1×10^6 benchmarks (vertical control points) and 3×10^5 horizontal control points, and there are millions of gravity values, although not all gravity points are monumented. With the advent of GPS there are a growing number of so called 3D points that have X, Y, and Z coordinates and hence latitude, longitude, *and* (ellipsoidal) height. Through the years this network has been built up by linking monuments using carefully planned observations. In the past these were mostly triangulation, trilateration, and other traditional control measurements. Today all horizontal geodetic control is established using GPS, and GPS is even being used to establish vertical control (see Section 10-8).

As a result of this transition between traditionally established geodetic control practice and GPS, the traditional standards established by the Federal Geodetic Control Subcommittee (FGCS) of the Federal Geographic Data Committee (FGDC) are in a state of flux. However, a draft document of the Federal Geodetic Control Committee (FGCC, now known as the FGCS) lists the information in Table 9-1 as the relative positioning accuracy standards for three-dimensional surveys using space system techniques. From this table it is clear that GPS is capable of determining baselines at accuracies of $1/10^8$.

TABLE 9-1 Geometric Relative Positioning Accuracy Standards for Three-Dimensional Surveys Using Space System Techniques

		(95 Percent Confidence Level)		
		Minimum Geometric Accuracy Standard		
Survey Categories	Order	Base Error	Line-Length Dependent Error	
		e (cm)	p (ppm)	a(1:a)
Global-regional geodynamics; deformation measurements	AA	0.3	0.01	1:100,000,000
National Geodetic Reference System, "primary" networks; regional-local geodynamics; deformation measurements	A	0.5	0.1	1:10,000,000
National Geodetic Reference System, "secondary" networks; connections to the "primary" NGRS network; local geodynamics; deformation measurements; high-precision engineering surveys	B	0.8	1	1:1,000,000
National Geodetic Reference System (terrestrial based); dependent control surveys to meet mapping, land information, property, and engineering requirements	(C)			
	1	1.0	10	1:100,000
	2-I	2.0	20	1:50,000
	2-II	3.0	50	1:20,000
	3	5.0	100	1:10,000

Note: For ease of computation and understanding, it is assumed that the accuracy for each component of a vector baseline measurement is equal to the linear accuracy standard for a single-dimensional measurement at the 95 percent confidence level. Thus the linear one-standard deviation (s) is computed by

$$s = \pm\left[\sqrt{e^2 + (0.1d \bullet p)^2}\right]/1.96$$

where *d* is the length of the baseline in kilometers.

9-3 Station Marks

Another major change in establishing control networks such as the NSRS is the placement of 3D monuments or marks. While one of the most stable marks is still a bronze marker cemented into solid bedrock, the use of 3D type monuments is now advocated by the NGS. It was found that concrete monuments in the form of cylinders or rectangular solids were subject to small but measurable frost movement, hence NGS designed the monument shown in Fig. 9-2. This monument was first used for maintaining height or elevation data, that is, for benchmarks, but when GPS became the technology of choice for horizontal control, such monuments were used for both horizontal and vertical control, hence the name 3D.

It is helpful and advisable to place an azimuth mark at a distance of not less than one-quarter mile from the station. Also, as a result of the expense of establishing such a station, it is advisable to have at least two reference marks so the station can be found more easily and/or reproduced if temporarily lost or destroyed. An azimuth or a reference mark consists of a bronze disk and bears an arrow pointing to the station mark.

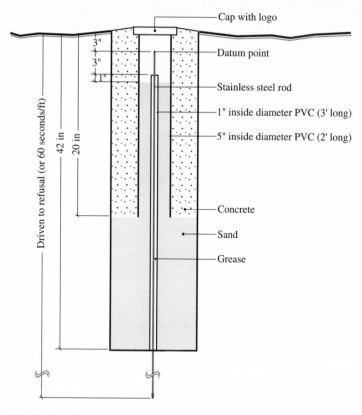

FIGURE 9-2 Diagram of an NGS 3D marker.

To ensure that a station will be of future use to other engineers, a very complete description of the station and its location is prepared.

This description should include the type of monument, the references used, and its location. The description of the location begins with the state and county and the distance and direction from the nearest town. It also includes the position in a particular quarter section (see Chapter 18) and the position with respect to the reference marks and nearby topographic features.

9-4 Baseline Measurement and Correction

Baselines are measured either by using invar tapes and related apparatus, by EDMs discussed in Chapter 2, or by GPS observations. The EDM shown in Fig. 2-20 is a long-range instrument suitable for the measurement of long baselines.

The principal corrections applied to field measurements of all baselines are for temperature, slope, and for reduction of the length to sea-level length. For a taped baseline the corrections for sag and tension are usually avoided by having the tape standardized under the conditions to which it will be subjected in the field.

Generally, all linear distances are reduced to their equivalent sea-level lengths. In Fig. 9-3, A and B are two points on the earth's surface at an average elevation h above sea level. Since the lengths of arcs are proportional to their radii,

$$\frac{s}{R} = \frac{D}{R+h} \quad \text{or} \quad s = \frac{DR}{R+h} \tag{9-1}$$

in which s is the sea level length, D is the measured length at an average elevation h, and R is the radius of curvature of the earth. When h is in feet, $R = 20,906,000$ ft; when h is in meters, $R = 6,372,200$ m. The more exact values of R are dependent on the mean latitude, the azimuth of the base, and the dimensions of the ellipsoid used to represent the sea-level surface (see Section 1-2).

For a baseline at an average elevation of 6100 ft and a measured length of 4324.186 ft, the sea-level length is, by Eq. (9-1),

$$s = \frac{4324.186 \times 20,906,000}{20,906,000 + 6100} = 4322.925 \text{ ft}$$

The use of EDMs eliminates the need for elaborate baseline preparation when taping such a line. The general requirement for a baseline that is to be measured by means of one of these instruments is a clear line of sight between the two ends of the base.

The time required to make the necessary instrument readings during the actual measuring operations is independent of the length of the baseline. When an EDM is used under normal conditions, the instrument can be set up and aligned, and all the readings can be made and recorded in minutes. There is a tremendous saving in time compared to taping, especially when baselines to 30 miles in length are considered.

The corrections for meteorological conditions and instrumental errors discussed in Sections 2-27 and 2-28 are applied to the measurements to obtain the corrected slope distance. To reduce the slope distance to sea level as required, either the reciprocal vertical angles are measured at the two ends of the line or else the elevations of the two ends of the line must be known. The methods of making the slope reductions are given in Sections 9-11 and 9-12.

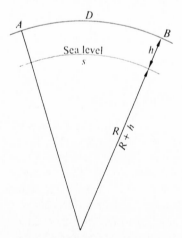

FIGURE 9-3 Reduction to sea level.

GPS lends itself to the rapid measurement of a baseline length as well as to the determination of the azimuth of the baseline. GPS is generally not used to measure baselines for triangulation or trilateration but is usually used in conjunction with existing control networks.

9-5 Spherical Excess

When a theodolite or total station is set up and leveled at a control station, the vertical axis of the instrument assumes a plumb position and the line of sight sweeps through a vertical plane. The vertical plane containing the line of sight to a distant station intersects the sea-level surface along an arc, which for the sake of this analysis is like the arc of a great circle. Thus in Fig. 9-4 a vertical plane containing the line of sight from A to B intersects the sphere (sea level) along arc AB. The direction reading on the theodolite's horizontal circle is represented by the line AB'. Also, a vertical plane containing the line of sight from A to C intersects the sphere along the arc AC. The direction reading on the theodolite's horizontal circle is represented by the line AC'. The horizontal angle $B'AC'$ obtained from the two direction readings is equal to the spherical angle BAC. The same situation would exist at stations B and C. Thus the theodolite effectively measures the spherical angles at the control stations forming the spherical triangle ABC.

The amount by which the sum of the three angles of a spherical triangle exceeds 180° is called the *spherical excess*. Before the closing error in the angles of a large triangle can be determined, the spherical excess must be known. The amount of this excess is dependent on the area of the triangle and on the latitudes of the vertexes. It is approximately equal to 1″ for every 75.5 mile² of area. A more exact value for the excess in seconds is

$$E'' = \frac{bc \sin A}{2R^2 \sin 1''} = mbc \sin A \tag{9-2}$$

FIGURE 9-4 Spherical excess.

(a)   (b)   (c)   (d)

FIGURE 9-5 Types of triangulation figures.

in which b, c, and A are the two sides and the included angle of the triangle, and R is the radius of curvature of the earth. Most tables intended for geodetic computations contain values of m for various latitudes. When the triangles are small, the mean radius of the earth can be used. For a triangle in latitude 45°, $m = 2.355 \times 10^{-10}$ when the dimensions are in feet and $m = 2.535 \times 10^{-9}$ when the dimensions are in meters. The excess is divided equally between the three angles of the triangle, the corrections being subtracted from the observed values.

9-6   Types of Figures in Triangulation

Although triangles are the basic figures in a triangulation net, they are not allowed to exist alone because they do not provide sufficient checks on the measurements and subsequent computations. Triangles are combined to form other figures that are considered as geometric entities. The most common figure used is the braced quadrilateral. The quadrilateral, shown in Fig. 9-5(a), comprises four triangulation stations in which there are eight measured angles. The term *braced* comes from the fact that the two diagonal lines are observed, thus bracing the four-sided outer figure.

The figure shown in view (b) is a center-point triangle, which is actually a quadrilateral with one of its corners lying inside the other three. If the center point is moved downward outside the triangle, for example, a braced quadrilateral results.

A center-point pentagon is shown in view (c), and a quadrilateral containing a center point is shown in view (d). Other types of figures can be identified in Fig. 19-1. These figures are combined either into chains or into area-wide nets.

9-7   Computation of a Quadrilateral

Using the quadrilateral shown in Fig. 9-6 as an example, two sides of each triangle are computed using the law of sines, since one side of the triangle is always known, either as a measured baseline or as a side computed from a previous triangle solution.

Table 9-2 shows the computations of determining the lengths of the example quadrilateral using the following adjusted angles:

$$a = 63° \ 17' \ 28.12''$$
$$b = 84° \ 18' \ 20.37''$$
$$c = 17° \ 52' \ 26.69''$$

$$d = 30° \ 41' \ 16.98''$$
$$e = 47° \ 07' \ 55.96''$$
$$f = 66° \ 35' \ 55.30''$$
$$g = 35° \ 34' \ 51.76''$$
$$h = 14° \ 31' \ 44.82''$$

The length of side AB is known to be 2899.06 ft.

The stations in triangle CBA are listed in the order determined by beginning with the unknown point C and going around the triangle in a clockwise direction. The adjusted angles are filled in as shown in Table 9-2. With the side BA known, the sides AC and CB are computed by using the law of sines.

$$\frac{AC}{\sin \ (c+d)} = \frac{BA}{\sin \ e} \quad \text{or} \quad AC + \frac{BA \sin \ (c+d)}{\sin \ e} \qquad (9\text{-}3a)$$

Also

$$\frac{CB}{\sin \ b} = \frac{BA}{\sin \ e} \quad \text{or} \quad CB = \frac{BA \sin \ b}{\sin \ e} \qquad (9\text{-}3b)$$

After the triangles are laid out as shown in Table 9-2, the adjusted angles are added to check that their sum is 180°. The length of the starting side BA of the first triangle is entered in the last column followed by the sines of the angles. The lengths of AC and CB are then computed by Eqs. (9-3a) and (9-3b) and shown in the table. The line $CA = AC$ is then used as the starting side of triangle DCA.

The sines of the angles are shown in Table 9-2 to clarify the steps involved in the triangle solutions. These need not be entered if the calculator generates the sines internally, as most calculators do.

The two triangles DBA and CBD in the second route are solved by starting from side BA and following a similar procedure. This route provides a check on the side DC and also shows consistency of the sides AD and CB.

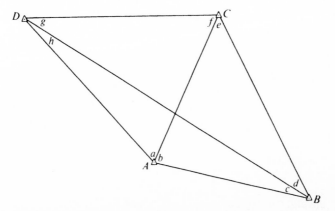

FIGURE 9-6 Quadrilateral.

9-8 Computation of Plane Coordinates

The objective of triangulation is to establish the horizontal positions of the triangulation stations in the network relative to one another and with respect to a horizontal datum. Chapter 11 contains a description of the state plane coordinate systems that should be the basis of any triangulation to be established on a plane coordinate system. The triangulation network must include at least one point, and preferably should include two or more points, the horizontal position of which is known with respect to the state plane coordinate system. In the network of Fig. 9-7(b), if A and B are points with known coordinates, the length and azimuth of the line AB can be computed by using Eqs. (8-12) and (8-13). If point C is also a point with known coordinates, then a check on the entire network is provided, because the discrepancy between the fixed position of C and its position as computed through the quadrilaterals can be determined.

If the coordinates of stations A and B in Fig. 9-6 are known, the length of line BA computed by Eq. (8-12) is used as the starting line in the computations for the lengths of the sides of the quadrilateral shown in Table 9-2. The azimuth from north of the line AB is determined by Eq. (8-13). The azimuth of the line AC is equal to the azimuth of AB minus angle b. The azimuth of BC is equal to the azimuth of BA plus angle $(c + d)$. Furthermore, when the azimuths of AC and BC have been computed, the difference between the two must be equal to angle e. If the figure has been adjusted, any discrepancy represents a mistake. These computations are shown in Table 9-3 for the two triangles CBA and DCA that will be used to compute the positions of C and D.

After the lengths of the lines AC and BC have been computed by solving the triangles, and the azimuths of the lines AC and BC have been determined, the latitudes and departures of these two lines are computed by Eqs. (8-3) and (8-4). Applying the latitudes and departures of AC and BC to the coordinates of A and B, respectively, in accordance with Eqs. (8-7) and (8-8), gives two sets of values of the coordinates of C. These two sets of values should not differ by more than 0.01 or 0.02 ft. The slight discrepancy is due either to rounding-off errors or to slight inconsistencies in the triangles.

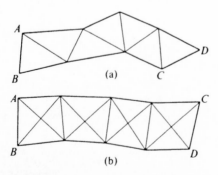

(a)

(b)

FIGURE 9-7 Triangulation nets.

TABLE 9-2 Computation of Lengths in Triangles

Sides	Stations	Angles	Adjusted Angles	Distances
BA				2899.06
	C	e	47° 07′ 55.96″	0.732 92548
	B	c + d	48° 33′ 43.67″	0.749 67381
	A	b	84° 18′ 20.37″	0.995 06538
AC			180° 00′ 00.00″	2965.31
CB				3935.94
CA				2965.31
	D	g + h	50° 06′ 36.58″	0.767 27890
	C	f	66° 35′ 55.30″	0.917 74558
	A	a	63° 17′ 28.12″	0.893 30193
AD			180° 00′ 00.00″	3546.82
DC				3452.35
BA				2899.06
	D	h	14° 31′ 44.82″	0.250 87197
	B	c	17° 52′ 26.69″	0.306 92610
	A	a + b	147° 35′ 48.49″	0.535 87391
AD			180° 00′ 00.00″	3546.82
DB				6192.52
BD				6192.52
	C	e + f	113° 43′ 51.26″	0.915 44565
	B	d	30° 41′ 16.98″	0.510 36357
	D	g	35° 34′ 51.76″	0.581 85393
DC			180° 00′ 00.00″	3452.35
CB				3935.95

TABLE 9-3 Computation of Azimuths of Lines in Triangles

Azimuth AB	=	104° 42′ 56.60″		Azimuth BA	=	284° 42′ 56.60″
− Angle b	=	− 84° 18′ 20.37″		+ Angle (c + d)	= +	48° 33′ 43.67″
Azimuth AC	=	20° 24′ 36.23″		Azimuth BC	=	333° 16′ 40.27″
		+ 360°				
		380° 24′ 36.23″				
		−333° 16′ 40.27″				
Angle e	=	47° 07′ 55.96″ (check)				
Azimuth AC	=	20° 24′ 36.23″		Azimuth CA	=	200° 24′ 36.23″
− Angle a	=	− 63° 17′ 28.12″		+ Angle f	= +	66° 35′ 55.30″
Azimuth AD	=	317° 07′ 08.11″		Azimuth CD	=	267° 00′ 31.53″
		−267° 00′ 31.53″				
Angle (g + h)	=	50° 06′ 36.58″ (check)				

TABLE 9-4 Position Computation

Station	Length	Cos Azimuth	Sin Azimuth	Y	X
A				622,516.21	1,442,416.25
	2965.31	+0.937 22075	+0.348 73667	+ 2,779.15	+ 1,034.11
C				625,295.36	1,443,450.36
	3935.94	−0.893 19764	+0.449 66429	− 3,515.57	+ 1,769.85
B				621,779.79	1,445,220.21
			Fixed	621,779.78	1,445,220.22
A				622,516.21	1,442,416.25
	3546.82	+0.732 76763	−0.680 47894	+ 2,598.99	− 2,413.54
D				625,115.20	1,440,002.71
	3452.35	+0.052 18331	+0.998 63752	+ 180.16	+ 3,447.65
C				625,295.36	1,443,450.36
			Fixed	625,295.36	1,443,450.36

The computation just discussed is known as a *position computation.* Table 9-4 shows the position computation for the quadrilateral in Fig. 9-6 when the angles given in Section 9-7 and the lengths determined in Table 9-2 are used. Each triangle is solved as if it were a two-sided traverse beginning on a known point and closing on a known point. The slight discrepancies are caused by rounding-off errors.

The coordinates of all the stations throughout the triangulation system are computed by using the strongest route of triangles in the network. When a station is reached whose coordinates are fixed, the positions of the intermediate stations can then be adjusted. If the triangulation system is of great extent and high precision, this adjustment should be made by an application of the least-squares principle. If the system is moderate in extent, an application of the compass rule described in Section 8-15 will give highly satisfactory results. In this case a traverse extending from one fixed point to another fixed point and including all the intermediate triangulation stations is selected in as direct a line as possible. This traverse is then adjusted by the compass rule.

9-9 Geodetic Positions

When the control survey system extends over a large area, as does the work of the National Geodetic Survey, the coordinates of the stations are given as the latitudes and longitudes on an adopted reference ellipsoid. These equations are necessarily complex. In computing these geodetic positions, the latitude φ_1 and the longitude λ_1 of one station must be known. The coordinates φ_2 and λ_2 of any adjacent station can be computed when the azimuth (reckoned clockwise from north) and the length of the line connecting the two stations are known. Note that the azimuth of the line at each point will in general be different. In the sequel the following notation is used:

a, b, major and minor semiaxes of the reference ellipsoid

f, flattening $= (a - b)/a$

Φ, geodetic latitude, positive north of the equator

L, difference in longitude, positive east

s, length of the geodesic (ellipsoidal distance between P_1 and P_2)

α_1, α_2, azimuths of the geodesic, clockwise from north, α_2 in the direction of $P_1 P_2$ produced

α, azimuth of the geodesic at the equator

$u^2 = \cos^2 \alpha (a^2 - b^2)/b^2$

U, reduced latitude, defined by $\tan U = (1 - f) \tan \Phi$

λ', difference in longitude on an auxiliary sphere

σ, angular distance $P_1 P_2$ on the sphere

σ_1, angular distance on the sphere from the equator to P_1

σ_m, angular distance on the sphere from the equator to the midpoint of the line

To determine the latitude and longitude of P_2 given Φ_1, λ_1, s, and α_1, the following steps are taken:

$$\tan \sigma_1 = \tan U_1 / \cos \alpha_1 \tag{9-4}$$

$$\sin \alpha = \cos U_1 \sin \alpha_1 \tag{9-5}$$

$$A = 1 + \frac{u^2}{256}[64 + u^2(-12 + 5u^2)] \tag{9-6}$$

$$B = \frac{u^2}{512}[128 + u^2(-64 + 37u^2)] \tag{9-7}$$

$$2\sigma_m = 2\sigma_1 + \sigma \tag{9-8}$$

$$\Delta\sigma = B \sin \sigma [\cos 2\sigma_m + \tfrac{1}{4} B \cos \sigma (-1 + 2 \cos^2 \sigma_m)] \tag{9-9}$$

where σ is first approximated by $\dfrac{s}{bA}$, then

$$\sigma_{new} = \frac{s}{bA} + \Delta\sigma \tag{9-10}$$

Therefore

$$\tan \Phi_2 = \frac{\sin U_1 \cos \sigma_{new} + \cos U_1 \sin \sigma_{new} \cos \alpha_1}{(1 - f)[\sin^2 \alpha + (\sin U_1 \sin \sigma_{new} - \cos U_1 \cos \sigma_{new} \cos \alpha_1)^2]^{\frac{1}{2}}} \tag{9-11}$$

$$\tan \lambda' = \frac{\sin \sigma_{new} \; \sin \alpha_1}{\cos U_1 \cos \sigma_{new} - \sin U_1 \sin \sigma_{new} \cos \alpha_1} \tag{9-12}$$

$$C = \frac{f}{16} \cos^2 \alpha [4 + f(4 - 3\cos^2 \alpha)] \tag{9-13}$$

$$L = \lambda' - (1 - C)f \sin \alpha \{\sigma_{new} + C \sin \sigma_{new}[\cos 2\sigma_m + C \cos \sigma_{new}(-1 + 2\cos^2 2\sigma_m)]\} \tag{9-14}$$

$$\lambda_2 = \lambda_1 + L \tag{9-15}$$

$$\tan \alpha_2 = \frac{\sin \alpha}{-\sin U_1 \; \sin \sigma_{new} + \cos U_1 \cos \sigma_{new} \cos \alpha_1} \tag{9-16}$$

In long lines Eqs. 9-8, 9-9, 9-10 are iterated until σ_{new} does not change. See the references at the end of the chapter for more information.

9-10 Precise Traverse

Under certain conditions a traverse of high precision may be required. In establishing control for a city survey, where many control points are needed, the traverse can frequently be used to advantage.

Electronic distance measuring systems are successfully employed in measuring the lengths of lines in precise traverses. Their use, however, is restricted to traverses composed of lines at least 500 ft in length. The amount of time needed to measure a line 500 ft long with the EDM is about the same as that required to measure the same line by using a tape. On shorter lines, taping the lines will prove more economical. Also, the inherent instrument errors, representing only a very small fraction of the length of a long line, are too large for lines under 500 ft in length.

GPS can be employed to measure the sides of the traverse as well as to determine the positions of the traverse stations directly, either as latitudes and longitudes or as plane coordinates on the state plane coordinate system (see Chapter 11). Thus it is not necessary to compute latitudes and departures as discussed in Chapter 8.

Traverses are classified as first order, second order, or third order, according to the degree of precision attained. The requirements for the various orders are shown in Table 9-5.

No check on the precision of the traverse is available until the traverse has been closed and the latitudes and departures and closures have been computed. The discrepancy is sometimes of such size that it is difficult to tell whether it is caused by a blunder or by an accumulation of small errors in the traverse. When a blunder has been made, it is often difficult to locate the incorrect value, and it may be necessary to rerun a large part or all of the traverse line. In triangulation, as each quadrilateral is completed in the field, there are checks on both the angles and the lengths. In GPS surveys of networks there are also additional checks available.

9-11 Reduction of Slope Distance by Vertical Angles

In Fig. 9-8 the positive angle α and the negative vertical angle β are measured from stations A and B, respectively. Vertical distances and angles have been greatly exaggerated in the diagram. The angle e is the refraction angle of the line of sight, assumed to be the same at both stations. It is further assumed that the height of the instrument above the ground and the height of the signal above the ground in each instance are the same, or that the two angles have been reduced to an equivalent situation (see Section 2-29).

The angle subtended at the center of the earth between the vertical lines through A and B is $\boldsymbol{\theta}$. In triangle OVB the angle at V is $90° - \boldsymbol{\theta}$, as shown, because the angle at B is $90°$.

The distance G is measured along the refracted line from A to B by the EDM. This distance, within the range of the instrument, can be taken as the straight-line slope distance AB. The difference between the two distances in 100 miles is less than a foot, decreasing to a negligible amount in the range of the instruments.

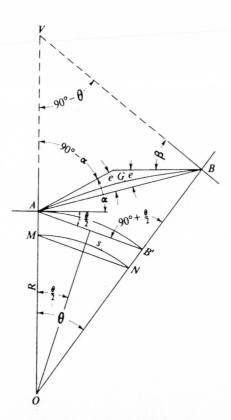

FIGURE 9-8 Reduction to sea-level distance by vertical angles.

TABLE 9-5 Classification of Traverses

Classification	First Order	Second Order		Third Order	
		Class I	Class II	Class I	Class II
Recommended spacing of principal stations	Network stations 10 to 15 km. Other surveys seldom less than 3 km.	Principal stations seldom less than 4 km except in metropolitan area surveys where the limitation is 0.3 km.	Principal stations seldom less than 2 km except in metropolitan area surveys where the limitation is 0.2 km.	Seldom less than 0.1 km in tertiary surveys in metropolitan area surveys. As required for other surveys.	
Horizontal directions or angles					
Instrument	0".2	0".2	0".2	1".0	1".0
Number of observations	16	8 or 12*	6 or 8*	4	2
Rejection limit from mean	4"	4" or 5"	4" or 5"	5"	5"
Length measurements					
Standard error	1 part in 600,000	1 part in 300,000	1 part in 120,000	1 part in 60,000	1 part in 30,000

*May be reduced to 8 and 4, respectively, in metropolitan areas.

Reciprocal vertical angle observations					
Number of and spread between observations	3D/R-10"	3D/R-10"	2D/R-10"	2D/R-10"	2D/R-20"
Number of stations between known elevations	4–6	6–8	8–10	10–15	15–20
Astro azimuths					
Number of courses between azimuth checks	5–6	10–12	15–20	20–25	30–40
Number of observations /night	16	16	12	8	4
Number of nights	2	2	1	1	1
Standard error	0".45	0".45	1".5	3".0	8".0
Azimuth closure at azimuth check point not to exceed	1".0 per station or 2" $\sqrt{N}$	1".5 per station or 3" $\sqrt{N}$. Metropolitan area surveys seldom to exceed 2".0 per station or 3" $\sqrt{N}$	2".0 per station or 6" $\sqrt{N}$. Metropolitan area surveys seldom to exceed 4".0 per station or 8" $\sqrt{N}$	3".0 per station or 10" $\sqrt{N}$. Metropolitan area surveys seldom to exceed 6" per station or 15" $\sqrt{N}$	8" per station or 30" $\sqrt{N}$
Position closure					
After azimuth adjustment	0.04 m $\sqrt{K}$ or 1:100,000	0.08 m $\sqrt{K}$ or 1:50,000	0.2 m $\sqrt{K}$ or 1:20,000	0.4 m $\sqrt{K}$ or 1:10,000	0.8 m $\sqrt{K}$ or 1:5,000

In the triangle AVB

$$(\beta + e) + (90° - \theta) + [(90° - \alpha) + e] = 180°$$

or

$$2e = \alpha + \theta - \beta$$

and

$$e = \frac{\alpha}{2} - \frac{\beta}{2} + \frac{\theta}{2}$$

In the triangle ABB'

$$\angle BAB' = \alpha - e + \frac{\theta}{2} = \alpha - \left(\frac{\alpha}{2} - \frac{\beta}{2} + \frac{\theta}{2}\right) + \frac{\theta}{2}$$

or

$$\angle BAB' = \frac{\alpha + \beta}{2} \tag{9-17}$$

Also the distance AB' is very nearly

$$AB' = AB \cos BAB' \text{ (approx)} \tag{9-18}$$

With the approximate value of AB', the value of θ can be obtained with sufficient accuracy by the relationship

$$\sin \frac{\theta}{2} = \frac{AB'}{2R} \tag{9-19}$$

in which R is the radius of the earth for the latitude of the area in the direction of the line.

In the triangle ABB'

$$\angle ABB' = 90° - \frac{\theta}{2} - \angle BAB' \tag{9-20}$$

The triangle ABB' can now be solved by the law of sines to give a more exact value of the distance AB'. Thus

$$AB' = \frac{AB \sin ABB'}{\sin[90° + (\theta/2)]} \tag{9-21}$$

The distance MN is the length of a chord connecting the sea-level positions of stations A and B, and is less than the length of the chord AB'. The amount to be subtracted from AB' is determined by an analysis of the sea-level correction given in Section 9-4. If C_1 denotes the amount by which the chord AB' must be reduced to obtain the chord MN, then

$$C_1 = AB' \frac{h_A}{R} \tag{9-22}$$

in which h_A is the elevation of station A, and R is the radius of the earth for the latitude of the area in the direction of the measured line.

The sea-level length s is greater than the corresponding chord length MN. The amount to be added to MN to obtain s is found as follows:

$$MN = 2R \sin \frac{\theta}{2} = 2R\left(\frac{\theta}{2} - \frac{\theta^3}{48} + \frac{\theta^5}{3840} - \cdots\right)$$

$$s = R\theta = 2R\frac{\theta}{2}$$

$$s - MN = 2R\frac{\theta}{2} - 2R\frac{\theta}{2} + 2R\frac{\theta^3}{48} - 2R\frac{\theta^5}{3840}$$

If the last term is neglected, the result is

$$C_2 = R\frac{\theta^3}{24} = \frac{\overline{MN}^3}{24R^2} \tag{9-23}$$

in which C_2 is the amount to be added to the chord length to obtain the sea-level length of the line, R is the radius of the earth expressed in the same units as the measured distance, and θ is the angle subtended by the line at the center of the earth, expressed in radians.

The value of C_2 is about 14 ft in a distance of 100 miles and decreases to 1 ppm of the measured length at about 20 miles. Thus, for distances less than 20 miles, the chord-to-arc correction given by Eq. (9-23) can be neglected.

The presentation given above for reducing slope distances by reciprocal vertical angles can be simplified by computing the effect of earth curvature as an angular value per 1000 ft of distance and then using this small angle in the reduction computations. In Fig. 9-9 the effect of curvature in 1 mile is shown to be 0.667 ft, in accordance with Eq. (3-1). The value 0.667 is computed for an average radius of curvature. The actual value is dependent on the radius of curvature of the earth along the line in question and will therefore vary a small amount from place to place.

The linear effect of curvature, BB' in Fig. 9-9, is converted to an angular value c by dividing BB' by the distance AB, giving

$$c'' = \frac{0.667}{5280 \text{ arc } 1''} = 26.06''/\text{mile}$$

or

$$e'' = 4.935''/1000 \text{ ft of distance} \tag{9-24a}$$

FIGURE 9-9 Angle of curvature.

In the metric system

$$c'' = \frac{0.0785}{1000 \text{ arc } 1''} = 16.19''/\text{km}$$ (9-24b).

In the triangle ABB' of Fig. 9-8 the angle c is the same as $\theta/2$. By Eq. (9-17), $\angle BAB'$ is the mean of the reciprocal vertical angles and the angle at B' is $90° + c$. Thus the angle at B' is $90° - c - \angle BAB'$, which corresponds to Eq. (9-20).

The triangle is solved by the law of sines to give the chord distance AB' which is then reduced to sea-level chord by Eq. (9-22), and to the sea-level length by Eq. (9-23).

EXAMPLE 9-1   The slope distance AB in Fig. 9-8, corrected for meteorological conditions and instrument constants, is 76,963.54 ft. The vertical angle at A is $+3° 02' 05''$, and at B is $-3° 12' 55''$. The elevation of A is 1056.67 ft. What is the sea-level length MN? (Note that these are the same data used in Example 3-1.)

Solution: The angle of curvature is, by Eq. (9-24a), $4.935 \times 76.963 = 379.8''$ or $6' 20''$. Angle $BAB' = (3° 02' 05'' + 3° 12' 55'')/2 = 3° 07' 30''$. The angle at B' = 90° $+ 06' 20'' = 90° 06' 20''$. The angle at B is $90° - (3° 07' 30'' + 0° 06' 20'') = 86° 46' 10''$. The distance AB' is, by the law of sines,

$$AB' = \frac{76,963.54 \sin 86° 46' 10''}{\sin 90° 06' 20''} = 76,841.36 \text{ ft}$$

Assuming a mean radius of curvature of 20,906,000 ft, the reduction of the chord AB to the chord MN is, by Eq. (9-22), $76,841.36 \times 1056.67/20,906,000 = 3.88$ ft. Thus $MN = 76,841.36 - 3.88 = 76,837.48$ ft. The corresponding sea-level distance is obtained by adding the value of C_2 given by Eq. (9-23) to MN. Thus

$$C_2 = \frac{76,837^3}{24 \times 20,906,000^2} = 0.04 \text{ ft}$$

which is quite negligible. The sea-level length is then

$$76,837.48 + 0.04 = 76,837.52 \text{ ft} \qquad \blacklozenge$$

If reciprocal vertical angles have not been measured, then a value for atmospheric refraction must be assumed. If this is taken to be about 14% of curvature, then the combined curvature and refraction angle can be taken as $(c + r) = 4.935 \times 0.86 = 4.24''/1000$ ft, or in the metric system $16.19 \times 0.86 = 13.92''/\text{km}$. Assuming the vertical angle at the lower station to have been measured, then angle $BAB' = \alpha + (c + r)$. If the vertical angle was measured at the upper station, then angle $BAB' = |\beta + (c + r)|$. The vertical angle β is always negative.

9-12   Reduction of Slope Distance by Station Elevations

Figure 9-10 shows the elevations of A and B. The distance AB is again assumed to be the straight-line distance measured with the EDM. This distance must be reduced to the chord distance AB'. The vertical distance BB' is the difference in elevation between A and B. Since the slopes are usually not great when trilateration is suitable, Eq. (2-6) can be used to determine the difference between the slope distance AB and the distance AC. Since $AB = AD$, then $AB - AC = CD$. By Eq. (2-6),

$$CD = \frac{\overline{BB'}^2}{2AB} = \frac{(\Delta h)^2}{2G} \qquad (9\text{-}25)$$

in which Δh is the difference in elevation, in feet or meters, between the two ends of the line, and G is the measured slope distance, also in feet or meters. If a line is 5 miles long and the difference in elevation between the two ends is 1000 ft, the last term in Eq. (2-5) amounts to about 0.07 ft. So, for shorter lines and large differences of elevation, this term should be evaluated.

The angle $\theta/2$ in Fig. 9-10 may be obtained by Eq. (9-19). But the value of AB' is not known. Its approximate value, however, is AC, which can be found by subtracting the distance CD computed by Eq. (9-25) from the measured distance G; and the distance AC can be used in Eq. (9-19) instead of AB'. Also

$$\sin \frac{\theta}{2} = \frac{B'C}{BB'} = \frac{B'C}{\Delta h}$$

and

$$B'C = \Delta h \sin \frac{\theta}{2} \qquad (9\text{-}26)$$

Then the distance $AB' = AB - (B'C + CD)$. Finally, the distance AB' is reduced to the sea-level distance by applying Eqs. (9-22) and (9-23).

The reduction using difference in elevation can be simplified by assuming that the angle $\theta/2$ in Fig. 9-10 is the curvature angle c'' given by Eq. (9-24), as was done in Section 9-11. Then the approximate value of AB' is

$$AB' = \sqrt{G^2 - \overline{BB'}^2} - BB' \sin c'' \qquad (9\text{-}27)$$

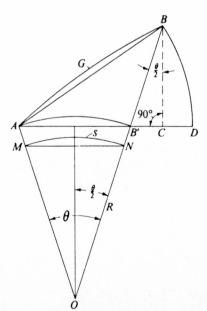

FIGURE 9-10 Reduction to sea-level distance by station elevations.

Since the value of c'' in the second part of the right side of Eq. (9-27) assumes that AB equals AB', the solution may require an iteration if the difference in elevation is substantial.

EXAMPLE 9-2   The slope distance between two points A and B, corrected for meteorological and instrumental effects, is 97,542.20 ft. The elevation of A is 4225.5 ft, and the elevation of B is 13,724.0 ft. What is the reduced sea-level distance?

Solution: The difference in elevation between the two points, BB' of Fig. 9-10 is 9498.5 ft. The curvature angle c'' is $4.935 \times 97.54 = 481'' = 0° \ 08' \ 01''$. The approximate value of AB' is, by Eq. (9-27),

$$AB' = \sqrt{97{,}542.20^2 - 9{,}498.5^2} - 9498.5 \sin 0° \ 08' \ 01''$$

or

$$AB' = 97{,}078.62 - 22.15 = 97{,}056.47 \text{ ft}$$

The approximate value of AB' is now used to obtain a second value of c''. Thus $c'' = 4.935 \times 97.06 = 479'' = 0° \ 07' \ 59''$. Then

$$AB' = 97{,}078.62 - 9498.5 \sin 0° \ 07' \ 59'' = 97{,}056.56 \text{ ft}$$

The difference between the first and second values of AB' is 0.09 ft, which is less than 1 ppm, of the measured distance.

The reduction to sea level is, by Eq. (9-22), $97{,}056.56 \times 4225.5/20{,}906{,}000 = 19.62$ ft and the reduced chord MN is $97{,}056.56 - 19.62 = 97{,}036.94$ ft. Since the distance is less than 20 miles, the chord-to-arc correction given by Eq. (9-23) can be neglected. The sea-level distance is thus 97,036.94 ft. ◆

9-13 Intersection

The method of intersection mentioned in Section 9-1 is applicable to a situation in which a point is to be established in the network of a triangulation system but in which the point is not occupied. The simplest case is that in which two angles are measured, one from each end of a known line. In Fig. 9-11 point P is observed from the two known points A and B where the angles α and β are measured. The azimuth of AP is then the azimuth of AB minus angle α. The azimuth of BP is the azimuth of BA plus angle β. The coordinates of P are then obtained by putting the two lines AP and BP in the form of Eq. (8-16) and solving the resulting pair of equations to obtain X_P and Y_P.

EXAMPLE 9-3   The coordinates of A in Fig. 9-11 are $X_A = 1{,}455{,}460.24$ and $Y_A = 640{,}276.08$; and the coordinates of B are $X_B = 1{,}463{,}876.62$ and $Y_B = 637{,}406.51$. Angle $\alpha = 74° \ 29' \ 16''$ and angle $\beta = 37° \ 46' \ 28''$. Compute the coordinates of P.

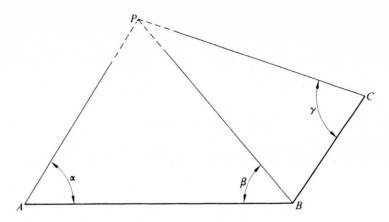

FIGURE 9-11 Intersection.

Solution: By Eq. (8-13),

$$\tan A_{AB} = \frac{X_B - X_A}{Y_B - Y_A}$$

$$
\begin{array}{ll}
X_B = 1{,}463{,}876.62 & Y_B = 637{,}406.51 \\
X_A = 1{,}455{,}460.24 & Y_A = 640{,}276.08 \\
\overline{X_B - X_A = \phantom{1{,}}8{,}416.38} & \overline{Y_B - Y_A = \phantom{1{,}}-2{,}869.57}
\end{array}
$$

and

$$\tan A_{AB} = \frac{8{,}416.38}{-2{,}869.57} = -2.93297602$$

$$
\begin{array}{ll}
A_{AB} = 108°\,49'\,36'' & A_{BA} = 288°\,49'\,36'' \\
-\alpha = -74°\,29'\,16'' & +\beta = +37°\,46'\,28'' \\
\overline{A_{AP} = 34°\,20'\,20''} & \overline{A_{BP} = 326°\,36'\,04''}
\end{array}
$$

Reducing X coordinates by 1,455,000 and Y coordinates by 637,000 for convenience gives $X'_A = 460.24$, $Y'_A = 3276.08$, $X'_B = 8876.62$, and $Y'_B = 406.51$. Then, by Eq. (8-16),

$$Y'_P - 3276.08 = \cot 34°\,20'\,20''(X'_P - 460.24)$$

$$Y'_P - 406.51 = \cot 326°\,36'\,04''(X'_P - 8876.62)$$

which reduce to

$$
\begin{array}{rl}
1.463810\,X'_P - Y'_P & = -2{,}602.38 \\
-1.516644\,X'_P - Y'_P & = -13{,}869.18 \\
\hline
2.980454\,X'_P & = 11{,}266.80 \\
X'_P & = 3780.23
\end{array}
$$

then

$$Y'_P = 8135.92$$

and

$$X_P = 1{,}455{,}000 + 3{,}780.23 = 1{,}458{,}780.23$$
$$Y_P = 637{,}000 + 8{,}135.92 = 645{,}135.92$$ ◆

Since there is no check on the position of the intersected point from observations only at A and B, a third point C is occupied, and the angle γ is measured. An intersection of lines AP and CP then gives an independent check on the computed position of the intersection point. If the intersection point is observed from more than the minimum of two control points, then its most probable position can be determined by the method of least squares. The variation of coordinates method is a convenient solution to a conditioned intersection. This is explained in Section A-6 of Appendix A.

9-14   Resection

The method of resection is a convenient way of determining the position of an unknown point by occupying the point and measuring the horizontal angles between at least three, and preferably more, control points. This is referred to as a *three-point* resection. The fact that the control points are not occupied, except perhaps to erect signals, makes the determination of the point somewhat risky, even when more than the minimum of three control stations are observed.

There are numerous approaches to the solution of the resection problem, and the interested reader should consult the readings at the end of this chapter for further study on the subject. The method presented here is but one of the available methods. The solution is set up for the hand or desk computer and is shown in Table 9-6.

In Fig. 9-12 A, B, and C are points whose positions (coordinates) are known and are visible from the unknown station P. If station P is occupied and the angles p'' and p' are measured, then these angles and the coordinates of the control points are sufficient to determine the position of P. The lengths and azimuths of the lines AB, BC, and CA are computed from the coordinates of these three points. The angle at A from C to B may be computed from the azimuths of the lines AC and AB. Let the angle at B from A to P be denoted by x and the angle at C from P to A be denoted by y, as shown in Fig. 9-12. Then, since the sum of the interior angles in a four-sided figure is 360°,

$$x + y + A + p'' + p' = 360°$$

Therefore

$$x + y = 360° - (A + p'' + p')$$

and

$$\tfrac{1}{2}(x + y) = 180° - \tfrac{1}{2}(A + p'' + p') \tag{9-28}$$

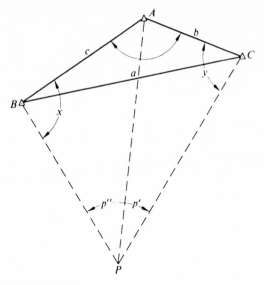

FIGURE 9-12 Three-point resection.

Furthermore $\sin x/AP = \sin p''/c$ and $\sin y/AP = \sin p'/b$. So

$$AP = \frac{c \, \sin \, x}{\sin \, p''} = \frac{b \, \sin \, y}{\sin \, p'}$$

Let $\sin y/\sin x = \tan Z$. Then

$$\tan Z = \frac{c \, \sin \, p'}{b \, \sin \, p''} \qquad (9\text{-}29)$$

When one is added to each side of the equation $\sin y/\sin x = \tan Z$, the result is

$$\frac{\sin \, y}{\sin \, x} + 1 = \tan Z + 1$$

or

$$\frac{\sin \, x + \sin \, y}{\sin \, x} = \tan Z + \tan 45° \qquad (9\text{-}30)$$

When each side of the same equation is subtracted from 1, the result is

$$1 - \frac{\sin \, y}{\sin \, x} = 1 - \tan Z$$

or

$$\frac{\sin \, x - \sin \, y}{\sin \, x} = 1 - \tan Z \tan 45° \qquad (9\text{-}31)$$

Dividing Eq. (9-31) by Eq. (9-30) gives

$$\frac{\sin\,x - \sin\,y}{\sin\,x + \sin\,y} = \frac{1 - \tan\,Z\,\tan\,45°}{\tan\,Z + \tan\,45°}$$

$$= \frac{1}{\tan(Z + 45°)} = \cot(Z + 45°)$$

The left side of Eq. (9-31) may be expressed in terms of the product of two functions, rather than in terms of the sum of two functions. Thus

$$\frac{2\,\cos\,\frac{1}{2}(x + y)\,\sin\,\frac{1}{2}(x - y)}{2\,\sin\,\frac{1}{2}(x + y)\,\cos\,\frac{1}{2}(x - y)} = \cot(Z + 45°)$$

or

$$\cot\,\frac{1}{2}\,(x + y)\,\tan\,\frac{1}{2}\,(x - y) = \cot(Z + 45°)$$

Hence

$$\tan\,\frac{1}{2}\,(x - y) = \cot(Z + 45°)\,\tan\,\frac{1}{2}\,(x + y) \qquad (9\text{-}32)$$

When the triangle ABC has been solved for the angle at A and the lengths b and c, the angle $\frac{1}{2}\,(x + y)$ is computed by Eq. (9-28). Then $\tan Z$ is computed by Eq. (9-29), and the angle $\frac{1}{2}\,(x - y)$ is computed by Eq. (9-32). Adding $\frac{1}{2}\,(x + y)$ and $\frac{1}{2}\,(x - y)$ gives x; subtracting $\frac{1}{2}\,(x - y)$ from $\frac{1}{2}\,(x + y)$ gives y. These two angles complete the data necessary to solve the triangles PBA and PAC, the length of the common side PA affording a check on the computations. The azimuths of the three lines BP, AP, and CP are determined from the known angles and azimuths. Their lengths are determined from the triangle solutions. The latitude and departure of any one of the lines are computed to determine the coordinates of P. The latitude and departure of another of the lines will afford a check on the coordinates of P.

A numerical example, indicating the manner in which the computations for solving a three-point resection may be tabulated, is shown in Table 9-6. Due regard must be given to the algebraic sign of $\tan\,\frac{1}{2}\,(x + y)$ and also of $\cot(Z + 45°)$ in the solution of Eq. (9-32). This determines whether $\frac{1}{2}\,(x - y)$ is positive or negative. If $\frac{1}{2}\,(x + y)$ is 90°, then the position of the unknown point is indeterminate since it lies on the circle passing through the three control stations.

In Fig. 9-13(a), $\frac{1}{2}\,(x + y) = 180° - \frac{1}{2}(A + p'' + p')$, as in Fig. 9-12. However, in Fig. 9-13(b) , $\frac{1}{2}\,(x + y) = \frac{1}{2}(A - p'' - p')$. Otherwise the solution for x and y is the same for the conditions in Fig. 9-13(b) as for those in Figs. 9-12 and 9-13(a).

A three-point resection can always be made to check mathematically, but *the mathematical checks do not provide a check on the field work or on the control points used.* The field angles may be checked by measuring the angles p'', p', and the exterior angle at P, since the sum of these three angles should be 360°.

The value of a three-point resection may be increased considerably by sighting on a *fourth* control point. The result is a conditioned three-point resection, since there are two independent solutions that should give the same position for the unknown point. If the discrepancy between the two computed positions of the point based on a conditioned resection is unreasonably large, then the cause may

TABLE 9-6 Computation of Three-Point Resection

<table>
<tr><td colspan="3" align="center">Determination of Angles</td></tr>
<tr><td>$c = 1642.83$ ft</td><td>$p' = 38°\ 12'\ 20''$</td><td>$\sin p' = 0.618\ 4846$</td></tr>
<tr><td>$b = 1076.44$ ft</td><td>$p'' = 49°\ 36'\ 10''$</td><td>$\sin p'' = 0.761\ 5697$</td></tr>
<tr><td></td><td>$A = 141°\ 28'\ 30''$</td><td></td></tr>
<tr><td></td><td>Sum $= \overline{229°\ 17'\ 00''}$</td><td></td></tr>
<tr><td></td><td>$\frac{1}{2}$ Sum $= 114°\ 38'\ 30''$</td><td></td></tr>
</table>

$$\tfrac{1}{2}(x + y) = 180° - \tfrac{1}{2}\ \text{sum} = 65°\ 21'\ 30'' \quad \tan \tfrac{1}{2}(x + y) = 2.179\ 9995$$

$$\tan Z = \frac{c\ \sin\ p'}{b\ \sin\ p''} = \frac{1642.83 \times 0.618\ 4846}{1076.44 \times 0.761\ 5697} = 1.239\ 4301$$

$$Z = 51°\ 06'09'' \quad Z + 45° = 96°\ 06'09'' \quad \cot(Z + 45°) = -0.106\ 9133$$

$$\tan \tfrac{1}{2}(x - y) = \cot(Z + 45°)\ \tan \tfrac{1}{2}(x + y) = -0.106\ 9133 \times 2.179\ 9995 = -0.233\ 0709$$

$$
\begin{aligned}
\tfrac{1}{2}(x + y) &= 65°\ 21'\ 30'' \\
\tfrac{1}{2}(x - y) &= \underline{-13°\ 07'\ 11''} \\
x &= 52°\ 14'\ 19''
\end{aligned}
$$

$$y = 78°\ 28'\ 41''$$

<table>
<tr><td colspan="4" align="center">Solution of triangles (see Table 9-2)</td></tr>
<tr><td colspan="4" align="center">Triangle PBA</td></tr>
<tr><td>Sides</td><td>Stations</td><td>Angles</td><td>Distances</td></tr>
<tr><td>$c = BA$</td><td></td><td></td><td>1642.83</td></tr>
<tr><td></td><td>$P = p''$</td><td>$49°\ 36'\ 10''$</td><td>0.761 5697</td></tr>
<tr><td></td><td>$B = x$</td><td>$52°\ 14'\ 19''$</td><td>0.790 5679</td></tr>
<tr><td></td><td>A</td><td>$(78°\ 09'\ 31'')$</td><td>0.978 7194</td></tr>
<tr><td>AP</td><td></td><td>$\overline{180°\ 00'\ 00''}$</td><td>1705.38</td></tr>
<tr><td>PB</td><td></td><td></td><td>2111.26</td></tr>
<tr><td colspan="4" align="center">Triangle PAC</td></tr>
<tr><td>Sides</td><td>Stations</td><td>Angles</td><td>Distances</td></tr>
<tr><td>$b = AC$</td><td></td><td></td><td>1076.44</td></tr>
<tr><td></td><td>$P = p'$</td><td>$38°\ 12'\ 20''$</td><td>0.618 4846</td></tr>
<tr><td></td><td>A</td><td>$(63°\ 18'\ 59'')$</td><td>0.893 4999</td></tr>
<tr><td></td><td>$C = y$</td><td>$78°\ 28'\ 41''$</td><td>0.979 8483</td></tr>
<tr><td>CP</td><td></td><td>$\overline{180°\ 00'\ 00''}$</td><td>1555.09</td></tr>
<tr><td>PA</td><td></td><td></td><td>1705.38 checks</td></tr>
</table>

be that the observer has sighted on an erroneously identified control point, the published position of a control point was wrong, a mistake was made in measuring the angles, or a mistake was made in the computations. If the control points are correct and the discrepancy is within reason, then its amount will give an indication of the accuracy of the field work. The least-squares adjustment of an overdetermined resection is explained in Section A-7 of Appendix A.

A resection can be performed by observing on two points of known position using a total station instrument. This is referred to as a *two-point resection.* In Fig. 9-14 reflectors are set up over the two known points A and B. The total station

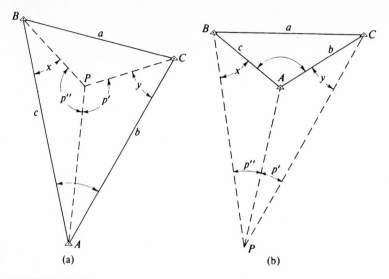

FIGURE 9-13 Three-point resection.

instrument occupies the unknown point P. The angle at P from A to B and the distances PA and PB are then measured. The microprocessor is instructed to inverse line AB according to Eqs. (8-12) and (8-13) to determine the length and azimuth of AB. Then, since the lengths of all of the sides are now known, the angles at A from B to P and at B from P to A can be computed using the law of sines, thus

$$\sin A = \frac{PB \, \sin \, P}{AB}$$

$$\sin B = \frac{PA \, \sin \, P}{AB}$$

The azimuths α_{AP} and α_{BP} are then (in this case)

$$\alpha_{AP} = \alpha_{AB} + \angle A$$

$$\alpha_{BP} = \alpha_{BA} - \angle B$$

A position computation is then made as explained in Section 9-8. All of the foregoing computations are performed in the computer of the total station instrument.

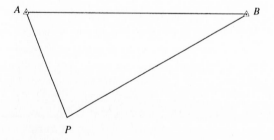

FIGURE 9-14 Two-point resection using total station instrument.

PROBLEMS

9-1. A 50-m invar tape is standardized under a 15-kg tension when supported at the 0-25-50 m points. What effect does a +1-kg systematic error in the applied tension have on a baseline that measures 10 km? $E = 2.0 \times 10^6 \, kg/cm^2$, $A = 0.040 \, cm^2$, $w = 25 \, g/m$.

9-2. The average elevation of a baseline is 985 m and the measured length of the base is 11,422.652 m. What is the length reduced to sea level?

9-3. What is the sea-level length of a baseline that measures 12,546.62 ft at an average elevation of 2850 ft?

9-4. A broken baseline is measured in two sections along two lines one of which deflects 3° 02′ 10″ from the other. The length of the first section is 7852.366 m, and the length of the second section is 6462.584 m. What is the length of the line joining the two baseline stations?

9-5. In Problem 9-4, if the deflection angle is 2° 24′ 28″ and the two sections are 8425.642 m and 1486.047 m, what is the length of the baseline?

9-6. The hypotenuse of a 30°-60°-90° triangle is 35 km long. The triangle is assumed to lie in latitude 45°. Compute the spherical excess to the nearest hundredth of a second.

9-7. What is the spherical excess to the nearest 0.01″ of a triangle in latitude 45° whose sides and the included angle are 39,000 ft, 86,000 ft, and 78° 26′ 28.2″, respectively.

9-8. In the quadrilateral shown in the accompanying illustration, all angles were measured with the same degree of precision, with the following results:

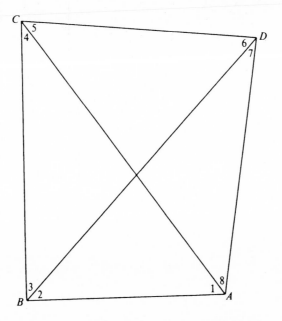

Angle	Value	Angle	Value
1	27° 02′ 13.0″	5	56° 11′ 57.4″
2	53° 41′ 08.3″	6	24° 31′ 23.8″
3	55° 25′ 23.9″	7	39° 53′ 02.7″
4	43° 51′ 14.9″	8	59° 23′ 36.1″

The coordinates of A are $X_A = 495{,}056.52$ ft and $Y_A = 504{,}574.66$ ft, and the coordinates of B are $X_B = 497{,}509.00$ and $Y_B = 501{,}524.50$ ft. Solve the remaining triangle sides.

9-9. Using triangle *CAB* of Problem 9-8, compute the coordinates of *C* and show a check.

9-10. Using triangle *DAC* of Problem 9-8, compute the coordinates of *D* and show a check.

9-11. In the quadrilateral shown in the illustration accompanying Problem 9-8, all angles were measured with the same degree of precision with the following results:

Angle	Value	Angle	Value
1	28° 00′ 29.4″	5	39° 17′ 30.0″
2	39° 39′ 46.0″	6	28° 22′ 45.4″
3	59° 59′ 55.6″	7	53° 10′ 42.0″
4	52° 19′ 49.0″	8	59° 09′ 02.6″

The coordinates of A are $X_A = 518{,}664.75$ ft and $Y_A = 473{,}340.01$ ft, and the coordinates of B are $X_B = 517{,}319.18$ ft and $Y_B = 474{,}602.18$ ft. Solve the remaining triangle sides.

9-12. Using triangle DAB of Problem 9-11, compute the coordinates of *D* and show a check.

9-13. Using the triangle *CDB* of Problem 9-11, compute the coordinates of *C* and show a check.

9-14. In the quadrilateral shown in the illustration accompanying Problem 9-8, all angles were measured with the same degree of precision with the following results:

Angle	Value	Angle	Value
1	19° 48′ 31.6″	5	19° 52′ 00.7″
2	27° 22′ 12.6″	6	27° 18′ 43.6″
3	103° 23′ 25.5″	7	61° 19′ 30.4″
4	29° 25′ 50.3″	8	71° 29′ 45.3″

The coordinates of A are $X_A = 522{,}680.20$ ft and $Y_A = 478{,}750.96$ ft, and the coordinates of B are $X_B = 519{,}708.08$ ft and $Y_B = 470{,}470.55$ ft. Solve the remaining triangle sides.

9-15. Using triangle *DAB* of Problem 9-14, compute the coordinates of *D* and show a check.

9-16. Using the triangle *CDB* of Problem 9-14, compute the coordinates of *C* and show a check.

9-17. The distance *AB* in Fig. 9-8 is determined by EDM measurement to be 28,565.50 ft. Angle α is $+ 2° 12′ 22″$, angle β is $-2° 16′ 06″$. The elevation of *A* is 1733.50 ft. Compute the sea-level length *MN*.

9-18. If only angle α had been measured in Problem 9-17, compute the sea-level length *MN*.

9-19. The distance *AB* of Fig. 9-8 is measured by EDM as 14,853.55 m. Angle α is $+1° 49′ 18″$, and angle β is $-1° 55′ 46″$. The elevation of *A* is 842.55 m. Compute the sea-level length *MN*.

9-20. If only angle β had been measured in Problem 9-19, compute the sea-level length *MN*.

9-21. Trilateration station *A* is at an elevation of 576.055 m above sea level, and station *B* is at an elevation of 1693.740 m. The slope distance from *A* to *B* as measured by EDM is 20,454.50 m. Reduce this distance to sea level.

9-22. Trilateration station A is at an elevation of 1065.80 ft above sea level, and station B is 3450.22 ft above sea level. The slope distance between the two stations measured by EDM is 9753.53 ft. Reduce this distance to sea level.

9-23. The coordinates of two control points A and B are X_A = 602,105.32 m and Y_A = 126,118.90 m; X_B = 601,048.82 m and Y_B = 125,613.48 m. The clockwise angle at A from B to an unknown point P is 52° 18′ 46.2″; the counterclockwise angle at B from A to P is 37° 28′ 16.0″. Compute the coordinates of P by intersection.

9-24. The coordinates of two control points A and B are X_A = 473,261.25 m and Y_A = 220,535.70 m; X_B = 473,224.68 m and Y_B = 221,381.86 m. The counterclockwise angle at A from B to P is 119° 52′ 20.4″; the clockwise angle at B from A to P is 23° 44′ 50.0″. Compute the coordinates of P by intersection.

9-25. The coordinates in feet of three control points are as follows:

Point	Y	X
A	98,202.66	38,762.50
B	110,002.65	61,252.84
C	89,102.32	78,565.12

(**a**) From an unknown point P_1 lying southerly from A, the clockwise angle from A to B is measured as 22° 30′ 14.6″ and that from B to C as 29° 48′ 50.6″. Compute the coordinates of P_1.

(**b**) From a point P_2 lying easterly from A and inside the triangle formed by the control points, the clockwise angle from A to B is measured as 102° 28′ 29.7″ and that from B to C as 86° 20′ 04.4″. Compute the coordinates of P_2.

(**c**) From a point P_3 lying northerly from B, the clockwise angle from C to B is measured as 12° 52′ 22.6″ and that from B to A as 53° 28′ 33.4″. Compute the coordinates of P_3.

BIBLIOGRAPHY   Adler, R. K., and Shmutter, B. 1971. Precise traverse in major geodetic networks. *Canadian Surveyor* 25:389.

Bodnar, A. N. 1990. National geodetic reference system statewide upgrade policy. *Tech. Papers Fall Conv. Amer. Cong. Surv. & Mapp.* p. A-71.

Bomford, Brigadier G. 1962. *Geodesy.* 3rd ed. London: Oxford Press.

Burke, K. F. 1971. Why compare triangulation and trilateration? *Proc. Ann. Meet. Amer. Cong. Surv. & Mapp.* p. 244.

Classification, Standards of Accuracy, and General Specifications of Geodetic Control Surveys. U.S. Department of Commerce, NOAA, 1974, reprinted 1980.

Danial, N. F. 1978. Another solution of the three-point problem. *Surveying and Mapping* 38:329.

Dracup, J. F. 1976. Tests for evaluating trilateration surveys. *Proc. Fall Conv. Amer. Cong. Surv. & Mapp.* p. 96.

Gale, P. M. 1969. Control surveys for the city of Houston. *Surveying and Mapping* 29:669.

Kesler, J. M. 1973. EDM slope reduction and trigonometric leveling. *Surveying and Mapping* 33:61.

Love, J. D. 1984. Publication of NAD 83. *Tech. Papers Fall Conv. Amer. Cong. Surv. & Mapp.* p. 377.

——*Manual of Second- and Third-Order Triangulation and Traverse.* Special Publication No. 145, U.S. Coast and Geodetic Survey, U.S. Government Printing Office.

Meade, B. K. 1967. High-precision transcontinental traverse surveys. *Surveying and Mapping* 28:41.

——1983. Latitude, longitude, and ellipsoidal height changes NAD 27 to predicted NAD 83. *Surveying and Mapping* 43:65.

——1983. Least-squares adjustment of a quadrilateral with six measured distances. *Surveying and Mapping* 43:385.

Mezera, D. F. 1983. Trilateration adjustment using unit corrections derived from least squares. *Surveying and Mapping* 43:315.

Mitchell, R. J., and Mitchell, D. J. 1987. The geometric strength of the three-point resection problem. *ACSM Bull. 107. Amer. Cong. Surv. & Mapp.* p. 19.

——*Proceedings Second International Symposium on Problems Related to the Redefinition of North American Geodetic Networks.* U.S. Department of Commerce, NOAA.

Reynolds, W. F. 1928. *Manual of Triangulation Computations and Adjustments.* Special Publication No. 138, U.S. Coast and Geodetic Survey, U.S. Government Printing Office.

Richardus, P. 1966. *Project Surveying.* New York: Wiley.

Spencer, J. F., Jr. 1988. Availability of NAD 83 products. *Tech. Papers Ann. Meet. Amer. Cong. Surv. & Mapp.* v. 1, p. 195.

Standards and Specifications for Geodetic Control Networks. Federal Geodetic Control Committee, NOAA 1984.

U.S. Coast and Geodetic Survey. *Formulas and Tables for the Computation of Geodetic Positions.* Special Publication No. 8, U.S. Government Printing Office.

Vincenty, T. 1975. Direct and inverse solutions of geodesics on the ellipsoid with application of nested equations. *Survey Review.* XXII, 176.

10 The Global Positioning System

10-1 Introduction

The navigation, positioning, and timing system known as the Global Positioning System (GPS) is a combination of several satellite navigation systems and concepts developed by or for the Department of Defense (DOD). The predecessors to GPS include the following satellite systems: 1) Transit, a navigation system developed by the Johns Hopkins Applied Physics Laboratory in the late 1950s and 1960s; 2) Timation, an experimental program developed for the Navy by the Naval Research Laboratory that demonstrated the ability to operate atomic clocks on board orbiting satellites and was used as a system concept for GPS; and 3) Project 621 B, an Air Force study program originated in 1964 by Aerospace Corporation and the Air Force's Space and Missile Organization. In 1968 a DOD Four-Service Executive Steering Group was established to investigate the development of a Defense Navigation Satellite System that would satisfy all of the DOD's satellite navigation requirements.

By 1972 the best characteristics of these programs had been determined to form the general system characteristics and initial design parameters for the system now known as the NAVSTAR Global Positioning System. By the late 1970s it was clear that both plane and geodetic surveying would be changed forever by this powerful space age technology. The surveying community is credited with discovering how to use the GPS signals for precise land measurements. The authors assert that no other technology has affected the surveying profession as profoundly as GPS. A number of excellent textbooks on GPS are referenced at the end of the chapter. The purpose of this chapter is to introduce the beginning surveying student to the principles of GPS so that the student can understand the limitations of the technology, attest to its accuracy and availability, and, of course, use it in various applications.

The technical and operational characteristics of GPS are organized into three distinct but somewhat arbitrary segments: the space segment, the operational control segment, and the user equipment segment. The GPS signals, which are broadcast by each satellite and carry data to both the user's equipment and ground control facilities, link the segments together into one system. Fig. 10-1 illustrates the segments of the GPS system and lists the transmitted signals. Fig. 10-2 shows the GPS satellite.

Each GPS satellite transmits signals centered on two microwave radio frequencies: 1575.42 megahertz, referred to as Link 1, or simply L1; and 1227.60 megahertz, referred to as L2. These channels lie in a band of frequencies known as the L-band, which starts just above the frequencies used by cellular telephones. The International Telecommunications Union, the radio regulation arm of the United Nations, has set aside special subbands within the L-band for satellite-based positioning systems. The L1 and L2 frequencies lie within these bands.

Such high frequencies are used for several reasons. The signals consist of a number of components. A bandwidth of about 20 MHz is required to transmit

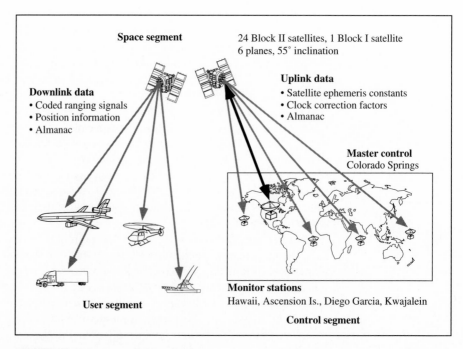

FIGURE 10-1 Segments of GPS system.

FIGURE 10-2 GPS satellite.

these components. A high, relatively uncluttered part of the radio spectrum is required for GPS-type signals.

GPS signals must provide a means for determining not only high-accuracy positions in real time but also velocities. Velocities are determined by measuring the slight shift in the frequency of the received signals due to the Doppler effect—essentially the same phenomenon associated with sound waves that gives rise to the change in pitch of a locomotive's whistle as a train passes in front of an observer at a level crossing. To measure velocities with centimeter-per-second accuracies, centimeter wavelength (microwave) signals are required.

Another reason for requiring such high frequencies is to reduce the effect of the ionosphere. The ionosphere affects the propagation speed of radio signals. The range between a satellite and a receiver derived by measuring travel times of the signal will therefore contain errors. The size of the errors gets smaller as higher frequencies are used. But even at the L1 frequency the error can amount to 30 meters for a signal arriving from directly overhead. For some GPS applications, an error of this size is tolerable. However, other applications require much higher accuracies. This is why GPS satellites transmit on two frequencies. If measurements made simultaneously on two well-spaced frequencies are combined, almost all of the ionosphere's effect can be removed.

Although high frequencies are desirable for the reasons just given, they should not be too high. For a given transmitter power, a received satellite signal becomes weaker as the frequency used becomes higher. The L-band frequencies used by GPS are therefore a good compromise between this so-called space loss and the perturbing effect of the ionosphere.

GPS signals, like most radio signals, start out in the satellites as pure sinusoidal waves or carriers. But pure sinusoids cannot be readily used to determine positions in real time. Although the phase of a particular cycle of a carrier wave can be measured very accurately, each cycle in the wave looks like the next one, so it is difficult to know exactly how many cycles lie between the satellite and the receiver. This *integer ambiguity* can be resolved using the differential techniques pioneered by surveyors and discussed in Section 10-9.

10-2   The Codes

For a user to obtain positions independently in real time, the signals must be altered in such a way that time-delay measurements can be made. This is achieved by modulating the carriers with pseudo random noise (PRN) codes.

The PRN codes consist of sequences of binary values (zeros and ones) that, at first sight appear to have been randomly chosen. But a truly random sequence can only arise from unpredictable causes over which, of course, there would be no control and which could not be duplicated. However, using a mathematical algorithm or special hardware devices called feedback shift registers, sequences can be generated that do not repeat until after some chosen interval of time. Such sequences are termed pseudo-random. The apparent randomness of these sequences makes them indistinguishable from certain kinds of noise such as the hiss heard when a radio is tuned between stations or the "snow" seen on a television screen. Although noise in a communications device is generally unwanted, in this case the noise is very beneficial.

Exactly the same code sequences are independently replicated in a GPS receiver. By aligning the replicated sequence with the received one and knowing the instant of time the signal was transmitted by the satellite, the travel time, and hence the range, can be computed. Each satellite generates its own unique codes, so a GPS receiver can easily identify which signal is coming from which satellite, even when signals from several satellites arrive at its antenna simultaneously.

Two different PRN codes are transmitted by each satellite: The C/A-code, or coarse/acquisition code, and the P-code, or precision code. The C/A-code is a sequence of 1023 binary digits, or chips, that is repeated every millisecond. This means that the chips are generated at a rate of 1.023 million per second and that one chip has a duration of about one microsecond. Each chip, riding on the carrier wave, travels through space at the speed of electromagnetic waves. A unit of distance can therefore be obtained by multiplying the time interval by this speed. One microsecond translates to approximately 300 meters. This is the wavelength of the C/A-code. Because the C/A-code is repeated every millisecond, a GPS receiver can quickly lock onto the signal and begin matching the received code with the one generated by the receiver.

The precision of a range measurement is determined in part by the wavelength of the chips in the PRN code. Higher precision can be obtained with shorter wavelengths. To get higher precision than those afforded by the C/A-code, GPS satellites also transmit the P-code. The wavelength of the P-code chips is only about 30 m, one-tenth the wavelength of the C/A-code chips: The rate at which the chips are generated is correspondingly ten times as fast, or 10.23 million per second.

The P-code is an extremely long sequence. The pattern of chips does not repeat until slightly more than 266 days, or about 2.35×10^{14} chips. Each satellite is assigned a unique seven-day segment of this code, which is initialized at midnight on Saturday each week.

The GPS PRN codes have additional useful properties. When a receiver is processing the signals from one satellite, it is essential that the signals received simultaneously from other satellites do not interfere. The GPS PRN codes have been specially chosen to be resistant to such interference. Also, the use of PRN codes results in a signal that is essentially impervious to unintentional or deliberate jamming from other radio signals, a possibility that the U.S. Department of Defense, the owner of the system, has to worry about.

The C/A-code is modulated onto the L1 carrier, whereas the P-code is modulated on both L1 and L2. This means that only users with dual-frequency GPS receivers can correct the measured ranges for the effect of the ionosphere. Users of single-frequency receivers must resort to models of the ionosphere that account for only a portion of the effect. Access to the lower-accuracy C/A-code is provided in the GPS *Standard Positioning Service* (SPS), the level of service authorized for civilian users. The *Precise Position Service* (PPS), which provides access to both the C/A-code and the P-code, is designed primarily for military uses. The SPS incorporates a further intentional degradation of accuracy, called Selective Availability (SA), which will be discussed later.

To convert the measured ranges between the receiver and the satellites to a position, the receiver must know where the satellites are. To do this in real time requires that the satellites broadcast this information. Accordingly, there is a message superimposed on both the L1 and L2 carriers along with the PRN codes. Each satellite broadcasts its own message, which consists of orbital information (the ephemeris) to be used in the position computation, the offset of its clock from GPS system time, and information on the health of the satellite and the expected accuracy of the range measurements.

The message also contains almanac data from the other satellites in the GPS constellation, as well as their health status and other information. The almanac data, a crude description of the satellite orbit, are used by the receiver to determine the location of each satellite. The receiver uses this information to acquire the signals quickly from satellites that are above the horizon but are not yet being tracked. So, once one satellite is tracked and its message decoded, acquisition of the signal from other satellites is quite rapid.

The broadcast message also contains another very important piece of information for receivers that track the P-code. As mentioned earlier, the P-code segment assigned to each satellite is seven days long. A GPS receiver with an

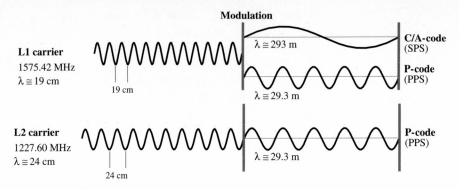

FIGURE 10-3 GPS satellite signals.

initially unsynchronized clock has to search through its generated P-code sequence to try to match the incoming signal. It would take many hours to search through just one second of the code, so the receiver needs some help. It gets this help from a special word in the message called the Hand-Over Word (HOW), which tells the receiver where in the P-code to start searching. Fig. 10-3 describes the signals.

10-3   Satellite Positioning

The GPS satellites provide the user with the capability of determining his position, expressed in Cartesian coordinates X_r, Y_r, Z_r or latitude, longitude, and height. Fundamentally this is accomplished using the principles of resection by measuring ranges. Think of the GPS satellites as frozen in space at a specific instant. The space coordinates X_s, Y_s, and Z_s, relative to the center of mass of the earth of each satellite, are known because the ephemeris (the orbit) of the satellites is determined by the operational control segment of the GPS system and broadcast to the users by each of the satellites.

If the ground receiver employed a clock that was set precisely to the GPS system time, the true distance (range) to each satellite could be accurately measured by recording the time required for the satellite signal to reach the receiver. Using this technique, ranges to only three satellites would be needed since the intersection of three spheres whose radii are the respective ranges yields the three unknowns X_r, Y_r, and Z_r and could be determined from three range equations. Fig. 10-4 illustrates the problem, and Eq. (10-1) expresses the solution.

$$R_j = [(X_{sj} - X_r)^2 + (Y_{sj} - Y_r)^2 + (Z_{sj} - Z_r)^2]^{\frac{1}{2}}, \ j \geq 3 \qquad (10\text{-}1)$$

Unfortunately there cannot be perfect synchronization between the satellite clocks and the receiver clock, and this clock offset or error introduces an unacceptable positional error. Moreover, as discussed later, both clocks are different

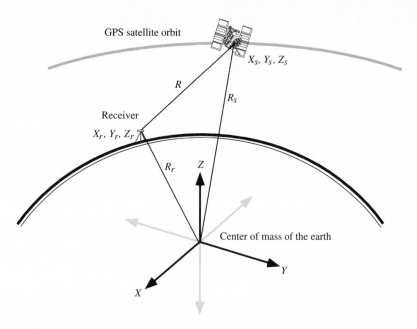

GPS satellite orbit

X_S, Y_S, Z_S

R

R_S

Receiver

X_r, Y_r, Z_r

R_r

Z

Center of mass of the earth

Y

X

FIGURE 10-4 Range from satellite to ground receiver.

than the GPS system time because of the usual errors (such as drift) associated with any timing device. The receiver clock problem can be circumvented by measuring simultaneously to at least four satellites and solving for this unknown clock error. These measured distances are called pseudoranges since they are the true ranges affected by the satellite and receiver clock error. In the case where only the satellite and receiver clock errors are considered, the mathematical model for the pseudoranges has the following form:

$$R_{rj} = \rho_{rj} + c(\Delta T_r - \Delta T_j), \ j \geq 4 \tag{10-2}$$

The subscript j refers to the number of satellites. The terms R_j, ρ_j, ΔT_r, ΔT_j are the measured pseudorange, the geometric range, the receiver and the satellite clock error, respectively, and c is the speed of light. The geometric distance ρ_j is, in Cartesian coordinates,

$$\rho_{rj} = [(X_j - X_r)^2 + (Y_j - Y_r)^2 + (Z_j - Z_r)^2]^{\frac{1}{2}} \tag{10-3}$$

The satellite clock error, ΔT_j, can be considered to be known from the ephemeris message or combined with ΔT_r and solved for as part of the solution depending on accuracy requirements and mode of operation. If there are four satellites and ΔT_j is known, a unique solution for X_r, Y_r, Z_r, and ΔT_r results, Fig. 10-5. This result can be easily converted to latitude, longitude and height with the appropriate equations.

User obtains pseudorange measurements

(R_1, R_2, R_3, R_4) to four satellites

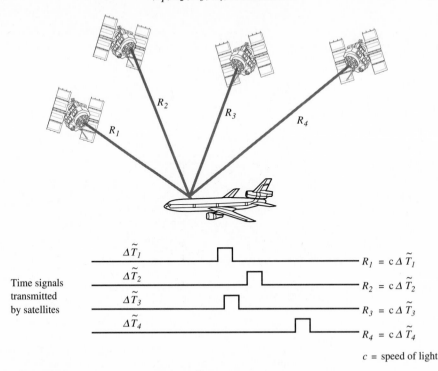

Pseudoranges	Position Equations
$R_1 = c \Delta \tilde{T}_1$	$(X_1 - X_r)^2 + (Y_1 - Y_r)^2 + (Z_1 - Z_r)^2 = (R_1 - c \Delta T_r)^2$
$R_2 = c \Delta \tilde{T}_2$	$(X_2 - X_r)^2 + (Y_2 - Y_r)^2 + (Z_2 - Z_r)^2 = (R_2 - c \Delta T_r)^2$
$R_3 = c \Delta \tilde{T}_3$	$(X_3 - X_r)^2 + (Y_3 - Y_r)^2 + (Z_3 - Z_r)^2 = (R_3 - c \Delta T_r)^2$
$R_4 = c \Delta \tilde{T}_4$	$(X_4 - X_r)^2 + (Y_4 - Y_r)^2 + (Z_4 - Z_r)^2 = (R_4 - c \Delta T_r)^2$

R_i = Pseudorange (i = 1, 2, 3, 4)

Pseudorange includes actual distance between satellite and user plus satellite clock bias, user clock bias, atmospheric delays, and receiver noise.

Satellite clock bias and atmospheric delays are compensated for by incorporation of deterministic corrections prior to inclusion into the solution.

X_i, Y_i, Z_i = satellite position (i = 1, 2, 3, 4)

Satellite position broadcast in navigation 50-Hz message

Receiver solves for:

X_r, Y_r, Z_r = receiver position

ΔT_r = user clock bias

FIGURE 10-5 Receiver position solution using signals from four satellites.

For many applications this simple "navigation" solution will suffice. These applications deal with facilities management (FM in the literature), which is the process of creating a spatially referenced inventory of land information features, such as light poles, stop signs, on-off ramps, and so on. It is critical that the modern surveyor understand the accuracy and limitation of this type of positioning, as it will become more important in the future as the GPS accuracy improves as a result of important future enhancements.

In considering the error sources in a single-point positioning from pseudoranges, it is first necessary to discuss two errors that are deliberately introduced by the DOD. Selective Availability (SA) is the deliberate degradation in the GPS satellite orbit information (the navigation message) and timing by the DOD. The total error introduced today (1996) is limited by policy to $[\sigma_x^2 + \sigma_y^2]^{\frac{1}{2}} \leq 50$ m. The other error introduced by the DOD is Anti-Spoofing (A-S), which is the encryption process used to deny unauthorized access to a modification of the P-code known as the Y-code. The A-S process prevents potential enemies from spoofing or fooling a friendly GPS user of the Y-code.

Other errors in the pseudoranges are listed in Table 10-1 and are briefly discussed below.

The radio waves emitted by the satellites pass through the earth's ionosphere and troposphere and create a delay in the arrival time of the signals and hence an error in the range. Errors exist in the clocks in the satellites, the ephemeris or orbit of the satellites, in the receiver whose position is to be determined, and because of a phenomenon called multipath. Multipath is caused by objects near the receiver antenna that reflect the satellite signals to the antenna after hitting the object. Those signals travel a longer path than those traveling directly to the antenna and hence cause an error. This is a condition that can largely be avoided by the surveyor in the field by placing the receiver antenna in a place free from reflected signals (for example, away from a building wall). Figure 10-6 shows a GPS receiver and antenna.

Another important aspect of GPS positioning with a single receiver is that in the future it is highly likely that the accuracy of such positioning will improve. Many experts believe that the instantaneous "navigation" accuracy of GPS will

TABLE 10-1   Errors in Single-Point Positioning

Error Source	Range Error (σ) in Meters
Selective availability (SA)	24.0
Atmospheric delay	
Ionospheric	7.0
Tropospheric	0.7
Clock and ephemeris error	3.6
Receiver noise	1.5
Multipath	1.2
Total range error	25.3
Resulting horizontal accuracy $\sqrt{\sigma_x^2 + \sigma_y^2}$	50.0

FIGURE 10-6 GPS receiver and antenna. Courtesy of Trimble.

improve to a level of 2–5 meters by the year 2005. This accuracy level will not satisfy the requirements for boundary or property surveying but will, as stated earlier, satisfy many requirements associated with facilities management and other tasks.

10-4   Differential GPS

By far the most significant application of GPS for surveyors is to accomplish highly accurate measurement of a vector (as in mathematics, both magnitude and direction) between two points, that is, ΔX, ΔY, and ΔZ. Usually one point is known with respect to the appropriate geodetic datum, and one or more additional points are determined with respect to the known point. In this fashion most of the errors listed in Table 10-1 tend to cancel out. For example, SA is nearly eliminated by differencing simultaneous measurements at two stations. This is because SA and other errors are common to both of the receivers, that is, the receiver occupying the known point and the receiver situated at the point to be determined.

The actual equations used to reduce GPS observations today are more complicated than those provided below. However, the basic principles can be grasped from the following discussion.

In Fig. 10-7, if we are only concerned with $X_2 - X_1$, $Y_2 - Y_1$, and $Z_2 - Z_1$, we can write equations differencing the pseudoranges to satellites A, B, C, and D from stations 1 and 2 as

$$R_{2A} - R_{1A} = (\rho_{2A} - \rho_{1A}) + c(\Delta T_2 - \Delta T_A) - c(\Delta T_1 - \Delta T_A) \qquad \text{(10-4)}$$

or

$$R_{2A} - R_{1A} = (\rho_{2A} - \rho_{1A}) + c(\Delta T_2 - \Delta T_1) \qquad \text{(10-5)}$$

which shows how the satellite clock error cancels out. Similarly

$$R_{2B} - R_{1B} = (\rho_{2B} - \rho_{1B}) + c(\Delta T_2 - \Delta T_1) \qquad \text{(10-6)}$$

$$R_{2C} - R_{1C} = (\rho_{2C} - \rho_{1C}) + c(\Delta T_2 - \Delta T_1) \qquad \text{(10-7)}$$

$$R_{2D} - R_{1D} = (\rho_{2D} - \rho_{1D}) + c(\Delta T_2 - \Delta T_1) \qquad \text{(10-8)}$$

where ρ_{rj} is given by Eq. 10-2, with $r = 1, 2$ and $j = A, B, C, D$. Since R_{1j} and R_{2j} are computed from the measurements at the two stations; and in this simplified example c, the speed of electromagnetic waves, is considered known; and the satellite coordinates X_j, Y_j, Z_j, $j = A, B, C, D, \ldots$; and the fixed-station coordinates, say X_1, Y_1, and Z_1, are considered known, the above four equations can be solved

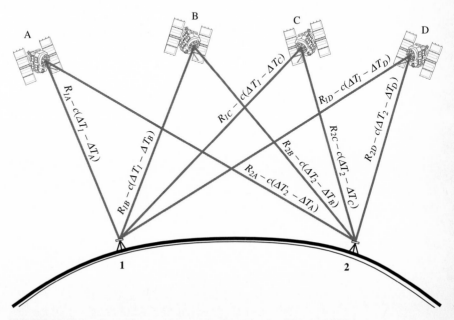

FIGURE 10-7 Satellite and ground receiver configuration for differential GPS.

for the four unknowns $(\Delta T_2 - \Delta T_1)$, X_2, Y_2, Z_2. The above technique is called single difference differential pseudoranging. This technique is common today and can be used to obtain accuracies of $\sigma = \pm 1$ m over distances of up to 50 km. In practice, of course, many more observations are taken on as many satellites as possible, resulting in the need for a least-squares solution to the problem.

Today's instruments are capable of making pseudorange measurements using either the C/A- or the P(Y)-code transition modulations. Computations of the precise vector between two points when the differential carrier phase measurement is used is, of course, quite similar to differential pseudoranging. Instead of differencing the C/A-code pseudoranges, a measurement comprised of a large number of whole (integer) carrier cycles plus a fractional cycle measurement is differenced. The integer portion of this measurement is unknown and must be estimated and ultimately solved for; this measurement can be thought of as an ambiguous pseudorange. Rewriting the right side of Eqs. 10-5 to 10-7 yields the following:

$$R_{2A} - R_{1A} = \Delta\Phi_{21A} - N_{21A} \qquad (10\text{-}9)$$

$$R_{2B} - R_{1B} = \Delta\Phi_{21B} - N_{21B} \qquad (10\text{-}10)$$

$$R_{2C} - R_{1C} = \Delta\Phi_{21C} - N_{21C} \qquad (10\text{-}11)$$

$$R_{2D} - R_{1D} = \Delta\Phi_{21D} - N_{21D} \qquad (10\text{-}12)$$

or

$$R_{2j} - R_{1j} = \Delta\Phi_{21j} - N_{21j} \quad ; \; j = A, \, B, \, C, \, D, \, \ldots \qquad (10\text{-}13)$$

where

$$R_{2j} - R_{1j} = (\rho_{2j} - \rho_{1j}) + c(\Delta T_2 - \Delta T_1) \qquad (10\text{-}14)$$

$$\Delta\Phi_{21j} = \Phi_{2j} - \Phi_{1j} \qquad (10\text{-}15)$$

$$N_{21j} = N_{2j} - N_{1j} \qquad (10\text{-}16)$$

and

$$N_{rj} = (n_c \lambda_c) \quad r = 1, 2 \qquad (10\text{-}17)$$

$$\Phi_{rj} = \left(\frac{\theta}{2\pi}\lambda_c\right)_{rj} \qquad j = A, \, B, \, C, \, D, \, \ldots \qquad (10\text{-}18)$$

and

$$n_c = \text{integral number of carrier wavelengths}$$
$$\lambda_c = \text{carrier wavelengths}$$
$$\theta = \text{phase angle of carrier}$$

Comparing Eqs. 10-9 to 10-12 and Eq. 10-14, it follows that the left side of Eqs. 10-9 to 10-12 contain three unknown coordinates (coordinates of station 2), four unknown biases (N_{21j}, $j = A$, B, C, D) and one unknown clock term $(\Delta T_2 - \Delta T_1)$, giving a total of eight unknowns.

These equations, called single-difference equations, are solved for by observing ranges at numerous epochs and the unknowns solved for by using least squares.

10-5   Kinematic GPS

With the advent of dual-frequency GPS receivers and sophisticated and clever techniques like those developed by NGS (see references), it is possible using differential GPS to obtain extremely high accuracies in real time while a rover receiver is moving. The trick is to be able to determine which wavelength has arrived at the rover. This is generally referred to as real-time ambiguity resolution on-the-fly (OTF). A number of sophisticated algorithms have been developed to do this, and it is generally possible to determine the correct wavelength (referred to as getting back "on lock") within seconds with only five satellites in view while the rover is moving.

The real-time technique would be common in current GPS surveying if it didn't cost the surveyor significantly more than post-processed results. As the cost of two-wavelength receivers comes down, it will become the practice of choice. Dramatic price reductions for all GPS receivers have occurred over the past few years.

The ambiguity resolution equations referred to in this section are beyond the scope of this textbook, however, the principles of satellite positioning outlined in this chapter are valid no matter what GPS method is used.

10-6   Accuracy of Differential Techniques

The errors shown in Table 10-1 take on a form similar to that of electronic distance measuring devices, that is,

$$\sigma_{GPS} = [\sigma_A^2 + \sigma_B^2(d)]^{\frac{1}{2}} \qquad (10\text{-}19)$$

where σ_{GPS} can be thought of as the standard error of a distance determined by GPS; σ_A is a constant error, usually 5 mm $\leq \sigma_A \leq$ 15 mm for high-quality two-wavelength devices; and σ_B is a distance (d) dependent error that ranges from $d \times \frac{1}{10^5}$ m to $d \times \frac{1}{10^8}$ m depending on the type of receiver used and the error model in postprocessing. Therefore, for medium length (10 km) distances, an average error of the distance between the base and rover station might be $\sigma_{GPS} = \pm 2$ cm under good conditions. Amazingly this result is true even if the receiver is used in a static or kinematic mode. Therefore GPS can provide the same or better accuracy than the best traditional surveying techniques over distances greater than 1000 m or so.

10-7   Transformation of GPS Results

Denoting the Cartesian (rectangular) coordinates of a point in space by X, Y, and Z, and assuming an ellipsoid of revolution with the same origin as the Cartesian coordinate system, the point can also be expressed by the ellipsoidal coordinates φ, λ, and h, as seen in Fig. 10-8. The relation between the Cartesian coordinates and the ellipsoidal coordinates is

$$X = (R_N + h) \cos \varphi \cos \lambda \qquad \text{(10-20a)}$$

$$Y = (R_N + h) \cos \varphi \sin \lambda \qquad \text{(10-20b)}$$

$$Z = \left(\frac{b^2}{a^2} R_N + h\right) \sin \varphi \qquad \text{(10-20c)}$$

where R_N is the radius of curvature in the prime vertical, thus

$$R_N = \frac{a^2}{[a^2 \cos^2 \varphi + b^2 \sin^2 \varphi]^{\frac{1}{2}}} \qquad \text{(10-21)}$$

and a and b are the semiaxes of the reference ellipsoid. The Cartesian coordinates are related to WGS-84, which is the datum (reference system) used by the GPS. There is little practical difference between WGS-84 and NAD 83.

Eqs. 10-20a, b, and c transform ellipsoidal coordinates φ, λ, and h into Cartesian coordinates X, Y, and Z. For GPS applications the inverse transformation is more important because the Cartesian coordinates are given and the ellipsoidal coordinates sought. Therefore the ellipsoidal coordinates φ, λ, and h must be computed from the Cartesian coordinates X, Y, and Z. Usually this problem is solved iteratively, although a solution in an explicit form is possible. This development is beyond the scope of this book and is fully explained in the references.

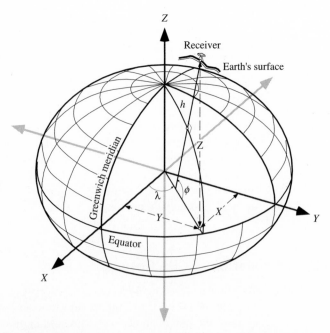

FIGURE 10-8 Coordinate reference system of GPS and conversion to latitude, longitude, and ellipsoidal heights.

10-8 Height Determination Using GPS

From the preceding section it is clear that the ellipsoid height h in Fig. 10-8 can be obtained using Eqs. 10-20a, b, and c, but the distance N in Fig. 10-9 must be known to determine the orthometric height H. Orthometric heights are obtained by leveling. The relationship is usually defined as

$$H = h - N$$

The X, Y, and Z coordinates determined by GPS must be transformed using the inverse of Eqs. 10-20a, b, and c but are still not useful when comparing orthometric heights that were determined by spirit leveling. The reason for this is the fact that leveling is affected by the earth's gravity field and the datum for leveling is usually an equipotential surface near or coinciding with mean sea level. This surface is called the geoid. The distance N is the separation between the ellipsoid and the geoid.

Today N has been modeled in the United States to $\sigma = \pm 10$ cm. These values $N(\varphi, \lambda)$ are available from the NGS. Moreover since the relative accuracy of the modeled geoid is very good and N changes slowly over short distances, the following procedure can be used:

1. First, occupy an existing (NAVD 88) benchmark with height H_{BM} and compare N_{GPS} from $N_{GPS} = h_{GPS} - H_{BM}$ with the modeled N, denoted by N_{MDL}
2. Determine a correction to the modeled N by $\Delta N = N_{GPS} - N_{MDL}$
3. If more than one benchmark is available nearby, determine ΔN at several points and average the result.
4. Use ΔN to correct the GPS determined heights using $H = h_{GPS} - (N_{MDL} + \Delta N_{AVG})$.

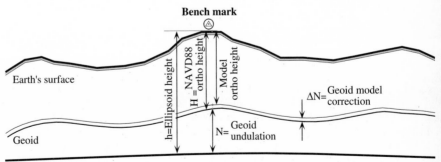

h = geometric height (height above ellipsoid)
H = orthometric height (height above "mean sea level")
N = geoid undulation

H = h - N

FIGURE 10-9 Relationship between ellipsoidal height and orthometric height.

10-9   Surveying with GPS

The most significant difference between GPS and traditional surveying techniques is the fact that GPS does not require line of sight between measured points. This can be taken into account advantageously to increase the efficiency and productivity of GPS surveying. For instance, when inexpensive dual-frequency GPS receivers are available, the baseline between the measured points may be on the order of 10–100 km. As a result very few *base stations* are needed to serve a very large area. With this scenario GPS surveys can be performed using only the rover receiver. The base station data can be downloaded through the network using different on-line services. When only single-frequency GPS receivers are available, the above scenario is more difficult since the baselines should be either shorter (approximately 10–20 km) or regional ionospheric corrections should be supplied with the base station data. This, however, is not the current practice because it is more convenient for GPS surveying companies to establish their own base stations for each project.

A typical GPS survey consists of the following steps:

1. Identifying sites and their accessibility.[†]
2. Recording the geometry of obstructions for each site.
3. Planning the GPS observations.
4. Conducting the GPS observations.
5. Estimating the baseline lengths.
6. Performing network adjustment.
7. Evaluating the adjustment results.
8. Performing format conversions according to user requests.

Currently 25 GPS satellites are in orbit (1 Block I and 24 Block II). As a result, if an observing site is clear of obstructions, planning will not be required because in this case the satellite coverage is good at all times, as shown in Fig. 10-10. This figure shows the number of visible satellites over a 24-hour period. The presence of obstructions can affect satellite visibility substantially, and therefore planning is necessary to determine the periods of best satellite coverage for making the GPS observations. In general, it is best to have a clear view of the sky above about 15° elevation above the horizon.

Leapfrogging techniques are common in GPS surveying. Currently the most common technique is to use three or more receivers in stop-and-go mode. In this mode one of the receivers is placed on the base station, preferably the one with the fewest obstructions. The other receivers (rover receivers) are placed on the points to be surveyed for the time required to determine the integer ambiguities necessary for cm-level surveying. The receiver firmware tracks internal indicators (for example, the quality of the geometry formed by the satellites being tracked, and called PDOP, with time of continuous tracking) for determining the length of time required to estimate the integer ambiguities successfully. Once the integer ambiguities are determined, the rest of the points can be surveyed with just one

---

[†]Access by road and easy intervisibility of the azimuth mark are important factors in planning a GPS survey.

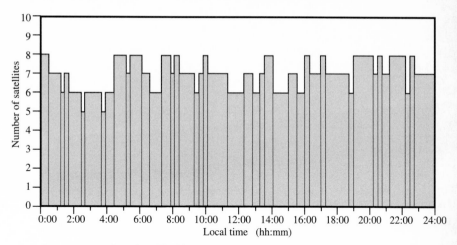

Location: The Ohio State University, Center for Mapping
Latitude: 39° 59' 57.80"   Longitude: 83° 02' 26.40"
Date: Feb. 12, 1996   Elevation angle ≥ 10°
Time zone: Eastern Standard Time (USA/CAN)   Local time: GMT - 5

FIGURE 10-10 Satellite visibility.

epoch of observations, assuming that lock to at least four satellites with good geometry is maintained. The receivers have indicators informing the users if a certain point should be reobserved for longer periods because lock to at least four satellites was not maintained.

A real-time implementation of this system referred to as Real-Time Kinematic (RTK) will resolve the integer ambiguities and perform cm-level positioning in real time. In this mode the ambiguity resolution is referred to as OTF ambiguity resolution. The advantage of the real-time system is that it does not require any postprocessing since the positions are computed in real time. The major disadvantage, however, is that it requires a direct communication link, which makes the system more expensive and more difficult to implement. The length of time required to perform ambiguity resolution and cm-level positioning depends primarily on the type of receivers used. With dual-frequency GPS receivers, the time required to perform OTF ambiguity resolution ranges from a few seconds to 1 or 2 minutes. With single-frequency receivers this time is much longer, on the order of 2 to 10 or 15 minutes, depending on the errors affecting the carrier phase measurements. However, by starting from a known baseline or performing an antenna swap, the ambiguity resolution time ranges from few seconds to one minute.

An antenna swap is performed by occupying the base station receiver on the base station and the roving receiver on a point nearby. After a few minutes of receiving signals at both points from at least four satellites, the two antenna are interchanged from one tripod to the other, care being taken to maintain lock on all four (or more) satellites. Data are collected for a few minutes, after which the two antenna are again returned to their original tripods, with lock maintained on the satellites. This process eliminates the integer ambiguity at the base station.

TABLE 10-2   Observational Strategies for cm-level Relative Positioning

Strategy	Description	Remarks
Static	Stationary.	Long observation time; yields baselines and ambiguities simultaneously (30–120 minutes).
Kinematic	One stationary, one moving.	Integer ambiguities are determined at start with static initialization. Maintain lock to at least four satellites.
Pseudostatic	One stationary, one moving. Moving receiver revisits points at least once at hourly intervals.	Baselines and integer ambiguities are determined simultaneously.
On-the-fly	One stationary, one moving.	Integer ambiguities are determined without static initialization (ambiguity resolution).

TABLE 10-3   Dual-Frequency Receiver List

Company	Model	Signal Tracked	Max. # Satellites Tracked	Size (in) (W x H x D)	Wt. (Lb.)	Power Use (Watts)	Price ($US in 1996)
Ashtech	Z-12 Field Surveyor	L1, CA and P-code L2, P-code	12	8.5 × 3.9 × 8.0	8.8	12	32,000
Leica	Kinematic GPS Receiver	L1, CA and P-code L2, P-code	9	8.0 × 7.0 × 3.0	7.5	8	17,845
Topcon America	Turbo SII	L1, CA and P-code L2, P-code	8	8.6 × 4.1 × 1.9	2.2	5	22,950
Trimble	Site Surveyor SSi	L1, CA and P-code L2, P-code	12	9.8 × 4.0 × 11.0	6.8	9	23,000

If lock to at least four satellites cannot be maintained, re-initialization is required either starting from a known baseline or by performing an antenna swap, which makes this mode of RTK difficult to implement in practice. Observational strategies that were common at the time of this writing are shown in Table 10-2. The student may want to speculate about the future from this table.

There are a variety of receivers available on the market today. Table 10-3 presents a summary of selected two-frequency receivers. These two-frequency devices are commonly used for highly precise work, but single-frequency receivers can accomplish a large number of surveying tasks.

BIBLIOGRAPHY   Bossler, J. D., and Hanson, R. H. 1984. The impact of VLBI and GPS on geodesy. *Surveying and Mapping* 44:105.

Chong, A. K. 1990. Quality control of stop and go kinematic GPS surveying. *Tech. Papers Fall Conv. Amer. Cong. Surv. & Mapp.* p. A-114.

Chrzanowski, A., Langley, R. B., Wells, D. E., and McLaughlin, J. D. 1983. A forecast of the impact of GPS on surveying. *Tech. Papers 43rd Ann. Meet. Amer. Cong. Surv. & Mapp.* p. 625.

Collected papers on global positioning system. *Tech. Papers Ann. Meet. Amer. Cong. Surv. & Mapp.* pp. 199–317.

Collins, J. 1983. Establishing first-order control by GPS satellite surveying instruments. *Tech. Papers 43rd Ann. Meet. Amer. Cong. Surv. & Mapp.* p. 449.

——1984. GPS Satellite surveying. *Tech. Papers 44th Ann. Meet. Amer. Cong. Surv. & Mapp.* p. 223.

——and Leick, A. 1985. Montgomery County (PA) GPS survey. *Tech. Papers 45th Ann. Meet. Amer. Cong. Surv. & Mapp.* p. 512.

——1994. GPS-controlled aerial mapping. *P.O.B.* 19:2.

——1991. GPS - the low-cost alternative. *P.O.B.* 16:3.

Georgiadou, Y., and Doucet, K. 1990. The issue of selective availability. Innovation column in *GPS World,* September/October. Metuchen, New Jersey: Aster Publishing Corp.

Georgiadou, Y., and Kleusberg, A. 1988. On carrier signal multipath effects in relative GPS positioning. *Manuscripta Geodaetica,* Springer-Verlag.

——1987. Ionospheric refractions and multipath effects in GPS carrier phase observations. *Presented at the IUGG XIX General Assembly,* Vancouver, B.C.

Goad, C. 1989. Kinematically surveying with GPS. *Tech. Papers Ann. Conv. Amer. Cong. Surv. & Mapp.* p. 299.

——1989. On the move with GPS. *P. O. B.* 14:4.

Grunthal, M. C. 1985. Analysis of short-term global positioning system observations. *Tech. Papers Fall Conv. Amer. Cong. Surv. & Mapp.* p. 980.

Hofmann–Wellenhof, B., Lichtenegger, H., and Collins, J. 1993. *GPS: Theory and Practice.* New York: Springer-Verlag Wien.

Janes, H. W., Langley, R. B. And Newby, S. P. 1991. Analysis of tropospheric delay prediction models: comparisons with ray tracing and implications for GPS relative positioning. *Bulletin Geodesique,* Springer-Verlag.

Klobuchar, J. A. 1991. Ionospheric effects on GPS. Innovation Column in *GPS World,* April. Metuchen, New Jersey: Aster Publishing Corp.

Lachepelle, G., and Wade, R. L. 1982. Navstar GPS single-point positioning. *Tech. Papers 42nd Ann. Meet. Amer. Cong. Surv. & Mapp.* p. 603.

Langley, R. B. 1995. A GPS glossary. Innovation Column in *GPS World,* October. Metuchen, New Jersey: Aster Publishing Corp.

——1991. The Mathematics of GPS. Innovation Column in *GPS World,* July/August. Metuchen, New Jersey: Aster Publishing Corp.

——1991. The Orbits of GPS satellites. Innovation Column in *GPS World,* Metuchen, New Jersey: Aster Publishing Corp.

——1990. Why is the GPS signal so complex?. Innovation Column in *GPS World,* May/June. Metuchen, New Jersey: Aster Publishing Corp.

Leick, A. 1990. *GPS Satellite Surveying.* New York: Wiley.

National Research Council, National Academy of Science, Committee on the Future of the Global Positioning System. 1995. *The Global Positioning System, a shared national asset.* Washington, D.C.: National Academy Press.

Reilly, J. P. 1994. GPS equipment survey. *P.O.B.* 19:5.

Remondi, B. W. 1986. Distribution of Global Position System ephemerides by the National Geodetic Survey. *Tech. Papers Ann. Conv. Amer. Cong. Surv. & Mapping* V. 2, p. 219.

——1986. Performing centimeter-level surveys in seconds with GPS carrier phase: Initial results. *Tech. Papers Ann. Conv. Amer. Cong. Surv. & Mapp.* V. 2, p. 302.

Rocken, C., and Meertens, C. 1991. Monitoring selective availability dither frequencies and their effect on GPS data. *Bulletin Geodesique.* Springer-Verlag.

Salsig, G. 1993. New frontiers for aerial control. *P.O.B.* 18:3.

Spilker, J. J., Jr. 1980. GPS signal structure and performance characteristics. *The Institute of Navigation: Global Positioning System,* Vol. 1, pp. 29–54.

Stowell, J. L., and Ott, G. E., Jr. 1985. Operational techniques of GPS as a survey instrument. *Tech. Papers Fall Conv. Amer. Cong. Surv. & Mapp.* p. 286.

Veatch, E., and Oswald, J. 1989. The kinematic GPS revolution "surveying on the move." *Tech. Papers Ann. Conv. Amer. Cong. Surv. & Mapp.* p. 288.

Vincenty, T. 1989. How to obtain elevations by GPS. *P.O.B.* 15:1.

Wegenast, D. A. 1989. A GPS high precision network for the state of Washington. *Tech. Papers Fall Conv. Amer. Cong. Surv. & Mapp.* p. 24.

Wells, D., et al. 1986. *Guide to GPS Positioning.* Ottawa, Canada: Canadian Institute of Surveying and Mapping.

Zilkoski, D. 1990. Minimum steps required when estimating GPS–derived orthometric heights. *Tech. Papers Fall Conv. Amer. Cong. Surv. & Mapp.* p. A-121.

11 State Plane Coordinate and Universal Transverse Mercator Systems

11-1 Introduction

In Chapter 8 the use of plane rectangular coordinates was discussed as applied to plane surveying, not only for defining the positions of survey stations, but also for solving such diverse problems as determining areas, locating the intersections of lines and curves, parting off land, and computing lengths and azimuths. No mention was made, however, of the extent of any given plane coordinate system. The methods of plane surveying are based on the assumption that all distances and directions are projected onto a horizontal plane surface that is tangent to the surface of the earth at one point within the area of the survey.

If two surveys are made independently of each other, then the measurements in each survey are referred to two different horizontal planes, which, of course, do not coincide. Furthermore, if the Y axis of a plane coordinate system for each of the two surveys is assumed to be parallel to the true meridian at one station of the survey, then even the Y axes of the two systems are not parallel with one another because of the convergence of the meridians. In a given system the farther the survey departs from the point of tangency, the more will the distances and angles as measured on the ground differ from the corresponding distances and angles as projected onto the horizontal plane. When this discrepancy becomes intolerable, then the limits of plane surveying and its inherent simplicity have been reached.

Chapter 9 briefly discusses the methods of geodetic surveying wherein all distances are reduced to a common reference surface conforming closely to the sea-level surface. Angles in triangles are considered as spherical (sometimes spheroidal) angles. Coordinates of points are computed with reference to parallels of latitude and meridians of longitude by using angles computed near the center of the earth rather than distances. Geodetic surveying is employed so that precise surveys can be extended over great distances in any direction without suffering the limitations of plane surveying methods. Geodetic surveying methods are more complex and more expensive, and they involve more difficult computations.

For more than a century, large surveying and mapping organizations of the federal government, notably the National Geodetic Survey, have established horizontal control monuments in the form of triangulation, GPS, traverse, and intersection stations. These stations have been located by methods of geodetic surveying. The network formed by these control points is being constantly filled in and added to by the same organizations. All the control points throughout the country bear a definite relationship, one to another, being referred to one common spheroidal surface.

Historically this surface has been the Clarke Spheroid of 1866 which is the basis of the North American Datum of 1927 (NAD 27). This horizontal control has recently been recomputed to take into account new surveys over the intervening years since the NAD 27 was established. The adjusted positions of the control points are based on the new North American Datum of 1983 (NAD 83), which is defined by the spheroid called the Geodetic Reference System of 1980 (GRS 80). Referring to Section 1-2, an oblate spheroid, also called an ellipsoid of revolution, is a solid obtained by rotating an ellipse on its shorter or minor axis. Both the Clarke 1866 spheroid and the GRS 80 spheroid are examples of ellipsoids of revolution. A comparison of these two spheroids is given here:

	Clarke Spheroid of 1866	GRS 80 Spheroid
Semimajor axis	6,378,206.4 m	6,378,137.0 m
Semiminor axis	6,356,583.8 m	6,356,752.314 m

The geographic positions of control points are referred to one or the other of these spheroids in terms of their latitude φ and longitude λ. The azimuths of lines between points are with respect to lines on the spheroid that pass through the axis of rotation, and are called geodetic azimuths (see Section 7-6).

The state plane coordinate systems have been devised to allow methods of plane surveying to be used over great distances, while maintaining the simplicity of plane rectangular X and Y coordinates and azimuths referred to meridians that are parallel with one another. At the same time a precision approaching that of geodetic surveying is maintained. As a result surveyors and engineers can incorporate the network of control established by geodetic surveying into their own surveys for purposes of coordination, checking, and reestablishing lost points.

If a land surveyor runs a traverse from one of these stations to the nearest corner of the land that he is to survey, he can calculate readily the state-coordinate position for each corner of the land and has in effect keyed his survey to all the stations of the National Geodetic Reference System (NGRS). As a result all stations of the NGRS become witnesses to the positions of the land corners whose coordinates on a state system are known. The material marks of the land corners—such as trees, stones, fence posts, or other monuments—may be destroyed, and yet the positions they occupied on the ground can be closely reproduced from any recoverable stations of the NGRS that are within practical distances of these corners. In this manner a survey station or land corner that is described in terms of a state coordinate system is practically indestructible.

The system is of great value to highway engineers and others whose work covers very large areas. Many surveys, which in the past would have been open traverse with few checks on their accuracy, can now start from one known point and end on another, thus providing a closure for the traverse.

The state coordinate systems are shown on many federal maps, particularly on topographic maps. These maps thus become useful to engineers who desire data for reducing their surveys to a state grid. At the same time their own surveys become available for transfer to the map.

11-2   Limits of State Plane Coordinate Systems

Just as a map of a considerable portion of the earth's surface is a compromise when compiled on a plane surface, namely, the map sheet, so must a state plane coordinate system be a compromise since it represents a spheroidal surface projected onto a curved surface that may be developed into a plane surface. However, this compromise does not prove to be at all serious since any distortion suffered through the projection of the spheroidal surface to the plane surface can be allowed for by appropriate computations.

If a state is not too large, the entire state can use one projection. If the state is of such size that the distortions in one projection would be too great, then the state is divided into *zones* of appropriate size. These zones are broken along county lines to permit an entire county to be included in the same zone.

Two types of projections are employed in developing the state plane coordinate systems, namely, the *transverse Mercator projection* and the *Lambert conformal projection*. The transverse Mercator projection employs a cylindrical surface, the axis of which is normal to the spheroid's axis of rotation and which intersects the surface of the spheroid along two ellipses equidistant from a meridian plane through the center of the area to be projected (see Fig. 11-1). The distortions occur in the east-west direction, and therefore the projection is used for states or zones that have relatively short east-west dimensions. An example is Vermont.

The Lambert conformal projection employs a conical surface, the axis of which coincides with the spheroid's axis of rotation and which intersects the surface of the spheroid along two parallels of latitude that are approximately equidistant from a parallel lying in the center of the area to be projected (see Fig. 11-5). The distortions occur in a north-south direction, and therefore the projection is used for states or zones with relatively short north-south dimensions. An example is Tennessee.

As originally designed, the size of the zones were kept small enough to keep the maximum distortion at any point to about one part in 10,000. Thus, if the distance in an east-west direction in a transverse Mercator projection is limited to about 158 miles or 254 km, or if the distance in a north-south direction in a Lambert projection is limited to about 158 miles, then the distortions will be such that a distance on the spheroid will not differ from the corresponding projected distance by any more than one part in 10,000. Thus it becomes necessary to assign more than one projection to some individual states. California, for example, has seven Lambert zones; Nevada has three transverse Mercator zones; New York has three transverse Mercator zones to cover all but Long Island, which is covered by one Lambert zone. Florida has two transverse Mercator zones covering the peninsula proper and one Lambert zone covering the western portion of the state. Alaska is covered by 10 zones: 8 are Lambert zones, 1 is a transverse Mercator zone, and 1 is an oblique Mercator zone, which is a Mercator projection with the axis of the cylinder oriented in an azimuth of about 323°.

Because of the adoption of the new NAD 83, there are now two separate and distinct state plane coordinate systems. One is a projection of NAD 27 and is referred to as the State Plane Coordinate System of 1927 (SPCS 27); the other is a projection of NAD 83 and is referred to as the State Plane Coordinate System of 1983 (SPCS 83). Since both of these systems are currently in use and will continue to be used for an indefinite period, this chapter will describe the use of both systems.

Table 11-1 shows the number and type of projections referred to SPCS 27. Table 11-2 shows the number and type of projections referred to SPCS 83. Comparison between the two tables indicates a few but significant changes. The states of Montana, Nebraska, and South Carolina are each covered by a single zone on SPCS 83. California has absorbed Zone 7 of SPCS 27 into Zone 5 of SPCS 83. Michigan changed from transverse Mercator to Lambert projections. The projections of Montana, Nebraska, and South Carolina exceed the 1 : 10,000 limit, but this is of little consequence because the distortions resulting from these larger zones are readily determined by computations.

11-3 Difference Between SPCS 27 and SPCS 83

From the standpoint of the practicing surveyor, the major differences between SPCS 27 and SPCS 83 are the projection of a more accurate datum, the different magnitude of numbers, and the use of the meter as the unit of measurement in the new system. Plane coordinate values were published in feet on SPCS 27, this being the survey foot defined in Section 1-4. Plane coordinate values are published in meters on SPCS 83, the meter also as defined in Section 1-4. As the states adopted SPCS 83, the choice was permitted between the use of the meter, the Survey Foot, and the International Foot in which 1 ft = 30.48 cm. exactly. At the time of this writing, of those states that have enacted or are in the process of enacting legislation, 9 states have adopted the Survey Foot together with the meter; 6 states have adopted the International Foot and the meter; the remaining 21 states have adopted the meter exclusively. Fourteen states have not at present begun any legislation toward adoption of SPCS 83.

Some of the projection constants have been changed as well as the ellipsoidal constants used in computation. The form and use of the tables prepared by NGS are different between the two systems. The 1 : 10,000 criterion has been preserved in all projections except in the Montana, Nebraska, and South Carolina projections.

TABLE 11-1   State Plane Coordinate Systems of 1927

State and Zone	Projection	State and Zone	Projection
Alabama		**Idaho**	
East	Transverse Mercator	East	Transverse Mercator
West	Transverse Mercator	Central	Transverse Mercator
		West	Transverse Mercator
Alaska[a]			
		Illinois	
Arizona		East	Transverse Mercator
East	Transverse Mercator	West	Transverse Mercator
Central	Transverse Mercator		
West	Transverse Mercator	**Indiana**	
		East	Transverse Mercator
Arkansas		West	Transverse Mercator
North	Lambert		
South	Lambert	**Iowa**	
		North	Lambert
California		South	Lambert
Zone 1	Lambert		
Zone 2	Lambert	**Kansas**	
Zone 3	Lambert	North	Lambert
Zone 4	Lambert	South	Lambert
Zone 5	Lambert		
Zone 6	Lambert	**Kentucky**	
Zone 7	Lambert	North	Lambert
		South	Lambert
Colorado			
North	Lambert	**Louisiana**	
Central	Lambert	North	Lambert
South	Lambert	South	Lambert
Connecticut	Lambert	**Maine**	
		East	Transverse Mercator
Delaware	Transverse Mercator	West	Transverse Mercator
Florida		**Maryland**	Lambert
East	Transverse Mercator		
West	Transverse Mercator	**Massachusetts**	
North	Lambert	Mainland	Lambert
		Island	Lambert
Georgia			
East	Transverse Mercator	**Michigan**[b]	
West	Transverse Mercator	East	Transverse Mercator
		Central	Transverse Mercator
Hawaii		West	Transverse Mercator
Zone 1	Transverse Mercator		
Zone 2	Transverse Mercator	**Minnesota**	
Zone 3	Transverse Mercator	North	Lambert
Zone 4	Transverse Mercator	Central	Lambert
Zone 5	Transverse Mercator	South	Lambert

[a]Alaska is divided into 10 zones although projection tables are not available. Zone 1 is an oblique transverse Mercator projection; Zones 2 through 9 are transverse Mercator projections; and Zone 10 is a Lambert projection.

[b]Michigan is also divided into three Lambert zones, the North, Central, and South. These projections were computed for an elevation of 800 ft rather than for sea level.

TABLE 11-1 (continued)

State and Zone	Projection	State and Zone	Projection
Mississippi		Pennsylvania	
East	Transverse Mercator	North	Lambert
West	Transverse Mercator	South	Lambert
Missouri		Rhode Island	Transverse Mercator
East	Transverse Mercator		
Central	Transverse Mercator	South Carolina	
West	Transverse Mercator	North	Lambert
Montana		South	Lambert
North	Lambert	South Dakota	
Central	Lambert	North	Lambert
South	Lambert	South	Lambert
Nebraska		Tennessee	Lambert
North	Lambert		
South	Lambert	Texas	
Nevada		North	Lambert
East	Transverse Mercator	North Central	Lambert
Central	Transverse Mercator	Central	Lambert
West	Transverse Mercator	South Central	Lambert
		South	Lambert
New Hampshire	Tranverse Mercator		
		Utah	
New Jersey	Transverse Mercator	North	Lambert
		Central	Lambert
New Mexico		South	Lambert
East	Transverse Mercator		
Central	Transverse Mercator	Vermont	Transverse Mercator
West	Transverse Mercator		
		Virginia	
New York		North	Lambert
Long Island	Lambert	South	Lambert
East	Transverse Mercator		
Central	Transverse Mercator	Washington	
West	Transverse Mercator	North	Lambert
		South	Lambert
North Carolina	Lambert		
		West Virginia	
North Dakota		North	Lambert
North	Lambert	South	Lambert
South	Lambert		
Ohio		Wisconsin	
North	Lambert	North	Lambert
South	Lambert	Central	Lambert
		South	Lambert
Oklahoma			
North	Lambert	Wyoming	
South	Lambert	Zone E	Transverse Mercator
		Zone EC	Transverse Mercator
Oregon		Zone WC	Transverse Mercator
North	Lambert	Zone W	Transverse Mercator
South	Lambert		

TABLE 11-2   State Plane Coordinate Systems of 1983

State and Zone	Projection	State and Zone	Projection
Alabama		Idaho	
East	Transverse Mercator	East	Transverse Mercator
West	Transverse Mercator	Central	Transverse Mercator
		West	Transverse Mercator
Alaska[a]			
		Illinois	
Arizona		East	Transverse Mercator
East	Transverse Mercator	West	Transverse Mercator
Central	Transverse Mercator		
West	Transverse Mercator	Indiana	
		East	Transverse Mercator
Arkansas		West	Transverse Mercator
North	Lambert		
South	Lambert	Iowa	
		North	Lambert
California		South	Lambert
Zone 1	Lambert		
Zone 2	Lambert	Kansas	
Zone 3	Lambert	North	Lambert
Zone 4	Lambert	South	Lambert
Zone 5	Lambert		
Zone 6	Lambert	Kentucky	
		North	Lambert
		South	Lambert
Colorado			
North	Lambert	Louisiana	
Central	Lambert	North	Lambert
South	Lambert	South	Lambert
		Offshore	Lambert
Connecticut	Lambert		
		Maine	
Delaware	Transverse Mercator	East	Transverse Mercator
		West	Transverse Mercator
Florida			
East	Transverse Mercator	Maryland	Lambert
West	Transverse Mercator		
North	Lambert	Massachusetts	
		Mainland	Lambert
Georgia		Island	Lambert
East	Transverse Mercator		
West	Transverse Mercator	Michigan	
		North	Lambert
		Central	Lambert
Hawaii		South	Lambert
Zone 1	Transverse Mercator		
Zone 2	Transverse Mercator	Minnesota	
Zone 3	Transverse Mercator	North	Lambert
Zone 4	Transverse Mercator	Central	Lambert
Zone 5	Transverse Mercator	South	Lambert

[a]Alaska is divided into 10 zones. Zone 1 is an oblique transverse Mercator projection; Zones 2 through 9 are transverse Mercator projections; and Zone 10 is a Lambert projection.

TABLE 11-2 (continued)

State and Zone	Projection	State and Zone	Projection
Mississippi		Pennsylvania	
East	Transverse Mercator	North	Lambert
West	Transverse Mercator	South	Lambert
Missouri		Rhode Island	Transverse Mercator
East	Transverse Mercator	South Carolina	Lambert
Central	Transverse Mercator		
West	Transverse Mercator	South Dakota	
		North	Lambert
Montana	Lambert	South	Lambert
Nebraska	Lambert	Tennessee	Lambert
Nevada		Texas	
East	Transverse Mercator	North	Lambert
Central	Transverse Mercator	North Central	Lambert
West	Transverse Mercator	Central	Lambert
		South Central	Lambert
New Hampshire	Transverse Mercator	South	Lambert
New Jersey		Utah	
(New York East)	Transverse Mercator	North	Lambert
		Central	Lambert
New Mexico		South	Lambert
East	Transverse Mercator		
Central	Transverse Mercator	Vermont	Transverse Mercator
West	Transverse Mercator	Virginia	
		North	Lambert
New York		South	Lambert
East (New Jersey)	Transverse Mercator		
Central	Transverse Mercator	Washington	
West	Transverse Mercator	North	Lambert
Long Island	Lambert	South	Lambert
North Carolina	Lambert	West Virginia	
		North	Lambert
North Dakota		South	Lambert
North	Lambert		
South	Lambert	Wisconsin	
		North	Lambert
Ohio		Central	Lambert
North	Lambert	South	Lambert
South	Lambert		
		Wyoming	
Oklahoma		East	Transverse Mercator
North	Lambert	East Central	Transverse Mercator
South	Lambert	West Central	Transverse Mercator
		West	Transverse Mercator
Oregon			
North	Lambert	Puerto Rico and	
South	Lambert	Virgin Islands	Lambert

The plane coordinates published for SPCS 83 are more accurate than those published for SPCS 27 because of the readjustment of the horizontal control network that defines NAD 83. This allows better ties between control points, and upgrades the reliability of all surveys tied to NAD 83 control. Coordinates on SPCS 27 are designated X and Y, while the corresponding coordinates on SPCS 83 are designated E for Easting and N for Northing.

Throughout the remainder of this chapter, the difference between the use of SPCS 27 and SPCS 83 will be discussed where appropriate.

11-4   Geodetic and Grid Azimuths

In placing a survey on a system of state plane coordinates, all bearings or azimuths are referred to the same meridian. Since the system is rectangular, all grid north-and-south lines are parallel to a *central meridian.* In general there is a difference between the grid azimuth or bearing of a line and its geodetic azimuth or bearing. This difference is caused by the convergence of meridians, the amount increasing with the latitude and the distance from the central meridian. When the grid coordinates of the two ends of a starting line are known, its grid azimuth may be computed by Eq. (8-13) and its grid length by Eq. (8-12).

The geodetic azimuth of a line is a computed azimuth based on triangulation, traverse and GPS observations, computations, and adjustments resulting from geodetic surveying. The difference between true or astronomical azimuth as defined in Section 7-6, and geodetic azimuth is relatively small and caused by the fact that the plumb line at a station is deflected from a true normal to the spheroid of computation for geodetic positions. The geodetic azimuth of a line is that published by the NGS along with the geodetic positions in triangulation, traverse, and GPS networks.

It is frequently important for the surveyor to know the difference between the geodetic and grid azimuths of a survey line for purposes of obtaining a check on a computed value or to provide a starting azimuth for a survey. On the Lambert grid in SPCS 27, the angle between the geodetic and grid meridians is designated by θ and is listed in the state projection tables. On the transverse Mercator grid, the angle $\Delta\alpha$, also employed in geodetic computations, represents the convergence of the meridians. This angle on either projection is referred to as the *mapping angle.*

In publications relating to SPCS 83, the difference between grid north and geodetic north is referred to as the *convergence angle* and is designated by the angle γ on both types of projections.

A line on the surface of the spheroid representing a line of sight between two points or, more exactly, representing the most direct path between the two points, can be thought of as a "great circle," as would be the case if the sea-level surface were a true sphere. This line is referred to as a *geodetic line.* The geodetic line projected onto the Lambert grid or the transverse Mercator would in general show a slight curvature. Thus if long lines are involved and utmost accuracy is to be obtained, the difference between geodetic and grid azimuths is slightly different from the value of θ and $\Delta\alpha$ or γ (see Sections 11-6 and 11-12).

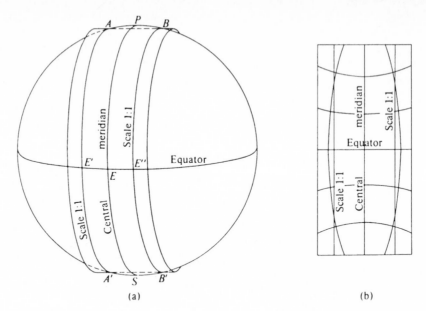

FIGURE 11-1 Transverse Mercator projection.

11-5 Transverse Mercator Projection

The transverse Mercator projection is represented diagrammatically in Fig. 11-1. View (a) shows the sea-level surface of the earth intersected by a cylindrical surface along the two ellipses $AE''A'$ and $BE''B'$, which are equidistant from a *central meridian PES*. This intersection along the two ellipses is effected by scaling down from a cylinder that would otherwise be tangent to the earth along the central meridian. View (b) shows a portion of the cylindrical surface developed into a plane surface on which the meridians and parallels of the earth's surface have been projected mathematically. On the development each ellipse appears as a straight line parallel to and equidistant from the central meridian. Along either of these two lines, a distance on the projection is the same as the corresponding distance on the sea-level surface.[†] Between the two lines, a distance is smaller than the corresponding distance on the sea-level surface. Outside the two lines, a distance on the projection is larger than the corresponding distance on the sea-level surface. The discrepancy between these corresponding distances depends on the position of the line being considered with respect to the central meridian. It is seen that the scale of a line running in an east-west direction varies from point to point. It is also seen, however, that a line on the projection parallel to the central meridian has a constant scale throughout its length, whether this scale is larger than, equal to, or less than that on the corresponding sea-level line.

[†]In this chapter the sea-level surface is assumed to coincide with the spheroidal surface. Thus a length at sea level is assumed to be the length on the spheroid. However, see Section 11-18.

In designing the transverse Mercator projection to a state or a zone, the width of the projection was limited to 158 miles or about 254 km and the lines along which the scale of the projection is exact are separated by about two-thirds this distance. At no point within these limits will the difference between a sea-level distance and the projected distance, called the *grid distance,* be greater than one part in 10,000.

11-6   Grid Azimuth on Transverse Mercator Projection of SPCS 27

With but two exceptions, a line run on the surface of the earth as a straight line does not have a constant azimuth with respect to the true or the geodetic meridians because of the convergence of these meridians. The exceptions are a straight line run along the true meridian from which the azimuth angle is measured and a line run along the equator. For example, a secant line run to establish a parallel of latitude, as described in Section 18-8, extends due east or due west at one point only. Immediately on leaving this point, the secant runs in a southeasterly or southwesterly direction.

In the transverse Mercator system of plane coordinates, the azimuth of any line is referred to the central meridian. This azimuth is also referred to any line parallel to the central meridian. The direction of each of this series of parallel lines is *grid north.* Thus a straight line on the transverse Mercator projection has a constant grid azimuth throughout its length. The angle between the grid north and the geodetic north at a point on this projection depends on two things: (1) the distance from the central meridian to the point, and (2) the latitude of the point.

In Fig. 11-2, P is a station whose latitude and longitude are, respectively, φ and λ. The line parallel to the central meridian passing through P is grid north. The curved line making the angle $\Delta\alpha$ with grid north at P is the geodetic meridian

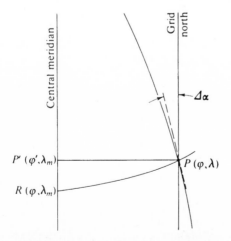

FIGURE 11-2 Relationship between geodetic and grid azimuths on transverse Mercator projection.

through P. The straight line PP', making an angle of $90°$ with the central meridian, is the projection of the arc of a great circle (nearly) passing through P. It intersects the central meridian at P', whose coordinates are φ' and λ_m. The parallel of latitudes through P is a curved line intersecting the central meridian at R. The latitude of R is φ. The value of $\Delta\alpha$ is

$$- \Delta\alpha = \Delta\lambda \; \sin \tfrac{1}{2}(\varphi + \varphi') \; \sec \tfrac{1}{2}(\varphi - \varphi') + \text{other terms} \qquad (11\text{-}1a)$$

Since the value of $\sec \frac{1}{2}(\varphi - \varphi')$ is practically unity, and since the other terms are small except for very exact determinations, the equation for $\Delta\alpha$ may be reduced to

$$-\Delta\alpha = \Delta\lambda \; \sin \tfrac{1}{2}(\varphi + \varphi') \qquad (11\text{-}1b)$$

in which $\Delta\lambda$ is the difference between the longitude of the station and the longitude of the central meridian for the given projection, and φ' is the latitude of the foot of the perpendicular from the point to the central meridian (P' in Fig. 11-2). The value of $\Delta\lambda$ is found from the relationship

$$\Delta\lambda = \lambda_m - \lambda \qquad (11\text{-}2a)$$

in which λ_m is the longitude of the central meridian, and λ is the longitude of the station.

If the station lies east of the central meridian, then $\Delta\lambda$ is positive $(+)$. If the station lies west of the central meridian, $\Delta\lambda$ is negative $(-)$. Because the direction in which $\Delta\lambda$ is computed by Eq. (11-2a) is opposite to that in which it is reckoned in Chapter 9, the algebraic sign of $\Delta\alpha$ as computed by Eq. (11-1b) is the same as the algebraic sign of $\Delta\lambda$. Thus, if the station lies east of the central meridian, $\Delta\alpha$ is positive, and if to the west, $\Delta\alpha$ is negative. The value of φ' is found by solving Eqs. (11-6) and (11-7).

Because of the different formulation for the solution of problems when using SPCS 83, the value of $\Delta\lambda$ is given by

$$\Delta\lambda = \lambda - \lambda_0 \qquad (11\text{-}2b)$$

in which λ_0 is the longitude of the central meridian.

The relationship between the geodetic azimuth and the grid azimuth of a line originating in P is given by the expression

$$\text{geodetic azimuth} - \text{grid azimuth} = \Delta\alpha \qquad (11\text{-}3)$$

In Fig. 11-3 lines AB and CD lie to the west of the central meridian, and EF and GH lie to the east of the central meridian. The grid lines joining the end points are straight lines in every case. The ground line, or geodetic line, joining any pair of end points, however, is curved when projected onto the grid. Furthermore, this curvature, which is greatly exaggerated in Fig. 11-3, increases outward from the central meridian. To allow for this curvature in going from the geodetic azimuth to the grid azimuth, or vice versa, when a line longer than 5 miles is involved, Eq. (11-3) must be modified by a small correction term. Thus

$$\text{geodetic azimuth} - \text{grid azimuth} = \Delta\alpha + \frac{(Y_2 - Y_1)(2X_1' + X_2')}{(6\rho_0^2 \sin 1'')_g} \qquad (11\text{-}4a)$$

in which X_1' and X_2' are the X coordinates of the origin and terminal of the line, respectively, *referred to the central meridian* [see Eq. (11-9)]; Y_1 and Y_2 are the Y coordinates; and ρ_0 is the mean radius of the earth in the vicinity of the projection. The subscript g indicates that the value of ρ_0 has been reduced by the scale factor mentioned in Section 11-4. Due regard must be given to the sign of the X' coordinates and the value $Y_2 - Y_1$. Notice that if the line is grid east-west, as is the line DF or CE in Fig. 11-3, no correction need be applied to $\Delta\alpha$ since $Y_2 - Y_1 = 0$.

The second term on the right side of Eq. (11-4a) is referred to as the t-T correction, in which t is the grid azimuth of the line and T is the grid azimuth of the geodetic line *projected onto the grid*. This is also referred to as the arc-to-chord correction, or the second term correction, and its symbol is δ.

In the SPCS 27 system, the coordinates are expressed in feet. Dividing the numerator and the denominator of the second term of Eq. (11-4a) by three gives

$$\delta = \frac{(Y_2 - Y_1)(2X_1' + X_2')/3}{(2\rho_0^2 \sin 1'')_g}$$

Let $\Delta Y = Y_2 - Y_1$ and $\Delta X = (2X_1' + X_2')/3$. Evaluating the denominator using the average radius of the ellipsoid for all zones gives

$$\delta = 2.36 \ \Delta X \ \Delta Y \times 10^{-10}$$

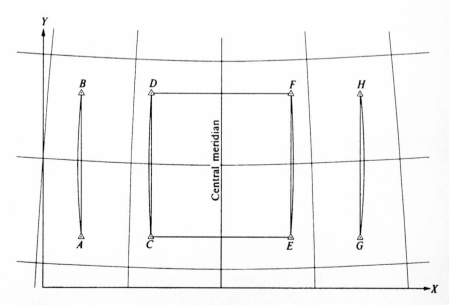

FIGURE 11-3 Curvature of geodetic lines on transverse Mercator projection.

with sufficient accuracy for all observable lines. Then Eq. (11-4a) becomes

$$\text{geodetic azimuth} - t = \Delta\alpha + 2.36\ \Delta X\ \Delta Y \times 10^{-10} \qquad (11\text{-}4b)$$

11-7 Geographic to Plane Coordinates on Transverse Mercator Projection of SPCS 27

The origin of coordinates on the transverse Mercator projection is taken far enough south and west of the zone covered to render all values of coordinates positive. To make all the X coordinates positive, the central meridian, which is the Y axis, is assigned a large X value denoted by X_0; usually $X_0 = 500{,}000$ ft. The X axis is a line perpendicular to the central meridian, where $Y = 0$.

The Y coordinate of a point on the central meridian is equal to the spheroidal or sea-level distance along the central meridian measured from the intersection of the X axis and the central meridian to the point and adjusted for the scale along the central meridian. The X coordinate of a point on the central meridian is, of course, the same as the X value of the central meridian.

Before the Y and X coordinates of any point whose latitude and longitude are known can be determined, the sea-level length of a line passing through the point and perpendicular to the central meridian must be computed. The latitude of the foot of this perpendicular on the central meridian must then be determined. In Fig. 11-4, P is a point whose latitude and longitude are φ and λ. The line $P'P$ is the sea-level line passing through P and perpendicular to the central meridian at P'. The azimuth (from south) of the line from P' to P is therefore 270°. The sea-level length of this line may be determined by the equation for $\Delta\lambda$ in the computation for geodetic positions (see Special Publication 241, U. S. Coast and Geodetic Survey). This length is given by Eq. (9-1):

$$\Delta\lambda = s\ \sin\ \alpha\ \sec\ \varphi \cdot A + \text{arcsin correction}$$

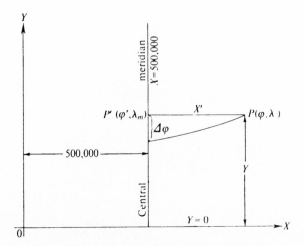

FIGURE 11-4 Rectangular coordinates on a transverse Mercator projection.

in which $\Delta\lambda$ is the difference in longitude between the point and the central meridian given by Eq. (11-2a), s is the sea-level length of the line $P'P$ in meters, α is the azimuth of $P'P$, A is the factor A for the known point P, and φ is the latitude of point P. Since the azimuth of $P'P$ is 270°, $\sin \alpha = -1$. Transposing terms then gives the following equation for the distance s in meters:

$$s = \frac{\Delta\lambda \, \cos \, \varphi}{A} - \text{arcsin correction} \tag{11-5}$$

In Eq. (11-5) $\Delta\lambda$ is positive if the point is east of the central meridian and negative if the point is west of the central meridian.

The latitude of P' is determined from the equation for $-\Delta\varphi$ in the computation for geodetic position, which is

$$-\Delta\varphi = Bs \, \cos \, \alpha + Cs^2 \sin^2 \alpha + (\delta\varphi)^2 \, D - Ehs^2 \sin^2 \alpha$$

in which $-\Delta\varphi$ is the minus difference in latitude going from P' to P, C is the factor C for the unknown point P', and α is the azimuth of the line $P'P$, which is 270°. Since $\cos \alpha = 0$, the first term drops out. Also, because the difference in latitude between P' and P can never be very large, the terms containing D and E are both neglected. So

$$-\Delta\varphi = Cs^2 \tag{11-6}$$

The distance s in meters is obtained from Eq. (11-5). The value of C should be obtained for the latitude of P'. However, since this is as yet unknown, the preliminary value of C for the known latitude of the point P is used. Then, when the preliminary value of φ' is determined, a better value of C is obtained to give a corrected value of $-\Delta\varphi$. It should be noted that the numerical value of $\Delta\varphi$ is always added to the latitude φ of point P to obtain φ'. Therefore

$$\varphi' = \varphi + \Delta\varphi \tag{11-7}$$

Once the value of φ' has been determined, the Y coordinate of P', and hence of P, is obtained by interpolation in the tables prepared for the particular projection by the NGS. This coordinate is a true length along the central meridian reduced for the scale along the central meridian. Therefore it may be tabulated for each latitude.

The length of the line $P'P$ on the plane projection is obtained by the equation

$$X' = s_g + \frac{s_g^3}{(6\rho_0^2)_g} \tag{11-8}$$

in which X' is the grid length of $P'P$ corresponding to s (the sea-level length of $P'P$), s_g is the distance s converted to feet and reduced by the overall scale reduction of the projection, and ρ_0 is the mean radius of curvature of the spheroid in the vicinity of the projection. This radius is assumed to be constant for any given projection and has also been reduced by the overall scale reduction.

The X coordinate of the point P may now be determined by adding the value of X' to the X coordinate of the central meridian. Thus

$$X = X_0 + X' \tag{11-9}$$

If $\Delta\lambda$ is negative, then s and consequently X' will also be negative and the value of X will be less than X_0.

To summarize, the equations are repeated here in their order of solution:

$$\Delta\lambda = \lambda_m - \lambda \tag{11-2a}$$

$$s = \frac{\Delta\lambda \cos \varphi}{A} - \text{arcsin correction} \tag{11-5}$$

$$-\Delta\varphi = Cs^2 \tag{11-6}$$

$$\varphi' = \varphi + \Delta\varphi \tag{11-7}$$

The value of Y is found by interpolating in an appropriate table with φ' as argument.

$$X' = s_g + \frac{s_g^3}{(6\rho_0^2)_g} \tag{11-8}$$

$$X = X_0 + X' \tag{11-9}$$

11-8   Tables for Transverse Mercator Projection of SPCS 27

The National Geodetic Survey has prepared and published tables for each projection in each state and makes them available through the U.S. Government Printing Office. These tables are used in place of Eqs. (11-1) through (11-9) and are convenient for use with computers or handheld calculators. The tables are based on the Clarke Spheroid of 1866 for points defined on NAD 27.

Referring to NAD 27, Table 11-3 shows a portion of the projection tables for Arizona. The function Y_0 is the Y coordinate of the intersection of the given parallel of latitude with the central meridian. The remainder of the quantities are used in the following equations, taken from Special Publication No. 257 for Arizona:

$$X = X' + 500,000 \tag{11-10}$$

which is the same as Eq. (11-9) in which $X_0 = 500,000$.

$$X' = H \cdot \Delta\lambda'' \pm ab \tag{11-11}$$

which performs the same function as Eq. (11-8).

$$Y = Y_0 + V\left(\frac{\Delta\lambda''}{100}\right)^2 \pm c \tag{11-12}$$

in which the middle term of the right side is essentially the same as Eq. (11-6).

$$\Delta\alpha'' = \Delta\lambda'' \sin \varphi + g \tag{11-13}$$

A comparison of Eq. (11-13) with Eq. (11-1b) shows that $\sin \varphi$ is the same as $\sin \frac{1}{2}(\varphi + \varphi')$. This is permissible since $\Delta\alpha$ is small, and $\Delta\varphi$ given by the middle term of Eq. (11-12) is also small. The g term is similar to the F term of Eq. (11-1a) and is usually quite small.

TABLE 11-3   Transverse Mercator Projection for Arizona East and Central Zones on SPCS 27

Latitude	Y_0 (ft)	ΔY_0 (per second)	a	H	ΔH (per second)	V	ΔV (per second)
33° 20′	848 779.94	101.064 33	−0.823	84.842 642	269.39	1.130 341	4.73
33° 21′	854 843.80	101.064 67	−0.821	84.826 479	269.51	1.130 625	4.73
33° 22′	860 907.68	101.065 00	−0.820	84.810 309	269.64	1.130 909	4.73
33° 23′	866 971.58	101.065 17	−0.819	84.794 131	269.74	1.131 193	4.72
33° 24′	873 035.49	101.065 33	−0.817	84.777 947	269.87	1.131 476	4.72

$\Delta\lambda''$	b	Δb	c
1000	+3.540	+0.315	−0.022
1100	+3.855	+0.305	−0.027
1200	+4.160	+0.294	−0.032
1300	+4.454	+0.282	−0.038
1400	+4.736	+0.268	−0.043

Table for $g\Delta\alpha'' = \sin\varphi(\Delta\lambda'') + g$

				$\Delta\lambda''$			
Latitude	0″	1000″	2000″	3000″	4000″	5000″	6000″
31°	0.00	0.00	0.02	0.08	0.19	0.37	0.64
32°	0	0	0.02	0.08	0.19	.038	0.65
33°	0	0	0.02	0.08	0.19	.038	0.65
34°	0	0	0.02	0.08	0.19	.038	0.65
35°	0	0	0.02	0.08	0.19	.038	0.65

Y	P	ΔP	X′	d
800,000	1.56187	1638	100,000	+0.02
900,000	1.57825	1649	150,000	+0.04

Table for e

Y \ X′	200,000	300,000	400,000	500,000	Y	East and Central Zones M	East and Central Zones ΔM
0	0	0.1	0.3	0.6	500,000	0.006 2436	664
1,000,000	0.0	0.1	0.4	0.8	600,000	0.006 3100	667
2,000,000	0.0	0.2	0.5	1.0	700,000	0.006 3767	672
					800,000	0.006 4439	676
					900,000	0.006 5115	680

Constants for East Zone: Central meridian 110°10′ 00.000″. Scale reduction (central meridian) 1 : 10,000. $(1/6\rho_0^2 \sin 1'')_g = 0.7872 \times 10^{-10}$.

Note that the functions Y_0, H, V, and a are based on the latitude of the point in question, whereas b, c, and g are based on the longitude.

The value of $\Delta\lambda$ needed for use in the tables is given by Eq (11-2a). The second term correction of $\Delta\alpha$ is given in Eq. (11-4a) or Eq. (11-4b).

EXAMPLE 11-1   The NAD 27 latitude and longitude of a control point in Arizona are $\varphi = 33° 22' 14.656''$ and $\lambda = 109° 52' 43.255''$. Compute the coordinates of the control point on the transverse Mercator East zone for Arizona. Compute $\Delta\alpha$, neglecting the correction term.

Solution: The difference in longitude is first computed by Eq. (11-2a). Thus

$$\Delta\lambda = 110° \ 10' \ 00.000'' - 109° \ 52' \ 43.255'' = +0° \ 17' \ 16.745'' = +1036.745''$$

The work then proceeds as follows:

$$(\Delta\lambda''/100)^2 = 107.484$$

$H = 84.806357$	(from Table 11-3)
$V = 1.130978$	(from Table 11-3)
$a = -0.820$	(from Table 11-3)
$b = +3.656$	(from Table 11-3)
$H \cdot \Delta\lambda = +87,922.57$	
$ab = \quad -3.00$	
$X' = \overline{+87,919.57}$	[by Eq. (11-11)]
$c = -0.02$	(from Table 11-3)
$V(\Delta\lambda''/100)^2 \pm c = 121.54$	
$Y_0(\text{minutes}) = 860,907.68$	(from Table 11-3)
$Y_0(\text{seconds}) = \quad +1,481.21$	(interpolation in Table 11-3)
$Y_0 = \overline{862,388.89}$	(Y value on central meridian for given latitude)
$X = 587,919.57$	[by Eq. (11-10)]
$Y = 862,510.43$	[by Eq. (11-12)]
$\Delta\alpha'' = +1036.745 \sin 33° \ 22' \ 15'' + 0 = +570.27''$	
$= +0° \ 09' \ 30.27''$ [by Eq. (11-13)]	

The values of the remaining functions given in the tables are needed to compute NAD 27 latitude and longitude from SPCS 27 X and Y coordinates according to the following equations:

$$Y_0 = Y - P\left(\frac{X'}{10,000}\right)^2 - d \qquad (11\text{-}14)$$

in which X', P, and d are obtained from the given coordinates. The value of X' is given by Eq. (11-10).

The latitude is obtained by interpolating in the tables using the value of Y_0 from Eq. (11-14). Then the value of H for this latitude can be interpolated.

$$\Delta\lambda''(\text{approx.}) = \frac{X'}{H} \qquad (11\text{-}15)$$

Then a is obtained from the known latitude and b from the approximate value of $\Delta\lambda''$. The final value for $\Delta\lambda''$ is then given by

$$\Delta\lambda'' = \frac{X' \mp ab}{H} \qquad (11\text{-}16)$$

and the value of $\Delta\alpha''$ is

$$\Delta\alpha'' = MX' - e \qquad (11\text{-}17)$$

Note that P and M are functions of the Y coordinate of the point, d is a function of the X coordinate of the point, and e depends on both X and Y.

EXAMPLE 11-2 The state plane coordinates of a point on the Arizona East system are X = 626,545.10 ft and Y = 868,452.25 ft. Determine the latitude and longitude of the point. Determine the value of $\Delta \alpha''$, neglecting the small correction term.

Solution: The value of X' is 626,545.10 – 500,000.00 = +126,545.10. Then, from Table 11-3, P = 1.57308 and d = +0.02. By Eq. (11-14),

$$Y_0 = 868,452.25 - 1.57308(12.65451)^2 - 0.02 = 868,200.32$$

Interpolating in the tables for φ gives

$$868,200.32 = Y_0$$
$$\underline{-866,971.58 \text{ (for } \varphi = 33° \ 23')}$$
$$1,228.74 \text{ (difference from tabular value)}$$

$$1,228.74/101.06517 = 12.158''$$

$$\varphi = 33° \ 23' \ 12.158''$$
$$H = 84.790852 \qquad \text{(from Table 11-3)}$$
$$\Delta \lambda''(\text{approx.}) = 126,545.10/84.790852 = +1492'' \quad \text{[by (Eq. 11-15)]}$$
$$a = -0.819 \text{ and } b = +4.984; \quad ab = -4.082$$
$$\Delta \lambda'' = (126,545.10 + 4.082)/84.790852 = +1492.486''$$
$$= +0° \ 24' \ 52.486'' \qquad \text{[by (Eq. 11-16)]}$$
$$\lambda = 109° \ 45' \ 07.514'' \qquad \text{[by (Eq. 11-2a)]}$$
$$M = 0.0064902 \quad \text{and} \quad e = 0$$
$$\Delta \alpha'' = 0.0064902 \times 126,545.10$$
$$= + \ 821.30''$$
$$= + \ 0° \ 13' \ 41.30'' \qquad \blacklozenge$$

The projection tables for Arizona as well as for the other states using the transverse Mercator projection also include a section for use in reducing sea-level distances to grid distances. Table 11-4 shows portions of the table containing the reduction factors included in the Arizona tables for the East and Central zones. Note that the scale factor varies as the distance east or west from the central meridian. The scale factor for the central meridian where X' = 0 is 0.999 9000, and changes from less than unity to greater than unity at $X' \approx \pm 295,832$ ft.

TABLE 11-4 Scale Factor for Arizona East and Central Zones on SPCS 27

X' (ft)	Scale Factor	X' (ft)	Scale Factor	X' (ft)	Scale Factor
0	0.999 9000	290,000	0.999 9963	500,000	1.000 1862
5,000	0.999 9000	295,000	0.999 9996	505,000	1.000 1919
10,000	0.999 9001	300,000	1.000 0030	510,000	1.000 1977
15,000	0.999 9003	305,000	1.000 0065	515,000	1.000 2036
20,000	0.999 9005	310,000	1.000 0100	520,000	1.000 2095

11-9   Grid Azimuth on Transverse Mercator Projection of SPCS 83

The grid azimuth of a line on SPCS 83 is computed by the same equations given in Section 11-6, except for the notation E instead of X, N instead of Y, and γ instead of $\Delta\alpha$. In the NGS publications for SPCS 83, the geodetic azimuth is denoted by the symbol α. Thus

$$\alpha - t = \gamma + \frac{(N_2 - N_1)(2E_1' + E_2')}{(6\rho_0^2 \sin 1'')_g} \tag{11-18a}$$

Applying the same reasoning given in Section 11-6, the second term (δ) on the right side of Eq. (11-18a) becomes

$$\delta = \frac{(N_2 - N_1)(2E_1' + E_2')/3}{(2\rho_0^2 \sin 1'')_g}$$

Let $\Delta N = N_2 - N_1$ and $\Delta E = (2E_1' + E_2')/3$, and evaluating the denominator using the average radius of the ellipsoid for all zones, remembering that N and E are in meters, gives

$$\delta = 25.4 \; \Delta N \; \Delta E \times 10^{-10}$$

with sufficient accuracy for all observable lines. Then Eq. (11-18a) becomes

$$\alpha - t = \gamma + 25.4 \; \Delta N \; \Delta E \times 10^{-10} \tag{11-18b}$$

If the SPCS 83 coordinates are expressed in feet, the Eq. (11-18a) becomes

$$\alpha - t = \gamma + 2.36 \; \Delta N \; \Delta E \times 10^{-10} \tag{11-18c}$$

11-10   Geographic to Plane Coordinates on Transverse Mercator Projection of SPCS 83

Two methods can be employed to convert from geographic to plane coordinates and vice versa when using SPCS 83. In the first method a series of mapping equations, which are given in NOAA Manual NOS NGS 5 *State Plane Coordinate System of 1983* (NGS 5), can be programmed in a computer with at least a ten-digit capacity. The constants for each zone of the transverse Mercator projection are also given in NGS 5. After the appropriate constants have been entered into the program, the latitude and longitude of a point are entered. The solution gives the N and E values of the point, the convergence angle, and the scale factor at the point. Refer to NGS 5 for a detailed discussion of the mapping equations. In Section 11-24 we explain how to use the Internet to compute plane coordinates and related quantities. If it is frequently necessary to convert many points from geographic to plane coordinates, or vice versa, use of the programmed mapping equations is the most desirable method.

The second method, which is advantageous when conversion is only occasionally necessary, is the use of tables prepared by NGS. These tables are rather cumbersome but give results comparable in accuracy to those of the mapping equations.

As an example of the use of these tables, consider a point lying in the Arizona East zone of SPCS 83. In the published tables for this zone, the following quantities are given:

Datum: NAD 83

Zone: Arizona East

Zone code: 0201

Ellipsoidal constants:

$$A = 6{,}378{,}137 \text{ m (semimajor axis of GRS 80 spheroid)}$$
$$F = 1/298.25722210 \text{ (flattening)}$$

Projection defining constants:

φ_B = 31° 00′ (latitude of grid origin)

λ_B = 110° 10′ (longitude of origin and central meridian, CM)

E_0 = 213,360 m (Easting coordinate of grid origin)

N_B = 0 m (Northing coordinate of grid origin)

K_0 = 0.999900 (scale along central meridian)

Latitude range from φ_{min} = 31° 00′ to φ_{max} = 37° 20′

Equations

$$\Delta\lambda = \lambda_B - \lambda \tag{11-19a}$$

$$p = 10^{-4}\,\Delta\lambda \tag{11-19b}$$

$$N = (I) + (II)p^2 + (III)p^4 \qquad \text{(grid Northing)} \tag{11-19c}$$

$$E' = (IV)p + (V)p^3 + (VI)p^5 \tag{11-19d}$$

$$E = E_0 + E' \qquad \text{(grid Easting)} \tag{11-19e}$$

$$\gamma = (XII)p + (XIII)p^3 \qquad \text{(plane from geodetic coordinates)} \tag{11-19f}$$

$$Q = 10^{-6}E' \tag{11-19g}$$

$$\gamma = (XIV)Q - (XV)Q^3 \qquad \text{(geodetic from plane coordinates} \tag{11-19h}$$

$$k = k_0[1 + (XVI)Q^2 + 0.00003Q^4] \tag{11-19i}$$

$$\lambda = \lambda_B - (IX)Q + (X)Q^3 + (XI)Q^5 \tag{11-19j}$$

$$\varphi' = \text{interpolation in (I) using } N \text{ as argument}$$

$$\varphi = \varphi' - (VII)Q^2 + (VIII)Q^4 \tag{11-19k}$$

The quantities (I) through (XVI) are given in Tables 11-5a through 11-5c and are functions of the geodetic latitude.

TABLE 11-5A Transverse Mercator Projection Tables for Arizona East Grid Coordinates from Geodetic Coordinates SPCS 83

Latitude	(I) (IV)	Diff. 1″	(II) (V)	Diff. 1″	(III) (VI)
31° 14′	25,867.4232	30.795150	3326.0491	0.016851	2.228
	264,614.9986	−0.774338	48.2920	−0.001037	−0.01
31° 15′	27,715.1322	30.795230	3327.0601	0.016832	2.227
	264,568.5383	−0.774711	48.2298	−0.001037	−0.01
31° 16′	29,562.8460	30.795310	3328.0701	0.016813	2.227
	264,522.0556	−0.775085	48.1675	−0.001037	−0.01
31° 17′	31,410.5646	30.795390	3329.0789	0.016795	2.227
	264,475.5505	−0.775458	48.1053	−0.001037	−0.010

Second Difference Corrections

00″	10″	20″	30″	00″	10″	20″	30″
60″	50″	40″	30″	60″	50″	40″	30″
(I)0.0000	−0.0003	−0.0005	−0.0006	(IV)0.0000	0.0016	0.0025	0.0028

TABLE 11-5B Transverse Mercator Projection Tables for Arizona East Geodetic Coordinates from Grid Coordinates SPCS 83

Latitude	(I) (IX)	Diff. 1″	(VII) (X)	Diff. 1″	(VIII) (XI)
31° 13′	24,019.7190	30.795070	1541.4680	0.016775	19.200
	37,784.1220	0.110514	268.8357	0.003259	3.521
31° 14′	25,867.4232	30.795150	1542.4745	0.016780	19.217
	37,790.7528	0.110606	269.0313	0.003263	3.526
31° 15′	27,715.1322	30.795230	1543.4813	0.016786	19.234
	37,797.3892	0.110698	269.2271	0.003267	3.530
31° 16′	29,562.8460	30.795310	1544.4885	0.016792	19.252
	37,804.0310	0.110790	269.4231	0.003271	3.535
31° 17′	31,410.5646	30.795390	1545.4960	0.016798	19.269
	37,810.6785	0.110883	269.6194	0.003275	3.540

Second Difference Corrections

00″	10″	20″	30″	00″	10″	20″	30″
60″	50″	40″	30″	60″	50″	40″	30″
(I)0.0000	−0.0003	−0.0005	−0.0006	(IX)0.0000	−0.0004	−0.0006	−0.0007

TABLE 11-5C   Transverse Mercator Projection Tables for Arizona East Convergence, Scale Factor, $(t - T)$ SPCS 83

Latitude	(I)	Diff. 1″	(XII)(XIV)	Diff. 1″	(XIII)(XV)	(XVI)
31°13′	24,019.7190	30.795070	5,182.7580	0.041458	3.014	
			19,582.5962	0.213950	218.240	0.012332
31°14′	25,867.4232	30.795150	5,185.2455	0.041451	3.014	
			19,595.4332	0.214026	218.460	0.012332
31°15′	27,715.1322	30.795230	5,187.7326	0.041444	3.015	
			19,608.2747	0.214101	218.680	0.012332
31°16′	29,562.8460	30.795310	5,190.2192	0.041436	3.015	
			19,621.1208	0.214176	218.901	0.012332
31°17′	31,410.5646	30.795390	5,192.7054	0.041429	3.015	
			19,633.9713	0.214251	219.122	0.012332

Second Difference Corrections

00″	10″	20″	30″	00″	10″	20″	30″
60″	50″	40″	30″	60″	50″	40″	30″
(I)0.0000	−0.0003	−0.0005	−0.0006	(XIV)0.0000	−0.0003	−0.0005	−0.0006

EXAMPLE 11-3   Compute the grid coordinates of a point lying in Arizona East zone of SPCS 83 whose geodetic coordinates are

$$\varphi = 31°\ 15'\ 23.54600'' \qquad \lambda = 111°\ 12'\ 43.75500''$$

Solution: By Eq. (11-19a),

$$\Delta\lambda = 110°\ 10' - 111°\ 12'\ 43.75500'' = -1°\ 02'\ 43.75500''$$

$$= -3763.75500''$$

By Eq. (11-19b),

$$p = -0.37637550$$

From Table 11-5a,

$$\text{tabular (I)} = 27,715.1322$$

$$23.54600 \times 30.795230 = 725.1045$$

$$\text{second difference} \quad -0.0006$$

$$\text{(I)} = \overline{28,440.2361}$$

$$\text{tabular (II)} = 3327.0601$$

$$23.54600 \times 0.016832 = 0.3963$$

$$\text{(II)} = \overline{3327.4564}$$

$$\text{tabular (III)} = 2.227$$

By Eq. (11-19c),

$$N = 28{,}440.2361 + 3327.4564p^2 + 2.227p^4 = \qquad 28{,}911.643 \text{ m}$$

$$
\begin{aligned}
\text{tabular (IV)} &= 264{,}568.5383\\
23.54600 \times (-0.774711) &= \quad\ -18.2413\\
\text{second difference} &= \qquad\ \ 0.0028\\
\text{(IV)} &= \overline{264{,}550.2998}\\
\text{tabular (V)} &= \qquad\ \ 48.2298\\
23.54600 \times (-0.001037) &= \qquad -0.0244\\
\text{(V)} &= \overline{\qquad\ \ 48.2054}\\
\text{tabular (VI)} &= \qquad -0.0100
\end{aligned}
$$

By Eq. (11-19d),

$$E' = 264{,}550.2998p + 48.2054p^3 - 0.0100p^5 = -99{,}572.821 \text{ m}$$

By Eq. (11-19e),

$$E = 213{,}360 - 99{,}572.821 = 113{,}787.179 \text{ m} \qquad \blacklozenge$$

EXAMPLE 11-4   Compute the convergence angle and the scale factor at the point in Example 11-3.

Solution:

$$
\begin{aligned}
\text{tabular (XII)} &= 5187.7326\\
23.54600 \times 0.041444 &= \quad\ 0.9758\\
\text{(XII)} &= \overline{5188.7084}\\
\text{tabular (XIII)} &= \qquad 3.015
\end{aligned}
$$

By Eq. (11-9f),

$$\gamma = 5188.7084p + 3.015p^3 = -1953.06''$$

By Eq. (11-19g),

$$Q = 10^{-6}E' = -0.09957282$$

$$\text{tabular (XVI)} = 0.012332$$

By Eq. (11-9i),

$$k = 0.999900(1 + 0.012332Q^2 + 0.00003Q^4) = 1.00002226 \qquad \blacklozenge$$

Equations (11-19e), (11-19g), (11-19h), (11-19i), (11-19j), and (11-19k) are used when converting from grid N and E coordinates to geodetic coordinates and for computing the convergence angle and scale factor at the point.

EXAMPLE 11-5   Using the results of Example 11-3, compute the geodetic coordinates of the point. The solution to this example shows the method of conversion and verifies the reliability and accuracy of the tables. The results of Example 11-3 are $N = 28{,}911.643$ m; $E = 113{,}787.179$ m.

Solution: By Eq. (11-19e),

$$E' = 113{,}787.179 - 213{,}360.000 = -99{,}572.821 \text{ m}$$

By Eq. (11-19g),

$$Q = -0.099572821$$

tabular value of (I) for 31° 15′ =	27,715.132 m
N =	28,911.643 m
diff. =	$\overline{1196.511}$ m
diff. 1″ =	30.795230 m
diff. in seconds =	38.85377″

$$\varphi' = 31° \ 15' \ 38.85377''$$

tabular (VII) =	1543.4813
38.85377 sec =	0.6522
(VII) =	$\overline{1544.1335}$

tabular (VIII) =	19.234
38.85377 sec =	0.011 (interpolated)
(VIII) =	$\overline{19.245}$

By Eq. (11-19k),

$$\varphi = 31° \ 15' \ 38.85377 - 1544.1335Q^2 + 19.245Q^4 = 31° \ 15' \ 23.54600''$$

tabular (IX) =	37,797.3892
38.85377 sec =	4.3010 (φ')
second diff. =	$\overline{-0.0006}$
(IX) =	37,801.6896

tabular (X) =	269.2271
38.85377 sec =	0.1269 (φ')
(X) =	$\overline{269.3540}$

tabular (XI) =	3.530
38.85377 sec =	$\overline{0.003}$ (φ')
(XI) =	3.533

By Eq. (11-19j),

$$\lambda = 110° \ 10' - 37{,}801.6896Q + 269.3540Q^3 + 3.533Q^5$$
$$\lambda = 111° \ 12' \ 43.75492''$$

The difference between the initial longitude and that obtained by the reverse computation is 0.00008″ or approximately 2 mm.

$$\text{tabular (XIV)} = 19{,}608.2747$$
$$38.85377 \text{ sec} = 8.3186 \; (\varphi')$$
$$\text{second diff.} = -0.0005$$
$$\overline{\text{(XIV)} = 19{,}616.5928}$$

$$\text{tabular (XV)} = 218.680$$
$$38.85377 \text{ sec} = 0.143 \; (\varphi')$$
$$\overline{\text{(XV)} = 218.823}$$

By Eq. (11-19h),

$$\gamma = 19{,}616.5928Q - 218.823Q^3 = -1953.06''$$

The solution for k is identical to that computed in Example 11-4, that is, by Eq. (11-19i),

$$k = 0.999900 \; (1 + 0.012332Q^2 + 0.00003Q^4) = 1.00002226 \qquad \blacklozenge$$

The State of Arizona has adopted the meter or the International Foot (IF) for use with SPCS 83. Thus, if the surveyor works with the foot, the results of Example 11-3 must be converted from meters to feet. From Example 11-3,

$$E_{\text{IF}} = 113{,}787.179/0.3048 = 373{,}317.516 \text{ ft}$$
$$N_{\text{IF}} = 28{,}911.643/0.3048 = 94{,}854.472 \text{ ft.}$$

If the second term on the right side of Eq. (11-18a) is to be evaluated, the values of E_1' and E_2' must be determined by subtracting the value of E_0 in feet from the value of E in feet. The value of E_0 in feet on Arizona East is

$$(E_0)_{\text{IF}} = 213{,}360/0.3048 = 700{,}000.000 \text{ ft}$$

The value of E' for the point in Example 11-3 is then

$$E_{\text{IF}}' = 373{,}317.516 - 700{,}000.000 = -326{,}682.484 \text{ ft}$$

If geodetic coordinates are to be computed from grid coordinates that are in IF units, the reverse of the above example must be performed (see Example 11-6).

EXAMPLE 11-6   The grid coordinates of a point on the Arizona East zone of SPCS 83 are $N = 132{,}354.233$ ft and $E = 792{,}825.414$ ft. Convert these coordinates to meters.

Solution: The value of

$$N_{\text{M}} = 132{,}354.233 \times 0.03048 = 40{,}341.570 \text{ m}$$
$$E_{\text{M}} = 792{,}825.414 \times 0.3048 = 241{,}653.186 \text{ m}$$

These results are then used as shown in Example 11-5 if geodetic coordinates are to be determined. $\blacklozenge$

As a further example of conversion from meters to feet or vice versa, consider the Indiana East transverse Mercator zone of SPCS 83. Indiana has adopted the Survey Foot (SF) and the meter. The grid origin for this zone is E_0 = 100,000 m and N_0 = 250,000 m.

EXAMPLE 11-7   The grid coordinates of a point in Indiana East zone of SPCS 83 are N = 1,382,464.500 ft, and E = 292,464.755 ft. Convert these to meters.

Solution:

$$N_M = 1{,}382{,}464.500 \times 12/39.37 = 421{,}376.022 \text{ m}$$
$$E_M = 292{,}464.755 \times 12/39.37 = 89{,}143.436 \text{ m}$$

The reverse of Example 11-7, that is, conversion of N and E from meters to Survey Feet, should be obvious.   ◆

11-11   Lambert Conformal Projection

The Lambert projection is represented diagrammatically in Fig. 11-5. View (a) shows the sea-level surface of the earth or the spheroid intersected by a cone along two parallels of latitude AL_1B and CL_2D. These are known as the *standard parallels* of the projection. View (b) shows a portion of the conical surface developed into a plane surface on which the meridians and parallels of the earth's surface have been projected mathematically. The developments of the two parallels appear as arcs of concentric circles with the center at the apex of the developed cone (not shown on the illustration). Along these two parallels the distances on the projection are the same as corresponding distances on the sea-level surface. Between the two parallels, a distance on the projection is smaller than the corresponding distance on the sea-level surface. Outside the two parallels a projected distance is larger. The discrepancy between these corresponding distances depends on the position of the line being considered with respect to the two standard parallels. It is seen that the scale of a line running in a north-south direction varies from point to point. It is also seen, however, that a due east-west line has a constant scale through its length, whether this scale be larger than, equal to, or less than that on the corresponding sea-level line.

To apply the Lambert conformal projection to a state or a zone, the width of the projection in a north-south direction is limited to 158 miles or 254 km, and the standard parallels are separated by about two-thirds this distance. At no point within these limits will the discrepancy between a sea-level distance and the grid distance be greater than 1 part in 10,000.

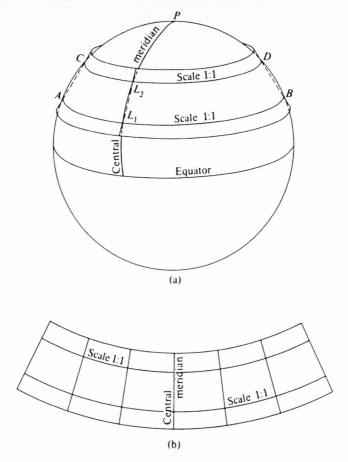

FIGURE 11-5 Lambert conformal projection.

11-12 Grid Azimuth on Lambert Conformal Projection of SPCS 27

The central meridian and all lines parallel to it on the Lambert projection are grid-north lines. Therefore the azimuth of any line on the plane projection is referred to any of these grid-north lines. The relationship between geodetic azimuth and grid azimuth is indicated in Fig. 11-6. The point P has geographic coordinates φ and λ. Two geodetic meridians are shown. Line AB is the central meridian of the projection, and PB is the meridian through P. The line through P, parallel to the central meridian, is the grid meridian or grid-north line at P. The point B is the apex of the developed cone. The angle at B between the meridian through P and the central meridian, which is designated θ, is known as the *mapping angle* for the point.

To compute the mapping angle for a point, first determine the difference in longitude $\Delta\lambda$ between the central meridian and the point by the equation

$$\Delta\lambda = \lambda_m - \lambda \tag{11-20}$$

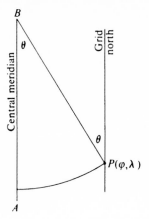

FIGURE 11-6 Relationship between geodetic and grid azimuths on Lambert projection.

Then multiply $\Delta\lambda$ by a constant of the given projection, which gives the relationship between $\Delta\lambda$ and θ. That is

$$\theta = l\,\Delta\lambda \qquad (11\text{-}21)$$

where l is the constant. If the point is east of the central meridian, then $\Delta\lambda$ and θ are positive. If the point is west of the central meridian, $\Delta\lambda$ and θ are negative.

As seen from Fig. 11-6, the relationship between the geodetic azimuth and the grid azimuth at a point is

$$\text{geodetic azimuth} - \text{grid azimuth} = +\;\theta \qquad (11\text{-}22)$$

Due regard must be paid to the algebraic sign of θ. Tables giving the value of θ for every minute of longitude are available for the various Lambert projections. These tables are discussed in Section 11-14.

In Fig. 11-7 the arc designated by φ_0 is a certain parallel of latitude lying slightly north of an arc midway between the standard parallels. The Y-coordinate of the intersection of this parallel with the central meridian is designated by Y_0, as shown. The line AB lies north of this parallel. The grid line joining the two points A and B is a straight line. The geodetic line joining the two points, when projected onto the grid, is seen to curve outward away from the parallel designated by φ_0. The geodetic line CD is also seen to curve outward away from this parallel, whereas the grid line between C and D is straight. To allow for this curvature in going from the geodetic azimuth to the grid azimuth, or vice versa, when a long line is involved, Eq. (11-22) must be modified by a small correction term. Thus

$$\text{geodetic azimuth} - \text{grid azimuth} = +\;\theta - \frac{X_2 - X_1}{2\rho_0^2 \sin 1''}\left(Y_1 - Y_0 + \frac{Y_2 - Y_1}{3}\right) \qquad (11\text{-}23)$$

in which X_1 and Y_1 and X_2 and Y_2 are the coordinates of the ends of the line, Y_0 is as defined above, and ρ_0 is the mean radius of curvature of the earth at latitude φ_0. The value of $1/(2\rho_0^2 \sin 1'')$ and the value of Y_0 are given in the projection tables for each zone.

FIGURE 11-7 Curvature of geodetic lines on Lambert projection.

Notice that if a line runs grid north or grid south, as does the line AC or BD in Fig. 11-7, there is no correction term to θ because $X_2 - X_1 = 0$. Also if a line other than a grid north-south line crosses the parallel designated by φ_0, its projection onto the grid may have a reverse curvature.

The second term in Eq. (11-23) is the $t - T$ correction discussed in Section 11-6 and designated δ. Let $\Delta X = (X_2 - X_1)$ and $\Delta Y = \left(Y_1 - Y_0 + \dfrac{Y_2 - Y_1}{3} \right)$, and evaluating the denominators using the average radius of the ellipsoid for all zones gives

$$\delta = 2.36 \; \Delta X \; \Delta Y \times 10^{-10}$$

with sufficient accuracy for all observable lines. Designating t for the grid azimuth of the line, Eq. (11-23) becomes

$$\text{geodetic azimuth} - t = \theta - 2.36 \; \Delta X \; \Delta Y \times 10^{-10} \tag{11-24}$$

11-13 Geographic to Plane Coordinates on Lambert Conformal Projection of SPCS 27

On a Lambert projection the central meridian is the Y axis, and the X axis is perpendicular to the central meridian. The origin of coordinates is far enough south and west to render all the coordinate values positive. The Y axis is assigned a large X value. It is usually, but not always, 2,000,000 ft. This large value is designated by X_0; that is, $X_0 = 2,000,000$ ft. Except for three projections, the Y value of the X axis is 0 ft.

In Fig. 11-8 P is a point whose geographic coordinates are φ and λ. The parallel of latitude through P is a circle whose radius is R. As the latitude increases, the corresponding value of R decreases. The value of R for any latitude may be

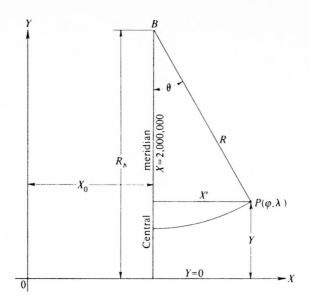

FIGURE 11-8 Rectangular coordinates on Lambert projection.

obtained from the tables to be discussed in the next section. The mapping angle θ for point P may be determined by Eq. (11-21), or it may be obtained by interpolation in the proper table for the given projection. The distance along the central meridian from the X axis to point B is designated by R_b. It is constant for a given projection. Thus the X axis is at a distance R_b south of B.

If the latitude of point P in Fig. 11-8 is known, R can be obtained. Also, if the longitude of the point P is known, θ can be obtained. Then the Y and X coordinates of P can be computed by means of the following equations:

$$Y = R_b - R \cos \theta \qquad (11\text{-}25)$$

$$X = X_0 + R \sin \theta \qquad (11\text{-}26)$$

Because of the large numbers involved, the calculator employed to solve Eqs. (11-25) and (11-26) must be able to generate sines and cosines to ten places after the decimal.

11-14 Tables for Lambert Projections of SPCS 27

The National Geodetic Survey has prepared tables for each projection in each state and makes them available through the U.S. Government Printing Office. The tables are based on the Clarke Spheroid of 1866 for points defined on NAD 27.

Table 11-6 shows a portion of Tables I and II for the North zone of the State of Ohio.

Table I gives for every minute of latitude the value of R, the Y coordinate of the intersection of the parallel of latitude with the central meridian, and the tabular difference for $1''$ of latitude for both R and Y'. The values of R decrease, while

TABLE 11-6   Lambert Projection for Ohio-North of SPCS 27

Table I

Latitude	R (ft)	Y′ Y Value on Central Meridian (ft)	Tabular Difference for 1″ of Latitude (ft)	Scale Expressed as Ratio
40° 51′	24,128,050.71	431,107.76	101.19850	0.9999463
40° 52′	24,121,978.80	437,179.67	101.19883	0.9999452
40° 53′	24,115,906.87	443,251.60	101.19883	0.9999443
40° 54′	24,109,834.94	449,323.53	101.19917	0.9999434
40° 55′	24,103,762.99	455,395.48	101.19950	0.9999426

Table II

1″ of Longitude = 0.65695032″ of θ

Longitude	θ
84° 46′	−1° 29′ 20.71461″
84° 47′	−1° 30′ 00.13163″
84° 48′	−1° 30′ 39.54865″
84° 49′	−1° 31′ 18.96567″
84° 50′	−1° 31′ 58.38269″

Constants

$$R_b = 24,559,158.47 \text{ ft}$$
$$X_O = C = 2,000,000 \text{ ft}$$
$$Y_O = 510,419.96 \text{ ft}$$
$$l = 0.65695032$$
$$\lambda_m \text{ (central meridian)} = 82° 30′ 00.000″$$
$$1/(2\rho_0^2 \sin 1″) = 2.3574 \times 10^{-10}$$

the values of $Y′$ increase. Table I also shows the amount by which a sea-level length must be increased or decreased to obtain the corresponding grid length. The values in the last column of Table I are the scale factors by which the sea-level lengths are multiplied to give the corresponding grid lengths. Note that the scale varies with the latitude.

Table II gives the value of θ for each full minute of longitude. To interpolate, the number of seconds of longitude is multiplied by the constant appearing in the heading, and the product is added to or subtracted from the tabular value. Table II is not really necessary, since $\Delta \lambda$ is readily obtained by Eq. (11-20). The value of θ can then be found by Eq. (11-21), in which l is the value given at the head of Table II or is included among the constants of the projection.

The constants following Table II are those discussed in Sections 11-11, 11-12, and 11-13.

EXAMPLE 11-8   The latitude and longitude of a control point in Ohio are $\varphi = 40° 52′ 44.602″$ and $\lambda = 84° 48′ 16.225″$. Compute the plane coordinates of the control point on the projection for the Ohio North zone.

Solution: The value of R for the point is obtained from Table I by interpolation, as follows:

$$R \text{ for } 40° \, 52' = 24{,}121{,}978.80 \text{ ft}$$

$$101.19883 \times 44.602'' = \underline{\qquad -4{,}513.67}$$

$$R \text{ for } 40° \, 52' \, 44.602'' = 24{,}117{,}465.13 \text{ ft}$$

The value of θ is obtained from Table II by interpolation.

$$\theta \text{ for } 84° \, 48' = -1° \, 30' \, 39.54865''$$

$$0.656\ 95032 \times 16.225'' = \underline{\qquad -10.65902''}$$

$$\theta \text{ for } 84° \, 48' \, 16.225'' = -1° \, 30' \, 50.20767''$$

Then the plane coordinates may be computed by Eqs. (11-25) and (11-26).

$$
\begin{array}{ll}
R_b = \quad 24{,}559{,}158.47 & X_0 = \quad 2{,}000{,}000.00 \\
R \cos \theta = -24{,}109{,}046.29 & R \sin \theta = \quad -637{,}190.13 \\
\hline
Y = \quad 450{,}112.18 \text{ ft} & X = \quad 1{,}362{,}809.87 \text{ ft}
\end{array}
$$

◆

EXAMPLE 11-9   A line joins the control point in Example 11-8 with a point whose coordinates are X = 1,414,516.20 ft and Y = 459,168.24 ft. Compute the geodetic azimuth of this line in the direction of the second point.

Solution: The grid azimuth is found by Eq. (8-13)

$$
\begin{array}{lll}
X_2 = 1{,}414{,}516.20 & Y_2 = 459{,}168.24 & \\
X_1 = 1{,}362{,}809.87 & Y_1 = 450{,}112.18 & \tan \alpha = \dfrac{+51{,}706.33}{+\ 9056.06} \\
\Delta X = \quad +51{,}706.33 & \Delta Y = +9{,}056.06 & \quad\quad = +5.70958342
\end{array}
$$

$$\text{grid } \alpha = 80° \, 03' \, 56.67''$$

The value of θ for the first point was determined in Example 11-8 by using the longitude of the point. However, in this example we will assume that only the plane coordinates of the ends of the line are known. It can be seen from Fig. 11-8 that

$$\tan \theta = \frac{X'}{R_b - Y} = \frac{X - X_0}{R_b - Y} \tag{11-27}$$

The computations for determining θ follow:

$$
\begin{array}{lll}
X = 1{,}362{,}809.87 & R_b = 24{,}559{,}158.47 & \tan \theta = \dfrac{-637{,}190.13}{24{,}109{,}046.29} \\
X_0 = 2{,}000{,}000.00 & Y = \quad 450{,}112.18 & \tan \theta = -0.02642\ 95038 \\
X' = -637{,}190.13 & R_b - Y = 24{,}109{,}046.29 & \theta = -1° \, 30' \, 50.21''
\end{array}
$$

The correction term in Eq. (11-23) is now computed:

$$X_2 - X_1 = +51,706.33 \qquad Y_1 = 450,112.18$$
$$Y_2 - Y_1 = +9,056.06 \qquad Y_0 = 510,419.96$$
$$Y_1 - Y_0 = -60,307.78$$
$$\frac{1}{2\rho_0^2 \sin 1''} = 2.357 \times 10^{-10} \qquad \frac{(Y_2-Y_1)/3 \qquad +3,018.69}{-57,289.09}$$

correction term $= 5.171 \times 10^4 \times 2.357 \times 10^{-10} \times -0.573 \times 10^5 = -0.70''$

Finally, the geodetic azimuth is determined by applying Eq. (11-23).

$$\text{grid azimuth} = 80° \ 03' \ 56.67''$$
$$+ \ \theta = \underline{-1° \ 30' \ 50.21''}$$
$$78° \ 33' \ 06.46''$$
$$- \text{correction term} \qquad \underline{+0.70''}$$
$$\text{geodetic azimuth} \qquad 78° \ 33' \ 07.16''$$

◆

EXAMPLE 11-10   The state plane coordinates of a point in the Ohio North zone are $X = 2,097,462.33$ ft and $Y = 440,028.92$ ft. Determine the latitude and longitude of the point.

Solution: The value of θ is obtained by Eq. (11-27) given in Example 11-9, as follows:

$$X = 2,097,462.33 \qquad R_b = 24,559.158.47 \qquad \tan \theta = \frac{+97,462.33}{24,119,129.55}$$

$$X_0 = 2,000,000.00 \qquad Y = \ \ \ \ 440,028.92 \qquad \tan \theta = +0.00404 \ 08726$$
$$X' = \ \ \ +97,462.33 \qquad R_b - Y = 24,119,129.55 \qquad \theta = +0° \ 13' \ 53.4853''$$
$$\theta = +833.4853''$$

By Eq. (11-21),

$$\Delta\lambda = \frac{+833.4853}{0.656 \ 95032} = +1268.719'' = +0° \ 21' \ 08.719''$$

By Eq. (11-20), the longitude is

$$\lambda = 82° \ 30' \ 00.000'' - 0° \ 21' \ 08.719''$$
$$\lambda = 82° \ 08' \ 51.281''$$

The value of R can be obtained by a rearrangement of Eq. (11-25). Thus

$$R = \frac{R_b - Y}{\cos \theta} = \frac{24,119,129.55}{0.99999 \ 18355} = 24,119,326.47$$

Finally, the latitude is obtained from the value of R by interpolating in Table I of the projection tables.

$$R \text{ for } 40° \, 52' = 24{,}121{,}978.80$$

$$R \text{ for point} = 24{,}119{,}326.47$$

$$\text{difference for seconds} = \overline{\quad 2{,}652.33 \quad}$$

$$\text{number of seconds} = \frac{2{,}652.33}{101.19883} = 26.209''$$

$$\varphi = 40° \, 52' \, 26.209''$$

11-15 Grid Azimuth on Lambert Conformal Projection of SPCS 83

Calculation of grid azimuths on SPCS 83 is essentially the same as that developed for SPCS 27. The only difference is in the notations used by the NGS. E is used in place of X, N in place of Y, γ in place of θ, and the symbol α for geodetic azimuth. Thus

$$\alpha - t = \gamma - \frac{E_2 - E_1}{(2\rho_0^2 \sin 1'')_g}\left(N_1 - N_0 + \frac{N_2 - N_1}{3}\right) \qquad (11\text{-}28)$$

Letting $\Delta E = (E_2 - E_1)$, $\Delta N = \left(N_1 - N_0 + \dfrac{N_2 - N_1}{3}\right)$, evaluating the denominator for the average radius of the spheroid, and remembering that the units are all in meters, Eq. (11-28) becomes

$$\alpha - t = \gamma - 25.4 \, \Delta E \, \Delta N \times 10^{-10} \qquad (11\text{-}29a)$$

with sufficient accuracy for all observable lines. The quantity t is the grid azimuth of the line.

If SPCS 83 coordinates are expressed in feet, Eq. (11-28) becomes

$$\alpha - t = \gamma - 2.36 \, \Delta E \, \Delta N \times 10^{-10} \qquad (11\text{-}29b)$$

11-16 Geographic to Plane Coordinates on Lambert Conformal Projection of SPCS 83

Geographic coordinates are converted to state plane coordinates either by the use of polynomials or by use of tables published by National Geodetic Survey in NOAA Manual NOS NGS5 *State Plane Coordinate System of 1983* (NGS5). Any surveyor or engineer using SPCS 83 must avail himself of this most valuable publication. The coefficients and constants used in this section are taken from that publication.

POLYNOMIAL METHOD

Given a point of known latitude φ and longitude λ, the conversion to plane coordinates N and E is performed quite readily by use of the following equations:

$$\Delta\varphi = \varphi - B_0 \qquad (\Delta\varphi \text{ in decimal degrees}) \qquad (11\text{-}30)$$

(B_0 is the latitude of the central parallel and is equivalent to φ_0 in SPCS 27.)

$$u = L_1 \, \Delta\varphi + L_2 \, \Delta\varphi^2 + L_3 \, \Delta\varphi^3 + L_4 \, \Delta\varphi^4 + L_5 \, \Delta\varphi^5 \qquad (11\text{-}31)$$

Some of the Lambert zones require only three terms, some need four terms, and a few need five terms. The coefficients are given for each Lambert zone of each state in NGS 5. The author of NGS 5 suggests the nested form of Eq. (11-31), thus

$$u = \Delta\varphi [L_1 + \Delta\varphi \{L_2 + \Delta\varphi (L_3 + \Delta\varphi (L_4 + L_5 \, \Delta\varphi))\}] \qquad \text{(11-32a)}$$

$$R = R_0 - u \qquad (R_0 \text{ is the mapping radius at } B_0) \qquad \text{(11-32b)}$$

$$\gamma = (L_0 - \lambda) \sin (B_0) \qquad \text{convergence angle } (L_0 = \text{longitude} \qquad \text{(11-32c)}$$
$$\text{of central meridian)}$$

$$E' = R \sin \gamma \qquad \text{(11-32d)}$$

$$N' = u + E' \tan \tfrac{\gamma}{2} \qquad \text{(11-32e)}$$

$$E = E' + E_0 \qquad \text{(Easting)} \qquad \text{(11-32f)}$$

$$N = N' + N_0 \qquad \text{(Northing)} \qquad \text{(11-32g)}$$

$$k = F_1 + F_2 u^2 + F_3 u^2 \qquad \text{Grid scale factor} \qquad \text{(11-32h)}$$

EXAMPLE 11-11   Given a point in latitude 37° 45′ N and longitude 121° 00′ W lying in Zone 3 of California SPCS 83. Compute the Northing and Easting of the point, the convergence angle, and scale factor at the point.

Solution: The following constants and coefficients are obtained from NGS 5 table for California Zone 3:

$$
\begin{aligned}
B_0 &= 37.7510694363° & L_1 &= 110{,}983.9104 \\
\sin B_0 &= 0.612232038295 & L_2 &= 9.43943 \\
R_0 &= 8{,}246{,}930.3684 \text{ m} & L_3 &= 5.64142 \\
L_0 &= 120° \, 30′ & L_4 &= 0.019048 \\
E_0 &= 2{,}000{,}000.0000 \text{ m} & F_1 &= 0.999929178853 \\
N_0 &= 638{,}844.8039 \text{ m} & F_2 &= 1.23137 \times 10^{-14} \\
& & F_3 &= 4.89 \times 10^{-22}
\end{aligned}
$$

Then by Eq. (11-30), $\Delta\varphi = 37.75° - 37.7510694363° = -0.0010694363°$
by Eq. (11-32a), $u = -118.6901784$ (student should verify this)
by Eq. (11-32b), $R = 8{,}246{,}930.3684 + 118.6902 = 8{,}247{,}049.058$ m
by Eq. (11-32c), $\gamma = (120.50° - 121.00°) \sin B_0 = -0.50° \times 0.612232038$
$\gamma = -0.306116019148°$
by Eq. (11-32d), $E' = 8{,}247{,}049.058 \sin (-0.306116019°)$
$E' = -44{,}061.566$ m
by Eq. (11-32e), $N' = -118.6901784$ m
$+ (-44{,}061.566) \tan(-0.1530580100)$
$= -0.985$ m
by Eq. (11-32f), $E = 2{,}000{,}000.000 - 44{,}061.566 = 1{,}955{,}938.433$ m
by Eq. (11-32g), $N = -0.985 + 638{,}844.804 = 638{,}843.819$ m
by Eq. (11-32h), $k = 0.999929179$ (student should verify this)   ◆

To convert from plane coordinates (E, N) to geographic coordinates (φ, λ), the following polynomial is used:

$$\Delta\varphi = \varphi - \varphi_0 = G_1 u + G_2 u^2 + G_3 u^3 + G_4 u^4 + G_5 u^5 \qquad (11\text{-}33)$$

or, in the nested form

$$\Delta\varphi = u[G_1 + u\{G_2 + u(G_3 + u(G_4 + G_5 u))\}] \qquad (11\text{-}34a)$$

$$N' = N - N_0 \qquad (11\text{-}34b)$$

$$E' = E - E_0 \qquad (11\text{-}34c)$$

$$R' = R_0 - N' \qquad (11\text{-}34d)$$

$$\gamma = \tan^{-1}\frac{E'}{R'} \qquad \text{(convergence angle)} \qquad (11\text{-}34e)$$

$$\lambda = L_0 - \frac{\gamma}{\sin B_0} \qquad \text{(longitude)} \qquad (11\text{-}34f)$$

$$u = N' - E' \tan\frac{\gamma}{2} \qquad (11\text{-}34g)$$

$$\varphi = B_0 + \Delta\varphi \qquad \text{(latitude)} \qquad (11\text{-}34h)$$

$$k = F_1 + F_2 u^2 + F_3 u^3 \qquad \text{(grid scale factor)} \qquad (11\text{-}34i)$$

EXAMPLE 11-12   Given a point with coordinates $E = 2,050,000.000$ m and $N = 640,000.000$ m on California SPCS 83 Zone 3. Compute the latitude, longitude, convergence angle, and grid scale factor for the point.

Solution: The coefficients G_1 through G_4 are given in NGS 5 as follows:

$$G_1 = 9.010315015 \times 10^{-6}$$
$$G_2 = -6.90503 \times 10^{-15}$$
$$G_3 = -3.71614 \times 10^{-20}$$
$$G_4 = -9.8819 \times 10^{-28}$$

The remainder of the coefficients and constant for this zone are given in Example 11-11.

$N' = 640,000,000 - 638,844.8039 = 1,155.1961$ m	[11-34b]
$E' = 2,050,000.000 - 2,000,000.000 = 50,000.000$ m	[11-34c]
$R' = 8,246,930.3684 - 1,155.1961 = 8,245,775.172$ m	[11-34d]
$\gamma = \tan^{-1} 50,000.000/8,245,775.172 = 0.347420807°$	[11-34e]
$\lambda = 120.50 - 0.347420807/0.612232038295 = 119.9325341°$	[11-34f]
$\lambda = 119° 55' 57.123''$	
$u = 1,155.1961 - 50,000.000 \tan 0.173710404°$	
$= 1003.604711$	[11-34g]
$\Delta\varphi = +0.009042788°$ (student should verify this)	[11-34a]
$\varphi = 37.75106944° + 0.009042788° = 37.76011222°$	
$= 37° 45' 36.4040''$	[11-34h]
$k = 0.99992919$ (student should verify this)	[11-34i]

TABLE METHOD

EXAMPLE 11-13 Given a point in latitude 37° 45′ N and longitude 121° 00′ W lying in Zone 3 of California SPCS 83. Compute the Northing and Easting of the point, the convergence angle, and the scale factor at the point.

Solution: Table 11-7 shows a portion of the Lambert projection table for California Zone 3 on SPCS 83. These tables are not contained in NGS 5 but are published separately for each state which is on the Lambert projection of SPCS 83. The constants necessary for this problem, however, are given in NGS 5 for all zones, and are as follows for California Zone 3.

$$\lambda_{CM} = 120° 30′ \qquad \text{(longitude of central meridian)}$$

$$\ell = 0.612232038294 = \sin \varphi_0$$

$$R_b = 8,385,775.1723 \text{ m} \quad \text{(mapping radius of grid origin)}$$

$$E_0 = 2,000,000.000 \text{ m} \quad \text{(Easting coordinate of CM)}$$

$$N_b = 500,000.000 \text{ m} \qquad \text{(Northing coordinate of origin)}$$

The equations for the conversion from φ, λ to N, E are

$$\gamma = (\gamma_{CM} - \lambda) \sin \varphi_0 \tag{11-35a}$$

$$E = R \sin \gamma + E_0 \qquad (R \text{ from table}) \tag{11-35b}$$

$$N = R_b - R \cos \gamma + N_b \tag{11-35c}$$

By Eq. (11-35a),

$$\gamma = (120.50° - 121.00°) \times 0.612232038 = -0.306116019°$$

$$\gamma = -0° 18′ 22.01767°$$

$$R \text{ for } 37° 45′ = 8,247,049.059 \text{ m} \qquad \text{(from Table 11-7)}$$

$$E = 8,247,049.059 \sin (-0.306116019°) + 2,000,000.000 \tag{11-35b}$$

$$E = 1,955,938.433 \text{ m}$$

$$N = 8,385,775.172 - 8,247,049.059 \cos (-0.306116019°) + 500,000,000 \qquad \text{[11-35c]}$$

$$N = 638,843.818 \text{ m}$$

$$k = 0.99992918 \qquad \text{(from Table 11-7)}$$

TABLE 11-7 Lambert Projection for California Zone 3 SPCS 83

Latitude	R (m)	Tabular Difference for 1″ of Latitude (m)	k (Scale Expressed as Ratio)
37° 42′	8,252,598.230	30.82865	0.99992957
37° 43′	8,250,748.511	30.82873	0.99992936
37° 44′	8,248,898.788	30.82882	0.99992923
37° 45′	8,247,049.059	30.82890	0.99992918
37° 46′	8,245,199.324	30.82899	0.99992922

EXAMPLE 11-14 Given a point on the California Zone 3 of SPCS 83 with coordinates $E = 2{,}050{,}000.000$ m and $N = 640{,}000.000$ m. Compute the latitude and longitude of the point, the convergence angle, and the grid scale factor.

Solution: The constants needed for this solution are given in Example 11-13. The equations for the conversion are as follows:

$$\tan \gamma = (E - E_0)/(R_b - (N - N_b)) \tag{11-36a}$$

$$R = (R_b - (N - N_b))/\cos \gamma \tag{11-36b}$$

$$\lambda = \lambda_{\mathrm{CM}} - \gamma/\ell \tag{11-36c}$$

φ is obtained from Table 11-7 by interpolation.

By Eq. (11-36a),

$$\tan \gamma = \frac{2{,}050{,}000.000 - 2{,}000{,}000.000}{8{,}385{,}775.1723 - (640{,}000 - 500{,}000.000)}$$

$$\gamma = + 0°\,20'\,50.71490''$$

By Eq. (11-36b),

$$R = \frac{8{,}385{,}775.1723 - (640{,}000 - 500{,}000)}{\cos\ 0°\,20'\,50.71490''}$$

$$R = 8{,}245{,}926.763 \text{ m}$$

By Eq. (11-36c),

$$\lambda = 120.50° - 0.347420806°/0.612232038 = 119.9325341°$$

$$\lambda = 119°\,55'\,57.123''$$

From Table 11-7,

$$R \text{ for } 37°\,45' = 8{,}247{,}049.059$$

$$R \text{ for } \varphi\quad = \underline{8{,}245{,}926.763}$$

$$\text{Difference}\quad = \quad\ \ 1{,}122.296$$

$$\text{Diff. for } 1'' \quad\ \ \ \ \ \overline{\quad\ 30.82890\quad} = 36.4040''$$

$$\varphi = 37°\,45'\,36.4040''$$

By interpolation in Table 11-7, the grid scale factor is 0.99992920. The grid coordinates given in Examples 11-12 and 11-14 are in meters. If the surveyor uses the SF or the IF as adopted by his state, he must make the appropriate conversion as outlined in Section 11-10. This conversion is the same as that performed for the transverse Mercator projections.

11-17 Conversion from SPCS 27 Coordinates to SPCS 83 Coordinates

The North American Datum of 1983 was developed for several reasons: (1) the adoption of the worldwide reference ellipsoid (GRS 80 spheroid) in place of the Clarke 1866 spheroid; (2) the accumulation of new observations over the years using increasingly more accurate survey instrumentation and methods; (3) the advent of the transit satellite system followed by the more accurate GPS system, which allowed the tying together of widely spaced points into a unified system of control points; (4) the inconsistency in closures between different portions of the 1927 datum points experienced by surveyors and engineers using the 1927 network; and (5) the power of large mainframe computers to handle the simultaneous adjustment of a quarter of a million points using over a million original geodetic observations.

Of primary concern to surveyors and engineers is the conversion of horizontal control defined by X and Y coordinates on SPCS 27 to E and N values on SPCS 83. Although several methods for this conversion have been proposed, two techniques are in use at the present time. The first method to be discussed is the most desirable and the most accurate. To use this method, the surveyor or engineer must select those control networks in which he has the highest degree of confidence, and which have been tied to the National Network established by NGS. Such a network may encompass a city, county, or municipality, a large construction project such as a power or transportation network, or a survey for a subdivision. The overriding consideration must be the integrity of each such network. Having selected those networks, the surveyor or engineer must then go to the *original* observations of angles and distances and submit them to NGS for processing through its National Geodetic Survey Data Base. Specifications and input formats for submitting these original observations are published by NGS in NOAA manual NOS NGS 2 titled *Input Formats and Specifications of the National Geodetic Survey Data Base—Volume I, Horizontal Control Data,* and must be rigorously followed. The data will be processed by NGS, and both the geodetic and grid coordinates of the points on NAD 83 will be furnished to the surveyor. Any prior or subsequent surveys tied to these points by the surveyor can then be adjusted to these points using the method of least squares (see Appendix A).

The second and highly practical method of conversion is by the application of the program NADCON, also developed by NGS. The agency has computed the shift in latitude and longitude from NAD 27 to NAD 83 for points on a grid covering the lower 48 states at 15-min intervals, covering Alaska at 7.5-min intervals, and covering Hawaii at 1.5-min intervals. These shifts were determined by computing the difference in latitude ($\delta\varphi$) and the difference in longitude ($\delta\lambda$) between NAD 27 and NAD 83 at all of the NGS first- and second-order stations in these regions based on the principle of minimum curvature (Dewhurst, 1990). The shift in latitude over the conterminous United States, expressed in meters, is shown in Fig. 11-9. The shift in longitude is shown in Fig. 11-10. To use the NADCON program, the surveyor must convert all X and Y coordinates on SPCS 27 to their latitude and longitude values on NAD 27 (see Examples 11-2 and 11-10). Each geodetic position defines a unique point on the above-mentioned grid. After φ and λ of NAD 27 are entered, the program searches for the grid element containing the point, computes the corrections $\delta\varphi$ and $\delta\lambda$, adds these to the 1927 values, and outputs φ and λ of 1983. The surveyor then converts these to N and E on SPCS 83.

The Army Corps of Engineers has combined NADCON with programs for conversion of geodetic positions to grid coordinates and vice versa. Thus the surveyor need not make these calculations. He inputs SPCS 27 coordinates directly, along with the appropriate zone designation, and receives SPCS 83 coordinates as the output from the program. This program, designated CORPSCON, is also available through NGS.

FIGURE 11-9 Latitude change from NAD 27 to NAD 83 in meters. By permission of National Geodetic Survey.

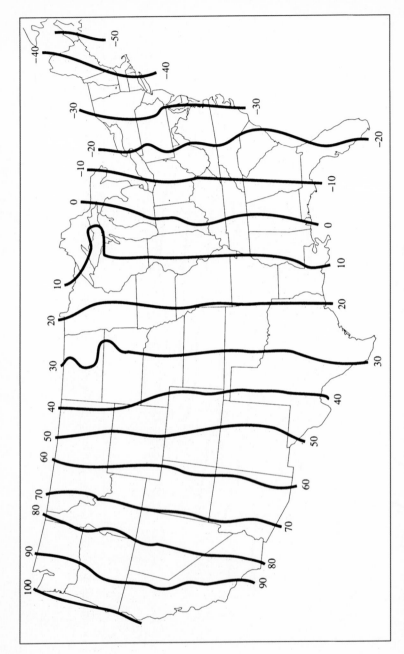

FIGURE 11-10 Longitude change from NAD 27 to NAD 83 in meters. By permission of National Geodetic Survey.

11-18   Application of the State Coordinate Systems

As mentioned in Section 11-1, the national network of horizontal control monuments is available to any engineer or surveyor who desires to connect his work with the federal system by methods of plane surveying. It is made usable through the state coordinate systems computed for each state, because the positions of these monuments are published both in geographic and state plane coordinates. If a point to be used is one whose grid coordinates are not known, they may be computed from the latitude and longitude of the point by the equations and tables discussed in previous sections.

When a federal monument has been established, it is the practice to set also a second monument, called an *azimuth mark*, that is far enough away to give an accurate point for backsighting. In addition to the azimuth marks so set, other control monuments in the vicinity can be used as backsights. The grid azimuths, as well as the geodetic azimuths, of these lines to azimuth marks are usually published along with the position data. If, however, no grid azimuth is available for a particular line whose geodetic azimuth is known, then the convergence angle γ or the angle $\Delta\alpha$ or the angle θ is computed as explained in Section 11-6 or Section 11-12 and the grid azimuth may be obtained by Eq. (11-3) or (11-4a) or by Eq. (11-22) or (11-23). If the plane coordinates of the two ends of the line are known, then the grid azimuth may be determined by Eq. (8-13) of Section 8-21.

When a traverse survey is to be related to the state plane coordinate system, all of the lengths of the traverse lines measured with the tape or the EDM, after correction for systematic errors, must be reduced to their sea-level or ellipsoidal lengths. The method for making this reduction for baseline and trilateration measurements using EDM was presented in Chapter 9.

The sea-level reduction can also be expressed approximately as a reduction factor derived from Eq. (9-1).

Thus

$$s = D\left(1 - \frac{h}{R}\right)$$

The sea-level reduction factor r_{SL} is given by

$$r_{SL} = \left(1 - \frac{h}{R}\right) \tag{11-37}$$

A horizontal distance D at elevation h is then reduced to its sea-level length s by multiplying by the sea-level reduction factor.

Table 11-8 gives the values of r_{SL} for an average radius of the sea-level surface of 20,906,000 ft when the elevation of the line is given in feet. If the elevation is expressed in meters, it must be converted to feet by multiplying by 39.37/12 before entering the table. If the radius of the earth in the project vicinity is substantially different from the average given above, the values given in Table 11-8 may not be sufficiently accurate. The reduction is then made using Eq. (11-37).

TABLE 11-8   Sea-Level Reduction Factors

Elevation h (ft)	Sea-Level Reduction Factor	Elevation h (ft)	Sea-Level Reduction Factor
500	0.9999 761	5,500	0.9997 369
1,000	0.9999 522	6,000	0.9997 130
1,500	0.9999 283	6,500	0.9996 891
2,000	0.9999 043	7,000	0.9996 652
2,500	0.9998 804	7,500	0.9996 413
3,000	0.9998 565	8,000	0.9996 173
3,500	0.9998 326	8,500	0.9995 934
4,000	0.9998 087	9,000	0.9995 695
4,500	0.9997 848	9,500	0.9995 456
5,000	0.9997 608	10,000	0.9995 217

◀ **EXAMPLE 11-15**   A line measures 22,456.72 ft at elevation 3260 ft. The average radius of curvature in the area is 20,847,000 ft. Compute the sea-level distance using Table 11-8 and also by Eq. (11-37).

Solution: By interpolation in Table 11-8,

$$r_{SL} = 0.9998\ 565 - 239 \times \frac{260}{500} = 0.9998\ 441$$

and

$$s = 22,456.72 \times 0.9998\ 441 = 22,453.22 \text{ ft}$$

By Eq. (11-37),

$$r_{SL} = 1 - \frac{3260}{20,847,000} = 0.9998\ 436$$

and

$$s = 22,456.72 \times 0.9998\ 436 = 22,453.21 \text{ ft}$$

The difference between the two values in this example is insignificant.   ◆

In Section 10-8, the concept of the geoidal separation was introduced when computing elevations from GPS observations. When using SPCS 27 coordinates, the geoidal separation is ignored since these coordinates are derived from geodetic coordinates referred to the Clarke spheroid of 1866. The Clarke spheroid generally conforms to the geoid throughout the conterminous United States, and thus the geoidal separation is rather insignificant. However, SPCS 83 coordinates are derived from geodetic coordinates referred to the GRS 80 spheroid in which the geoidal separation *does* become significant. If the accuracy of the survey warrants, the geoidal separation N should be taken into account when reducing measured lengths to the GRS 80 spheroid (see footnote on page 378). In the conterminous United States, the value of N is negative since the geoid lies below the GRS 80 spheroid. Values of the geoidal separation are published by NGS together with NAD 83 coordinate information.

EXAMPLE 11-16 A line measures 6500 m at an elevation of 2500 m above sea level. The geoidal separation in the area is − 31 m. What is the effect of ignoring the geoidal separation?

Solution: The radius of the earth is 20,906,000 × 0.3048 = 6,372,150 m. Ignoring the geoidal separation, the sea-level reduction factor is

$$r_{SL} = \left(1 - \frac{2500}{6,372,150}\right) = 0.99960767$$

and the reduced length is 6,497.450 m.

Taking into account the geoidal separation, the sea-level reduction factor is

$$r_{SL} = \left(1 - \frac{2469}{6,372,150}\right) = 0.99961253$$

and the reduced length is 6497.481 m.

The difference between the two values is about 1 part in 209,700, which may or may not be significant. ◆

If a traverse is conducted in hilly or mountainous country, a higher accuracy is obtained by reducing the length of each line using the average elevation of the two ends of the line as the value of h in Eq. (11-37) or in Table 11-8. This refinement is usually not necessary except for highest accuracy over long traverses.

The sea-level lengths of the lines in the survey must be further reduced to their corresponding grid lengths by application of the grid scale factors given in the projection tables. When working on the transverse Mercator projection, the grid scale factor is a function of the distance east or west of the central meridian, as shown in Table 11-4 for Arizona. Note that if the area is less than about 295,000 ft or 90,000 m east or west of the central meridian ($X' = \pm$ 295,000 ft), the grid lengths will be less than the sea-level lengths, whereas beyond this distance, the grid lengths will be greater than the sea-level lengths.

When working on the Lambert conformal projection, the grid scale factor is a function of the latitude of the area as shown in Table I for Ohio North (Table 11-6). The factor is thus constant for east-west lines but changes in the north-south direction. The value of X' needed for determining the grid scale factor on the transverse Mercator projection is obtained by Eq. (11-9). When using the Lambert conformal projection, the latitude of the area must be known in order to determine the scale factor. This can be obtained either by the method shown in Example 11-10 or else by scaling it from a map of the area.

If the traverse contains long lines running generally east-west on the transverse Mercator projection or north-south on the Lambert conformal projection, an added refinement can be made to enhance the accuracy of the survey. This consists of obtaining the grid scale factor for both ends of a line and for its midpoint and applying the equation

$$k = \frac{k_1 + 4k_m + k_2}{6}$$

(11-38)

in which k is a weighted scale factor for reducing the sea-level to the corresponding grid length. Usually, however, it is sufficiently accurate to compute the average of the scale factors at the ends of each line.

In an area of limited extent, it is often desirable and expedient to assign a combination factor for sea-level and grid lengths to an area whose elevation does not fluctuate too much. For example, if the average elevation in a given area is 2600 ft above sea level, the sea-level reduction factor is 0.9998756. If the area is in such a latitude on a Lambert projection that the grid scale factor is 0.9999000, the combination factor in this area would be $(0.9998756)(0.9999000) = 0.9997756$. A ground length of 10,000 ft in this area would therefore correspond to a grid length of 9997.756 ft.

Figure 11-11 illustrates the application of the sea-level reduction and grid scale corrections to the length of a line. The line AB lying at an elevation h is projected to the sea-level length $A'B'$ by the sea-level reduction factor. The trace of the grid projection surface is shown at five different positions. If the grid surface lies below sea level as shown at position 1, the grid scale factor is less than unity and so is the combination factor. If the grid surface generally coincides with sea level as shown at position 2, the grid scale factor is unity and the combination factor is less than unity.

If the grid surface lies above the sea-level surface as at positions 3, 4, and 5, the grid scale factor is greater than unity. At position 3 the combination factor is less than unity. At position 4 the grid surface is coincident with the ground surface, hence the combination factor is unity. At position 5 the grid surface lies above the ground surface, and the combination factor is greater than unity.

The lengths of traverse lines are usually short enough to allow the methods given in Sections 8-4, 8-5, and 8-6 to be used for computing azimuths from measured angles. After the measured lengths have been reduced to their appropriate grid lengths and the azimuths have been computed and adjusted, the traverse is then computed and adjusted on the state coordinate system by the methods presented in Chapter 8. If, however, the lines in the traverse are long, as is quite feasi-

FIGURE 11-11 Relationship between sea-level length, grid length, and lengths of lines at different elevations.

ble using EDM, then the azimuths of the lines must be modified by the small correction term given in either Eq. (11-4a) or (11-23).

In Fig. 11-3 assume that the grid azimuth of CD is known. This is the direction of the straight line from C to D. If a theodolite is set up at C and a backsight is made to D, the grid projection of the line of sight is the curved line CD. If the theodolite is now sighted on E, the measured angle would be larger than the plane angle DCE. If this measured angle were to be added to the grid azimuth of CD to obtain the grid azimuth of CE, the resulting value would be too large by the small angle between the straight line CD and the curved line CD. The small angle is the correction term of Eq. (11-4a) or Eq. (11-23). The same analysis can be made of the angle at A from B to C shown in Fig. 11-7. Thus when the traverse contains very long lines, a preliminary solution for the coordinates of the points will allow an evaluation of the correction term at each end of each line. The corrected grid azimuths are then used to compute the final coordinates of the traverse stations.

In the following two examples, the unit is the foot, but meters could be used just as well. Also, X refers to Easting and Y refers to Northing.

EXAMPLE 11-17   Referring to Fig. 11-12, a traverse is run from a fixed point A to a fixed point E. The grid azimuth from A to an azimuth mark P is 18° 52′ 42.20″, and the grid azimuth from E to another fixed control station Q is 8° 48′ 52.30″, both measured clockwise from north. The plane coordinates of the Ohio North Lambert projections of SPCS 27 are shown in the table below.

Point	Y (ft)	X (ft)
A	446,145.60	2,113,664.23
E	430,164.08	2,162,796.14
Q	454,494.86	2,166,569.06

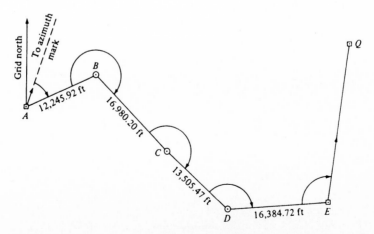

FIGURE 11-12 Traverse on Lambert projection, Ohio North zone of SPCS 27.

The ground lengths measured with the EDM together with the elevations of the stations and the measured angles to the right as shown in Fig. 11-12 are given in Table 11-9. Compute the adjusted coordinates of B, C, and D.

Solution: The reductions of the measured length to their sea-level or spheroidal lengths are performed by Eq. (11-37), using as h the average of the elevations at the ends of each line. Preliminary coordinates must be computed in order to determine the correct grid scale factor for each line and to compute the small correction to each measured angle. These preliminary coordinates are computed using unadjusted azimuths obtained from the measured angles (see Section 8-6) and a grid scale factor for $\varphi = 40° 53'$ obtained from Table 11-6. The published geodetic position of A is $\varphi = 40° 53' 25.950''$N and $\lambda = 81° 35' 20.000''$ W. The preliminary grid azimuths are computed as shown in Table 11-10.

TABLE 11-9   Data for Traverse Computations

Point	Measured Length (ft)	Elevation (ft)	Angle to Right
A		1246	48° 18′ 42.15″
	12,245.92		
B		988	249° 22′ 16.36″
	16,980.20		
C		959	176° 50′ 03.30″
	13,505.47		
D		1162	133° 31′ 44.46″
	16,384.72		
E		1280	101° 53′ 18.20″

TABLE 11-10   Preliminary Azimuths

Line	Preliminary Grid Azimuth
A Az mark	18° 52′ 42.20″
+ ∠A	+ 48° 18′ 42.15″
AB	67° 11′ 24.35″
BA	247° 11′ 24.35″
+ ∠B	+ 249° 22′ 16.36″
BC	136° 33′ 40.71″
CB	316° 33′ 40.71″
+ ∠C	+ 176° 50′ 03.30″
CD	133° 23′ 44.01″
DC	313° 23′ 44.01″
+ ∠D	+ 133° 31′ 44.46″
DE	86° 55′ 28.47″
ED	266° 55′ 28.47″
+ ∠E	+ 101° 53′ 18.20″
EQ	8° 48′ 46.67″

TABLE 11-11 Preliminary Grid Lengths

Line	Average Elevation (ft)	Sea-Level Reduction Factor	Sea-Level Lengths (ft)	Preliminary Grid Scale Factor	Preliminary Grid Length (ft)
AB	1117	0.99994657	12,245.26	0.9999443	12,244.58
BC	974	0.99995341	16,979.41	0.9999443	16,978.46
CD	1061	0.99994925	13,504.78	0.9999443	13,504.02
DE	1221	0.99994160	16,383.76	0.9999443	16,382.85

TABLE 11-12 Preliminary Coordinates

Point	Preliminary Azimuth	Preliminary Grid Length (ft)	Latitude (ft)	Departure (ft)	Y	X
A					446,145.60	2,113,664.23
	67° 11′ 24.35″	12,244.58	+4,746.92	+11,287.01		
B					450,892.52	2,124,951.24
	136° 33′ 40.71″	16,978.46	−12,328.24	+11,674.02		
C					438,564.28	2,136,625.26
	133° 23′ 44.01″	13,504.02	−9,277.68	+9,812.40		
D					429,286.60	2,146,437.66
	86° 55′ 28.47″	16,382.85	+878.95	+16,359.25		
E					430,165.55	2,162.796.91

Since the agreement between the preliminary grid azimuth and the fixed azimuth of *EQ* is good, no adjustment is made. Reduction to preliminary grid lengths are shown in Table 11-11. Computation of preliminary coordinates are shown in Table 11-12.

The agreement between the preliminary coordinates and the fixed coordinates of *E* is sufficient to permit the use of the preliminary coordinates in computing the scale factors and the reduction of measured angles to plane angles. It is to be noted that if the traverse stations could be identified with sufficient accuracy on a series of U.S. Geological Survey quadrangle maps, their approximate coordinates could be scaled directly from the quadrangle maps since these maps contain state coordinate grid ticks. These scaled coordinates, together with the geodetic latitude, which could also be scaled, would then be sufficient for determining grid scale factors and angle reduction.

The small corrections to be applied to each angle in order to convert from spherical angles (see Fig. 9-4) to plane angles are computed using the last term of Eq. (11-23). Figure 11-13 shows the situation at points *A, B,* and *C.* At *A* the sight to the azimuth mark is too short to have any effect, but the sight to *B,* represented by the dashed line, in which the curvature is greatly exaggerated, the measured angle is too large. Thus it must be reduced by the last term of Eq. (11-23) (see Fig. 11-7), which amounts to 0.17″ in this example. At *B* the measured spherical angle must be reduced because of the curvature, on the grid projection, of both lines of sight *BA* and *BC.* Example 11-9 shows the computation of this correction. The corrected angles are shown in Table 11-13.

These plane angles are now used to compute the grid azimuths of the traverse lines as shown in Table 11-14.

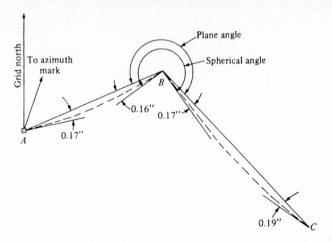

FIGURE 11-13 Reduction from spherical to plane angles in traverse computations.

TABLE 11-13   Correction to Plane Angles

Point	Measured Angle to Right	Corrections	Plane Angles
A	48° 18′ 42.15″	− 0.17″	48° 18′ 41.98″
B	249° 22′ 16.36″	− 0.16″− 0.17″	249° 22′ 16.03″
C	176° 50′ 03.30″	− 0.19″− 0.17″	176° 50′ 02.94″
D	133° 31′ 44.46″	− 0.18″− 0.31″	133° 31′ 43.97″
E	101° 53′ 18.20″	− 0.31″− 0.03″	101° 53′ 17.86″

TABLE 11-14   Adjusted Grid Azimuths

Line	Unadjusted Grid Azimuth	Correction	Adjusted Grid Azimuth
A-Az mark	18° 52′ 42.20″		
+ ∠A	+48° 18′ 41.98″		
AB	67° 11′ 24.18″	+1.46″	67° 11′ 25.64″
BA	247° 11′ 24.18″		
+ ∠B	+249° 22′ 16.03″		
BC	136° 33′ 40.21″	+2.93″	136° 33′ 43.14″
CB	316° 33′ 40.21″		
+ ∠C	+176° 50′ 02.94″		
CD	133° 23′ 43.15″	+4.39″	133° 23′ 47.54″
DC	313° 23′ 43.15″		
+ ∠D	+133° 31′ 43.97″		
DE	86° 55′ 27.12″		
ED	266° 55′ 27.12″	+5.86″	86° 55′ 32.98″
+ ∠E	+101° 53′ 17.86″		
EQ	8° 48′ 44.98″	+7.32″	8° 48′ 52.30″
Fixed EQ	8° 48′ 52.30″		
	$e = -7.32″$		

If the geographic latitudes of the points have been scaled from quadrangle maps, the grid scale factor for each point can be obtained by interpolation in Table 11-6. Otherwise, the simplest method is to compute R for each point, as shown in Example 11-10 using the preliminary coordinates and interpolated in Table 11-6. The calculation of R for point B is given here:

$$X_B = \quad 2,124,951.24 \qquad R_b = 24,559,158.47 \qquad \tan \theta = \frac{124,951.24}{24,108,265.95}$$

$$\begin{array}{lll} X_0 & -2,000,000.00 & Y_B & -450,892.52 & \theta = \quad 0.296956845° \\ \overline{X'} & 124,951.24 & \overline{R_b - Y} = 24,108,265.95 & \theta = 0° \; 17' \; 49.04464'' \end{array}$$

By Eq. (11-21), $\qquad R = \dfrac{24,108,265.95}{\cos \theta} = 24,108,589.75$

This computation is made most efficiently with a handheld calculator by temporarily storing $R_b - Y$, computing θ in degrees, recalling $R_b - Y$, and then dividing by $\cos \theta$.

The values of R and the interpolated grid scale factors are given in Table 11-15. Computation of coordinates based on the adjusted grid azimuths given in Table 11-14, and correct grid lengths are shown in Table 11-16. The correct grid lengths are obtained by multiplying the sea-level length given in Table 11-11 by the

TABLE 11-15 Grid Scale Factors

Point	R (ft)	Grid Scale Factor	Line	Average Grid Scale Factor
A	$\theta = 40° \; 53' \; 25.950'' \, N$	0.9999439	AB	0.9999436
B	24,108,589.75	0.9999432	BC	0.9999442
C	24,120.981.13	0.9999452	CD	0.9999460
D	24,130,316.21	0.9999467	DE	0.9999467
E	24,129,542.10	0.9999466		

TABLE 11-16 Final Coordinate Computation

Point	Adjusted Grid Azimuth	Adjusted Grid Length (ft)	Latitude	Departure	Y	X
A					446,145.60	2,113,664.23
	67° 11′ 25.64″	12,244.57	+4,746.84	+11,287.03		
B					450,892.44	2,124,951.26
	136° 33′ 43.14″	16,978.46	−12,328.38	+11,673.87		
C					438,564.06	2,136,625.13
	133° 23′ 47.54″	13,504.05	− 9,277.87	+9,812.26		
D					429,286.19	2,146,437.39
	86° 55′ 32.98″	16,382.89	+878.59	+16,359.31		
E					430,164.78	2,162,796.70
				Fixed	430,164.08	2,162,796.14
					$e_L = +0.70$	$e_D = + 0.56$

average grid scale factors given in Table 11-15. Finally, the coordinates are adjusted as shown in Chapter 8. ◆

The traverse computations in Example 11-17 have been refined to take into account the average elevation of each individual line, the average scale factor for each individual line, and the small corrections used to reduce the measured angles to their corresponding plane angles. These refinements assume that the traverse measurements were of a very high order of accuracy. Under most conditions, especially for shorter lines, the traverse computations can be greatly simplified by using an average elevation of all of the traverse stations, a grid reduction factor based on an average value for φ, and ignoring the small corrections to the measured angles.

◆**EXAMPLE 11-18** Given the measured data in Example 11-17, compute the coordinates of B, C, and D using an average elevation, an average scale factor, and neglecting the reduction to plane angles.

Solution: The average elevation of the traverse stations is 1127 ft. By Eq. (11-37), the sea-level reduction factor is 0.99994609. The average latitude, scaled from quadrangle maps is 40° 52′ 20″ for which the grid scale factor is 0.9999449. The combination factor is then 0.99994609 × 0.9999449 = 0.99989099. The measured lines reduced to the grid are given in Table 11-17.

Referring to the preliminary azimuths computed in Table 11-10 by using the measured angles, the angular error is − 5.63″. Distributing this error over the azimuths of the five lines gives the adjusted azimuths shown in Table 11-18. This is similar to the adjustments shown in Table 11-14 for the more refined computations.

Finally, the coordinates are computed as shown in Table 11-18, and are adjusted as shown in Chapter 8. If, for example, the compass rule of adjustment were to be applied to the traverse computed by the more refined method, and by the simplified method, the greatest discrepancy in this example is 0.05 ft in the X coordinate of point D. The student can verify this by making the adjustments. ◆

Triangulation of moderate extent may be executed and computed by using grid coordinates on the state coordinate systems. A line makes an ideal baseline when the coordinates of both ends are known beforehand, since both the grid azimuth and the grid distance may be computed from the coordinates by inversing the line. Plane angles are used to solve the triangles and to carry grid azimuths through the triangulation system.

·**TABLE 11-17** Grid Lengths of Traverse Lines

Line	Measured Length (ft)	Grid Length (ft)
AB	12,245.92	12,244.59
BC	16,980.20	16,978.35
CD	13,505.47	13,504.00
DE	16,384.72	16,382.93

TABLE 11-18 Final Coordinate Computation

Point	Adjusted Grid Azimuth	Adjusted Grid Length (ft)	Latitude	Departure	Y	X
A					446,145.60	2,113,664.23
	67° 11′ 25.48″	12,244.59	+4,746.86	+11,287.04		
B					450,892.46	2,124,951.27
	136° 33′ 42.96″	16,978.35	−12,328.29	+11,673.81		
C					438,564.17	2,136,625.08
	133° 23′ 47.39″	13,504.00	−9,277.83	+9,812.23		
D					429,286.34	2,146,437.31
	86° 55′ 32.97″	16,382.93	+878.60	+16,359.35		
E					430,164.94	2,162,796.66
					430,164.08	2,162,796.14
					$e_L = +0.86$	$e_D = +0.52$

In performing trilateration on the state coordinate system, the measured lengths are all reduced to sea level by the methods outlined in Sections 9-11 and 9-12. They are then reduced to grid lengths for use in computing the coordinates of the trilateration stations by applying the appropriate grid scale factors.

Because of the inferior accuracy and reliability in the determination of the position of a point by the methods of intersection and resection, there is little or no justification for applying any azimuth correction to the lines. If the positions of the known points are on the state coordinate system, then the computations are straightforward as outlined in Sections 9-13 and 9-14, and the new point is automatically on the state coordinate system.

A problem frequently arises in which a large survey extends from one coordinate-projection zone into an adjacent zone. The solution requires a conversion from plane coordinates of one zone to geodetic coordinates, as illustrated in Examples 11-2 and 11-10, and then a conversion from geodetic coordinates to plane coordinates of the next zone, as illustrated in Examples 11-1 and 11-8. The transition usually requires that at least two points in each zone near the common boundary of the two zones be converted so a line of known grid azimuth can be established for each zone.

11-19 Land Areas from State Coordinates

If the state coordinates of the corners of a parcel of land are known, the grid area encompassed within the boundary may be computed by Eq. (8-25) or (8-26). Since this grid area does not, as a rule, correspond to the actual ground area, it should be *divided* by the *square* of the combination factor used to determine grid lengths from ground lengths. For example, if the grid area of a parcel of land lying in an area whose combination factor is 0.9997750 was computed by Eq. (8-25) to be 45, 675 ft^2, then the actual ground area is $(45,675)/(0.9997750)^2$ = 45,696 ft^2.

11-20 Project Lengths from State Coordinates

Engineering projects require the field location of construction lines and survey points. Since most large projects will be established with reference to control points on the state coordinate system, the construction plans and drawings will contain state coordinates and grid lines. If the state coordinates of two ends of a construction line are used to inverse the line, the grid length and grid bearing or azimuth will result. The proper length to be laid off, however, is the actual ground length. The difference between the grid length and the ground length can be taken into account by one of two methods. The first method, which is quite practical, is to ignore the difference. This method assumes that the errors in layout measurements are larger than the errors introduced by neglecting the difference between the grid and the ground distance. For example, in an area whose elevation is 2500 ft, the sea-level reduction factor is 0.9998804. Suppose the grid scale factor is 0.9999000 or 1 part in 10,000. The combination factor is then 0.9997804. This means that the difference between the ground length and the grid length is only 0.02 ft/100 ft or 0.22 ft/1000 ft. This small discrepancy, amounting to about 1 part in 5000, could be ignored on most construction projects. In high country the sea-level reduction factor becomes significant, particularly if the project lies in an area of the state projection where the grid scale factor is less than unity.

The second method is to compute the combination sea-level grid factor, and then to *divide* all grid distances indicated on, or derived from, the plans in order to obtain the correct ground or project lengths. To accomplish this without misunderstanding between the surveyor, engineer, and contractor, a note should accompany each construction drawing giving explicit instructions to the user. The note could possibly read, "All distances shown on this set of plans (this drawing) or derived from plane coordinates shown on the plans (drawing) are grid distances on the ___Coordinate System, Zone___. To obtain ground distances for laying out construction lines, divide grid distances by 0.9998940."

11-21 Lost and Obliterated Corners

When the coordinates of a land corner on a state system are known and the corner cannot be found by the usual methods, its ground position can be determined by locating the point corresponding to those coordinates. This is accomplished by running a traverse from other stations, for which the state coordinate data are available, to a new point in the immediate vicinity of the point sought for. The coordinates of the new point and the sought-for point are compared, and the azimuth and distance from the new point to the old one are laid out on the ground. The point thus established is the ground position corresponding to the coordinates of the required corner. If any remnants of the original corner remain in position, they may be found by a careful and detailed examination of the ground within a short distance of the restored position.

Marks that are otherwise difficult to find, and which if found are apt to be considered as of doubtful origin, may be recovered and proved authentic by the use of the state coordinate system. A fragment of stone or a piece of decayed wood may constitute a satisfactory recovery if found in the ground position corresponding to the coordinates of the landmark. If no marks are found in the position indicated by the coordinates, it may safely be concluded that the monument has been destroyed, and the coordinate position may be accepted as primary evidence of the ground location.[†]

As the accuracy of the restored position reflects the errors of both the original survey and the restoration survey, care must be taken to maintain a high order of accuracy in the restoration survey, especially when the restoration survey is longer than the original one. Although it is not necessary to employ the same stations in the resurvey that were used in the original survey, it is desirable to use stations that were in the same traverse or survey net. A station on some other traverse may not be so well related in position to the lost mark because of the methods that were used in the previous adjustment.

11-22   Descriptions by Coordinates

A legal description of a parcel of land identifies the land for title purposes and provides information necessary for locating the land on the ground. A description that is satisfactory for title purposes may be wholly inadequate for a field location of the property. One of the simplest forms of land description for title purposes is afforded by the public land method of surveying, in which a description by reference to township, range, and section can apply to one and only one parcel of land. Thus "the northwest quarter of the southeast quarter ($NW\frac{1}{4} SE\frac{1}{4}$) of Section Ten (10), Township Two South (T2S), Range Six East (R6E) of the Meridian of Michigan" is satisfactory for title purposes. It is also satisfactory for survey purposes as long as the controlling monuments are in existence. For the sake of perpetuating the corners, it would be helpful to add the coordinates of the center of the section, its south and east quarter-corners, and its southeast corner. If a corner monument is later obliterated, these coordinates will provide an effective means of recovering and identifying any part of the monument that may remain in position. If the original monument is entirely destroyed, the coordinates become strong evidence of where it stood and make possible an accurate replacement.

When the description is by metes and bounds, it may be simplified by reference to a plat that is made a part of the conveyance. The azimuths and lengths of the sides of the parcel, as well as the coordinates of the place of beginning, can be shown on this plat.

Where no plat is available, the following form taken from the Second Progress Report of the Joint Committee on the Real Property Division, American Bar Association, and the Surveying and Mapping Division, American Society of Civil Engineers, Proceedings, A.S.C.E., June, 1941, may be used:

[†]Refer to Chapter 18 for a discussion of the relocation of lost corners on the U.S. Public Land Survey System.

_____ situated in the Town of _____, County of _____, State of _____, and bounded as follows:

"Beginning at a drill hole in a stone mound which is set in the corner of a stone wall on the north line of Farm Road at the southwest corner of land of Peter L. Prince and at the southeast corner of land hereby conveyed, the coordinates of which monument referred to the Massachusetts Coordinate System, Mainland Zone, are $x = 417{,}603.29$, $y = 316{,}042.17$. Thence, on an azimuth of 81° 39′ 30″, 123.39 feet along the northerly line of Farm Road to an iron pin at the southwest corner of the tract hereby conveyed; thence, on an azimuth of 181° 47′ 30″, 145.82 feet along the easterly line of land of Arthur C. Hicks to an iron pin in a stone wall at the northwest corner of the tract hereby conveyed; thence, on an azimuth of 276° 32′ 00″, 62.04 feet along a stone wall on the southerly line of Peter L. Prince to a drill hole in a stone bound in the wall at the northeast corner of land hereby conveyed; thence, on an azimuth of 335° 10′ 20″, 133.09 feet along a stone wall on the westerly line of land of said Prince, to the place of beginning."

Zero azimuth is grid south in the Massachusetts Coordinate System, Mainland Zone.

This description was written June 1, 1939, from data secured by survey made by Fred L. Connor, Civil Engineer, in March and April, 1939.

11-23   Universal Transverse Mercator (UTM) System

The universal transverse Mercator (UTM) system shown in Fig. 11-14 employs 60 zones numbered 1 through 60 commencing at $\lambda = 180°$, each of which embraces 6° of longitude, and lies between 80° north latitude and 80° south latitude. The central meridian for each zone lies at the center of that zone. Coordinates are expressed in meters. The geometry of each zone of the UTM is identical with that of the state transverse Mercator zone described in Sections 11-5 through 11-10.

The zone number of any point can be determined from the following equations

$$N = (180° + \lambda_E)/6$$
$$N = (180° - \lambda_W)/6 \qquad (11\text{-}39)$$

in which N is the zone number and λ_E and λ_W denote east and west longitude. If the values in parentheses are not exactly divisible by 6, then N is the next higher zone.

EXAMPLE 11-19   Compute the zone number for a point P whose longitude is 25° 42′ 16″ E. Compute the zone number for a point Q whose longitude is 123° 45′ 36″ W.

Solution: By Eq. (11-39),

$$N_P = (180° + 25° \ 42' \ 16'')/6 = 34 + 1 = 35$$
$$N_Q = (180° - 123° \ 45' \ 36'')/6 = 9 + 1 = 10$$

The origin of coordinates for each zone is the intersection of its central meridian and the equator. The coordinates of the origin in the northern hemisphere are $X_0 =$

FIGURE 11-14 Universal transverse Mercator (UTM) system. The zones are numbered from 1 to 60 eastward from 180° west longitude.

500,000 m and $Y_0 = 0$ (see Section 11-7). In the southern hemisphere $X_0 = 500,000$ m and $Y_0 = 10,000,000$ m. The scale factor along the central meridian is 0.9996. The ellipses of intersection of the enveloping cylinder (see Fig. 11-1) with the ellipsoid along which the scale is unity lie 180,000 m east and 180,000 m west of the central meridian. The scale at the outer east-west limits of each zone is about 1.0010 at the equator.

The equations for making the conversion between geodetic positions and grid coordinates mentioned in Section 11-10 are equally valid for these conversions using UTM. In Section 11-24 we explain how to use the Internet to compute UTM coordinates.

The UTM system is employed by the U.S. Department of Defense in its worldwide mapping programs. Surveys of many military and defense-oriented installations are often required to be expressed in UTM grid coordinates. Because the UTM grid embraces a much wider area than the transverse Mercator or Lambert grids used to define the state plane coordinate systems, it is found quite useful to those responsible for planning over large areas embracing one or more states.

11-24 Conversion of Coordinates Using PC-Based Software

Based on software provided by the NGS, we[†] have developed PC-based software to perform conversion between geodetic, state plane, and UTM coordinates. An easy-to-use interface will facilitate student usage. The software can be downloaded over an Internet connection with any Web browser.

The downloadable software includes the executable file of the program (TRANSFOR.EXE), relevant data files, as well as input file for examples 11-1, 11-2, 11-3, 11-5, 11-8, 11-10, 11-11, and 11-12 respectively. This program performs coordinate transformation between the following coordinate frames:

- Geodetic
- State Plane Coordinate System 27
- State Plane Coordinate System 83
- Universal transverse Mercator

For geodetic and universal transverse Mercator, the user can choose between NAD 27 and NAD 83 datums. Also the user can choose the linear units and/or the angular format of the input/output data when applicable.

The hardware required for installation is a personal computer with a 386DX or higher processor running Microsoft Windows version 3.1 or later. A mouse or compatible pointing device is also required.

To install this program, do the following:

1. Make a directory for the program (for example, C:\transfor) and for the examples (for example C:\ex).
2. Use a Web browser to connect to the URL "http: //www.awl.com/cseng/ titles/0-673-99752-9/"
3. Extract the program files by executing program exe.
4. Extract the examples by executing examples. exe
5. In the program manager create a new program group and item for the executable file (TRANSFOR.exe) as follows:
   a. In the program manager Menu select FILE.
   b. In the File submenu select NEW.
   c. Select New Group, and then type the name for this group, TRANSFOR.
   d. In the File submenu select NEW again.
   e. Select New Item. Specify the name, the command line, the working directory, for example,

   TRANSFOR

   c:\transfor\transfor

   c:\ex

[†]The authors are grateful to Dr. Ayman Habib for his work in converting this software.

To execute the program, click on the icon created by step 5. Then you need to specify the input and the output files. At this stage the program will sequentially activate the controls that you need to specify, namely

- Input/output coordinate frame
- Datum, NAD 27 or NAD 83 (only if you choose geodetic or Universal Transverse Mercator)
- Output angular format (if required)
- Input/output linear units (if required)

Only after all the necessary parameters are input will the TRANSFOR button be activated. At any point you can click the CANCEL button to clear all the parameters, allowing you to start from scratch.

PROBLEMS

11-1. The latitude and longitude of a point on the Arizona East projection of SPCS 27 are, respectively, 33° 22′ 19.6432″ N and 110° 32′ 16.8250″ W. Compute the SPCS 27 state plane coordinates of the point.

11-2. Solve Problem 11-1 for a point in latitude 33° 23′ 16.2422″ N and longitude 109° 48′ 16.2323″ W.

11-3. Solve Problem 11-1 for a point in latitude 33° 21′ 54.5628″ N and 110° 28′ 27.9640″ W.

11-4. The plane coordinates of a point on the Arizona East projection of SPCS 27 are $X = 402,562.20$ ft and $Y = 861,496.88$ ft. Compute the latitude and longitude of the point. Compute $\Delta\alpha$ at the point.

11-5. Solve Problem 11-4 for a point whose coordinates are $X = 392,851.66$ ft and $Y = 860,021.95$ ft.

11-6. Solve Problem 11-4 for a point whose coordinates are $X = 398,560.20$ ft and $Y = 865,471.08$ ft.

11-7. The plane coordinates of a point on the Arizona East projection of SPCS 27 are $X = 202,156.85$ ft and $Y = 757,525.20$ ft. Compute the scale factor at the point.

11-8. Solve Problem 11-7 for a point whose coordinates are $X = 515,562.78$ ft and $Y = 861,444.48$ ft.

11-9. Solve Problem 11-7 for a point whose coordinates are $X = 802,496.11$ ft and $Y = 808,472.92$ ft.

11-10. Compute the angle $\Delta\alpha$ at the point given in Problem 11-1

11-11. Compute the angle $\Delta\alpha$ at the point given in Problem 11-2.

11-12. Compute the angle $\Delta\alpha$ at the point given in Problem 11-3.

11-13. Three points on the Arizona East projection of SPCS 27 have the following coordinates:

Point	Y (ft)	X (ft)
A	928,122.75	538,476.16
B	950,472.15	502,102.25
C	992,159.25	526,304.15

Compute the plane angle at B from A to C. Compute the spherical angle at B from A to C (see Fig. 11-3).

11-14. Given the plane coordinates of Problem 11-13, compute the plane and spherical angles at A from C to B.

11-15. Given the plane coordinates of Problem 11-13, compute the plane and spherical angles at C from B to A.

11-16. The latitude and longitude of a point on the Arizona East projection of SPCS 83 are, respectively, 31° 15′ 42.5628″ N and 108° 15′ 43.8621″ W. Compute the SPCS 83 coordinates of the point, in meters. Convert these to coordinates expressed in International Feet. Compute the scale factor at the point.

11-17. Solve Problem 11-16 for a point in latitude 31°16′43.5682″ N and longitude 110° 04′ 25.4628″ W.

11-18. Solve Problem 11-16 for a point in latitude 31° 14′ 20.0266″ N and longitude 110° 45′ 42.6846″ W.

11-19. The plane coordinates of a point on Arizona East projection of SPCS 83 are N = 28,202.466 m and E = 298,506.922 m. Compute the latitude and longitude of the point. Compute the convergence angle γ.

11-20. Solve Problem 11-19 for a point whose coordinates are N = 27,045.232 m and E = 131,462.560 m.

11-21. Solve Problem 11-19 for a point whose coordinates are N = 31,912.244 m and E = 273,462.865 m.

11-22. Compute the scale factor at the point in Problem 11-19.

11-23. Compute the scale factor at the point in Problem 11-20.

11-24. Compute the scale factor at the point in Problem 11-21.

11-25. Compute the convergence angle γ at the point in Problem 11-16.

11-26. Compute the convergence angle γ at the point in Problem 11-17.

11-27. Compute the convergence angle γ at the point in Problem 11-18.

11-28. Three points on the Arizona East projection of SPCS 83 have the following coordinates:

Point	N (m)	E (m)
P	26,476.815	204,184.822
Q	31,488.028	203,910.050
R	27,572.468	223,246.719

Compute the plane angle at P from Q to R. Compute the spherical angle at P from Q to R.

11-29. Given the coordinates of Problem 11-28, compute the plane and spherical angles at Q from R to P.

11-30. Given the coordinates of Problem 11-28, compute the plane and spherical angles at R from P to Q.

11-31. The latitude and longitude of a point on the Ohio North projection of SPCS 27 are, respectively, 40° 52′ 16.2642″ N and 83° 48′ 15.4020″ W. Compute the SPCS 27 plane coordinates of the point. Compute the scale factor at the point.

11-32. Solve Problem 11-31 for a point in latitude 40° 54′ 18.2686″ N and longitude 84° 00′ 02.5686″ W.

11-33. Solve Problem 11-31 for a point in latitude 40° 53′ 54.4896″ N and longitude 81° 06′ 02.8466″ W.

11-34. The plane coordinates of a point on the Ohio North projection of SPCS 27 are X = 1,946.258.05 ft and Y = 447,246.85 ft. Compute the latitude and longitude of the point. Compute the scale factor at the point.

11-35. Solve Problem 11-34 for a point with coordinates X = 2,256,422.15 ft and Y = 436,020.90 ft.

11-36. Solve Problem 11-34 for a point with coordinates X = 1,848,500.22 ft and Y = 440,910.02 ft.

11-37. The coordinates of two points A and B on the Ohio North projection of SPCS 27 are X_A = 1,978,223.20 ft and Y_A = 444,426.88 ft; X_B = 2,125,456.87 ft and Y_B = 439,181.08 ft. Compute the geodetic azimuth of AB and BA. Use Eq. (11-23).

11-38. Three points on the Ohio North projection of SPCS 27 have the following coordinates:

Point	Y (ft)	X (ft)
P	450,121.85	1,858,462.15
Q	188,182.08	1,860,121.24
R	181,729.59	2,104,565.67

Compute the plane angle and the spherical angle at Q from P to R.

11-39. Given the plane coordinates of Problem 11-38, compute the plane angle and the spherical angle at P from R to Q.

11-40. Given the plane coordinates of Problem 11-38, compute the plane angle and the spherical angle at R from Q to P.

11-41. The latitude and longitude of a point on California Zone 3 projection of SPCS 83 are, respectively, 37° 43′ 02.5613″ N and 120° 45′ 18.2244″ W. Compute the plane coordinates of the point, in meters, using both polynomials and tables. Convert the coordinates to Survey Feet. Compute the convergence angle γ and the scale factor.

11-42. Solve Problem 11-41 for a point in latitude 37° 45′ 02.0030″ N and longitude 121° 44′ 20.5656″ W.

11-43. Solve Problem 11-41 for a point in latitude 37° 44′ 42.2866″ N and longitude 119° 04′ 28.5652″ W.

11-44. The plane coordinates of a point on California Zone 3 of SPCS 83 are N = 632,880.946 m and E = 1,962,463.180 m. Compute the latitude and longitude of the point using both polynomials and tables. Compute the convergence angle γ and the scale factor.

11-45. Solve Problem 11-44 for a point whose coordinates are N = 638,195.262 m and E = 2,122,464.580 m.

11-46. Solve Problem 11-44 for a point whose coordinates are N = 635,454.181 m and E = 1,918,422.822 m.

11-47. Three points on the California Zone 3 projection of SPCS 83 have the following coordinates:

Point	N (m)	E (m)
A	634,015.263	2,015,004.156
B	639,262.121	1,931,464.223
C	633,181.006	1,910,876.850

Compute the plane angle and the spherical angle at A from B to C.

11-48. Given the coordinates in Problem 11-47, compute the plane angle and spherical angle at B from C to A.

11-49. Given the coordinates in Problem 11-47, compute the plane angle and spherical angle at C from A to B.

11-50. The area of a tract of land is computed from the state plane coordinates of its boundary corners and is found to be 5.6562 acres. The scale factor in the area is 0.9999716. The average elevation is 1850 ft above sea level. What is the actual area in this elevation?

BIBLIOGRAPHY
Buckner, B. 1996. We've missed the point on state plane coordinates. ACSM Bulletin No. 159. *Amer. Cong. on Surv. & Mapp.* p. 22.

Burkholder, E. F. 1980. The Michigan scale factor. *Proc. Ann. Meet. Amer. Cong. Surv. & Mapp.* p. 180.

Department of the Army. 1958. *Universal Transverse Mercator Grid.* Technical Manual TM5-241-8. Department of Defense, Washington, D.C.

——. 1958. *Universal Transverse Mercator Grid Tables for Latitude 0°–80°: Clarke 1866 Spheroid (Meters): Vol. II: Transformation from Geographic to Grid.* Technical Manual TM5-241-4/1, Department of Defense, Washington, D.C.

——. 1958. *Universal Transverse Mercator Grid Tables for Latitude 0°–80°: Clarke 1866 Spheroid (Meters): Vol. II: Transformation from Grid to Geographic.* Technical Manual TM5-241-4/2, Department of Defense, Washington, D.C.

Dewhurst, W. T. 1990. *NADCON—the Application of Minimum-Curvature Derived Surfaces in the Transformation of Positional Data from the North American Datum of 1927 to the North American Datum of 1983.* NOAA Technical Memorandum NOS NGS-50. Rockville, MD.

Gale, P. 1971. Practical uses of coordinates—or, control surveys in operation. *Proc. Fall Conv. Amer. Cong. Surv. & Mapp.* p. 45.

Gergen, J. G. 1982, The North American Datum and the surveyor. *Tech Papers 42nd Ann. Meet. Amer. Cong. Surv. & Mapp.* p. 421.

Hintz, R. J., and Onsrud, H. J. 1989–1990. Issues for the Bureau of Land Management to consider in converting from NAD 27 to NAD 83. *P.O.B.* 15:2.

Mitchell, H. C., and Simmons, L. G. 1945. *The State Coordinate System (a Manual for Surveyors).* Special Publication No. 235, U.S. Coast and Geodetic Survey, U.S. Government Printing Office. Washington, D.C.

Moffitt, F. H., & Jordan, S. 1987. Conversion of state plane coordinates from NAD 27 to NAD 83 by second degree conformal transformation. *Bulletin 106 Amer. Cong. on Surv. & Mapp.* p. 13.

Pace, G. M. 1984. An ordinary surveyor looks at the state plane coordinate system. *Surveying and Mapping* 44:23.

Read, J. R. 1981. A coordinate system for North America based upon the 6° U.T.M. Zone. *Surveying and Mapping* 41:23.

Stem, J. E. 1984. Status of legislative action to enact the "State Plane Coordinate System of 1983."*ACSM Bull. 86, Amer. Cong. Surv. & Mapp.* p. 33.

——. 1990. *State Plane Coordinate System of 1983.* NOAA Manual NOS NGS 5, U.S. Department of Commerce. Rockville, MD.

Stephenson, R. B. 1978. Using your state coordinate system. *Proc. Fall Conv. Amer. Cong. Surv. & Mapp.* p. 378.

Stoughton, H. W. 1984. Analysis of various algorithms to compute scale factors on the Lambert conformal conic projection. *Tech. Papers Fall Conv. Amer. Cong. Surv. & Mapp.* p. 136.

———. (no date). The North American Datum of 1983. Amer. Assn. for Geodetic Surveying Monograph No. 2. *Amer. Cong. Surv. & Mapp.*

Vincenty, T. 1984. Proposal for unified state plane coordinate system. *ACSM Bull. No. 92, Amer. Cong. Surv. & Mapp.* p. 17.

———. 1985. Precise determination of the scale factor for Lambert conical projection coordinates. *Surveying and Mapping* 45:315.

———. 1986. Use of polynomial coefficients in conversions of coordinates on the Lambert conformal conic projection. *Surveying and Mapping* 46:15.

Wade, E. B. 1987. Datum transformation from NAD 27 to NAD 83. *Tech. Papers Ann. Conv. Amer. Cong. Surv. & Mapp.* p. 27.

12 Practical Astronomy

12-1 Uses of Astronomy

The engineer or surveyor needs a knowledge of practical astronomy to make the observations and computations necessary for the determination of latitude, time, longitude, and azimuth. This chapter presents, in condensed form, methods of observation and computation that can be used when a survey is being conducted with an ordinary engineer's transit or a theodolite. For the development of the formulas used and for more exact methods, reference can be made to the readings at the end of the chapter.

The various heavenly bodies are considered to be located on the surface of a sphere of infinite radius. The center of this sphere is the center of the earth. The field observations usually consist of the measurement of horizontal and vertical angles to some heavenly body and the time of observation. The computations involve the solution of the spherical triangle formed by the observer's zenith, the celestial north pole, and the heavenly body.

Where only moderate precision is required, the field observations can be made either with a sextant or with a transit that is equipped preferably with a full vertical circle and a prismatic eyepiece. A good watch or a time module is needed for determining the instant at which the observation is made. An ephemeris is required for the computations. The *Astronomical Ephemeris,* which is printed each year by the U.S. Government Printing Office, contains tables that furnish the positions of the principal heavenly bodies. The *Almanac for Computers* is published by the U.S. Naval Observatory for use by those who wish to generate the positions of the heavenly bodies by computers. The Sokkia Company publishes the *Sokkia Celestial Observation Handbook and Ephemeris,* which is indispensable to the surveyor who contemplates making astronomical observations. It contains several examples of the use of the tables for a variety of applications of astronomical observations. The *Sokkia Ephemeris* is authored by Drs. Richard L. Elgin, David R. Knowles, and Joseph E. Senne of the firm Elgin, Knowles & Senne, Inc., of Fayetteville, AR. Table 12-1 is the ephemeris data for the sun and Polaris (the pole star) for the month of May 1996 as it appears in the *Sokkia Ephemeris* for 1996. This table should be referred to when studying many of the examples given in this chapter.

TABLE 12-1   Ephemeris of the Sun and Polaris for May 1996 0^h Universal Time

Day	GHA (Sun)	Declination	Equation of Time APPT—Mean	Semi-Diam.	GHA (Polaris)	Declination	Greenwich Transit
	° ′ ″	° ′ ″	M S	′ ″	° ′ ″	° ′ ″	H M S
1 W	180 43 16.3	15 05 46.5	02 53.09	15 53.9	182 29 43.2	89 14 46.26	11 48 05
2 TH	180 45 02.8	15 23 46.6	03 00.18	15 53.7	183 28 46.8	89 14 45.92	11 44 09
3 F	180 46 41.2	15 41 31.4	03 06.75	15 53.4	184 27 48.7	89 14 45.58	11 40 14
4 SA	180 48 11.4	15 59 00.6	03 12.76	15 53.2	185 26 49.0	89 14 45.25	11 36 18
5 SU	180 49 33.3	16 16 13.9	03 18.22	15 52.9	186 25 48.3	89 14 44.95	11 32 23
6 M	180 50 46.6	16 33 11.0	03 23.11	15 52.7	187 24 47.3	89 14 44.66	11 28 28
7 TU	180 51 51.4	16 49 51.7	03 27.42	15 52.5	188 23 46.8	89 14 44.39	11 24 32
8 W	180 52 47.3	17 06 15.6	03 31.16	15 52.3	189 22 47.3	89 14 44.14	11 20 37
9 TH	180 53 34.5	17 22 22.4	03 34.30	15 52.0	190 21 48.9	89 14 43.88	11 16 42
10 F	180 54 12.8	17 38 11.8	03 36.85	15 51.8	191 20 51.4	89 14 43.62	11 12 46
11 SA	180 54 42.2	17 53 43.6	03 38.81	15 51.6	192 19 54.1	89 14 43.34	11 08 51
12 SU	180 55 02.7	18 08 57.4	03 40.18	15 51.4	193 18 56.4	89 14 43.04	11 04 55
13 M	180 55 14.3	18 23 52.9	03 40.95	15 51.1	194 17 57.5	89 14 42.73	11 01 00
14 TU	180 55 17.0	18 38 29.9	03 41.14	15 50.9	195 16 57.1	89 14 42.41	10 57 04
15 W	180 55 11.1	18 52 48.0	03 40.74	15 50.7	196 15 55.1	89 14 42.09	10 53 09
16 TH	180 54 56.4	19 06 47.0	03 39.76	15 50.5	197 14 51.5	89 14 41.77	10 49 14
17 F	180 54 33.3	19 20 26.6	03 38.22	15 50.3	198 13 46.7	89 14 41.47	10 45 19
18 SA	180 54 01.7	19 33 46.5	03 36.12	15 50.1	199 12 41.3	89 14 41.19	10 41 24
19 SU	180 53 21.9	19 46 46.5	03 33.46	15 49.9	200 11 35.6	89 14 40.92	10 37 29
20 M	180 52 34.0	19 59 26.3	03 30.26	15 49.7	201 10 30.1	89 14 40.66	10 33 34
21 TU	180 51 38.1	20 11 45.6	03 26.54	15 49.6	202 09 25.1	89 14 40.42	10 29 39
22 W	180 50 34.3	20 23 44.2	03 22.29	15 49.4	203 08 20.6	89 14 40.18	10 25 44
23 TH	180 49 23.0	20 35 21.8	03 17.53	15 49.2	204 07 16.7	89 14 39.95	10 21 49
24 F	180 48 04.1	20 46 38.2	03 12.28	15 49.0	205 06 13.1	89 14 39.71	10 17 54
25 SA	180 46 38.0	20 57 33.0	03 06.53	15 48.9	206 05 09.6	89 14 39.46	10 13 58
26 SU	180 45 04.7	21 08 06.2	03 00.32	15 48.7	207 04 06.0	89 14 39.21	10 10 03
27 M	180 43 24.5	21 18 17.5	02 53.64	15 48.6	208 03 01.7	89 14 38.95	10 06 08
28 TU	180 41 37.6	21 28 06.6	02 46.50	15 48.4	209 01 56.3	89 14 38.68	10 02 13
29 W	180 39 44.0	21 37 33.4	02 38.93	15 48.3	210 00 49.6	89 14 38.41	9 58 18
30 TH	180 37 44.1	21 46 37.7	02 30.94	15 48.1	210 59 41.2	89 14 38.14	9 54 24
31 F	180 35 37.9	21 55 19.2	02 22.53	15 48.0	211 58 31.1	89 14 37.88	9 50 29

Although the surveyor is ordinarily interested primarily in the azimuth of a certain line, most azimuth methods in use require the surveyor to know the instant at which the observation is made and the latitude. Certain observations require precise timing (see Sections 12-15 and 12-17), while others require the time accurate to a few minutes (see Sections 12-14 and 12-16). Latitudes are obtained by scaling from reliable maps or else by conversion from grid coordinates to geodetic coordinates, as discussed in Chapter 11.

12-2   Definitions

Figure 12-1 shows the *astronomical* or *celestial triangle PZS* and the principal points and planes on the celestial sphere. The *celestial poles P* and *P′* are the points of intersection, with the celestial sphere, of the axis of rotation of the earth, if that axis were extended indefinitely. The *celestial equator E′S″WVE* is the

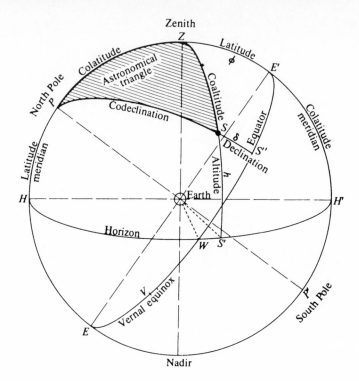

FIGURE 12-1 Celestial sphere.

intersection of the celestial sphere with the plane passing through the center of the earth and perpendicular to the axis of rotation.

If a vertical line at any point on the earth's surface is extended, it intersects the upper portion of the celestial sphere at the *zenith Z* and the lower portion of the sphere at the *nadir*. A *vertical circle* is a great circle passing through the zenith and the nadir; ZSS' is an arc of a vertical circle. The *meridian plane* $HPZE'H'P'E$ contains a vertical circle passing through the celestial poles, the *meridian* being the intersection of this plane with the celestial sphere. An *hour circle* is a great circle passing through the north and south celestial poles; PSS'' is an arc of an hour circle.

The *horizon* is the intersection of the celestial sphere with the plane that passes through the center of the earth and is perpendicular to a vertical line; $HWS'H'$ is half of the horizon.

The *azimuth* of a heavenly body S is the angular distance $HH'S'$ measured clockwise along the horizon from the north point of the meridian to the vertical circle through the heavenly body. The azimuth is considered positive when measured clockwise. In Fig.12-1 the angle Z is shown to be 360° minus the azimuth of the body.

The *altitude h* of a heavenly body S is the angular distance $S'S$, measured upward from the horizon along a vertical circle. The *coaltitude*, or *zenith dis-*

tance, ZS equals 90° minus the altitude. The altitude corresponds to the vertical angle to the body. The zenith distance corresponds to the zenith angle to the body (see Section 4-19).

The *latitude* φ of a point on the earth's surface is the angular distance $E'Z$ measured along the meridian from the equator to the zenith of that point. It is numerically equal to the altitude HP of the pole measured at the point. Latitudes north of the equator are considered positive and those south of the equator, negative. The *colatitude* PZ equals 90° minus the latitude.

The *vernal equinox V* is that intersection of the plane of the earth's orbit with the equator at which the sun appears to cross the equator from south to north. This occurs about March 21. The *right ascension* of a heavenly body S is the angular distance VS'' measured eastward along the equator from the vernal equinox to the hour circle through the body. The *declination* δ of a heavenly body S is the angular distance $S''S$ measured from the equator to the body along the hour circle through the body. If the body is north of the equator, the declination is considered positive; if south, negative. The *codeclination,* or *polar distance, PS* equals 90° minus the declination.

The *hour angle t* of a heavenly body S is the angular distance $E'S''$ measured westward along the equator from the meridian to the hour circle through the heavenly body. Hour angles measured to the west are positive and to the east, negative. Three hour angles are of particular importance. The *local hour angle* (LHA) is the hour angle of a heavenly body with respect to a local meridian, as shown in Fig. 12-1.

The LHA is designated t and is the angle at P in the celestial triangle. The Greenwich hour angle (GHA) is the hour angle with respect to the meridian of Greenwich. The *sidereal hour angle* (SHA) of a heavenly body is its angular distance, measured westward along the equator, from the vernal equinox to the hour circle through the heavenly body.

If the equatorial plane of Fig. 12-1 is viewed along the line PP', the hour circles become lines radiating outward from P, as shown in Fig. 12-2. The line PG is the hour circle passing through Greenwich Observatory. The line PL is the local meridian. The longitude of the local meridian is thus the angle GPL or arc GL. The line PSS'' is the hour circle passing through the heavenly body S and corresponds to the hour circle PSS'' of Fig. 12-1. Point V is the position of the vernal equinox which lies on the equator. The local hour angle of the heavenly body is $\angle LPS''$ or arc LS''; the Greenwich hour angle of the body is $\angle GPS''$ or arc GLS''; the SHA is the arc $VGLS''$ measured westward; and the right ascension of the body is the arc VS'' measured eastward. Thus it is seen that the SHA of a body is equal to 360° minus the right ascension of the body.

The *transit* of a heavenly body is its passage over the observer's meridian. The *upper branch* of the meridian is the half of the meridian, from the north celestial pole to the south celestial pole, that contains the observer's zenith, as arc $PZE'H'P'$ in Fig. 12-1. The *lower branch* of the meridian is the half containing the nadir. An *upper transit* of a body occurs at the instant at which the body is on the upper branch of the meridian. A *lower transit* occurs at the instant at which the body is on the lower branch of the meridian.

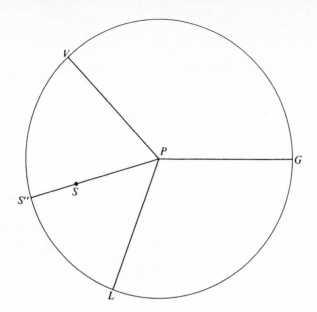

FIGURE 12-2 Horizontal angles and right ascension.

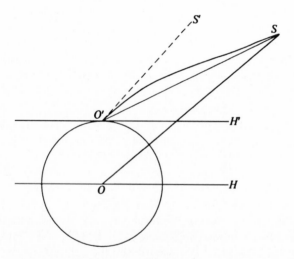

FIGURE 12-3 Refraction and parallax.

The *north meridian* is the arc of the meridian from the observer's zenith to the nadir and containing the north celestial pole. This is the arc *ZPHE-nadir* in Fig. 12-1. The *south meridian* contains the south celestial pole.

The *refraction* of a body is the angular increase in its apparent altitude due to the refraction of light. In Fig. 12-3 the angle $S'O'H'$ is the measured altitude, and angle $S'O'S$ is the refraction. The *correction* for refraction C_r is always negative, as seen from the figure. It is a maximum when the altitude is small and decreases to zero when the altitude is 90°. Values of the correction for refraction in degrees can be obtained by the following equation:

$$C_r = \frac{0.27306b}{(460 + F)\tan h} \tag{12-1a}$$

in which b is the barometric pressure in inches of mercury, F is the temperature in °F, and h is the measured altitude (vertical angle) to the sun or star. The quantity b can be obtained from Fig. 2-23 if the elevation of the point is known or it can be computed by the following formula:

$$b = \text{inverse log} \frac{92,670 - \text{elevation}}{62,737}$$

in which the elevation is in feet.

Owing to the uncertainty in the correction when the altitude is small, observations on heavenly bodies within 20° of the horizon should be avoided if possible when the measured altitude is required (see Section 12-16). However, if the altitude need not be measured, as when observing the sun for azimuth using the hour angle method described in Section 12-17, the closer the sun is to the horizon, the better.

EXAMPLE 12-1   The observed altitude to a star is 33° 15′ 00″ when the temperature is 72°F and the barometer reads 28.0 in. What is the refraction correction?

Solution: By Eq. (12-1a),

$$C_r = \frac{0.27306 \times 28.0}{(460 + 72)\tan 33°\ 15'\ 00''} = -0°\ 01'\ 19''$$

◆

EXAMPLE 12-2   The observed altitude to the sum's limb is 27° 30′ when the temperature is 48°F, the elevation of the point of observation is 4850 ft, and the barometer reading is unknown. What is the refraction correction?

Solution: The barometric pressure obtained from Fig. 2-23 for an elevation of 4850 ft is 25.0 in. Then, by Eq. (12-1a),

$$C_r = \frac{0.27306 \times 25.0}{(460 + 48)\tan 27°\ 30'} = -0°\ 01'\ 33''$$

The *correction for parallax* (in altitude) is the small angular increase in the apparent altitude of a heavenly body to allow for the fact that the observations are made at the surface of the earth instead of at its center. In Fig. 12-3 the angle $SO'H'$ is the angle (altitude) at the surface corrected for refraction. The angle SOH is the true altitude, or the altitude that would be measured at the center of the earth. Observations on the sun, the moon, and the planets must be corrected for parallax. Observations on the stars need no parallax correction because any star is virtually at an infinite distance from the earth. Parallax correction is always added. Parallax of the sun in seconds is computed by the following formula:

$$C_p = 8.794 \cos h \tag{12-1b}$$

◆

12-3   Azimuth Angle from Spherical Trigonometry

Formulas expressing the azimuth angle at Z in the PZS spherical triangle are derived from the formulas found in spherical trigonometry. The three spherical angles are $P = t$, Z, and S. The three spherical sides opposite the angles are, respectively, $90° - h$, $90° - \delta$, and $90° - \varphi$.

By the law of sines in a spherical triangle whose angles are A, B, and C, and whose opposite sides are a, b, and c,

$$\frac{\sin A}{\sin a} = \frac{\sin B}{\sin b} = \frac{\sin C}{\sin c} \tag{12-2}$$

By the law of cosines,

$$\cos a = \cos b \cos c + \sin b \sin c \cos A \tag{12-3a}$$

$$\cos b = \cos a \cos c + \sin a \sin c \cos B \tag{12-3b}$$

Rearranging Eq. (12-3b) gives

$$\sin a \sin c \cos B = \cos b - \cos a \cos c \tag{12-3c}$$

and substituting the value of $\cos a$ from Eq. (12-3a) into Eq. (12-3c) gives

$$\sin a \sin c \cos B = \cos b - \cos b \cos^2 c - \sin b \sin c \cos c \cos A$$

but $\cos^2 c = 1 - \sin^2 c$, and dividing by $\sin c$ gives

$$\sin a \cos B = \cos b \sin c - \sin b \cos c \cos A \tag{12-4}$$

By Eq. (12-2),

$$\frac{\sin Z}{\sin (90° - \delta)} = \frac{\sin t}{\sin (90° - h)}$$

or

$$\sin Z = \frac{\sin t \cos \delta}{\cos h} \tag{12-5}$$

By Eq. (12-3a),

$$\cos Z = \frac{\sin \delta - \sin h \sin \varphi}{\cos h \cos \varphi} \tag{12-6}$$

From Eq. (12-5),

$$\sin Z \cos h = \sin t \cos \delta \tag{12-7}$$

By Eq. (12-4),

$$\cos h \cos Z = \sin \delta \cos \varphi - \cos \delta \sin \varphi \cos t \tag{12-8}$$

Dividing Eq. (12-7) by Eq. (12-8) gives

$$\tan Z = \frac{\sin t \cos \delta}{\sin \delta \cos \varphi - \cos \delta \sin \varphi \cos t} \tag{12-9}$$

Dividing both numerator and denominator of Eq. (12-9) by cos δ gives

$$\tan Z = \frac{\sin t}{\tan \delta \cos \varphi - \sin \varphi \cos t} \qquad (12\text{-}10)$$

Eq. (12-5), (12-6), and (12-10) and their variations are used to compute the azimuth of the sun or a star.

The altitude h of a star can be determined by Eq. (12-3a):

$$\sin h = \sin \varphi \sin \delta + \cos \varphi \cos \delta \cos t \qquad (12\text{-}11)$$

The above equations or their variation are repeated in the appropriate sections for convenience of the student.

12-4 Time

There are three kinds of time that may be involved in astronomical observations, namely, *sidereal time, apparent time* or *true solar time,* and *mean solar time.* A *sidereal day* is the interval of time between two successive upper transits of the vernal equinox over the same meridian. The sidereal day therefore begins at the instant at which the vernal equinox is on the upper branch of the meridian. The sidereal time at any instant, referred to a particular meridian, is the hour angle of the vernal equinox referred to the same meridian. The sidereal time for the local meridian shown in Fig. 12-2 is the arc *LV.*

An *apparent solar day* is the interval of time between two successive lower transits of the sun's center over the same meridian. The mean sun is a fictitious body which travels at a uniform rate along the equator. Because of the obliquity of the plane of the earth's orbit (which is about 23° 26.5′) and because of the variation in the earth's angular velocity about the sun in its elliptical orbit, the interval between two successive lower transits of the sun's center over the same meridian varies from day to day. Since it would not be feasible to keep time on clocks and watches if the apparent solar day were used, the mean solar day is used.

A *mean solar day* is the interval of time between two successive lower transits of the mean sun over the same meridian. The mean sun is a fictitious body which travels at a uniform rate along the equator. The mean sun is sometimes ahead of and sometimes behind the apparent, or true, sun. There are 24 hr in a day. The mean solar time at any instant for a given meridian is the hour angle of the mean sun referred to the same meridian plus 12 hr. At midnight, the hour angle of the mean sun is 12 hr. At mean solar noon, the hour angle of the mean sun is 0 hr.

The difference between apparent solar time and mean solar time is called the *equation of time,* or apparent – mean = equation of time. Its value for each day of the year is given in the ephemeris.

A *solar year* contains 365.2422 solar days. Because the earth rotates on its axis each day and orbits around the sun once a year, any point on the earth faces the sun one time less than it faces any of the stars and the vernal equinox in a period of one year. Therefore the solar year contains 366.2422 sidereal days. The mean solar day is longer than the sidereal day by $3^{\mathrm{m}}\,55.909^{\mathrm{s}}$ of mean solar time or $3^{\mathrm{m}}\,56.555^{\mathrm{s}}$ of sidereal time.

A solar interval of time can be converted to a corresponding sidereal interval by adding 9.856^S for each hour in the solar interval. A sidereal interval of time can be converted to a corresponding solar interval by subtracting 9.830^S for each hour in the sidereal interval.

12-5   Standard Time

Standard time is the mean solar time referred to meridians 15° apart, measured from Greenwich, which lies at 0° longitude. Since there are 24 hr in the mean solar day, then $24^h = 360°$ and $1^h = 15°$. Therefore the standard meridians are 1 hr apart. Eastern standard time (EST) is the time of the 75th meridian west of Greenwich; Central standard time (CST), of the 90th meridian; Mountain standard time (MST), of the 105th meridian; and Pacific standard time (PST), of the 120th meridian.

Conversion from degrees, minutes, and seconds to hours, minutes, and seconds, or the reverse, is based on the following relationships:

$$
\begin{array}{rclcrcl}
360° & = & 24^h & \qquad & 24^h & = & 360° \\
1° & = & 4^m & & 1^h & = & 15° \\
1' & = & 4^s & & 1^m & = & 15' \\
1'' & = & 0.067^s & & 1^s & = & 15''
\end{array}
$$

In Fig. 12-4 the mean sun, which travels westward along the equator, is shown to about 10° 30′ or 42^m beyond the 90th meridian. Central standard time is therefore 12:42 P.M. At the Greenwich meridian at the same instant, the time is 6^h later. So the Greenwich standard time is 6:42 P.M. The standard time referred to the Greenwich meridian is called Greenwich civil time (GCT), Greenwich mean time (GMT), or universal time (UT). The mean sun has not yet reached the 120th meridian, and the Pacific standard time is therefore before noon. Since the difference between CST and PST is 2^h, the Pacific standard time, or time based on the 120th meridian, is 10^h42^m A.M.

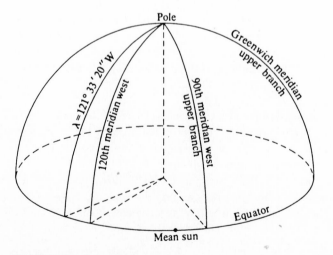

FIGURE 12-4 Time. (mean).

12-6 Local Time

Local time is the time based on the observer's meridian. *Local civil time* (LCT), which is based on mean solar time, is the hour angle of the mean sun measured from the local meridian plus 12 hr. *Local apparent time* (LAT), which is based on the true sun, is the hour angle of the sun's center plus 12 hr. For example, when the actual sun's center is on the observer's meridian, the time is local apparent noon. *Local sidereal time* is the hour angle of the vernal equinox measured from the observer's meridian.

At the meridian in longitude 121° 33′ 20″, standard time is based on the 120th meridian. For the instant shown in Fig. 12-4, the time is 10:42 A.M., PST. Local civil time is earlier by an amount corresponding to 1° 33′ 20″, or 6^m 13.3^s. The local civil time in longitude 121° 33′ 20″ is therefore 10:35:46.7 A.M. The local civil time at longitude 72° 03′ 14″ for the instant shown in Fig. 12-4 is found by adding to the Central standard time the amount corresponding to 90° − 72° 03′ 14″. The time interval for 90° − 72° 03′ 14″, or 17° 56′ 46″, is 1^h 11^m 47.1^s, and the LCT is 1^h 53^m 47.1^s P.M.

12-7 Time by Radio

If a shortwave receiver is available, time may be determined several times daily by picking up the signals sent out by the U.S. National Institute of Standards and Technology radio station WWV on 2.5, 5, 10, 15, and 20 MHz. These time signals are based on time kept by atomic clocks, and the time is called Coordinated Universal Time (UTC). There is a slight difference, up to about 0.7^s, between UTC and UT1, the latter based on travel of the mean sun and used to reduce astronomical observations. Thus, if the time signals are to be used for the precise timing of an astronomical event, they must be corrected to UT1 by a correction referred to as the DUT correction. Then UT1 = UTC + DUT. The value of the DUT correction is superimposed on the Bureau of Standards radio time signal. [†]In Canada, UTC is broadcast on radio station CHU at 3.33, 7.335, and 14.67 MHz.

12-8 Observing the Sun with the Theodolite

Observations on the sun are made to determine the altitude or the azimuth of the sun's center or both. Altitude or zenith distance is read on the vertical circle of a theodolite, whereas azimuth is related to horizontal-circle readings. Because it is quite difficult to set the intersection of the cross hairs directly on the center of the sun, the cross hairs are usually made to become tangent to the edges of the sun, and then the necessary amounts are added to or subtracted from the circle readings to reduce the observation to the sun's center.

Figure 12-5 shows the design of a glass reticule that can be used to great advantage when a survey requires many solar observations. The usual horizontal

[†]An excellent discussion of the conversion between UTC and UT1 for astronomical observations and the use of the HP-41CX Time Module is given in an article by Professor David R. Knowles of the University of Arkansas. The article appears in Volume 11, No. 2 December 1985–January 1986 of *Point of Beginning* (P.O.B.) published by Business News Publishing Co., 755 W. Big Beaver Road, Ste. 1000, Troy, MI 48084.

FIGURE 12-5 Reticle for sighting sun's center.

and vertical cross hairs and the stadia hairs are augmented by a pair of vertical tick marks having the same spacing as the stadia hairs. Inside the stadia hairs and these two tick marks is inscribed a solar circle with an angular radius of 15′ 45″. This is slightly smaller than the *semidiameter* of the sun, which is approximately 0° 16′. The sun's image can thus be centered on this circle with a very high degree of accuracy.

If a total station instrument of the type shown in Fig. 4-34 is used for solar observations, an objective lens filter must be used to protect the electronic components of the EDM. If the EDM is separately mounted, as shown in Fig. 2-15, and if it is not detached for solar observations, its objective lens must be capped to protect its components.

Although the sun can be observed directly through the telescope if the appropriate kind of dark filter is attached to the eyepiece, the usual practice is to view the image of the sun projected onto a white card held behind the eyepiece of the theodolite, as shown in Fig. 12-6. Two methods will be discussed in this section, each using the white-card technique. The first method is referred to as the center-tangent or disappearing-segment method. The second is known as the quadrant-tangent method.

When the center-tangent method is used, the observation is always made on the trailing edge, or *limb,* of the sun. During the morning hours, the sun's altitude is increasing, and the lower limb is observed for altitude (vertical-circle reading). During the afternoon hours, the sun's altitude is decreasing, and the upper limb is observed for altitude. Throughout all hours of the day, the sun moves from east to west, and the east or trailing limb is observed for azimuth (horizontal-circle reading).

When the theodolite is to be leveled over a point for an astronomical observation to measure the altitude or zenith distance of a heavenly body, the bubble of the telescope level must be used as described in Section 4-18. This procedure brings the vertical axis of the instrument more perfectly vertical than would be possible by using the plate levels. Vertical circle compensators eliminate this problem.

With the theodolite set up and leveled, the telescope is pointed in the direction of the sun, and the telescope bubble is centered. The vertical circle is read to determine the index error (see Section 4-19). If the theodolite is equipped with a pendulum compensator, there will be no index error. The telescope is raised and

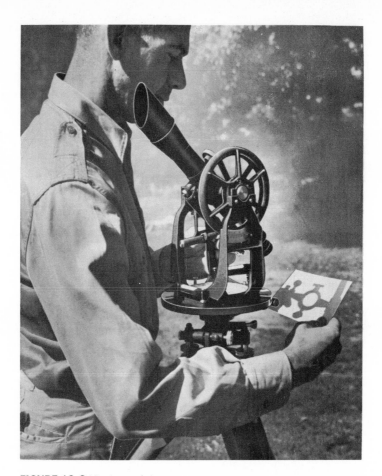

FIGURE 12-6 Viewing sun's image.

pointed as nearly as possible toward the sun without actually looking through the telescope. A white card is held from 3 to 4 in. behind the eyepiece, and the telescope is rotated both in altitude and azimuth until it casts a circular shadow on the card (see Fig. 12-6). The cross hairs are then focused on the card. The telescope is now focused to make the image of the sun clear and sharp on the card. The image of the sun produced by an erecting telescope will be upside down when the observer looks directly at the card. Therefore the lower edge of the sun's image is actually the upper limb of the real sun, and the upper edge of the image is the lower limb of the real sun. The right edge of the sun's image, as seen on the card, is the western or leading limb of the real sun; the left edge of the image is the eastern or trailing limb of the real sun.

In the morning, when it is desired to make the horizontal cross hair tangent to the lower limb of the sun, the vertical-motion tangent screw is turned so that the horizontal cross hair cuts the sun's image in the position indicated by the dashed lines in Fig. 12-7(a). The altitude of this cross hair now remains stationary. As the sun's image moves downward and to the right, as indicated by the arrow, the small

FIGURE 12-7 Image of sun as seen on card held behind eyepiece of erecting telescope.

segment becomes smaller. The observer moves the vertical cross hair by means of the upper horizontal-motion tangent screw so that this cross hair continuously bisects the disappearing segment. He stops at the instant the edge of the sun's image becomes tangent to the horizontal cross hair. At this point the vertical cross hair passes through the center of the sun and the horizontal-circle reading needs no correction. However, the horizontal cross hair is tangent to the lower limb, and the vertical angle must be increased by one-half the angle subtended by the sun. This is the sun's semidiameter and is about $0° 16'$. Its value is given for different times of the year in the ephemeris. The vertical angle is corrected for index error, parallax, and refraction before the semidiameter is added. The observation is repeated with the telescope reversed, as explained in Section 12-16.

In the afternoon, when it is desired to make the horizontal cross hair tangent to the sun's upper limb, the horizontal cross hair is first set as shown in Fig. 12-7(c). Also, the semidiameter must be subtracted from the vertical-circle reading after corrections for index error, parallax, and refraction have been applied.

If it is desired to make the vertical cross hair tangent to the trailing limb of the sun, either in the morning or in the afternoon, the vertical cross hair should be set to cut the sun's image in the position indicated by the dashed lines in Fig. 12-7(b) or (d). The horizontal cross hair is then moved so as to bisect the disappearing segment continuously until tangency is reached. At this point the horizontal cross hair passes through the center of the sun, and the vertical-circle reading need be corrected only for index error, parallax, and refraction. The clockwise horizontal-circle reading, however, is too small (too large if the sun is north of the observer) by an amount found by multiplying the semidiameter by sec h, where h is the altitude or vertical angle. This amount is added to (or subtracted from) the horizontal-circle reading.

In Fig. 12-8, Z is the observer's zenith, P is the pole, arc ZPM is part of the observer's meridian, S is the sun's center, and angle MOS' is the azimuth of the sun's center. The arc ZEE' is part of the vertical circle tangent to the sun's eastern limb. This arc lies in the vertical plane containing the line of sight of the theodolite when it is brought tangent to the eastern limb. Thus the azimuth of the sun's eastern limb is the angle MOE', which is smaller than the angle MOS' by the small angle e. In the triangle ZSE the angle at Z is e, the angle at E is $90°$, side ZS is the

FIGURE 12-8 Correction to horizontal-circle reading for sun's semidiameter.

coaltitude of the sun, or $90° - h$, and side SE is the sun's semidiameter d. By the law of sines of spherical trigonometry for the triangle ZSE,

$$\frac{\sin Z}{\sin SE} = \frac{\sin E}{\sin ZS} \quad \text{or} \quad \frac{\sin e}{\sin d} = \frac{\sin 90°}{\sin(90° - h)} = \frac{1}{\cos h} = \sec h$$

Since the semidiameter d is small, it may be assumed that

$$e = d \sec h \qquad (12\text{-}12)$$

in which e is the amount by which the horizontal-circle reading to the sun's eastern limb must be increased. Both d and e are expressed in minutes of arc.

EXAMPLE 12-3   On the morning of May 13, 1996, the horizontal cross hair of a theodolite was brought tangent to the lower limb of the sun, as indicated in Fig. 12-7(a), and the vertical circle read + 40° 22′. The vertical circle read + 0° 02′ when the telescope bubble was centered. If the refraction is 1′ 09″ (see Example 12-1) what was the true altitude of the sun's center?

Solution: The computations may be arranged as follows:

vertical-circle reading	= + 40° 22′
index correction	= − 0° 02′
	+ 40° 20′
refraction	= − 1′ 09″
	40° 18′ 51″
sun's parallax	= + 7″ [from Eq. (12-1b)]
	40° 18′ 58″
semidiameter for May 13	= + 15′ 51″ (from ephemeris)
true altitude of sun's center =	40° 34′ 49″

EXAMPLE 12-4   On the afternoon of May 18, 1996, the vertical cross hair of a theodolite was brought tangent to the eastern limb of the sun, as indicated in Fig. 12-7(d). The clockwise horizontal circle read 322° 44′ 30″ and the vertical circle read + 27° 52′. The index error of the vertical circle was determined to be zero. What was the corrected horizontal-circle reading, and what was the altitude of the sun's center?

Solution: The calculations follow:

horizontal-circle reading	=	322° 44′ 30″
semidiameter × sec h	= +	17′ 54″ (semidiameter from ephemeris)
corrected horizontal- circle reading	=	323° 02′ 24″
vertical-circle reading	= +	27° 52′
refraction	= −	1′ 48″ (previously computed)
		+ 27° 50′ 12″
parallax	= +	8″ [from Eq. (12-1b)]
true altitude of sun	= +	27° 50′ 20″

◆

When the position of the sun is observed in the manner just described, several observations should be taken in rapid succession, and the mean values of the corrected horizontal-circle readings and the vertical angles should be used. Each of the observed values should be corrected for index error, refraction and parallax, semidiameter, and d sec h. The watch time of each observation should be noted to the nearest second, whether or not the times are exactly correct. Then, after the several horizontal-circle readings and vertical angles have been corrected, both the values of the corrected horizontal-circle readings and the values of the corrected vertical angles should be plotted versus the time. Each plot should be very nearly a straight line. Values that are not close to the adopted straight line should be rejected before the mean values for the group of observations are computed (see Section 12-16).

When the sun is observed by the quadrant-tangent method, the white card is used in the manner previously described. Figure 12-9 shows the motion of the sun's image as it would be projected through an erecting telescope onto the card. In the morning the first part of the observation is made as indicated in Fig. 12-9(a). The vertical cross hair is always kept tangent to the sun's western, or leading, limb by means of the upper horizontal tangent screw. The vertical motion of the theodolite is not disturbed, and thus the horizontal cross hair remains at the same altitude. At the instant at which the sun's lower limb (the upper edge of the sun's image) becomes tangent to the horizontal cross hair, as shown by the solid circle, horizontal motion of the theodolite is stopped. At the same instant, the watch time is noted.

FIGURE 12-9 Image of sun as seen on card held behind eyepiece of erecting telescope (quadrant-tangent method).

Next, the vertical cross hair is held at the same azimuth, as shown in Fig. 12-9(b), while the horizontal cross hair is kept tangent to the sun's upper limb (the lower edge of the sun's image) by means of the vertical slow-motion screw. At the instant at which the eastern limb becomes tangent to the vertical cross hair, as shown by the solid circle, the theodolite motion is stopped and the watch time is noted.

In the afternoon the sun's upper limb is first made tangent, as indicated in Fig. 12-9(c), and then the eastern limb is made tangent as shown in Fig. 12-9(d).

When the quadrant-tangent method is used, the observer must watch for tangency at two points of the sun's disk, instead of at only one point in the center-tangent method. It is therefore more difficult to make the observation. Presumably, a pair of observations, one carried out as in Fig. 12-9(a) and the other as in (b), could be made in quick succession, and the readings could be averaged to eliminate the semidiameter corrections to both the horizontal-circle and vertical-circle readings. Moreover, if another pair of observations is made with the telescope reversed, the index error of the vertical circle could also be eliminated. However, a set of solar observations should never be used for computation until all the corrections have been made to the circle readings and the resulting angles have been plotted as a function of time. This plot uncovers any possible mistakes that would otherwise escape notice if all observations are presumed valid and the results are averaged. In each instance shown in Fig. 12-9, the corrections would consist of index correction, refraction and parallax, semidiameter, and $d \sec h$. Section 12-16 includes a discussion of mistakes in observations and the ways in which defective observed values may be rejected on a valid basis.

An ingenious device, which is known as *Roelof's solar prism* and which can be fitted to the objective of the telescope, produces four overlapping images of the sun to give the solar-disk pattern shown in Fig. 12-10. By direct viewing through the theodolite the small diamond-shaped area in the center of the pattern can be bisected simultaneously by the horizontal and vertical cross hairs with a high degree of accuracy. Because of the pattern symmetry, such a pointing is made directly on the sun's center, and the semidiameter corrections are thus eliminated. Because of the direct-viewing feature, the observations can be made quite rapidly. Of course, corrections for index error, refraction, and parallax must be applied to the vertical-circle reading. There is, however, no correction to the horizontal-circle reading.

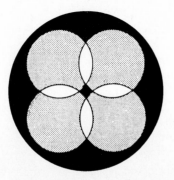

FIGURE 12-10 Pattern of sun's image produced by Roelof's solar prism.

12-9   Observing a Star with the Theodolite

Unlike the sun, a star appears as a point of light in the heavens when viewed through a theodolite telescope. It is therefore quite easy to observe the instant at which the star's image crosses either the horizontal cross hair when the star's altitude is being measured or the vertical cross hair when a horizontal angle to the star is being measured. Because observations are taken on stars under such conditions that the amount of natural light coming through the telescope usually is not sufficient to allow the observer to see the cross hairs, artificial light must be provided. A good method is to shine a flashlight obliquely across the objective lens at such an angle that there is sufficient light to make the cross hairs visible, but not enough to blot out the image of the star. Some instruments have internal illumination, the brightness of which can be controlled by a rheostat.

If a star is to be observed for determining the azimuth of a line or a backsight has to be made on a ground station during the observation for any other reason, then the ground station itself must be illuminated. Suitable light can usually be provided by holding a flashlight behind the mark when the station is being observed.

Preparations for star observations should be made in daylight or during twilight hours, if possible, simply for convenience. Some stars may actually be observed in the twilight hours. This condition is, of course, most desirable, because the observer and the notekeeper then have at least some natural light in which to work.

12-10   Latitude

The latitude of a point on the earth's surface can be obtained by scaling it from a map, by measuring the altitude of the sun at local apparent noon, or by measuring the altitude of a circumpolar star (see Section 12-12). Where the latitude is to be used for the computation of azimuth, it is generally accurate enough to scale the latitude from a good map. The final calculated results will probably be as good as can be obtained with the engineer's transit, for they will not be seriously affected by a slight error in the latitude used in the computations.

12-11   Latitude by Sun's Altitude at Noon

Where a meridian has been determined, the latitude can be obtained by measuring the altitude of the sun's lower or upper limb at the instant the sun crosses the meridian. Where the meridian is unknown, this altitude may be measured when the sun reaches the highest point in its path. The observed angle is corrected for the index error of the vertical circle, for refraction and parallax, and for the semidiameter of the sun. The sun's declination at the instant of apparent noon and the semidiameter of the sun can be obtained from the ephemeris. From Fig. 12-11, which is a section taken in the plane of the meridian, it is evident that the latitude, which is measured by the arc ZE, will be

$$\varphi = 90° - (h - \delta) \qquad (12\text{-}13)$$

North declination is considered positive and south declination is negative. Instrumental errors and the correction for semidiameter can be eliminated if one

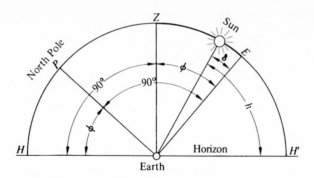

FIGURE 12-11 Latitude by sun's altitude at noon.

vertical angle is measured to the sun's lower limb with the telescope normal and another is measured immediately afterward to the upper limb with the telescope inverted. If the second observation is made within 1 or 2 min of apparent noon, the sun's altitude will not have changed appreciably during that interval.

EXAMPLE 12-5 An observation on the sun was taken on May 1, 1996, when the sun was on the meridian. The Pacific daylight savings time (PDST) of observation was 11:35 A.M. The measured vertical angle to the sun's lower limb is 58° 22′ 10″. Temperature is 76°F, and barometer is 27.6 in. of mercury. Index error is 0. Compute the latitude of the place of observation.

Solution:

instant of observation	$= 11^h\ 35^m$ PDST
	$= 10^h\ 35^m$ PST
longitude of standard meridian	$=\ \ \underline{8^h\ 00^m}$
universal time (UT) of observation	$= 18^h\ 35^m$
declination of sun at 0^h UT on May 2	$= +15°\ 23′\ 46.6″$ (from ephemeris)
declination of sun at 0^h UT on May 1	$= \underline{+15°\ 05′\ 46.5″}$ (from ephemeris)
change in declination for 24^h	$= +\ \ 0°\ 18′\ 00.1″$

change in declination to time of

$$\text{observation } 0°\ 18′\ 00.1″ \times \left(\frac{18^h\ 35^m}{24^h}\right) \qquad = +\ \ 0°\ 13′\ 56.3″$$

declination at time of observation	$= +15°\ 19′\ 42.8″ = \delta$
measured vertical angle to sun's lower limb	$58°\ 22′\ 10″$
semidiameter at time of observation	$=\qquad 15′\ 53.7″$
refraction	$= -\qquad\quad 31.2″$ [from Eq. (12-1a)]
parallax	$= +\qquad\quad\ \underline{4.6″}$ [from Eq. (12-1b)]
corrected vertical angle to sun's center	$=\ \ \overline{58°\ 37′\ 37.1″} = h$

By Eq. (12-13),

$$\varphi = 90° - (58°\ 37′\ 37.1″ - 15°\ 19′\ 42.8″) = 46°\ 42′\ 06″$$

12-12 Latitude by Observation on Circumpolar Star

A circumpolar star is one that never goes below the observer's horizon. Thus, as shown in Fig. 12-11, the codeclination or polar distance *PH* must be less than the latitude of the place. Also from Fig. 12-11 it will be noted that the altitude of the pole is also a measure of the latitude of the observer. Latitude can be determined by measuring the vertical angle to a circumpolar star when it is on the meridian. The star most frequently used by engineers is Polaris (α Ursae Minoris), because it is quite close to the pole and it can be readily identified. The brighter stars about the North Pole are shown in Fig. 12-12. Polaris can usually be identified without difficulty by first locating the Great Dipper (Ursa Major). The edge of the dipper that is formed by the stars α and β points toward Polaris.

If the meridian is unknown, the observation can be made when the star reaches either the highest or the lowest point in its path. These positions are known as *upper culmination* (UC) and *lower culmination* (LC). Upper culmination occurs when the star is on the upper branch of the meridian, whereas lower culmination occurs when the star is on the lower branch of the meridian. The time

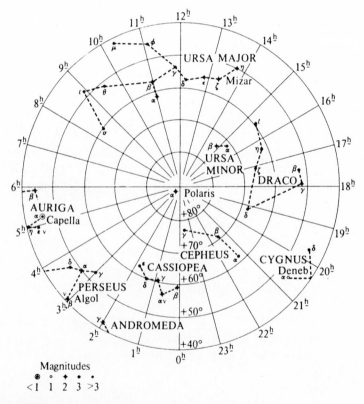

FIGURE 12-12 Stars about the North Pole.

of culmination, which occurs approximately $3^m 56^s$ earlier each day, may be found with the aid of special tables in the ephemeris.

In the Sokkia ephemeris, the universal time of upper culmination for each day is listed as the Greenwich Transit (see Table 12-1). This is the instant at which Polaris is on the upper branch of the Greenwich meridian. Since the sidereal day is shorter than the solar day, to find the local time of upper culmination at any other place, subtract 10^s for each 15° the place of observation is west of Greenwich, or add 10^s for each 15° the place is east of Greenwich. If standard time is desired, add or subtract 4 min for every degree of longitude by which the place is west or east of the standard time meridian (60°, 75°, 90°, 105°, and so on). To determine the time of lower culmination, add $11^h 58.0^m$ to the time of upper culmination for the same date if the time of upper culmination is less than $11^h 58.0^m$; or subtract $11^h 58.0^m$ if the time of upper culmination is greater than $11^h 58.0^m$.

EXAMPLE 12-6   What is the Pacific daylight saving time of upper culmination of Polaris on May 7, 1996, at a place in longitude 121° 14′ 56″ W? What is the PDST of lower culmination on the same date?

Solution: The computations may be arranged as follows

UC for Greenwich meridian May 7	=	$11^h 24^m 32^s$ (from
corr. to longitude of place (121° 14′ 56″/15) × 10^s =		$-1^m 21^s$ ephemeris)
UC for place of observation	=	$11^h 23^m 11^s$
reduction to PST, 1° 14′ 56″ converted to time	=	$+ 5^m 00^s$
PST of UC at place of observation	=	$11^h 28^m 11^s$ A.M. May 7
PDST of UC at place of observation	=	$12^h 28^m 11^s$ P.M. May 7
interval between UC and LC	=	$-11^h 58^m$
PDST of LC at place of observation	=	$0^h 30^m 11^s$ A.M. May 7 ◆

Since the latitude of Polaris changes very slowly, a more accurate value of the latitude will be obtained if a number of vertical angles are measured in rapid succession, half of them with the telescope normal and the others with it inverted. The mean of the vertical angle is corrected for refraction. The declination of the star is obtained from the ephemeris. If the observation is made at lower culmination, the latitude is found from the relationship

$$\varphi = h + (90° - \delta) = h + p \qquad (12\text{-}14)$$

If the observation is made at upper culmination, the relationship is

$$\varphi = h - (90° - \delta) = h - p \qquad (12\text{-}15)$$

in which p is the polar distance obtained from the declination given in the ephemeris.

EXAMPLE 12-7   The altitude of Polaris was measured as 38° 14′ 40″ at lower culmination at the place and date given in Example 12-6. The temperature was 62°F; the barometric pressure was 29.2 in.; and there was no index error. What is the latitude of the place of observation?

Solution: The declination of Polaris for May 7, 1996 at the time of observation, interpolated from Table 12-1 is 89° 14′ 44″. Thus, by Eq. (12-14),

$$
\begin{aligned}
\text{measured } h &= 38°\ 14′\ 40″ \\
\text{refraction correction [by Eq. (12-1a)]} &= -\quad 1′\ 10″ \\
h &= \overline{38°\ 13′\ 30″} \\
(90° - \delta) = p &\quad +\ 0°\ 45′\ 16″ \\
\varphi &= \overline{38°\ 58′\ 46″}
\end{aligned}
$$

12-13   Latitude from Polaris at Any Hour Angle

During part of the summer, both upper and lower culmination of Polaris can occur during daylight hours. If the time is known within a few minutes, the altitude can be observed whenever the star is visible, and the latitude can be computed without serious error by the relationship

$$\varphi = h - p\cos t + \tfrac{1}{2}p^2 \sin^2 t \tan h \sin 1″ \tag{12-16}$$

in which h is the true altitude, obtained by correcting the observed altitude for index error and refraction; t is the hour angle; and p is the polar distance, expressed in seconds.

The nearer Polaris is to culmination, the smaller is the error due to errors in time. For hour angles of 3, 9, 15, and 21^h, an error of 1^m in time causes an error of about 12″ in the computed latitude. For hour angles of 6 and 18^h, an error of 1^m in time causes an error of about 17″ in the computed latitude.

From an examination of the ephemeris, it would appear that the Greenwich Hour Angle (GHA) of Polaris changes by only slightly less than 1° in a 24-hr period. The actual change is almost 361°. Thus, between May 15 and May 16, the actual change is 197° 14′ 51.5″ − 196° 15′ 55.1″ + 360° = 360° 58′ 56.4″. This fact must be borne in mind when determining the GHA at any time other than at 0^h universal time.

The method of calculating the latitude will be apparent from the example in Table 12-2.

12-14   Azimuth from Observation on Polaris at Elongation

The two most common methods of determining the azimuth of a line are by observations on circumpolar stars and by solar observations. The simplest method is to observe Polaris either at the apparently most westerly point in its path (western elongation) or at the apparently most easterly point (eastern elongation). The azimuth of Polaris at elongation may be computed from Eq. (12-2). Thus

TABLE 12-2   Computation for Latitude from Polaris
at Any Hour Angle

Date: May 22, 1996
Longitude: 110° 37′ 01″ W.
Time: $8^h 22^m 20^s$ P.M., MDST.
Observed vertical angle on Polaris: 42° 37′ 30″

Mountain daylight saving time	=	$20^h\ 22^m\ 20^s$
Mountain standard time	=	$19^h\ 22^m\ 20^s$
Longitude west from Greenwich	+	$7^h\ 00^m\ 00^s$
Greenwich time, May 22	=	$26^h\ 22^m\ 20^s$
UT interval since 0^h May 23	=	$2^h\ 22^m\ 20^s$
Greenwich Hour Angle (GHA) at 0^h UT on May 24	=	205° 06′ 13.1″
GHA on 0^h UT on May 23	=	204° 07′ 16.7″
		0° 58′ 57.4″
		+ 360°
		360° 58′ 57.4″

Change from 0^h May 23 to time of observation

$$360° 58′ 57.4″ \left(\frac{2^h\ 22^m\ 20^s}{24^h} \right) \qquad = \qquad 35° 40′ 49.6″$$

GHA at time of observation $= 239° 48′ 06.3″$
− longitude $ -110° 37′ 01.0″$
Hour angle of Polaris, t $ = 129° 11′ 05″$

$$\varphi = h - p \cos t + \tfrac{1}{2} p^2 \sin^2 t \tan h \sin 1″$$

Observed altitude $= \quad$ 42° 37′ 30″ $\qquad\qquad$ 90° 00′ 00″
Refraction $\quad = \quad - \quad$ 01′ 02″ $\qquad \delta = 89° 14′ 40″$
$h \qquad\qquad = \quad$ 42° 36′ 28″ $\qquad p = $ 0° 45′ 20″ $= 2720″$

$p \cos t = 2720 \cos 129° 11′ 05″ = -1719″ = -0° 28′ 39″$

$\tfrac{1}{2} p^2 \sin^2 t \tan h \sin 1″ = \tfrac{1}{2} \times 2720^2 \sin^2 129° 11′ 05″ \tan 42° 36′ 28″ \sin 1″ = 10″$

$\varphi = 42° 36′ 28″ - (-0° 28′ 39″) + 0° 00′ 10″ = 43° 05′ 17″$

$$\sin Z = \frac{\sin S \cos \delta}{\cos \varphi}$$

but since S is 90° when Polaris is at elongation,

$$\sin Z = \frac{\cos \delta}{\cos \varphi} = \sin p \sec \varphi \qquad (12\text{-}17)$$

in which Z is the azimuth measured from the north, p is the polar distance, δ is the declination, and φ is the latitude.

The time of elongation can be determined from the time of upper culmination by adding or subtracting the proper interval given in Table 12-3.

If the exact time is unknown, the star can be followed with the vertical cross hair in the theodolite. When it appears to move vertically on the cross hair, it is at elongation. The horizontal angle is then measured from Polaris to some fixed point, or the line of sight is brought to the ground, a point on the line is established, and the angle is measured the next morning. If Polaris was at eastern elongation, its azimuth is added to the clockwise-measured angle to obtain the required azimuth of the line. If the star was at western elongation, its azimuth is subtracted from the clockwise-measured angle.

TABLE 12-3 Mean Time Interval Between Upper Culmination and Elongation[a]

Latitude	Time Interval	Latitude	Time Interval	Latitude	Time Interval
10°	5^h 58.3^m	40°	5^h 55.7^m	52°	5^h 53.9^m
15°	5^h 58.0^m	42°	5^h 55.4^m	54°	5^h 53.6^m
20°	5^h 57.6^m	44°	5^h 55.2^m	56°	5^h 53.1^m
25°	5^h 57.2^m	46°	5^h 54.9^m	58°	5^h 52.7^m
30°	5^h 56.7^m	48°	5^h 54.6^m	60°	5^h 52.1^m
35°	5^h 56.2^m	50°	5^h 54.3^m	62°	5^h 51.6^m

[a]Eastern elongation precedes and western elongation follows upper culmination.

12-15   Azimuth from Observation on Polaris at any Hour Angle

If the time is accurately known, the observations on Polaris can be made whenever the star is visible, and the azimuth can be calculated by either of the two following relationships:

$$\sin Z = -\sin t \cos \delta \sec h \qquad (12\text{-}18)$$

$$\tan Z = \frac{\sin t}{\sin \varphi \cos t - \cos \varphi \tan \delta} \qquad (12\text{-}19)$$

in which Z is the azimuth of Polaris, measured from the north; t is the hour angle, obtained from the time, as in Section 12-13; δ is the declination, taken from the ephemeris; h is the altitude, obtained by correcting the observed altitude for index error and refraction; and φ is the latitude.

If Z is positive, Polaris is east of north. If Z is negative, Polaris is west of north, in which case adding 360° gives the azimuth clockwise from north.

If the latitude is unknown, the altitude is measured and the first equation is used. When the altitude is measured, the same observations can be used for the determination of both azimuth and latitude. The signs of the trigonometric functions must be taken into account in using these equations. When the star is near elongation, the effect of errors in time on the computed azimuth will be negligible. When the star is near culmination, an error of 1^m in the time at a latitude of about 40° will cause an error of about 25″ in the computed azimuth.

The advantage of observing Polaris at any hour angle is that the surveyor can choose the time of observation rather than having to observe eastern or western elongation at a fixed time, which might not be convenient.

EXAMPLE 12-8   The following notes were recorded for an observation on Polaris to determine the azimuth of the line AB of Fig. 12-13.

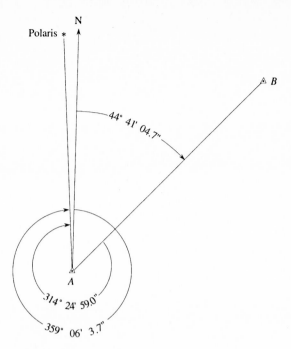

FIGURE 12-13 Azimuth of line by Polaris observation.

May 12, 1996
Latitude 37° 48′ 50.2″
Longitude 119° 38′ 12.6″
Watch set to UT
Instrument at *A*, backsight to *B*

Object	*D* or *R*	Circle	Time UT
			May 13, 1996
B	D	0°00′ 12″	
Polaris	D	314°25′ 28″	2ʰ 16ᵐ 21.5ˢ
Polaris	R	134°24′ 58″	2ʰ 19ᵐ 36.1ˢ
B	R	180°00′ 16″	

Compute the azimuth of the line *AB*.

Solution:

GHA of Polaris at 0ʰ May 13	=	194° 17′ 57.5″
GHA of Polaris at 0ʰ May 14	=	195° 16′ 57.1″
declination of Polaris at 0ʰ May 13	=	89° 14′ 42.73″
declination of Polaris at 0ʰ May 14	=	89° 14′ 42.41″
average time of observation in UT	=	2ʰ 17ᵐ 58.8ˢ

GHA at time of observation   =   228° 53′ 18.7″ (see Table 12-2)

longitude   − 119° 38′ 12.6″

LHA at time of observation on May 12 = t =   109° 15′ 06.1″

declination δ =   89° 14′ 42.7″

By Eq. (12-19),

$$\tan Z = \frac{\sin 109° \ 15' \ 06.1''}{\sin 37° \ 48' \ 50.2'' \cos 109° \ 15' \ 06.1'' - \cos 37° \ 48' \ 50.2'' \tan 89° \ 14' \ 42.7''}$$

$Z = -0° \ 53' \ 56.3'' = 359° \ 06' \ 03.7''$

angle direct   =   314° 25′ 16″

angle reverse   =   314° 24′ 42″

avg. angle   = 314° 24′ 59.0″

azimuth Polaris   359° 06′ 03.7″

azimuth of AB 44° 41′ 04.7″   ◆

Averaging the times of observation and using a single value for the declination are justified because these two quantities change very little during a normal set of observations. If additional sets are observed, then of course a new average time will be obtained for each new set. If the sets are taken in fairly quick succession, a single value for the declination can be used, based on the average of the initial and final times of observation.

12-16   Azimuth from Sun's Altitude

If the time is known to within 1 or 2 min, the azimuth of a line can be determined by measuring the horizontal angle from the line to the sun's center and measuring the altitude of the sun's center. A total of eight measurements, four with the telescope direct and four with it reversed and taken in quick succession, will give results reliable to within 15″ when a 30-second transit is used.

As shown in Fig. 12-14, the instrument is set up and leveled over a selected point, such as station TT41, and the line of sight is directed toward the sun. The telescope bubble is centered and the vertical circle is read to determine the index error, which is recorded. The horizontal circle is set to read 0° 00′ 00″, and a backsight is taken on another selected point, such as station TT42. The upper clamp is loosened, the telescope is pointed at the sun, and the image is obtained on a white card as described in Section 12-7. Four observations are taken with the image located as indicated in Fig. 12-7 or 12-9. In the morning these positions would correspond to those in views (a), (b), (a), and (b). In the afternoon the positions would be as in views (c), (d), (c), and (d). For each observation the times to the nearest second at the instant of tangency, the horizontal-circle reading, and the vertical angle are recorded.

The telescope is then reversed, pointed at the sun again, and adjusted to give an image on the card. This pointing is made with the upper clamp of the theodolite loosened. Four observations are taken with the telescope reversed, the positions

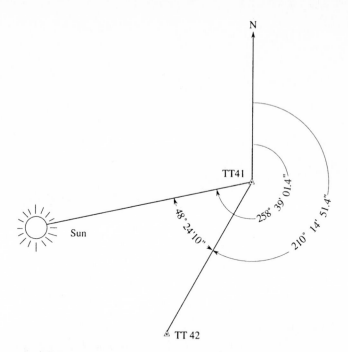

FIGURE 12-14 Azimuth of line by solar observation.

of the image being the same as those with the telescope direct, and the corresponding readings are recorded.

Finally, the upper clamp is loosened and a sight is taken back to station TT42. The horizontal-circle reading, which should be 180° 00′ 00″, is recorded. The index error should be determined after all the observations are made to see whether or not it has changed.

The measured horizontal and vertical angles are corrected to give the true horizontal and vertical angles to the sun's center, as explained in Section 12-8 and as demonstrated in Examples 12-3 and 12-4. To determine the validity of the observations, a plot is prepared showing corrected horizontal angles and corrected vertical angles as functions of time. If the observations have been made within a period of about 10 min or less, the two functions should plot as straight lines.

In Fig. 12-15 eight observations are plotted for an afternoon observation. The horizontal-angle function is shown as a straight line, but the vertical-angle function is shown as two separate straight lines that are parallel. The offset e between these two lines is analyzed as twice the unaccounted-for systematic error of the instrument. If each vertical angle read with the telescope direct were reduced by $e/2$ and each vertical angle read with the telescope reversed increased by $e/2$, the resulting eight values would plot on one straight line. This, however, is not necessary, since the mean of the eight measured vertical angles is free from the systematic error.

In Fig. 12-16 eight morning observations have been plotted. The second vertical angle measured with the telescope reversed is seen to be wrong by the amount e. This error is obviously caused by a mistake in reading the vertical circle. Before the means of the observed values are found, both the horizontal angle and the

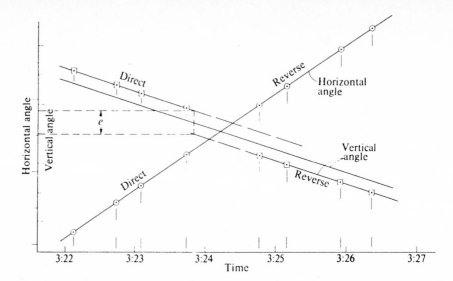

FIGURE 12-15 Plot of horizontal and vertical angles versus time.

vertical angle for the second reverse readings and the horizontal and vertical angles for the corresponding symmetrical direct readings would be rejected. These latter angles would be those for the third direct readings. This rejection is necessary to eliminate systematic instrument errors to the fullest possible extent. Note that the corresponding two instants of time would also be rejected. The means of the remaining six values of horizontal angles, vertical angles, and times would be used in the computation for azimuth.

Occasionally the plotted results for a series of solar observations will leave some doubt in regard to the validity of the corrected vertical angles. Since the sun's altitude is a function of time, the change of vertical angle with respect to time can be determined by expressing the altitude h in terms of the sun's hour angle t using Eq. (12-11). If the resulting equation is differentiated for h with respect to t, the slope of the curve for vertical angle versus time can be expressed as follows:

$$\frac{dh}{dt} = -\frac{\cos \varphi \cos \delta \sin t}{\cos h} \qquad (12\text{-}20)$$

The mean corrected vertical angle for the set of observations gives h. Angle t is obtained by converting the mean watch time of observation first to Greenwich civil time, then to Greenwich apparent time by applying the equation of time, and, finally, to local apparent time by subtracting the west longitude of the place of observation (or adding if the longitude is east). The result is the hour angle of the sun in hours, minutes, and seconds. This angle is converted to degrees, minutes, and seconds for use in Eq. (12-20). In the morning the hour angle is considered negative; in the afternoon, positive. If dh/dt is to be expressed as angular minutes of altitude per minute of time, then Eq. (12-20) must be modified to give the following relationship:

$$\frac{dh}{dt} = -\frac{15 \cos \varphi \cos \delta \sin t}{\cos h} \qquad (12\text{-}21)$$

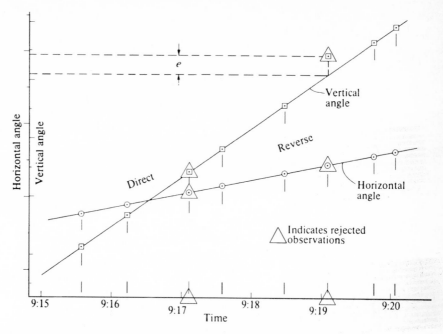

FIGURE 12-16 Rejection of observations by analysis of angle-time plot.

This slope can then be plotted on the diagram for vertical angle versus time, and a line having this slope should be parallel with the line passing through all valid vertical angles. Observations can then be rejected on this basis. If an observation for a vertical angle is rejected, it is necessary to reject also the corresponding horizontal angle. Furthermore, a symmetrical set of observations must also be rejected, as previously mentioned.

The azimuth of the sun is computed from either of the following relationships:

$$\tan \tfrac{1}{2}Z = \sqrt{\frac{\sin(s - \varphi)\,\sin(s - h)}{\cos s\,\cos(s - p)}} \tag{12-22}$$

$$\cos Z = \frac{\sin \delta - \sin \varphi \sin h}{\cos \varphi \cos h} \tag{12-23}$$

in which Z is the azimuth of the sun, measured from north clockwise in the morning and counterclockwise in the afternoon; s is $\tfrac{1}{2}(\varphi + h + p)$; φ is the latitude; h is the true altitude, obtained from the mean of the corrected vertical angles; p is the polar distance, or $(90° - \delta)$; and δ is the declination, obtained for the mean instant of time from the ephemeris. The best results will be obtained if the observations are made when the local mean time is between 8 and 10 A.M. or between 2 and 4 P.M. Except when the observation is made early in the morning or late in the afternoon, Eq. (12-22) should be used to compute the azimuth of the sun.

The uncertainty in the determination of an accurate azimuth by the sun's altitude method results from (1) the uncertainty in the refraction at low sun angles in early morning and late afternoon, particularly at high latitudes in winter, and (2) the fact that between 10 A.M. and 2 P.M. local time, a small error in the sun's altitude causes a large error in its azimuth. If the sun must be observed at those times, the hour angle method, discussed in Section 12-17 should be used.

The horizontal angle between the backsight and the position of the sun represented by the mean of the corrected angles is used with the azimuth of the sun to obtain the azimuth of the line. Computations showing the use of both formulas to determine the azimuth of the line from station TT41 to station TT42 in Fig. 12-14 are given in Table 12-4.

12-17 Azimuth from Hour Angle of the Sun

If the instant at which the sun's center or one of its limbs crosses the vertical cross hair of the theodolite is accurately timed (see Section 12-7), the sun's azimuth can be computed by the following relationship:

$$\tan Z = \frac{-\sin t}{\cos \varphi \tan \delta - \sin \varphi \cos t} \qquad (12\text{-}24)$$

in which Z is the sun's azimuth, measured from the south meridian. The procedure for making the observation is essentially that discussed in Section 12-16 except that the vertical circle is not read. Because the accuracy of the hour angle method is dependent on precise time, the instant of observation must be noted by the

---

TABLE 12-4 Computations for Azimuth from Altitude of Sun

Date: May 14, 1996. Instrument at station TT41. Latitude 37° 52′ 30″N.

Mean corrected horizontal angle to sun	=	48° 24′ 10″
Mean corrected vertical angle to sun = h	=	44° 54′ 49″
Mean instant of observation	=	4ʰ 12ᵐ 56ˢ PDST
	=	3ʰ 12ᵐ 56ˢ PST
Longitude of standard meridian		8ʰ 00ᵐ 00ˢ
		11ʰ 12ᵐ 56ˢ
		+12ʰ
Greenwich civil time of observation (UT)	=	23ʰ 12ᵐ 56ˢ
Declination of sun at 0ʰ UT May 14, 1996	= +	18° 38′ 29.9″
Change in declination (see Example 12-5)	= +	0° 13′ 50.1″
Declination at instant of observation	= +	18° 52′ 20.0″

$$\tan \tfrac{1}{2} Z = \sqrt{\frac{\sin (s - \varphi)\,\sin (s - h)}{\cos s \,\cos (s - p)}}$$

$p =$	71° 07′ 40.0″	$Z =$ 101° 20′ 58.6″	(counterclockwise from north)
$\varphi =$	37° 52′ 30.0″	$Z =$ 258° 39′ 01.4″	(clockwise from north)
$h =$	44° 54′ 49.0″	Horizontal angle = − 48° 24′ 10″	
	2)153° 54′ 59.0″	Azimuth TT41 to TT42 = 210° 14′ 51.4″	
$s =$	76° 57′ 29.5″		
$s - h =$	32° 02′ 40.5″		
$s - \varphi =$	39° 04′ 59.5″	$\cos Z = \dfrac{\sin \delta - \sin \varphi \sin h}{\cos \delta \cos h}$	
$s - p =$	5° 49′ 49.5″	$Z =$ 101° 20′ 58.6″	(counterclockwise from north)
		$Z =$ 258° 39′ 01.4″	(clockwise from north)
		Horizontal angle = − 48° 24′ 10.0″	
		Azimuth TT41 to TT42 = 210° 14′ 51.4″	

observer rather than by the note keeper since the observer can anticipate when to stop the stop watch or the time module.

After the backsight has been made on the ground azimuth mark, a number of observations are timed with the telescope in the direct position followed by the same number of reversed observations, reading the horizontal circle at each instant of time. Finally, a sight is then taken on the azimuth mark with the telescope still reversed to provide a check. Each observation is then reduced, and a plot of the sun's azimuth versus time should be made to reveal any mistakes in reading either the horizontal circle or the instant of observation.

The observation should be made on the sun's center by means of Roelof's solar prism or a reticule such as that shown in Fig. 12-5. If the eastern limb of the sun is observed by the method shown in Figs. 12-7(b) and (d), the timing is quite accurate. Since the horizontal angle from the azimuth mark is measured to the trailing limb of the sun, this angle must be increased (clockwise) to take into account the sun's semidiameter as given by Eq. (12-12). Note that the vertical circle need not be read to determine the altitude h since all of the information for its computation is available using Eq. (12-11). That is,

$$\sin h = \sin \varphi \sin \delta + \cos \varphi \cos \delta \cos t$$

If the observation is made by the method shown in Fig. 12-7(a) or (c), the timing is not as accurate, but the sun's semidiameter need not be computed. The average of the two observations, one taken as shown in Fig. 12-9 (a) and (b) or (c) and (d) will give essentially the correct azimuth. However, in this case an error in azimuth as much as 6″ can occur when averaging left and right limb pointings, since the changing altitude also changes the semidiameter corrections.

Because of the availability of accurate time that can be carried in the time modules associated with handheld calculators, the hour angle method is preferred over the altitude method. There is no need to measure the vertical angle using the hour angle method. This eliminates the effect of refraction and the necessity to observe horizontal and vertical angles simultaneously. Furthermore, the hour angle method can be used for longer periods of the day, whereas the altitude method results in substantial errors in azimuth when the sun is near the meridian.

As an example of the hour angle method, a single observation on the sun's eastern or trailing limb using a time module set to UT1 is shown in Table 12-5 and diagrammed in Fig. 12-17.

12-18   Order of Making Observations

In most azimuth observations, the time and the latitude will be known with sufficient accuracy to permit the use of solar observations, or of observations on Polaris at any hour angle. The time is obtained from a good watch, or by comparing a fair one with the time furnished by radio time signals, or by use of a time module. The latitude can usually be scaled with sufficient accuracy from an existing map.

If it is necessary to make observations where no preliminary information is available, an approximate latitude can be obtained by measuring the maximum

TABLE 12-5   Computations for Azimuth from Hour Angle of Sun

Date: May 19, 1996. Instrument at station 17; backsight on station 18.
Latitude 37° 52′ 20″ N, longitude 119° 12′ 43″ W (from map)
Horizontal angle 265° 24′ 22″ from station 18 clockwise to sun's trailing limb
Instant of time of observation $17^h 02^m 56^s$ GCT (UTC corrected to UT1)

GHA 0^h (May 19) $= 180° 53′ 21.9″$
GHA 24^h (May 20) $= 180° 52′ 34.0″$
Declination 0^h (May 19) $+ \;\; 19° 46′ 46.5″$
Declination 24^h (May 20) $+ \;\; 19° 59′ 26.3″$

$$\text{GHA} = 180° 53′ 21.9″ + (180° 52′ 34.0″ - 180° 53′ 21.9″ + 360°)\left(\frac{17^h\,02^m\,56.2^s}{24^h}\right)$$

$$\text{GHA} = 436° 36′ 50.9″$$
$$= 76° 36′ 50.9″$$
$$\underline{-119° 12′ 43.0″} \text{ (longitude)}$$
$$\text{LHA} = -42° 35′ 52.1″ = t \text{ (east)}$$

$$\delta = 19° 46′ 46.5″ + (19° 59′ 26.3″ - 19° 46′ 46.5″)\left(\frac{17^h 02^m 56.2^s}{24^h}\right)$$

$$\delta = 19° 55′ 46.2″$$

$$\tan Z = \frac{-\sin(-42° 35′ 52.1″)}{\cos 37° 52′ 20″ \tan 19° 55′ 46.2″ - \sin 37° 52′ 20″ \cos(-42° 35′ 52.1″)}$$

$Z = -76° 14′ 39.2″$ (east of south)
Azimuth of sun $= 103° 45′ 20.8″$ clockwise from north
$\sin h = 37° 52′ 20″ \sin 19° 55′ 46.2″ + \cos 37° 52′ 20″ \cos 19° 55′ 46.2″$
$\qquad \times \cos(-42° 35′ 52.1″)$
$h = 49° 04′ 21.2″$
semidiameter 15′ 49.9″ (from ephemeris)
semidiameter 15′ 49.9″ sec 49° 04′ 21″ = 24′ 10.0″

Clockwise angle station 18 to sun's center $= \;\; 265° 24′ 22″ + 0° 24′ 10.0″$
$= \;\; 265° 48′ 32.0″$
Azimuth of sun $= \;\; 103° 45′ 20.8″$
$\underline{+360°}$
$463° 45′ 20.8″$
$\underline{-265° 48′ 32.0″}$
Azimuth station 17 to station 18 $= \;\; 197° 56′ 48.8″$

altitude of the sun. The approximate local apparent time would be obtained at the same observation, since the maximum altitude will occur at the instant of local apparent noon. A more exact latitude could be obtained by measuring the maximum or the minimum altitude of Polaris or any other convenient star.

From the observed latitude, a very close approximation to the true meridian could be established by observations on Polaris at elongation. Local time can be obtained by an observation on the sun at local apparent noon, the previously located meridian being used, or by measuring the altitude of the sun in the manner described in Section 12-8.

By repeating the observations and using the data obtained from each preceding observation, the location of the meridian, the local time, and the latitude can eventually be determined with considerable exactness.

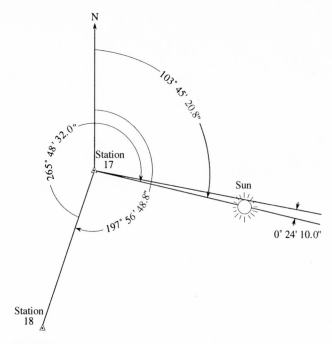

FIGURE 12-17 Azimuth of line by hour angle method.

The easiest way of determining the longitude is by comparing the local mean time with some standard time, sent out by a radio transmitter. With the small, compact shortwave receivers now available, there should be few times when this method cannot be used. Books on astronomy contain methods by which the approximate longitude can be obtained by observations on the moon.

PROBLEMS[†]

12-1. Determine the sun's declination and the local apparent time for the instant of 10:42:50 A.M., CST, on Oct. 12 at a place whose longitude is 89° 25′ 25″ W.

12-2. What is the PST of local apparent noon on Dec. 12 at a place whose longitude is 121° 35′ W? What is the sun's declination at this instant?

12-3. What is the sun's hour angle on Aug. 12 at a place whose longitude is 76° 22′ W at the instant of 5:15:20 A.M., PST?

12-4. When the altitude of the sun's lower limb is measured at local apparent noon on Sept. 5, the vertical-circle reading is + 44° 25′. The index error is + 0° 03′. The temperature is 80°F and the barometric pressure is 28.98 in.Hg. What is the latitude of the place of observation if the longitude is 85° 20′ W?

[†]The ephemeris for the current year is to be used in these problems.

12-5. Determine the PST of eastern and western elongation of Polaris on April 5 at a place whose longitude is 122° 40′ W and whose latitude is 42° 32′ N.

12-6. Polaris was observed at upper culmination in longitude 105° 04′ 30″W on the night of Dec. 19, and the vertical angle was + 38° 30′. The temperature is 40°F and the barometric pressure is 29.10 in.Hg. What was the Mountain standard time? What is the latitude of the place of observation?

12-7. What is the azimuth of Polaris at 8:55:10 P.M., EST, on Feb. 10, at a place whose latitude is 44° 20′ and whose longitude is 74° 21′ 30″ W?

12-8. From the following observations made on Polaris at longitude $7^h 22^m 43^s$ west, determine the latitude of the place and the azimuth of the line *AB*.

Date: Sept. 16. Transit at station *A* at elevation of 6110 ft.
Temperature: 75°F.

MST P.M.	Vernier A	Vernier B	Vertical Angle	Telescope	Pointing
	0° 00′ 00″	180° 00′ 00″		Direct	NW to station *B*
$7^h 12^m\ 16^s$	54° 24′ 00″	234° 24′ 00″	+ 42° 58′ 30″	Direct	N to Polaris
7 14 42	234° 24′ 30″	54° 25′ 00″	42° 59′ 00″	Reversed	N to Polaris
	180° 00′ 00″	0° 00′ 30″		Reversed	NW to station *B*

12-9. The following observations were made on the sun in latitude 42° 16′30″ N.

Date: Aug. 12. Transit at station *C* at elevation of 905 ft. Temperature: 82°F.

EST A.M.	Vernier A	Vernier B	Vertical Angle	Telescope	Pointing
	0° 00′ 00″	180° 00′ 00″		Direct	SW to station *D*
$10^h 24^m 32^s$	268° 48′ 30″	88° 49′ 00″	+ 49° 57′	Direct	Sun's lower limb
25 20	268° 38′ 00″	88° 38′ 30″	50° 20′	Direct	Sun's eastern limb
26 04	269° 16′ 00″	89° 16′ 00″	50° 11′	Direct	Sun's lower limb
28 12	89° 28′ 00″	269° 27′ 30″	50° 48′	Reversed	Sun's eastern limb
28 58	90° 07′ 00″	270° 07′ 30″	50° 39′	Reversed	Sun's lower limb
29 41	89° 54′ 00″	269° 54′ 30″	51° 02′	Reversed	Sun's eastern limb
	180° 00′ 00″	0° 00′ 30″		Reversed	SW to station *D*

The longitude of the place of observation is 82° 02′ 20″ W. Correct the horizontal and vertical angles to the sun's center, and plot the corrected values versus the corresponding times of observation. Compute the azimuth of the line *CD*.

12-10. The eastern limb of the sun was observed at $23^h 15^m 20.0^s$ UT on the afternoon of March 6 at a place whose latitude is 37° 58′20″ N and whose longitude is 120° 15′44″ W. What is the azimuth of the sun's center?

12-11. The eastern limb of the sun was observed on the morning of October 20 at $10^h 52^m 14.6^s$ EST at a place in latitude 43° 14′20″ N, and longitude 77° 42′16″ W. What is the azimuth of the sun's center?

BIBLIOGRAPHY   1989. *Almanac for Computers*. Washington, D.C.: Nautical Almanac Office, U.S. Naval Observatory.

Boucher, P. 1983. Azimuth determination by solar observation. New perspectives on an old problem. *Surveying and Mapping* 43:307.

Brinker, D. M. 1985. Descriptive geometry on the celestial sphere. *Tech. Papers 45th Ann. Meet. Amer. Cong. Surv. & Mapp.* p. 539.

——. 1986. Descriptive geometry of the celestial sphere, Part 1. *ACSM Bull. 101. Amer. Cong. Surv. & Mapp.* p. 19.

——. 1987. Descriptive geometry of the celestial sphere, Part 2. *ACSM Bull. 110, Amer. Cong. Surv. & Mapp.* p. 13.

Brown, G. D. 1984. Solar azimuth leap-frog surveys. *Tech. Papers 44th Ann. Meet. Amer. Cong. Surv. & Mapp.* p. 88.

Buckner, R. B. 1975. Reasons and methods for accurate direction in land surveys. *Surveying and Mapping* 35:305.

——. 1984. *A Manual on Astronomic and Grid Azimuth*. Rancho Cordova, CA: Landmark Enterprises.

Burtch, R. 1982. Time and time frequency standards. *Tech. Papers 42nd Ann. Meet. Amer. Cong. Surv. & Mapp.* p. 361.

Chait, B. 1970. Azimuth and longitude from near meridian star observations by theodolite. *The Canadian Surveyor* 24:215.

Elgin, R. L., Knowles, D. R., and Senne, J. H. 1991. *Celestial Observation Handbook and Ephemeris*. Lietz Sokkia, Overland Park, KS.

Hoskinson, A. J., and Duerksen, J. A. 1947. *Manual of Geodetic Astronomy*. Special Publication No. 237, U.S. Coast and Geodetic Survey, Washington, D.C.: U.S. Government Printing Office.

Peterson, W. E. 1986. Solar observations with an electronic theodolite. *Tech. Papers Ann. Meet. Amer. Cong. Surv. & Mapp.* p. 177.

Savlan, J. A. 1985. Solar hour angle: A consistent solution for astronomical bearing. *Tech. Papers 45th Ann. Meet. Amer. Cong. Surv. & Mapp.* p. 546.

Senne, J. H. 1987. The use of internal ephemeris generation for astronomical azimuth computation. *Tech. Papers Fall Conv. Amer. Cong. Surv. & Mapp.* p. 182.

Solaric, N., Spoljaric, S. D., and Lukinac, Z. 1994. Accuracy of automatic grid azimuth determination by astronomical methods with Leica-Kern E2 theodolite. *Surveying and Land Information Systems* 54:1.

13 Horizontal and Vertical Curves

13-1 Introduction

The horizontal alignment of a highway, railroad, or canal consists of a series of straight lines and curves. The straight portions are called *tangents*. As shown in Fig. 13-1, successive tangents change direction by deflection angles designated Δ_i. The change in direction at each intersection is distributed along a curve or series of curves to make a smooth transition. Figure 13-1 shows simple circular curves with radii designated R_i and central angles Δ_i. Horizontal transition curves can also be made up of multiple circular curves of different radii called *compound curves* or of circular curves joined to the tangents by spirals of third degree.

The grade line on a profile of any length is likewise made up of straight lines and curves, as indicated in Fig. 13-2. Intersections of the grade lines are designated V_i, grades are designated g_i, and the lengths of the curves joining the grade

FIGURE 13-1 Horizontal alignment consisting of tangents joined by simple circular curves.

FIGURE 13-2 Grade lines on a profile joined by vertical curves. The vertical scale is exaggerated.

lines are designated L_i. The curves joining the grade lines are called *vertical curves* and are generally parabolas (second-degree curves). Their function is to make a smooth transition from one grade to another—in other words, to spread out the change in grade over a distance L.

This chapter is designed to give the reader an understanding of the basic geometry of horizontal and vertical curves and how they are located or constructed in the field. Those interested in a more thorough treatment of the subject should consult texts on route location.

All curves that appear in this chapter are measured in feet. However, all of the equations to be developed are just as valid when using the metric system as long as a station is understood to be 100 m. Thus the designation m can be used wherever the designation ft is used, except where noted.

13-2   Notation for Circular Curves

The notation commonly used on circular curves is shown in Fig. 13-3. The point at which the two tangents to the curve intersect is called the *vertex,* which is designated V, or the *point of intersection* (PI). The deflection angle between the tangents, which is equal to the angle at the center of the curve, is denoted by Δ. If the survey (shown in Fig. 13-3) is progressing to the right, the straight line to the left of the PI is the *back tangent* and the one to the right is the *forward tangent.* The beginning point of the curve is called the *point of curvature* (PC). This is sometimes referred to as the *tangent-to-curve point* (TC), or the *beginning of curve* (BC). The end of the curve is the *point of tangency* (PT). This is also referred to as the *curve-to-tangent point* (the CT), or the *end of curve* (EC).

The distance from the PI to the PC or the PT is the *tangent distance T.* The distance from the PI to the middle point of the curve, measured along the bisector of the central angle, is the *external distance E.* The distance from the middle point of the curve to the middle point of the chord joining the PC and the PT is the *middle ordinate M.* The distance along the line joining the PC with the PT is the *long chord C* (see Fig. 13-6). The angle subtended at the center of the curve by a 100-ft chord is the *degree of curve D,* by *chord* definition. The angle subtended at the center of the curve by a 100-ft arc is the *degree of curve D,* by *arc* definition. The *radius* of the curve is designated by R.

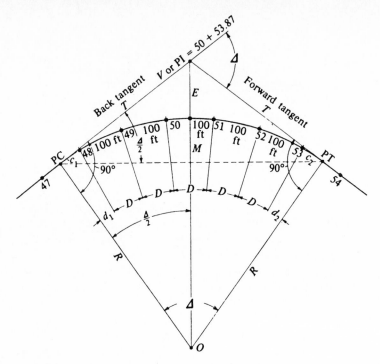

FIGURE 13-3 Circular curve.

As the station method is commonly used in recording the horizontal distances, and as the PC of a curve will rarely be at a full station, the distance c_1 from the PC to the first full station on the curve will be, in general, less than 100 ft. The central angle subtended by this distance is d_1. The distance from the last full station on the curve to the PT is c_2, and the corresponding central angle is d_2.

A curve can be designated by either the radius or the degree of curve. The designation by the radius is finding widespread use in highway practice. Usually some integral multiple of 50 ft is used as the radius of the curve. For the purpose of computing elements needed to lay out the curve in the field, the *length of the curve* is defined as the *difference in stationing along the curve between the PC and the PT of the curve.* By arc definition this corresponds to the length of the curve measured along the actual arc. By chord definition it does not correspond either to the length measured along the arc or to the length measured along a series of chords unless the curve is, by chance, composed of full 100-ft chords only. The length L of a curve herein defined is a mathematical value used to compute other elements used in laying out the curve in the field.

If the physical length of a curve measured along the actual arc is needed for property description, but the curve is defined by the chord definition, its length is obtained by $R\Delta$, in which Δ is the central angle in radians obtained with the aid of Table C-1 in Appendix C.

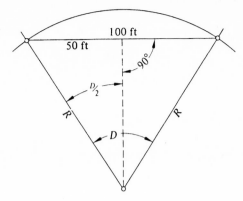

FIGURE 13-4 Relation between R and D by chord definition.

13-3 Radius and Degree of Curve

The relationship between the radius and the degree of curve by chord definition is shown in Fig. 13-4. In either of the two right-angled triangles formed by bisecting the central angle D,

$$\sin \tfrac{1}{2}D = \frac{50}{R} \tag{13-1}$$

from which

$$R = \frac{50}{\sin \tfrac{1}{2}D} \tag{13-2}$$

The relationship between the radius and the degree of curve by arc definition is shown in Fig. 13-5. When D is expressed in degrees,

$$\frac{100}{D} = \frac{2\pi R}{360°}$$

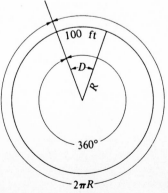

FIGURE 13-5 Relation between R and D by arc definition.

from which

$$R = \frac{5729.578}{D}$$

(13-3)

or

$$D = \frac{5729.578}{R}$$

(13-4)

Whether the degree of curve is by arc or by chord definition, the angle D subtends a full station on the curve.

From Eqs. (13-2) and (13-3), if the degree of curve is 1°, the radius of the chord definition curve would be 5729.651 ft, whereas the radius of the arc definition curve would be 5729.578 ft. The difference between the two values would be of no significance in laying out the curves. However, for sharper curves with more curvature and shorter radii, the difference between the two definitions is quite obvious. Suppose that $D = 20°$. Then by chord definition the radius is 287.94 ft, whereas by arc definition it is 286.48 ft.

13-4 Equations for Circular Curves

The relationships involving the radius R of a circular curve, the deflection angle Δ between the tangents, and other elements of the circular curve are shown in Fig. 13-6. The following equations apply to both the chord definition and the arc definition:

$$T = R \tan \tfrac{1}{2}\Delta$$

$$E = R \sec \tfrac{1}{2}\Delta - R = R(\sec \tfrac{1}{2}\Delta - 1)$$

(13-5)

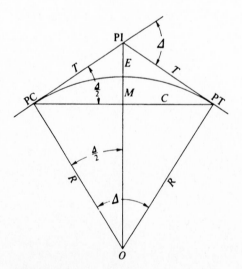

FIGURE 13-6 Elements of a circular curve.

or

$$E = R \text{ exsec } \tfrac{1}{2}\Delta$$

$$M = R - R \cos \tfrac{1}{2}\Delta = R(1 - \cos \tfrac{1}{2}\Delta) \tag{13-6}$$

or

$$M = R \text{ vers } \tfrac{1}{2}\Delta \tag{13-7}$$

$$C = 2R \sin \tfrac{1}{2}\Delta \tag{13-8}$$

The length of the curve, or the *difference in stationing* between the PC and PT, is computed by the relationship

$$L = \frac{\Delta}{D} \text{ (expressed in stations)}$$

$$L = 100 \, \frac{\Delta}{D} \text{ (expressed in feet or meters)} \tag{13-9}$$

13-5   Selection of Curve

Any two given tangents can be connected by an infinite number of circular arcs. The curve to be used in a particular case is determined by assuming D, R, T, E, or L. Since all these quantities are not independent, only one can be assumed. The other values are calculated from the relationships developed in the preceding section.

Field conditions frequently decide which quantity should be assumed. Thus, when the survey is following the bank of a stream, the external distance may be the limiting factor. On a winding road the tangent lengths may be restricted. On high-speed modern pavements and railroads, an attempt is made to keep the degree of curve below a given maximum. Wherever possible a radius of more than 1000 ft (300 m) is adopted. This radius corresponds to a degree of curve of about 5° 44'.

13-6   Stations of PI, PC, and PT

When a road or railroad is planned, its location is first made on a map sheet to fit the existing conditions of man-made and natural features, culture, and topography. This is referred to as a *paper location.* The coordinates of the PIs are carefully scaled from the map (see Section 8-33). The coordinates can be used in one of two different ways. First, each tangent can be inversed using Eqs. (8-11) and (8-12), from which all the tangent lengths and deflection angles are determined. The line is then staked out and stationed by the methods discussed in Sections 6-6, 6-7, and 6-8.

The alternative way is to run a random control traverse through the project area and then compute the lengths and bearings of tie lines by inversing between the points on the control traverse and the PIs, as discussed in Section 8-24. The tie lines are run in the field to locate the PIs. The tangents are then stationed by surveying between the PIs.

In the latter method for locating the PIs and stationing the tangents, there are inevitable errors in the location survey. To ensure that the curves will fit the tangents as located, the deflection angles at each PI between the back tangent and the forward tangent must be measured by double centering.

The net result of the field location of the tangents is a set of lines stationed from the point of beginning, throughout the project, together with the deflection angles between the lines. A section of such a field location is shown in Fig. 13-7(a). The distance to the first PI is 814.72 ft, to the second PI is 1506.54 ft, and so on. The station of the first PI is thus 8 + 14.72 and the station of the second PI is 15 + 06.54.

In order to make the stationing continuous along the tangents and the curves throughout the project, the line must be restationed to reflect the difference between the straight-line distance of Fig. 13-7(a) and the combination straight-line and curve distances of Fig. 13-7(b). In this restationing, the new stations of all the PCs and PTs are required.

The station of the first PC is obtained by the following:

$$PC = PI - T \tag{13-10}$$

in which T is the tangent distance of the first curve. In the first curve of Fig. 13-7(b), the PC station is

$$
\begin{aligned}
PI &= 8 + 14.72 \\
T &= -3 + 03.13 \\
\hline
PC &= 5 + 11.59
\end{aligned}
$$

The PT station is then computed by the following:

$$PT = PC + L \tag{13-11}$$

(a)

(b)

FIGURE 13-7 "Old" stationing along straight lines and "new" stationing along tangents and curves.

in which L is the length of the curve computed by Eq. (13-9). The station of the PT for the first curve of Fig. 13-7(b) is

$$
\begin{array}{rl}
PC = & 5 + 11.59 \\
L = & 5 + 92.78 \\
\hline
PT = & 11 + 04.37
\end{array}
$$

The second and subsequent PC stations are determined as follows. The distance S of Fig. 13-7(b) is the difference between the length of the line joining the two PIs and the sum of the T_1 and T_2. Then

$$PC_2 = PT_1 + S_{1-2} \tag{13-12}$$

In this example the length of the line between the two PIs is $1506.54 - 814.72 = 691.82$, and $S = 691.82 - (303.13 + 245.96) = 142.73$. Then by Eq. (13-12),

$$
\begin{array}{rl}
PT_1 = & 11 + 04.37 \\
S_{1-2} = & 1 + 42.73 \\
\hline
PC_2 = & 12 + 47.10
\end{array}
$$

The station of the PT for the second curve is obtained by Eq. (13-11), which is $17 + 32.31$. the remainder of the curves in the project are treated in the same manner as the second curve.

The difference between the "old" stationing along the straight lines and the "new" stationing along the tangents and curves is accounted for where required by means of an equation. For example, a point on the located line may be defined as old station $56 + 42.70$ = new station $51 + 17.66$. The meaning of this equation should be obvious.

13-7   Central Angle and Chord to First Curve Station

In Fig. 13-8, the PC station is $20 + 54.00$, the first curve station is $21 + 00$, and the second curve station is $22 + 00$. By either the chord or the arc definition, the angle at the center subtending a full station is equal to the degree of curve D. Let c_1' equal the *difference in stationing* between the PC and the first station on the curve expressed in feet. Then by direct proportion

$$\frac{d_1}{D} = \frac{c_1'}{100}$$

or

$$d_1 = \frac{c_1'D}{100} \quad \text{and} \quad \frac{d_1}{2} = \frac{c_1'D}{200} \tag{13-13}$$

In Fig. 13-8 let $D = 15°$. Then since $c_1' = 46.00$ ft, $d_1 = 46.00 \times 15/100 = 6.90° = 6° \, 54'$, and $d_1/2 = 3° \, 27'$.

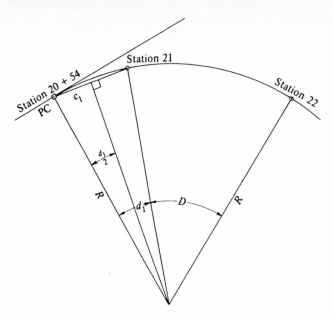

FIGURE 13-8 Relation between central angle and deflection angle.

The *actual* chord length between the PC and station 21 + 00, designated c_1 in Fig. 13-8, is used to lay out the point on the ground. It is found from the relationship

$$\frac{\frac{1}{2}c_1}{R} = \sin \tfrac{1}{2}d_1$$

from which

$$c_1 = 2R \sin \tfrac{1}{2}d_1 \tag{13-14}$$

The value of c_1 in Fig. 13-8 will depend on whether the degree of curve is determined by the chord or arc definition, since the value of R and consequently T will be different for the two definitions for a given D.

13-8   Central Angle and Chord from Last Curve Station to PT

If c_2' denotes the *difference in stationing* between the last station on the curve and the PT, then the central angle d_2 between these two points is

$$d_2 = \frac{c_2'D}{100} \quad \text{and} \quad \frac{d_2}{2} = \frac{c_2'D}{200} \tag{13-15}$$

The chord length c_2 between the last curve station and the PT is

$$c_2 = 2R \sin \tfrac{1}{2}d_2 \tag{13-16}$$

13-9   Central Angle and Chord Between Any Two Curve Points

From Sections 13-7 and 13-8 it is obvious that the angle at the center subtended between any two points on the curve is proportional to the *difference in stationing* between the two points. Let c' be the difference in stationing between any two points on the curve, let d be the central angle between the two points, and let D be the degree of curve, either by chord or arc definition. Then

$$d = \frac{c'D}{100} \text{ and } \frac{d}{2} = \frac{c'D}{200} \qquad (13\text{-}17)$$

The *actual* chord distance c between the same two points, either by chord or arc definition, is given by the relationship

$$c = 2R \sin \tfrac{1}{2} d \qquad (13\text{-}18)$$

EXAMPLE 13-1   The PC of an arc definition curve is at station 17 + 73.35; the length of the curve, by Eq. (13-9), is 788.96 ft; the degree of curve D is 7° 30′. What is the value of d_1 and d_2? What are the chord lengths c_1 between the PC and the first curve station, and c_2 between the last curve station and the PT? What is the chord length between station 19 + 00 and station 23 + 50?

Solution: By Eq. (13-11), the PT station is

$$1773.35 + 788.96 = 25 + 62.31$$

Then

$$c_1' = 1800.00 - 1773.35 = 26.65 \text{ ft}$$

$$c_2' = 2562.31 - 2500.00 = 62.31 \text{ ft}$$

By Eq. (13-13),

$$\frac{d_1}{2} = 26.65 \times \frac{7.5}{200} = 0.99938° = 0° \, 59' \, 58''$$

By Eq. (13-15),

$$\frac{d_2}{2} = 62.31 \times \frac{7.5}{200} = 2.33663° = 2° \, 20' \, 12''$$

By Eq. (13-3),

$$R = \frac{5729.58}{7.5} = 763.94 \text{ ft}$$

Then by Eq. (13-14),

$$c_1 = 2 \times 763.94 \sin 0° \, 59' \, 58'' = 26.65 \text{ ft}$$

and by Eq. (13-16)

$$c_2 = 2 \times 763.94 \sin 2° \, 20' \, 12'' = 62.29 \text{ ft}$$

Note the c_1 and c_1' are equal to one another, and c_2 differs from c_2' by only 0.02 ft.

By Eq. (13-17), one-half the central angle between station 19 and station 23 + 50 is

$$\frac{d}{2} = 450.00 \times \frac{7.5}{200} = 16.8750° = 16°\ 52'\ 30''$$

By Eq. (13-18), the chord joining these two points on the curve is

$$2 \times 763.94 \sin 16°\ 52'\ 30'' = 443.52 \text{ ft} \qquad \blacklozenge$$

EXAMPLE 13-2   What is the chord length between any two successive stations of the curve of Example 13-1? Between any two half stations? Between any two quarter stations?

Solution: The central angle between any two successive stations is D; between any two half stations it is $D/2$; and between any two quarter stations it is $D/4$. Then by Eq. (13-18),

$$c_{100} = 2R \sin \frac{D}{2} \qquad \text{(13-19)}$$

$$c_{50} = 2R \sin \frac{D}{4} \qquad \text{(13-20)}$$

$$c_{25} = 2R \sin \frac{D}{8} \qquad \text{(13-21)}$$

in which c_{100}, c_{50}, and c_{25} are the respective chord lengths desired. Thus

$$c_{100} = 2 \times 763.94 \sin 3°\ 45' \qquad = 99.93 \text{ ft}$$
$$c_{50} = 2 \times 763.94 \sin 1°\ 52'\ 30'' = 49.99 \text{ ft}$$
$$c_{25} = 2 \times 763.94 \sin 0°\ 56'\ 15'' = 25.00 \text{ ft} \qquad \blacklozenge$$

Example 13-2 shows that the chord lengths c_{100}, c_{50}, and c_{25} are all less than, if not equal to, the corresponding difference in stationing when using the arc definition. However, using a chord-definition curve, $c_{100} = 100$ ft, by definition; but c_{50} is greater than 50 ft and c_{25} is greater than 25 ft. This can be seen by an inspection of Fig. 13-9. The chord between station 23 and station 24 is 100 ft, by definition. The chord between station 23 and station 23 + 50 is longer than 50 ft because it is the hypotenuse of a right triangle one side of which is 50 ft. Also, the chord between station 23 and station 23 + 25 is longer than 25 ft.

13-10   Deflection Angles to Points on Curve

The angle formed between the back tangent and a line from the PC to a point on the curve is the *deflection angle* to the point. This deflection angle, measured at the PC between the tangent and the line to the point, is one-half the central angle subtended between the PC and the point. This relationship comes directly from the geometry of a circle. An angle between a tangent and a chord is measured by one-half the intercepted arc, whereas the central angle is measured by the whole arc. Thus in Fig. 13-10

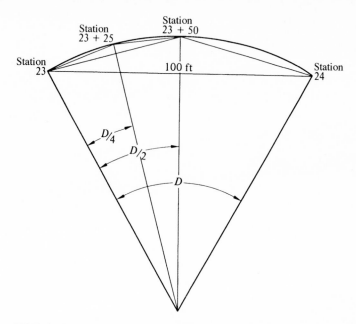

FIGURE 13-9 Chord length on chord-definition curve.

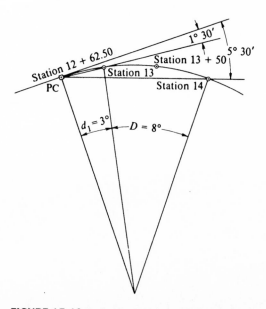

FIGURE 13-10 Deflection angles to points on curve.

the central angle d_1 between the PC at station $12 + 62.50$ and the first full station on the curve is $3°$, and the deflection angle $\frac{1}{2}d_1$ between the back tangent and the line directed to station $13 + 00$ is $1° 30'$. Similarly, the central angle between the PC and the second full station on the curve is $d_1 + D = 3° + 8° = 11°$ and the deflection angle, which is $\frac{1}{2}d_1 + \frac{1}{2}D$, is $5° 30'$. If each half station were to be located, then the deflection angle to station $13 + 50$ in Fig. 13-10 would be $\frac{1}{2}d_1 + \frac{1}{4}D = 3° 30'$.

The deflection angle to each succeeding *full* station is obtained by adding the value $\frac{1}{2}D$ to the preceding deflection angle. To locate *half* stations, $\frac{1}{4}D$ is added; to locate *quarter* stations, $\frac{1}{8}D$ is added. The last deflection is obtained by adding $d_2/2$ to the next to last calculation. This is the deflection angle to the PT and must equal $\Delta/2$ as shown in Fig. 13-3. This provides a check on the calculations.

EXAMPLE 13-3   Compute the deflection angles to each half station of the curve given in Example 13-1.

Solution: The value of c_2' in this case is $2562.31 - 2550.00 = 12.31$ ft, and $d_2/2$ is $(12.31 \times 7.5)/200 = 0° 27' 42''$. The value of Δ is, by Eq. (13-9), $59° 10' 20''$. The calculations for the deflection angles are as follows:

Station	Deflection Angle
PC = station 17 + 73.35	0° 00'
+ d_1/2	+ 0° 59' 58''
18 + 00	0° 59' 58''
+ D/4	+ 1° 52' 30''
18 + 50	2° 52' 28''
+ D/4	+ 1° 52' 30''
19 + 00	4° 44' 58''
+ D/4	+ 1° 52' 30''
19 + 50	6° 37' 28''
.	.
.	.
.	.
25 + 00	27° 14' 58''
+ D/4	+ 1° 52' 30''
25 + 50	29° 07' 28''
+ d_2/2	+ 0° 27' 42''
PT = station 25 + 62.31	29° 35' 10'' = Δ/2

In highway engineering, the radius R is usually given as a round number rather than the degree of curve D. The deflection angles to the curve stations can then be computed by first computing the deflection angle per foot of curve and then multiplying by the distance (the difference in stationing) between the PC and the curve point. In Fig. 13-11, the angle subtending an arc equal to the radius is 1 radian or $57.29578°$. The deflection angle from the tangent at the PC to the end of the arc is thus $28.64789°$. Therefore the deflection angle per foot is $28.64789°/R$.

FIGURE 13-11 Deflection angle to a point at an arc length R from the PC.

EXAMPLE 13-4 Given an arc definition curve of radius 600 ft and a central angle of 42° 34′. The PC is at station 14 + 62.20. Compute the deflection angles to each one-half station on the curve. Compute the chords c_1, c_2, and c_{50}.

Solution: The length of the curve is $R\Delta$ or $600 \times 0.74292849 = 445.76$ ft. PT station is 1462.20 + 445.76 = 1907.96 ft or station 19 + 07.96. The deflection per foot is 28.64789°/600 = 0.047746°.

The calculations for the deflection angles are as follows:

Station	Distance from PC	Deflection Angle	Deflection Angle
PC = 14 + 62.20	0°	0°	0°
15 + 00	37.80	1.80480°	1° 48′ 17″
15 + 50	87.80	4.19210°	4° 11′ 32″
16 + 00	137.80	6.57940°	6° 34′ 46″
16 + 50	187.80	8.96670°	8° 58′ 00″
17 + 00	237.80	11.35400°	11° 21′ 14″
17 + 50	287.80	13.74130°	13° 44′ 29″
18 + 00	337.80	16.12860°	16° 07′ 43″
18 + 50	387.80	18.51590°	18° 30′ 57″
19 + 00	437.80	20.90320°	20° 54′ 12″
PT = 19 + 07.96	445.76	21.28326°	21° 17′ 00″ = $\Delta/2$

$c_1 = 2 \times 600 \times \sin 1° 48′ 17″ = 37.79$ ft
$c_2 = 2 \times 600 \times \sin(7.96 \times 0.047746)° = 7.96$ ft
$c_{50} = 2 \times 600 \times \sin(50 \times 0.047746)° = 49.99$ ft

13-11 Curve Location

To locate the position of a simple curve in the field, the intersection angle Δ is measured at the PI with the theodolite. Based on an assumed value for one of the elements of the curve, the other elements and the stationing of the PC and the PT are computed. A table of deflection angles is prepared for locating full stations, half stations, or quarter stations. With the theodolite still at the PI, the PC and the PT are located by laying off the tangent distance T.

The instrument is next set up at the PC, and a backsight is taken along the back tangent toward the PI. The first deflection angle is turned off, and the actual chord distance c_1 is laid off on the line of sight to locate the first curve station. When full stations are located, the second deflection angle is turned off by setting the horizontal circle to this reading. The chord distance between the first and second curve stations is laid off by measuring from the first curve station, and the forward end of the chord is brought on the line of sight. This procedure establishes the second curve station. Each station is located by a line of sight from the PC and a chord distance from the preceding station, until the end of the curve is reached or until a setup on the curve becomes necessary. The previously located position of the PT provides a check on the accuracy of the field work.

The complete computations for the curve shown in Fig. 13-3 are given in Table 13-1. The order in which these computations are made is indicated by the numbers in parentheses. The deflection angles are computed more exactly than the angles can be turned off in the field. This is done to check the value of the final deflection angle, since it must agree with $\frac{1}{2}\Delta$. The curve is based on the chord definition for degree of curve.

The complete computations for an arc definition curve are shown in Table 13-2. Every half station is to be located by chords and deflection angles. Since the radius is assumed to be an integral number of feet, the degree of curve must be computed to at least five places after the decimal as shown in the table. The 1″ discrepancy in the check on deflection angles is due to rounding-off errors and is of no consequence.

TABLE 13-1 Computations for Circular Curve Full Stations, Chord Definition

$\Delta = 23°\,18'$ $\frac{1}{2}\Delta = 11°\,39'$	D (assumed) $= 4°\,00'$ chord definition

PI = station 50 + 53.87			$c_1' = 41.52$	(6)
$R = 50/\sin 2°\,00' = 1432.69$	(1)		$\frac{1}{2}d_1 = \dfrac{41.52 \times 4°}{200} = 0°\,49'\,49''$	(7)
$T = 1432.69 \tan 11°\,39' = 295.39$	(2)		$c_1 = 2 \times 1432.69 \times \sin 0°\,49'\,49'' = 41.52$	(8)
$L = 100 \times 23.3/4 = 582.50$	(3)		$c_2' = 40.98$	(9)
			$\frac{1}{2}d_2 = \dfrac{40.98 \times 4}{200} = 0°\,49'\,11''$	(10)

$$\begin{array}{rl}
\text{PI} = & 50 + 53.87 \\
T = & \underline{2 + 95.39} \\
\text{PC} = & 47 + 58.48 \qquad (4)\\
L = & \underline{5 + 82.50} \\
\text{PT} = & 53 + 40.98 \qquad (5)
\end{array}$$

$c_2 = 2 \times 1432.69 \times \sin 0°\,49'\,11'' = 40.99$ (11)

Station	Deflection angle	(12)
PC = 47 + 58.48	0° 00′ 00″	
48	0° 49′ 49″	
49	2° 49′ 49″	
50	4° 49′ 49″	
51	6° 49′ 49″	
52	8° 49′ 49″	
53	10° 49′ 49″	
PT = 53 + 40.98	11° 39′ 00″	

(checks $\frac{1}{2}\Delta$)

TABLE 13-2 Computations for Half Stations, Arc Definition

$\Delta = 28°\,46'$ $\frac{1}{2}\Delta = 14°\,23'$	R (assumed) $= 850$ ft arc definition	
$D = 5729.58/850 = 6.74068°$ (1)	$D/4 = 1°\,41'\,07''$ (1)	
$T = 850 \tan 14°\,23' = 850 \times 0.25645$	$c_1' = 47.43$ (6)	
$\qquad = 217.98$ ft (2)	$c_2' = 29.33$ (7)	
$L = \dfrac{100 \times 28.76667}{6.74068} = 426.76$ (3)		
PI $= 26 + 20.55$	Station	Deflection angle (13)
$T = \underline{\ 2 + 17.98\ }$	PC $= 24 + 02.57$	$0°\,00'\,00''$
PC $= 24 + 02.57$ (4)	$24 + 50$	$1°\,35'\,55''$
$L = \underline{\ 4 + 26.76\ }$	25	$3°\,17'\,02''$
PT $= 28 + 29.33$ (5)	$25 + 50$	$4°\,58'\,09''$
$\frac{1}{2}d_1 = \dfrac{47.43 \times 6.74068}{200} = 1°\,35'\,55''$ (8)	26	$6°\,39'\,16''$
	$26 + 50$	$8°\,20'\,23''$
$\frac{1}{2}d_2 = \dfrac{29.33 \times 6.74068}{200} = 0°\,59'\,19''$ (9)	27	$10°\,01'\,30''$
	$27 + 50$	$11°\,42'\,37''$
$c_1 = 2 \times 850 \sin 1°\,35'\,55'' = 47.43$ (10)	28	$13°\,23'\,44''$
$c_2 = 2 \times 850 \sin 0°\,59'\,19'' = 29.33$ (11)	PT $= 28 + 29.33$	$14°\,23'\,03''$
$c_{50} = 2 \times 850 \sin (6.74068/4)° = 49.99$ (12)		(checks $\frac{1}{2}\Delta$)

13-12 Moving Up on Curve

It frequently happens that because of the length of the curve or because of obstacles on the line, the entire curve cannot be staked out with the transit at the PC of the curve. If the transit is properly manipulated, it can be moved forward along the curve as many times as may be necessary, and the deflection angles previously computed can be used in staking out the remaining portions of the curve.

A portion of the curve for which the computations are tabulated in Table 13-1 is shown in Fig. 13-12 to a distorted scale. Let it be assumed that the theodolite is moved to station 50, which has been located by turning off a deflection angle of

FIGURE 13-12 Moving up on a curve.

4° 49′ 49″ from the tangent at the PC. If a backsight is taken on the PC with the circle reading 180°, it is apparent that the circle will read 4° 49′ 49″ when the line of sight is directed forward along the tangent at station 50. Since the angle between this tangent and the chord to station 51 is $\frac{1}{2}D$, or 2° 00′, the circle will read 6° 49′ 49″ when the line of sight is directed along the chord to station 51. This reading of 6° 49′ 49″ is the previously computed deflection angle for station 51. The reading of 8° 49′ 49″, previously computed for station 52, will be the reading when the line of sight is directed along the chord from station 50 to station 52.

Thus if the circle is set at 180° and a backsight is taken on the PC, the previously computed deflection angles can be used in locating stations beyond the one occupied by the theodolite. If the instrument is in good adjustment, the circle can be set to 0° 00″ and the backsight made with the telescope upside down. Then when the telescope is plunged to its direct position, the line of sight is oriented along the line marked 0° 00″ in Fig. 13-12.

If a second or subsequent curve setup is required, the backsight circle reading then corresponds to the computed deflection angle for the station on which the backsight is made. For example, if a setup is made at station 52 of Fig. 13-12 and a backsight is taken to station 50, the circle reading should read 180° + 4° 49′ 49″ when the backsight is made with the telescope direct or 4° 49′ 49″ when the backsight is made with the telescope upside down. The remainder of the curve stations are then set using the deflection angles as computed.

13-13   Location of Curve by Inversing

In modern transportation engineering, all of the surveys are placed on a plane coordinate system. If horizontal control for the location of tangents and curves has been established, the coordinates of all of the PIs, PCs, and PTs and the curve stations can be computed, based on the final alignment of the center line with respect to the control points and on the geometry of each of the curves. By inversing between a control point occupied by an instrument and any point on the tangents or curves, the direction and distance to the point can be computed. This direction and distance can be laid out using a combination theodolite-EDM or else by use of the total station instrument. In the former case a listing of all angles measured from a control line together with the corresponding distances are precomputed and furnished to the survey party. When using the total station instrument, the coordinates of the control point, the backsight point, and all the points to be located are loaded into the memory of the instrument's microprocessor. As each point is called up, the direction and distance for that point are computed. The point is then located as discussed in Section 4-24.

Although different approaches can be taken to locate a curve and tangents by the inversing technique, modified by such considerations as clearance, topography, preliminary surveys, type of instrument, and availability of field personnel, the following example is presented to give the student some insight into the procedure.

Suppose that station 0 + 00 of Fig. 13-7 has the following set of coordinates (kept reasonably small to simplify the discussion): X = 15,926.14 ft, Y = 26,244.20 ft. Further assume that the azimuth of the line joining station 0 + 00 and station $8 + 14.72$ is $92°\ 12'$. A traverse can be computed commencing at station $0 + 00$; to PC_1 which is at station $5 + 11.59$; to the center of the curve (point O of Figure 13-13); to PT_1; to PT_2, and so on. Latitudes and departures are computed for these traverse lines, following which the coordinates are determined, as shown in Table 13-3.

A check on the calculations can be made at PT_1 by comparing the coordinates obtained in Table 13-3 with those obtained by traversing from station $0 + 00$ to station $8 + 14.72$ (PI_1), then along the forward tangent a distance equal to tangent distance T_1. This is shown in Table 13-4. This check is made to detect mistakes and to avoid the effect of accumulated rounding-off errors.

In this example, assuming an arc-definition curve,

$$D = \frac{5729.578}{1150} = 4.9822° = 4°\ 58'\ 56'' \qquad \text{[by Eq. (13-4)]}$$

$$d_1 = \frac{88.41 \times 4.9822°}{100} = 4.4048° = 4°\ 24'\ 17'' \qquad \text{[by Eq. (13-13)]}$$

$$d_2 = \frac{4.37 \times 4.9822°}{100} = 0.2177° = 0°\ 13'\ 04'' \qquad \text{[by Eq. (13-15)]}$$

The azimuth of all of the radii between the center of the curve at 0 and the curve stations 6 through 11 of Fig. 13-13 can be computed using the above angles. The latitudes and departures of these radii from O to the curve stations are then computed and added to the coordinates of O to obtain the coordinates of the curve

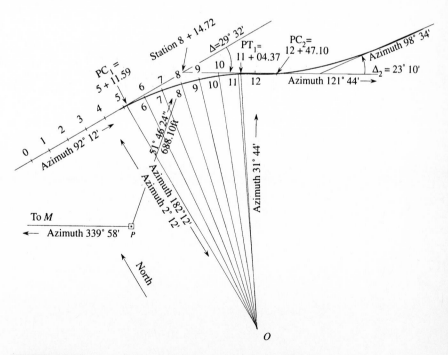

FIGURE 13-13 Location of curve and tangent points by inversing.

TABLE 13-3   Calculation of Latitudes, Departures, and Coordinates for Curve Location

Station	Azimuth	Length	Latitude[a]	Departure[a]	Y	X
0 + 00					26,244.20	15,926.14
	92° 12′	100.00	−3.839	+99.926		
1 + 00					26,240.36	16,026.07
	92° 12′	100.00	−3.839	+99.926		
2 + 00					26,236.52	16,125.99
	92° 12′	100.00	−3.839	+99.926		
3 + 00					26,232.68	16,225.92
	92° 12′	100.00	−3.839	+99.926		
4 + 00					26,228.84	16,325.84
	92° 12′	100.00	−3.839	+99.926		
5 + 00					26,225.01	16,425.77
	92° 12′	11.59	−0.445	+11.582		
PC$_1$ = 5 + 11.59					26,224.56	16,437.35
	182° 12′	1150.00	−1,149.152	−44.146		
Point O					25,075.41	16,393.21
	31° 44′	1150.00	+978.081	+604.862		
PT$_1$ = 11 + 04.37					26,053.49	16,998.07
	121° 44′	95.63	−50.298	+81.334		
12 + 00					26,003.19	17,079.40
	121° 44′	47.10	−24.773	+40.059		
PC$_2$ = 12 + 47.10					25,978.42	17,119.46

[a]Carried an extra place to reduce accumulated rounding-off error.

TABLE 13-4   Calculations to Check the Coordinates of PT$_1$

Station	Azimuth	Length	Latitude	Departure	Y	X
0 + 00					26,244.20	15,926.14
	92° 12′	814.72	−31.28	+814.12		
8 + 14.72					26,212.92	16,740.26
	121° 44′	303.13	−159.44	+257.81		
11 + 04.37					26,053.48	16,998.07

stations. Calculation of latitudes and departures of the radii is shown in Table 13-5. The coordinates of the curve stations are shown in Table 13-6.

Let point P of Fig. 13-13 be a control point whose coordinates are X_p = 16,180.61, Y_p = 25,751.88. This point is occupied to set stations on the tangents and curves of Fig. 13-13. A backsight is taken on control point M with the azimuth of MP set to read 339° 58′ on the horizontal circle. A listing of directions and distances from P to the station points have been computed by inversing each line. For example, to set station 8, the inverse calculations are

$$\tan (\text{azimuth } P - 8) = \frac{16{,}721.16 - 16{,}180.61}{26{,}177.66 - 25{,}751.88} \qquad A = 51° \, 46′ \, 24″$$

$$\text{distance } (P - 8) = [(16{,}721.16 - 16{,}180.61)^2 + (26{,}177.66 - 25{,}751.88)^2]^{1/2}$$

$$= 688.10 \text{ ft}$$

TABLE 13-5 Calculation of Latitudes and
Departures of Radii

Line	Azimuth	Length	Latitude	Departure
O-6	6° 36′ 17″	1150.00	+1142.37	+132.27
O-7	11° 35′ 13″	1150.00	+1126.56	+230.98
O-8	16° 34′ 09″	1150.00	+1102.25	+327.95
O-9	21° 33′ 05″	1150.00	+1069.60	+422.44
O-10	26° 32′ 01″	1150.00	+1028.87	+513.73
O-11	31° 30′ 57″	1150.00	+980.37	+601.14

TABLE 13-6 Coordinates of
Curve Stations

Station	Y	X
6	26,217.78	16,525.48
7	26,201.97	16,624.19
8	26,177.66	16,721.16
9	26,145.01	16,815.65
10	26,104.28	16,906.94
11	26,055.78	16,994.35

The inverse computations can be performed in the office if it is assumed that point *P* will be the occupied station. The advantage of the modern total station instrument is the ability for making these computations directly at the instrument. When the "one-person" system described in Section 4-24 is used, the calculations are made directly at the reflector. Thus, when using the total station instrument, *any* control point can be used to locate the curve and tangent stations.

13-14 Intersection of Curve and Straight Line

One of the common problems in right-of-way surveys is the intersection of a straight line with a curve. Thus in Fig. 13-14 *DG* represents a property line intersected by the curve *ABC*. When the curve is flat, the intersection *B* can be located on the ground by setting points at *E* and *F* on the curve on both sides of the intersection and close enough together so that no appreciable error will be introduced by considering the curve to be a straight line between these two points. The point *B* is found at the intersection of the chord *EF* and the property line *DG*.

The intersection can be obtained mathematically if the distance *AV*′ and the angle *GV*′*V* are measured on the ground. The figure *V*′*BOA* can be considered as a traverse with two missing quantities, the length *V*′*B* and the direction of the line *BO*. These values can be calculated by the methods of Section 8-30. From the computed direction, the central angle *AOB* and the station of *B* on the curve can be calculated.

A second mathematical solution will be apparent from the following relationships: In Fig. 13-14,

$$\text{angle } GV'V = \text{angle } DV'A = \text{angle } IOA = a$$

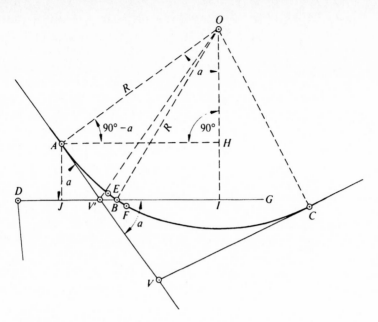

FIGURE 13-14 Intersection of curve and straight lines.

In triangle AOH,

$$AH = R \sin a \text{ and } OH = R \cos a$$

In triangle $AV'J$,

$$V'J = AV' \cos a \text{ and } AJ = AV' \sin a$$

Also,

$$V'I = AH - V'J$$

$$OI = OH + AJ$$

$$\cos IOB = \frac{OI}{R}$$

$$\tan IOV' = \frac{V'I}{OI}$$

$$BI = R \sin IOB$$
$$V'B = V'I - BI$$

and

$$\text{angle } AOB = \text{angle } AOI - \text{angle } IOB$$

When these computations have been completed, B may be located by measuring the computed distance $V'B$ along $V'G$ or by means of a deflection angle and a chord measured from the PC of the curve. The length of the chord AB is $2R \sin \frac{1}{2} AOB$.

The problem of intersecting a line with a curve can be solved by the methods given in Section 8-21 if the line and the curve are both on a coordinate system.

13-15   Vertical Curves

When the grade line of a highway or a railroad changes, provision must be made for a vehicle to negotiate this transition smoothly and to provide vision over the crest of a hill far enough ahead to give the operator of the vehicle ample time to react to a dangerous situation. The parabola is most commonly used for connecting two different grades to provide for this transition. It is easy to compute elevations on a parabola, and such a curve also provides a constant rate of change of grade. The results of the vertical-curve computations are the grade elevations at selected points along a route from the beginning of the curve to its end. These elevations are used, in turn, to control grading operations when the roadbed is to be brought to the desired grade by excavation and the construction of embankments.

The length of a vertical curve in 100-ft stations is designated as L and is measured along the horizontal. The two grades in the direction of stationing are g_1 and g_2. The total change in grade is $(g_2 - g_1)$. The rate of change of grade per station, designated as r, is found by dividing the total change in grade by the length of the curve in stations. Thus

$$r = \frac{g_2 - g_1}{L} \tag{13-22}$$

in which r is the rate of change of grade per station; g_1 is the initial grade, in percent; g_2 is the final grade, in percent; and L is the length of the curve, in stations. When r is specified, the required length L is found by the relationship

$$L = \frac{g_2 - g_1}{r} \tag{13-23}$$

If two given grades are to be connected by a vertical curve, then either r or L must be assumed and the other value is computed. The sharpness of a vertical curve in railroad location is usually defined by the allowable rate of change of grade.

When a vertical curve is laid out so that the intersection of the grade lines, called the *point of grade intersection* and designated as V, lies midway between the two ends of the curve measured horizontally, then the vertical curve is called an *equal-tangent* parabolic vertical curve.

Two methods are available for computing the elements of a vertical curve. The first method, presented in Section 13-16, treats the curve analytically. By this method problems involving high and low points on the curve, vertical clearance, and curve intersections can be dealt with most efficiently. Also, the analytic method can be adapted to computer programming very easily. The second method, presented in Section 13-17, is somewhat more easily applied for simple vertical-curve problems. It takes advantage of the geometric properties of the parabola. The solution of a complicated curve problem, however, is more difficult by the geometric method.

13-16 Vertical Curves by Equation of Parabola

In Fig. 13-15, the beginning of a vertical curve is abbreviated as BVC, the intersection of the tangents as V, and the end of the vertical curve as EVC. The initial grade is g_1 and the final grade is g_2. The y values are elevations in feet, and the x values are stations beyond the BVC. The y axis is passed through the BVC, and the x axis lies on the datum. The equation of a parabola with the axis in the y direction is

$$y = ax^2 + bx + c \qquad (13\text{-}24)$$

When $x = 0$, y is the elevation of the BVC. Therefore c is the elevation of the BVC, and Eq. (13-24) becomes

$$y = ax^2 + bx + (\text{elevation of BVC}) \qquad (13\text{-}25)$$

The first derivative of y with respect to x from Eq. (13-24) is

$$\frac{dy}{dx} = 2ax + b \qquad (13\text{-}26)$$

When $x = 0$, the slope of the curve, is g_1. Since this slope equals the first derivative of the curve at $x = 0$, it follows that $b = g_1$. Equation (13-24) then becomes

$$y = ax^2 + g_1x + (\text{elevation of BVC}) \qquad (13\text{-}27)$$

The second derivative is the rate of change of slope or grade of the curve. So $2a = r$ and Eq. (13-24) becomes

$$y = \frac{r}{2}x^2 + g_1x + (\text{elevation of BVC}) \qquad (13\text{-}28)$$

which is the equation of the equal-tangent parabolic vertical curve used to connect two grades. In Eq. (13-28) y is the elevation of a point on the curve, and x is the distance in stations between the BVC and the point. If the elevations of the points

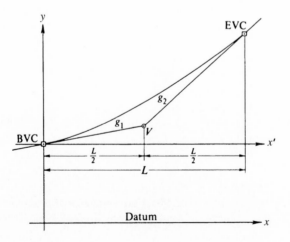

FIGURE 13-15 Parabola with rectangular coordinate axes.

above or below the BVC are desired, the x' axis becomes the y' axis, as shown in Fig. 13-15, and c becomes 0. For these conditions,

$$y' = \frac{r}{2}x^2 + g_1x \qquad (13\text{-}29)$$

The value of r must be assigned its proper algebraic sign. Equation (13-22) gives the algebraic sign directly. In Fig. 13-16 it is seen that if the vertical curve opens upward, r is positive; and if it opens downward, r is negative.

EXAMPLE 13-5   Two grades, for which $g_1 = +1.25\%$ and $g_2 = -2.75\%$, intersect at station 18 + 00, and the elevation of the intersection is 886.10 ft. If the length of the curve is to be 600 ft, what are the elevations of the BVC, the EVC, and all full stations on the curve?

Solution: See Fig. 13-17. The BVC is at station 15, and the EVC is at station 21. The elevation of the BVC is obtained by going backward along grade g_1 from the point of intersection for a distance of 300 ft. The elevation of the EVC is obtained

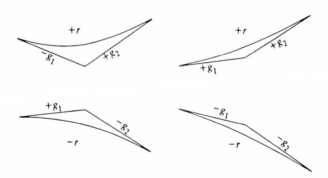

FIGURE 13-16 Algebraic sign of r.

FIGURE 13-17 Parabolic vertical curve.

by going forward from the point of intersection along grade g_2 for 300 ft. The computations follow:

$$\text{elevation of BVC} = 886.10 - 3 \times 1.25 = 882.35 \text{ ft}$$

$$\text{elevation of EVC} = 886.10 - 3 \times 2.75 = 877.85 \text{ ft}$$

By Eq. (13-22), the change of grade per station is

$$r = \frac{-2.75 - 1.25}{6} = -0.667\%$$

The equation of the curve is $y = -0.333x^2 + 1.25x + 882.35$.   ◆

The elevations of the points on the curve are computed by preparing the accompanying table. Note that the elevation of a point is the sum of the values in the fourth, fifth, and sixth columns of Table 13-7.

13-17   Vertical Curves by Tangent Offsets from Grade Lines

Three properties of an equal-tangent vertical curve are as follows: (1) The offsets from the tangent to the curve at a point are proportional to the squares of the horizontal distances from the point; (2) offsets from the two grade lines are symmetrical with respect to the point of intersection of the two grade lines; (3) the curve lies midway between the point of intersection of the grade lines and the middle point of the chord joining the BVC and the EVC. Proof of these properties is left to the reader.

In Fig. 13-18, by the first property, $bB = 4aA$; $cC = 9aA$; $OV = 16aA$; $eE = 4fF$; $dD = 9fF$; $OV = 16fF$. By the second property, $aA = fF$; $bB = eE$; $cC = dD$. By the third property, $MO = OV$, or $OV = \frac{1}{2}MV$. The distance MV is obtained by subtracting the elevation of V from the elevation of M. The elevation of M is the mean of the elevations of the BVC and the EVC, since it lies at the middle point of the chord joining the two points. The elevations of points A, B, C, D, E, and F, the BVC, and the EVC are computed from the assigned grades of the two tangents. By employing the properties of the curve, the elevations of the points on the curve can then be computed.

TABLE 13-7   Computations for Vertical Curve by Equation of Parabola

Station	x	x^2	$\frac{r}{2}x^2$	$g_1 x$	Elevation BVC	Elevation Curve
BVC = 15	0	0	0	0	882.35	882.35
16	1	1	−0.33	+1.25	882.35	883.27
17	2	4	−1.33	+2.50	882.35	883.52
18	3	9	−3.00	+3.75	882.35	883.10
19	4	16	−5.33	+5.00	882.35	882.02
20	5	25	−8.33	+6.25	882.35	880.27
EVC = 21	6	36	−12.00	+7.50	882.35	877.85

EXAMPLE 13-6 For the two grade lines of Fig. 13-18, $g_1 = -6\%$ and $g_2 = +2\%$. The elevation of the intersection of the grade lines at station 50 is 550.97 ft. The length of the curve is 800 ft. Compute the stations and elevations of the BVC and the EVC and the elevations at all other stations on the curve.

Solution: The computations may be arranged as follows:

$$\text{station of BVC} = 50 - 4 = \text{station } 46$$

$$\text{elevation of BVC} = 550.97 + 4 \times 6 = 574.97 \text{ ft}$$

$$\text{station of EVC} = 50 + 4 = \text{station } 54$$

$$\text{elevation of EVC} = 550.97 + 4 \times 2 = 558.97 \text{ ft}$$

$$\text{elevation of middle point of chord} = \frac{574.97 + 558.97}{2} = 566.97 \text{ ft}$$

$$\text{offset to curve at intersection} = \frac{566.97 - 550.97}{2} = 8.00 \text{ ft} = VO$$

$$\text{offset at } A \text{ and } F = (\tfrac{1}{4})^2 \times 8.00 = 0.50 \text{ ft}$$

$$\text{offset at } B \text{ and } E = (\tfrac{2}{4})^2 \times 8.00 = 2.00 \text{ ft}$$

$$\text{offset at } C \text{ and } D = (\tfrac{3}{4})^2 \times 8.00 = 4.50 \text{ ft}$$

$$\text{elevation of } A = 574.97 - 6.00 = 568.97 \text{ ft}$$

$$\text{elevation of } B = 568.97 - 6.00 = 562.97 \text{ ft}$$

$$\text{elevation of } C = 562.97 - 6.00 = 556.97 \text{ ft}$$

$$\text{elevation of } V = 556.97 - 6.00 = 550.97 \text{ ft}$$

$$\text{elevation of } F = 558.97 - 2.00 = 556.97 \text{ ft}$$

$$\text{elevation of } E = 556.97 - 2.00 = 554.97 \text{ ft}$$

$$\text{elevation of } D = 554.97 - 2.00 = 552.97 \text{ ft}$$

$$\text{elevation of } V = 552.97 - 2.00 = 550.97 \text{ ft (check)}$$

$$\text{elevation of } a = \text{station } 47 = 568.97 + 0.50 = 569.47 \text{ ft}$$

$$\text{elevation of } b = \text{station } 48 = 562.97 + 2.00 = 564.97 \text{ ft}$$

$$\text{elevation of } c = \text{station } 49 = 556.97 + 4.50 = 561.47 \text{ ft}$$

$$\text{elevation of } o = \text{station } 50 = 550.97 + 8.00 = 558.97 \text{ ft}$$

$$\text{elevation of } d = \text{station } 51 = 552.97 + 4.50 = 557.47 \text{ ft}$$

$$\text{elevation of } e = \text{station } 52 = 554.97 + 2.00 = 556.97 \text{ ft}$$

$$\text{elevation of } f = \text{station } 53 = 556.97 + 0.50 = 557.47 \text{ ft}$$

The foregoing computations should be arranged as shown in Table 13-8. ◆

13-18 Intermediate Points on Vertical Curves

Occasions will frequently arise where the elevations of points on vertical curves must be computed at intervals of 50, 25, or even 10 ft. Also, the elevations of random points on the curve are quite frequently necessary. The most direct way of

FIGURE 13-18 Properties of the vertical curve.

TABLE 13-8 Computations for Vertical Curve by Offsets from Both Tangents

Station	Tangent Elevation	Offset from Tangent	Elevation Curve
BVC = 46	574.97	0	574.97
47	568.97	+0.50	569.47
48	562.97	+2.00	564.97
49	556.97	+4.50	561.47
50	550.97	+8.00	558.97
51	552.97	+4.50	557.47
52	554.97	+2.00	·556.97
53	556.97	+0.50	557.47
EVC = 54	558.97	0	558.97

computing the elevations of these intermediate points is by use of Eq. (13-28) where x is the station or plus beyond the BVC and y is the elevation of the point in feet. In Example 13-5, if the elevation of station 17 + 22.33 is desired, then $x = 2.2233$ stations beyond the BVC. The elevation of station 17 + 22.33 is therefore

$$y = (-0.333)(2.2233)^2 + (1.25)(2.2233) + 882.35 = 883.48 \text{ ft}$$

The elevation of an intermediate point can be computed by the tangent offset method, although not quite so readily as by using the equation of the curve. In Example 13-6, the elevation of station 49 + 52 is obtained by first computing the tangent elevation at the station and then by computing the tangent offset at the station. The tangent elevation is 574.97 – (3.52)(6) = 553.85 ft. The tangent offset is $(3.52/4)^2 \times 8.00 = 6.20$ ft. Therefore the elevation of the curve at station 49 + 52 is 553.85 + 6.20 = 560.05 ft.

13-19   Location of Highest or Lowest Point

When g_1 and g_2 have opposite algebraic signs, either a high point or a low point will occur between the BVC and the EVC. Furthermore, this point may not fall on a full station or on a previously selected point. Sometimes it is necessary to determine the station and elevation of the high point or the low point to locate a clearance point, a drainage structure, or some other feature. The tangent to the curve at this point will be a horizontal line; that is, the slope of the curve will be zero. The position of the point can be determined then by equating to zero the first derivative of Eq. (13-28). For the high or low point,

$$\frac{dy}{dx} = rx + g_1 = 0$$

or

$$x = -\frac{g_1}{r} \tag{13-30}$$

The value of x in Eq. (13-30) is the distance in stations from the BVC to the high or low point. The elevation of the point is found by substituting the value of x obtained from Eq. (13-30) in Eq. (13-28) and solving for y.

The high point of the curve of Example 13-5 occurs at $x = 1.25/0.666 = 1.875$ stations beyond the BVC. Thus the summit is at station $16 + 87.5$. The elevation at the point is 883.52 ft.

In Example 13-6 the value of r found by Eq. (13-22) is $[2 - (-6)]/8 = +1.00$, and the low point occurs at $x = 6/1.00 = 6$ stations beyond the BVC. It is at station 52 where the elevation was previously computed.

13-20   Passing a Vertical Curve Through a Fixed Point

Occasionally, because of geometric constraints such as meeting a fixed grade elevation at a particular station, a vertical curve with given grades and intersection station must be computed. Such a situation will likely occur at a road intersection, a railroad crossing, or a grade separation structure.

Using the grades, intersection station, and elevation given in Example 13-5, let it be required to force the vertical curve to pass through station $18 + 25$ at an elevation of 881.20 ft. Obviously the curve must be lengthened. Thus a new value of L must be computed, and a new value for r results.

In Eq. (13-28),

$$\frac{r}{2} = \frac{-2.75 - 1.25}{2L}, \qquad x = \frac{L}{2} + 0.25; \text{ and}$$

$$\text{elevation of BVC} = 886.10 - \frac{g_1 L}{2}$$

Thus

$$881.20 = \left[\frac{-2.75 - 1.25}{2L}\left(\frac{L}{2} + 0.25\right)^2\right] + \left[1.25\left(\frac{L}{2} + 0.25\right)\right] + \left[886.10 - \frac{1.25L}{2}\right]$$

This reduces to the quadratic equation

$$L^2 - 9.425L + 0.25 = 0$$

Solving for L gives

$$L = 9.3984 \text{ stations and } \frac{L}{2} = 4.6992 \text{ stations}$$

The elevation of the BVC is

$$886.10 - 1.25 \times 4.6992 = 880.23 \text{ ft}$$

The station of the BVC is

$$18 - 4.6992 = 13.3008 \text{ stations}$$

The value of x is

$$18.25 - 13.3008 = 4.9492$$

and

$$\frac{r}{2} = \frac{-2.75 - 1.25}{2 \times 9.3984} = -0.2128 \text{ \%/station}$$

Then

$$y = -0.2128 \times 4.9492^2 + 1.25 \times 4.9492 + 880.23 = 881.20 \text{ ft}$$

13-21 Unequal Tangent Vertical Curve

The situation may arise in which it is not possible to construct a vertical curve having equal tangents because of limiting distances on one side or another of the point of intersection of two grade lines. Such a situation is shown in Fig. 13-19 in which

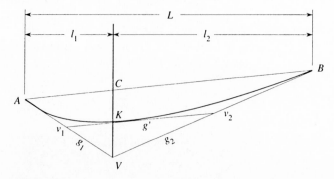

FIGURE 13-19 Unequal tangent vertical curve.

l_1 is less than l_2. A line v_1v_2 is drawn parallel with AB such that $Av_1 = v_1V$ and $Vv_2 = v_2B$. This constitutes two equal tangent parabolic vertical curves, the first from A to K and the second from K to B. The elevation of v_1 is the average of that of A and V; the elevation of v_2 is the average of that of V and B. Thus $CK = KV$. The vertical distance CV is, by derivation:

$$CV = \frac{l_1l_2}{L}(g_2 - g_1)$$

in which l_1, l_2 and L are distances in stations, and g_2 and g_1 are percent grades. Then, since $CK = KV$,

$$CK = \frac{l_1l_2}{2L}(g_2 - g_1)$$

The grade g' of v_1v_2 is the same as that of AB. Thus

$$g' = \frac{\text{elev } B - \text{elev } A}{L}$$

EXAMPLE 13-7   In Fig. 13-19, $g_1 = -4\%$, $g_2 = +3.5\%$, $l_1 = 250$ ft, $l_2 = 400$ ft, and the elevation of V is 450.00 ft. Point V is at station 55 + 00. Compute the elevations of each half station between A and B.

Solution: The elevation of A is $450.00 + 2.50 \times 4 = 460.00$ ft.
The elevation of B is $450.00 + 4.00 \times 3.5 = 464.00$ ft.

$$g' = \frac{464.00 - 460.00}{6.50} = +0.615\%$$

elevation of $K = 460.00 - 1.25 \times 4 + 1.25 \times 0.615 = 455.77$

elevation of $B = 455.77 + 2.00 \times 0.615 + 2.00 \times 3.5 = 464.00$ ft (check)

For the vertical curve from A to K,

$$r = \frac{0.615 - (-4.000)}{2.50} = 1.846\%/\text{station} \qquad \text{[By Eq. (13-22)]}$$

elevation of BVC at $A = 460.00$ ft

By Eq. (13-28), the equation of the curve is

$$y = 0.923x^2 - 4x + 460.00$$

For the vertical curve from K to B,

$$r = \frac{3.5 - 0.615}{4.00} = +0.721\%/\text{station} \qquad \text{[By Eq. (13-22)]}$$

elevation of BVC at $K = 455.77$

By Eq. (13-28), the equation of the curve is

$$y = 0.361x^2 + 0.615x + 455.77$$

TABLE 13-9 Unequal Tangent Vertical Curve by Equation of Parabola

Station	x	x^2	$r/2x^2$	$g_1 x$	Elev. BVC	Elev. Curve
52 + 50	0	0	0	0	460.00	460.00
53 + 00	0.5	0.25	+0.23	−2.00	460.00	458.23
53 + 50	1	1	+0.92	−4.00	460.00	456.92
54 + 00	1.5	2.25	+2.08	−6.00	460.00	456.08
54 + 50	2	4	+3.69	−8.00	460.00	455.69
55 + 00	2.5	6.25	+5.77	−10.00	460.00	455.77
55 + 50	0.5	0.25	+0.09	+0.31	455.77	456.17
56 + 00	1	1	+0.36	+0.62	455.77	456.75
56 + 50	1.5	2.25	+0.81	+0.92	455.77	457.50
57 + 00	2	4	+1.44	+1.23	455.77	458.44
57 + 50	2.5	6.25	+2.26	+1.54	455.77	459.57
58 + 00	3	9	+3.25	+1.85	455.77	460.87
58 + 50	3.5	12.25	+4.42	+2.15	455.77	462.34
59 + 00	4	16	+5.78	+2.46	455.77	464.00

The computations for the two curves are shown in Table 13-9. The low point on this curve is at station 54 + 66.68 at which the elevation is 455.67. ◆

13-22 Minimum Length of Vertical Curve

The length of a vertical curve on a highway should be ample to provide a clear sight that is sufficiently long to prevent accidents. The American Association of State Highway and Transportation Officials (AASHTO) has developed criteria for the distance required to pass another vehicle traveling in the same direction on a vertical curve and also for the distance required to stop a vehicle in an emergency. The former distance, called the *safe passing sight distance* and designated as S_{sp}, is based on the assumptions that the eyes of the driver of a vehicle are about 3.75 ft above the pavement and the top of an oncoming vehicle is about 4.50 ft above the pavement. The latter required distance, called the *safe stopping sight distance* and designated as S_{np}, is based on the assumption that an obstruction ahead of the vehicle is 0.50 ft above the pavement. The values of S_{sp} and S_{np} are given in Table 13-10.

As stated in Section 13-17, vertical offsets from a tangent to a parabola are proportional to the squares of the distances from the point of tangency. From this fact and Fig. 13-20, it can be shown that

$$h_1 : \frac{(g_1 - g_2)L}{800} = S_1^2 : \left(\frac{L}{2}\right)^2$$

$$S_1 = \sqrt{\frac{200 h_1 L}{g_1 - g_2}}$$

Similarly,

$$S_2 = \sqrt{\frac{200 h_2 L}{g_1 - g_2}}$$

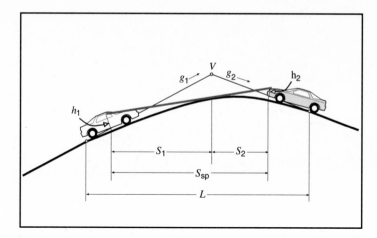

FIGURE 13-20 Computing minimum length of vertical curve.

If S_{sp} is substituted for $S_1 + S_2$, then

$$L = \frac{(g_1 - g_2)S_{sp}^2}{200(\sqrt{h_1} + \sqrt{h_2})^2} = \frac{(g_1 - g_2)S_{sp}^2}{3293} \qquad (13\text{-}31)$$

where g_1 and g_2 are expressed in percent of grade.

In terms of ratios for g_1, g_2

$$L = \frac{S_{sp}^2 (g_1 - g_2)}{33} \qquad (13\text{-}31a)$$

This is the case when the passing sight distance is less than the length L of the curve.

The heights h_1 and h_2 are shown as perpendicular to the road surface in Fig. 13-20, but the derivation assumes they are vertical. The error introduced is negligible.

If it is assumed that S_{sp} is longer than the curve, then

$$L = 2S_{sp} - \frac{33}{g_1 - g_2} \qquad (13\text{-}32)$$

in which distances are in feet and g_1 and g_2 are expressed as ratios.

If the safe stopping sight distance is less than the length of the curve, then

$$L = \frac{S_{np}^2(g_1 - g_2)}{14.0} \qquad (13\text{-}33)$$

If S_{np} is longer than the curve, then

$$L = 2S_{np} - \frac{14.0}{g_1 - g_2} \qquad (13\text{-}34)$$

On a highway with four or more traffic lanes, the safe stopping sight distance can be used to determine the required length of a vertical curve, because there is little probability of meeting oncoming vehicles in the passing lane. On a two-lane highway, however, the safe passing sight distance must be used if a vehicle is permitted to pass another one traveling in the same direction on the vertical curve. Use of the safe passing sight distance results in excessive lengths of vertical curves, and in most instances causes excessive excavation (see Chapter 17). Two methods are employed on two-lane roads to allow the safe stopping sight distance to be used for computing L. One method is to prohibit passing on crests and to indicate the restriction by appropriate center-line marking. The second method is to widen the pavement at the crest to permit two lanes in both directions for a sufficient distance.

◆EXAMPLE 13-8   The grades at a crest are $g_1 = +2\%$ and $g_2 = -3\%$, and the design speed is 60 mph. Compute the lengths of the vertical curves required for the safe passing sight distance and the safe stopping sight distance recommended by the AASHTO.

Solution: By Eq. (13-31) and Table 13-10

$$L = \frac{2100^2 \times 0.05}{33} = 6682 \text{ ft}$$

Also, By Eq. (13-32),

$$L = 2 \times 2100 - \frac{33}{0.05} = 3540 \text{ ft}$$

The length required for safe passing is therefore 6682 ft, since S_{sp} is less than L.
   By Eq. (13-33),

$$L = \frac{475^2 \times 0.05}{14} = 806 \text{ ft}$$

Also, By Eq. (13-34),

$$L = 2 \times 475 - \frac{14}{0.05} = 670 \text{ ft.}$$

The length required for safe stopping is thus 806 ft, since S_{np} is less than L.

   In this example it would not be necessary to apply Eq. (13-32) or (13-34) after it is found by Eq. (13-31) or (13-33) that the required sight distance is less than L.   ◆

TABLE 13-10 AASHTO Sight-Distance Recommendations

Design Speed (mph)	S_{sp} (ft)	S_{np} (ft)
30	1100	200
40	1500	275
50	1800	350
60	2100	475
70	2500	600
80	2700	750

13-23   Compound Curves

A compound curve consists of two or more consecutive circular arcs, the PT of one curve being the PC of the next and the centers of the curves being on the same side of the curve. Such a curve is shown in Fig. 13-21. Although it is beyond the scope of this text to go deeply into the problems of compound curves, it may be pointed out that many of the compound-curve equations can be developed by considering the polygon 1-2-3-4-5 in Fig. 13-21 as a five-sided traverse.

The values for this traverse are shown in Table 13-11.

Since the traverse is a closed one, the algebraic sums of the latitudes and departures must equal zero. From the departures,

$$T_s + T_l \cos \Delta - R_l \sin \Delta + (R_l - R_s) \sin \Delta_s = 0 \tag{13-35}$$

From the latitudes, north latitudes being considered negative for convenience, $-R_s + T_l \sin \Delta + R_l \cos \Delta - (R_l - R_s) \cos \Delta_s = 0$. If R_l is added and subtracted, this equation can be written

$$(R_l - R_s) - (R_l - R_s) \cos \Delta_s + T_l \sin \Delta - (R_l - R_l \cos \Delta) = 0 \tag{13-36}$$

TABLE 13-11   Basis of Equations for Compound Curves

| Side | Azimuth | Length | Departures | | Latitudes | |
			E	W	N	S
1–2	0°	R_s	0		R_s	
2–3	90°	T_s	T_s			0
3–4	90° + Δ	T_l	$T_l \cos \Delta$			$T_l \sin \Delta$
4–5	180° + Δ	R_l		$R_l \sin \Delta$		$R_l \cos \Delta$
5–1	Δ_s	$R_l - R_s$	$(R_l - R_s) \sin \Delta_s$		$(R_l - R_s) \cos \Delta_s$	

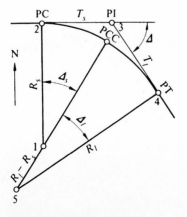

FIGURE 13-21 Compound curve.

From Eqs. (13-35) and (13-36) and the relation $\Delta = \Delta_s + \Delta_l$, the values of T_s, Δ_l, and Δ_s can be found when Δ, T_l, R_s, and R_l are known. For a complete discussion of compound curves, a text on route surveying should be consulted.

13-24 Reversed Curves

A reversed curve is composed of two simple curves turning in opposite directions, as shown in Fig. 13-22. The point of reverse curve, PRC, is the PT of the first curve and the PC of the second one.

If the angles Δ_1 and Δ_2 and the distance between intersection points PI_1 and PI_2 have been measured in the field, one radius, or one degree of curve, can be assumed and the other calculated. If R_1 is assumed, the first tangent distance T_1 is computed from the relationship:

$$T_1 = R_1 \tan \tfrac{1}{2} \Delta_1$$

This length subtracted from the total distance between the intersection points gives the value of T_2. Then

$$R_2 = T_2 \cot \tfrac{1}{2} \Delta_2$$

The use of reversed curves on railroads is limited to sidings and crossovers. The necessity of elevating the outer rail on a railroad, and the outer edge of a highway prevents the use of reversed curves except where very low speeds are encountered.

13-25 Easement Curves

The amount of superelevation, in feet per foot of width, on a curve can be determined from the equation

$$e = \frac{v^2}{32.2R} = 0.067\frac{V^2}{R} \tag{13-37}$$

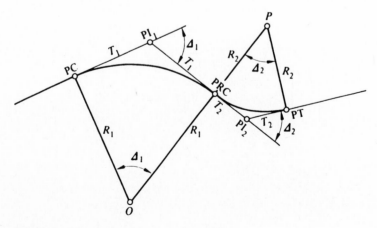

FIGURE 13-22 Reversed curve.

in which v is the velocity, in feet per second; V is the velocity, in miles per hour; and R is the radius of the curve, in feet.

On a straight track or pavement, the two edges are at the same elevation. On a circular arc the outer edge is elevated the proper amount for the radius of the curve and for the speed expected. These two requirements lead to an impossible condition at the PC of a simple curve, since the PC is both on the curve and on the tangent and at the same point there should be superelevation for the curve and none for the tangent.

The introduction of a spiral easement, or transition, curve between the tangent and the circular arc, as indicated in Fig. 13-23, permits the gradual elevation of the outer edge. There are many curves that can be used as easement curves. However, the one recommended by the American Railway Engineering Association (AREA) is probably as simple as any. With the aid of tables found in railroad engineering handbooks, the computations can be made almost as quickly as the computations for a simple curve.

The AREA spiral is based on the assumption that the superelevation is to change at a uniform rate along the spiral. The equation of this curve is

$$y = \frac{l^3}{6RL}$$

in which y is the tangent offset at any point whose distance from the beginning point TS (tangent to spiral) of the spiral is l, R is the radius of the circular arc, and L is the length of the spiral. The radius of this curve decreases from infinity at the TS to R at the SC (spiral to curve). Between the circular arc and the forward tangent is a second spiral whose ends are the CS (curve to spiral) and the ST (spiral to tangent).

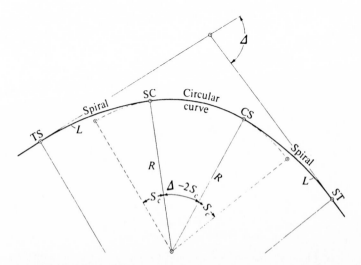

FIGURE 13-23 Spiral easement curves.

The principal differences between the circular curve and the spiral are as follows: The length of the spiral is twice the length of a circular arc with radius R and central angle S_c. In the case of the usual spiral, it is accurate enough for practical purposes to assume that the deflection angle from the tangent at the TS to any point on the spiral is one-third (instead of one-half) the central angle subtended by the chord from the TS to the point. The deflection angles are proportional to the squares of the distances along the spiral (rather than to the distances themselves, as in the case of the circular curve).

For the development of spirals and their applications, the reader should refer to Meyer and Gibson's *Route Surveying and Design* or Skelton's *Route Surveys*. These texts contain tables that simplify the field computations.

PROBLEMS

13-1. The degree of curve is 3° 30′. Compute the radius of the curve on the basis of (**a**) chord definition and (**b**) arc definition.

13-2. The degree of curve is 2° 00′. Compute the radius of the curve on the basis of (**a**) chord definition and (**b**) arc definition.

13-3. The degree of curve is 16° 00′. Compute the radius of the curve on the basis of (**a**) chord definition and (**b**) arc definition.

13-4. What is the degree of curve, arc definition of a 1500-ft radius curve?

13-5. What is the degree of curve, arc definition of a 350-ft radius curve?

13-6. A circular curve is to connect two tangents that intersect at an angle Δ of 16° 42′ at station 25 + 63.40. The tangent must be less than 500 ft. Using the arc definition, determine the degree (to the nearest 30′) of the flattest possible curve that will satisfy the conditions. Compute the deflection angles to each full curve station.

13-7. Using the information from Problem 13-6, compute the value of c_1 and c_2 given in Eqs. (13-14) and (13-16).

13-8. The deflection angle between two tangents is 18° 52′. The maximum permissible tangent distance is 550 ft. Using the arc definition, determine the degree of the flattest possible curve (nearest 30′) that will fit the conditions.

13-9. The PI of Problem 13-8 is at station 42 + 25.85. Compute the deflection angles to each full curve station.

13-10. Solve Problems 13-8 and 13-9 on the basis of chord definition.

13-11. A circular curve is to join two tangents that intersect at an angle of 21° 14′ 30″ at station 24 + 52.55. Using the arc definition, compute the radius (to a full 100 ft) of the flattest possible curve for which the arc length of the curve does not exceed 750 ft. Compute the deflection angles to each one-half station. Compute the values of c_1, c_2, and c_{50}.

13-12. The conditions are as stated in Problem 13-11, except that the minimum length of curve is 900 ft. Perform the same computations.

13-13. The external distance of a curve joining two tangents is not to be less than 35 ft. The deflection angle Δ is 38° 32′ at station 62 + 48.25. On the basis of arc definition, compute the degree of curve (nearest 30′) of the sharpest curve that will satisfy the limitation. Compute R, T, M, and C. Compute c_{100} for this curve.

13-14. The conditions are as stated in Problem 13-13, except that the external distance is not to be less than 20 ft. Perform similar computations.

13-15. Compute the chord length to the nearest 0.001 ft joining two adjacent full stations on a circular curve (**a**) when the degree by arc definition is 18°, (**b**) when it is 14°, (**c**) when it is 6°.

13-16. Compute the chord length to the nearest 0.001 ft joining two adjacent quarter stations on a circular curve (**a**) when the degree by arc definition is 16°, (**b**) when it is 12°, (**c**) when it is 8°.

13-17. Solve Problem 13-16 on the basis of chord definition.

13-18. The PC of a circular curve is located at station 42 + 08.25. the degree of curve by arc definition is 5° 00′. The curve is to be located by deflection angles. Stations 43, 44, and 45 can be located with the instrument set up at the PC. Because of an obstruction beyond station 45 the instrument must be moved up to occupy station 45 in order to locate the remainder of the curve. If a backsight of 0° 00′ is taken on the PC with the telescope inverted and the telescope is then plunged back to normal, what deflection angles should be turned to locate the remainder of the stations to the PT at station 51 + 83.20?

13-19. What is the central angle of the curve of Problem 13-18 (nearest 30″)?

13-20. The PC of a circular curve is located at station 174 + 48.18. The degree of curve by chord definition is 4° 00′. The curve is to be located by deflection angles. Stations 175 through 177 can be located from the PC instrument setup. The instrument is advanced to station 177 and a backsight of 0° 00′ is taken to the PC with the telescope inverted. If the telescope is then plunged back to normal, what deflection angles should be turned to locate the remainder of the stations to the PT at station 182 + 18.40?

13-21. What is the central angle of the curve of Problem 13-20 (nearest 30″)?

13-22. A series of horizontal circular curves are to join a series of tangents that have been located and stationed in the field from the beginning point designated station 0 + 00. The following data are given for each point of intersection:

PI Number	Original PI Station	Deflection Angle	Degree of Curve (Arc Definition)
1	12 + 12.44	8° 12′ R	2° 00′
2	21 + 76.15	20° 46′ R	2° 30″
3	31 + 50.05	18° 14′ L	3° 00′
4	49 + 21.40	22° 30′ R	3° 30′
5	65 + 18.92	End of line	

Compute the "new" stationing of the PCs, PTs, and the end of the line.

13-23. Compute the "new" stationing of Problem 13-22 on the basis of the chord definition.

13-24. In Fig. 13-14 $\Delta = 41° 44′, D = 4° 00′$ (chord definition), the PC at A is station 48 + 02.50, $AV' = 210.80$ ft, and angle $GV'V$ is 52° 20′. Compute $V'B$ and the stationing of B on the curve.

13-25. In Fig. 13-14 $\Delta = 48° 16′, D = 5° 30′$ (arc definition), the PC at A is station 36 + 14.95, $AV' = 261.85$ ft, and angle $GV'V$ is 39° 22′. Compute $V'B$ and the stationing of B on the curve.

13-26. A vertical curve joining two grade lines is 1800 ft long. Also, $g_1 = + 2.0\%$ and $g_2 = - 4.00\%$. The intersection of the grade lines at station 85 is at elevation 976.26 ft. Compute the elevation of each half station on the curve.

13-27. Compute the station and elevation of the summit of the curve of Problem 13-26.

13-28. A 1400-ft vertical curve joins two grades in which $g_1 = +2.25\%$ and g_2 is -0.90%. The intersection of the two grades at station 42 is at elevation 1065.14 ft. Compute the elevations of the full curve stations.

13-29. Compute the station and elevation of the summit of the curve of Problem 13-28.

13-30. A 1300-ft vertical curve joins two grades for which $g_1 = -2.25\%$ and $g_2 = +3.10\%$. The intersection of the grades is at station 50, and at elevation 575.66 ft. Compute the elevations of full curve stations.

13-31. Compute the station and elevation of the low point of the curve of Problem 13-30.

13-32. At what station or stations on the curve of Problem 13-30 is the elevation 586.00 ft?

13-33. A grade g_1 of -2.00% passes station 26 at an elevation of 861.22 ft, and a grade g_2 of $+2.35\%$ passes station 53 at an elevation of 848.86 ft. Compute the station and elevation of the point of intersection of these two grades. (The point-slope form of the equation of a line can be used.) Compute the elevations of the BVC, the EVC, and each full station on a 1200-ft vertical curve joining these two grades.

13-34. If g_1 is $+3.50\%$ and g_2 is -2.00%, compute the minimum length of a vertical curve necessary to provide safe passing sight distance for a design speed of 40 mph. Compute the length necessary to provide the safe stopping sight distance for a design speed of 60 mph.

13-35. If g_1 is $+2.75\%$ and g_2 is -4.10%, compute the minimum length of a vertical curve necessary to provide safe passing sight distance for a design speed of 50 mph. Compute the length necessary to provide the safe stopping sight distance for a design speed of 70 mph.

13-36. A grade g_1 of $+1.85\%$ intersects a grade of -3.20% at station 45. The distance back to the BVC is 500 ft and the distance forward to the EVC is 900 ft. The elevation of station 45 is 978.22 ft. Compute the elevations of the full stations on this unequal tangent vertical curve.

13-37. What is the station and elevation of the summit of the vertical curve of Problem 13-36?

13-38. A grade of -3.25% intersects a grade of $+2.40\%$ at station 80 whose elevation is 426.25 ft. The distance back to the BVC is 400 ft, and the distance forward to the EVC is 700 ft. Compute the elevations of the full stations on this unequal tangent vertical curve.

13-39. What is the station and elevation of the low point of the vertical curve of Problem 13-38?

13-40. In Fig. 13-21 $T_s = 325$ ft, $R = 450$ ft, $\Delta_s = 17°\ 40'$, $\Delta = 41°\ 20'$, and the PI is at station $136 + 22.45$. Using the arc definition for degree of curve, compute the T_l, R_l, and Δ_l and the stationing of the PCC and the PT.

13-41. In Fig. 13-21 $T_s = 380$ ft, $R_s = 500$ ft, $\Delta_s = 21°\ 15'$, $\Delta = 39°\ 50'$, and the PI is at station $68 + 42.85$. Using the chord definition for degree of curve, compute T_l, R_l, and Δ_l and the stationing of the PCC and the PT.

13-42. In Fig. 13-22 the distance from PI_1 to PI_2 is 1346.24 ft. The intersection angles are $\Delta_1 = 49°\ 50'$ and $\Delta_2 = 37°\ 15'$. The degree of the first curve, by chord definition, is $D_1 = 4°\ 30'$. Compute R_2 and D_2 necessary to fit the given conditions.

13-43. A pavement 60 ft wide is to be superelevated to allow safe negotiation of a $3°\ 00'$(arc definition) circular curve at a design speed of 60 mph. What will be the theoretical difference in elevation between opposite edges of the pavement?

13-44. If the gauge width between rails of a railroad track is 4 ft, $8\frac{1}{2}$ in., what is the superelevation of the outer rail for a design speed of 55 mph and a degree of curve (chord definition) of $2°\ 30'$?

BIBLIOGRAPHY   Danial, N. F. 1985. Computation of constrained vertical curves. *Tech. Papers Fall Conv. Amer. Cong. Surv. & Mapp.* p. 113.

Meyer, C. F., and Gibson, D. W. 1980. *Route Surveying and Design.* New York: Harper & Row.

Pryor, W. T. 1975, Metrication. *Surveying and Mapping* 35:229.

Skelton, R. R. 1949. *Route Surveys.* New York: McGraw-Hill.

14 Tacheometry

14-1 Use of Tacheometry

The term *tacheometry* in surveying is used to denote the procedures for obtaining horizontal distances and differences in elevation by rapid indirect methods, which are based on the optical geometry of the instruments employed. The procedure is sometimes referred to as stadia, optical distance measurement, or telemetry. The instruments employed are the engineer's transit or theodolite and the leveling rod or stadia rod, the telescopic alidade and the leveling rod or stadia rod, the theodolite and the subtense bar, and the self-reducing theodolite and the leveling rod. Horizontal distances are obtained with each of these combinations of instruments without resorting to direct taping. Differences in elevation can be determined indirectly with most of the combinations.

Tacheometry is used to measure the lengths of traverse sides, to check the more accurate taped distances in order to uncover gross errors or mistakes, to determine differences of elevation between points, and to carry lines of levels where a relatively low order of accuracy is permissible. Its most general use is found in the compilation of planimetric and topographic maps by field methods alone, by which distances, elevations, and directions to points are to be determined from field control points whose positions have been established by a higher order of accuracy.

The principles of stadia measurement using the transit or theodolite and the leveling rod will be developed thoroughly in this chapter. The other methods of tacheometry will then be discussed. The similarity and the difference in the principles of the various techniques can thus be better understood.

14-2   Principle of Stadia Measurements

For the measurement of stadia distances, the reticle in the telescope of the transit or theodolite is equipped with three horizontal cross hairs, the upper and lower hairs being called the *stadia hairs*. Three styles are shown in Fig. 14-1. The actual vertical separation of the upper and lower stadia hairs in the reticle is designated as i. In Fig. 14-2 points a and b represent the positions of the upper and lower stadia hairs with a spacing $ab = i$. By the laws of optics, a ray of light that is parallel with the optical axis of a lens will pass through the *principal focus* of the lens on the opposite side. In Fig. 14-2 a transit with a movable objective lens is used to illustrate the principles of stadia. The stadia equations developed from Fig. 14-2 will be modified later on to take into account the internally focused telescope described in Chapter 3, which has all but replaced the externally focusing telescopes found on older surveying instruments. Referring to Fig. 14-2, point F is the principal focus of the objective lens, and it is located on the optical axis in front of the movable objective lens at a distance f from the center of the lens. The distance f is known as the *focal length* of the lens, and it has a fixed value for a given lens.

Imagine a ray of light to be directed along the line aa', which is parallel with the optical axis of the telescope. This ray of light will be bent by the lens and will continue in the direction $a'FB$. Similarly, a ray along bb' will travel in the direction $b'FA$ on the other side of the lens. The vertical distance AB can be measured on a graduated rod. It is obtained by subtracting the stadia-hair reading at B from the stadia-hair reading at A. The distance AB, designated as s, is called the *stadia interval*.

FIGURE 14-1 Reticles with stadia hairs.

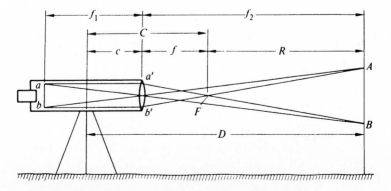

FIGURE 14-2 Principle of transit stadia measurement.

The distance $a'b'$ is equal to ab and is therefore equal to i. By similar triangles, $a'b'/f = AB/R$, in which R is the distance from the principal focus in front of the lens to the graduated rod. So the distance R may be found from the equation

$$R = \frac{f}{i} s \qquad (14\text{-}1\text{a})$$

in which f is the focal length, which is fixed for a given telescope; i is the spacing of the stadia hairs, which is also fixed for a given telescope; and s is the vertical distance on the rod between the upper stadia-hair reading and the lower stadia-hair reading. The ratio f/i is called the *stadia interval factor* and is designated as K. Eq. (14-1a) thus becomes

$$R = Ks \qquad (14\text{-}1\text{b})$$

The manufacturer can space the stadia hairs with relation to the focal length to obtain any convenient value of K desired. The most common value of f/i or K is 100, and this is the assumed value in this textbook.

The horizontal distance (D in Fig. 14-2) between the center of the instrument and the point at which the rod is held is $R + f + c$, in which c is the distance from the center of the instrument to the lens. This small distance c varies slightly when focusing the objective lens. Its exact value depends on the distance from the instrument to the rod, but the variation is not enough to affect the accuracy of stadia measurements. If C is substituted for $f + c$, then the horizontal distance D from the center of the instrument to the rod is given by the equation

$$D = Ks + C \qquad (14\text{-}2\text{a})$$

The distance C is called the *stadia constant*. Equation (14-2a) is the stadia equation for a horizontal line of sight using a movable objective telescope. It is applicable for lines of sight inclined by as much as 3° where the distance to be determined is not critical. For this reason, when a stadia interval is measured, it is permissible and desirable to set the lower cross hair on the nearest foot mark to facilitate reading the interval, even though the line of sight is not quite horizontal.

The value of C will vary from about 0.6 to 1.4 ft in the older transits. These instruments, which contain the movable objective lens, have for the most part been replaced by instruments with the internal-focusing lens. The geometry of the optics of the newer telescopes is such that the value of the stadia interval factor changes slightly when focusing the internal lens. This change is almost compensated for throughout the range of focus by a corresponding change in the value of C. This fact allows the constant C to be neglected; that is, $C = 0$. The value of the stadia interval factor is then assumed to be constant throughout the focusing range.

If the older instrument is used, the value of C can be assumed to be 1 ft. When K is 100, or nearly so, Eq. (14-2a) can be written as follows:

$$D = K(s + 0.01) \qquad (14\text{-}2\text{b})$$

in which the interval s is simply increased by 0.01 ft.

For long sights when C can be entirely neglected and when the internal-focusing instrument is employed, Eq. (14-2b) can be reduced to the simple form

$$D = Ks \tag{14-2c}$$

14-3   Determination of Stadia Constant

If the instrument is externally focusing, the stadia constant is readily determined by setting up the transit, focusing on a distant point at least 1000 ft away, and measuring the distance between the objective lens and the cross-hair ring. This is the focal length f. The telescope is then focused on an object between 200 and 300 ft away. The distance between the objective lens and the center of the instrument is then measured. This is the distance c in Fig. 14-2. The constant C is the sum of f and c.

If the instrument contains the fixed-objective lens, then the value of C can be assumed to be zero.

14-4   Determination of Stadia Interval Factor

If it is desired to determine the stadia interval factor $f/i = K$, a straight line from 400 to 800 ft in length is run on ground that is as nearly level as practicable. The instrument is set on this line. If the instrument is externally focusing, a point P is located on the line at a distance $f + c = C$ from the center of, and in front of, the instrument. If the instrument is internally focusing, then the position of P is directly under the instrument.

Points spaced about 50 ft apart are set on line, and the distances R_1, R_2, R_3, and so on are measured from point P to the successive points on line. A leveling rod equipped with two targets is held at each point in succession, and the intervals s_1, s_2, s_3, and so on are determined to the nearest 0.001 ft. Successive values of K are determined by the relationships $K_1 = R_1/s_1$, $K_2 = R_2/s_2$, $K_3 = R_3/s_3$, and so on, and the mean of the values of K thus obtained is taken as the stadia interval factor of the instrument.

14-5   Inclined Stadia Measurements

Inclined stadia measurements are more frequent than are horizontal measurements. Each inclined measurement is reduced to give the horizontal distance from the instrument to the rod and the difference in elevation between the telescope axis at the center of the instrument and the point at which the middle cross hair strikes the rod. This point on the rod is referred to as the middle cross-hair reading. In Fig. 14-3 an internal-focusing instrument in which $C = 0$ is located at M. The apex of the angle subtended by the upper and lower stadia hairs is assumed to fall at the middle of the telescope. The rod is located at O, the horizontal distance between the instrument and the rod is $MN = EG = H$, and the difference in elevation between the telescope axis at E and the middle cross-hair reading at R is $RG = V$. Thus V represents the total rise (or fall in the case of a minus vertical angle) of the line of sight from the instrument to the rod with the telescope inclined by the vertical angle a. The angle AEB is only about $0°\,34'$, and therefore the angles $RA'E$ and $RB'E$ are considered right angles. Therefore $A'B' = AB \cos a = s'$. The slope distance S is ER. By the ratio $f/i = K$, remembering that AB is the stadia interval s,

$$S = Ks' = Ks \cos a$$

The horizontal distance $H = S \cos a$. Thus

$$H = Ks \cos a \cos a$$

or

$$H = Ks \cos^2 a \tag{14-3a}$$

in which K is the stadia interval factor f/i, s is the stadia interval, and a is the vertical angle of the line of sight read on the vertical circle of the theodolite. If zenith angles are observed, the horizontal distance is given by

$$H = Ks \sin^2 Z \tag{14-3b}$$

The distance RG, which equals $S \sin a$, is the vertical distance V between the telescope axis and the middle cross-hair reading. Thus V is given by the equation

$$V = Ks \sin a \cos a \tag{14-4}$$

or

$$V = \tfrac{1}{2} Ks \sin 2a = \tfrac{1}{2} Ks \sin 2Z \tag{14-5}$$

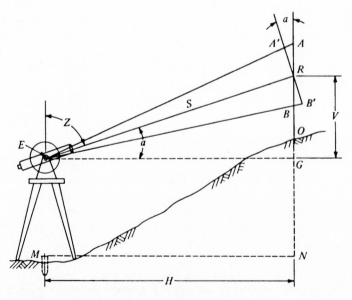

FIGURE 14-3 Inclined stadia measurement $(C = 0)$.

EXAMPLE 14-1 Stadia readings were made from an instrument set up to three different points. The value of K is 100. The recorded values are $s_1 = 0.848$ m and $a_1 = +3°\ 44'$; $s_2 = 1.455$ m and $a_2 = +0°\ 20'$; $s_3 = 1.820$ m and $a_3 = -12°\ 20'$. Compute the horizontal distance and the rise or fall of the line of sight.

Solution: By Eq. (14-3a),

$$H_1 = 100 \times 0.848 \cos^2 3°\ 44' = 84.44 \text{ m}$$
$$H_2 = 100 \times 1.455 \cos^2 0°\ 20' = 145.50 \text{ m}$$
$$H_3 = 100 \times 1.820 \cos^2 12°\ 20' = 173.70 \text{ m}$$

By Eq. (14-5),

$$V_1 = 50 \times 0.848 \sin 7°\ 28' = +5.51 \text{ m}$$
$$V_2 = 50 \times 1.455 \sin 0°\ 40' = +0.85 \text{ m}$$
$$V_3 = 50 \times 1.820 \sin(-24°\ 40') = -37.98 \text{ m}$$

◆

14-6   Reading a Stadia Interval

When using a food rod, the most convenient method of measuring a stadia interval is to begin by reading the vertical angle and the middle cross hair, and then set the lower cross hair on the nearest whole foot mark. Then read the upper stadia hair and get the difference, which is the stadia interval. When using a metric rod, the lower stadia hair is set to the nearest decimeter mark before reading the upper stadia hair.

14-7   Difference in Elevation Between Two Points

When the stadia methods are used to determine differences in elevation between points, three situations will prevail. In the first situation, which is associated with traversing and mapping details about a point, the difference in elevation between the point on the ground over which the instrument is set up and the point on the ground at which the rod is held is the desired value. In Fig. 14-4 the instrument occupies point M, and the rod is held at point O. A leveling rod or stadia rod is held alongside the instrument, and the height of the telescope axis above point M is measured. This height is recorded in the notes as the HI, which for this situation is understood to be the height of the instrument *above the ground.* The line of sight is directed at the rod held at O, and the middle cross-hair reading at R and the corresponding vertical angle a are recorded. The stadia interval is then read and recorded. The difference in elevation between points M and O is equal to $+ ME + GR - RO$. But the distance ME equals the HI, and the distance GR is the vertical distance V computed by Eq. (14-5). RO is the middle cross-hair reading, which will be referred to simply as the *rod reading.* Therefore the difference in elevation between M and O is (+HI + V − rod reading) if the vertical angle is positive, or (+HI − V − rod reading) if the vertical angle is negative.

EXAMPLE 14-2   The HI of the instrument of Example 14-1 is measured as 1.52 m. The middle cross-hair readings for the three observations are $R_1 = 1.42$ m; $R_2 = 0.44$ m; and $R_3 = 2.00$ m. Compute the difference in elevation between the point over which the instrument was set up and the points where the rod was held.

Solution:
$$DE_1 = +1.52 + 5.51 - 1.42 = +5.61 \text{ m}$$
$$DE_2 = +1.52 + 0.85 - 0.44 = +1.93 \text{ m}$$
$$DE_3 = +1.52 - 37.98 - 2.00 = -38.46 \text{ m}$$

It is convenient whenever possible to read the vertical angle when the rod reading equals the HI. The difference in elevation is then $+ \text{HI} \pm V - \text{HI} = \pm V$ directly. This procedure, however, is not always possible on account of obstructions on the line of sight.

In the second situation, which is sometimes associated with mapping details about a point, the difference in elevation between the telescope axis and the point on the ground at which the rod is held is the desired value. Again refer to Fig. 14-4, and let it be assumed that the elevation of point M has been established. The distance ME added to the elevation of M gives the elevation of the telescope axis *above the datum*, or the HI as understood in differential leveling and defined in Section 3-25. Then if the elevation of point O is desired, the difference in elevation between points E and O is added to the HI. In this instance the difference in elevation between E and O is $(+V - \text{rod reading})$. If the vertical angle is negative, the desired difference is $(-V - \text{rod reading})$.

In the third situation, which is associated with running a line of levels by stadia, the difference in elevation between two rod stations is desired. The elevation of the ground at the instrument setup is of no consequence. In Fig. 14-5 the difference in elevation between the backsight station B and the foresight station F is desired. The instrument is set up at A, and a stadia interval, vertical angle a_B, and rod reading R_B are observed on point B and recorded. Then an interval, vertical angle a_F, and rod reading R_F are observed on point F and recorded. The vertical distances $O_B R_B = V_B$ and $O_F R_F = V_F$ are computed by Eq. (14-5). The difference in

FIGURE 14-4 Difference in elevation between instrument ground station and rod station.

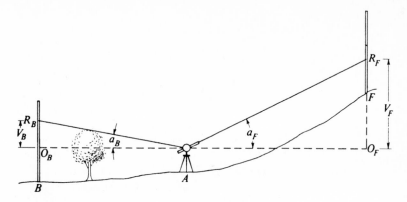

FIGURE 14-5 Difference in elevation between two rod stations.

elevation in this instance is (+ backsight rod reading $- V_B + V_F -$ foresight rod reading). Sketches should be drawn of other situations of this nature, in which the backsight and foresight vertical angles are, respectively, + and −, + and +, − and −, and − and +. The values to be added and subtracted to arrive at the difference in elevation between the two rod stations will become apparent from the sketches.

EXAMPLE 14-3   In Fig. 14-5, $s_B = 1.144$ m, $a_B = +0°\ 18'$, $R_B = 2.14$ m; $s_F = 1.458$ m, $a_F = +3°\ 16'$, and $R_F = 2.05$ m. What is the difference in elevation from B to F?

Solution: By Eq. (14-5),

$$V_B = 50 \times 1.144 \sin 0°\ 36' = 0.60 \text{ m}$$
$$V_F = 50 \times 1.458 \sin 6°\ 32' = 8.29 \text{ m}$$

The difference in elevation from B to F is then

$$DE = +2.14 - 0.60 + 8.29 - 2.05 = +7.78 \text{ m}$$ ◆

14-8   Stadia Traverse

A stadia traverse is sometimes run to obtain supplementary control based on existing higher-order control. This supplementary control is used to furnish instrument stations for the compilation of field data needed for the production of planimetric and topographic maps, or to provide picture points for controlling photogrammetric mapping. The traverse work may be performed concurrently with the gathering of data, or it may be performed separately from the map-compilation phase.

Three quantities are obtained from the results of the stadia traverse: (1) the length of each traverse side, (2) the azimuth or bearing of each traverse side, and (3) the difference in elevation between the ground stations at the ends of each traverse side. The lengths of the sides are obtained by the stadia reduction equations for horizontal distance given in Sections 14-2 and 14-5. The azimuths of the sides are most conveniently determined by carrying a traverse by azimuth as described in Section 8-7. Differences in elevation are obtained by the method discussed as the first situation of Section 14-7.

The most satisfactory results are obtained if observations for each line of the traverse are taken in both directions. This procedure reveals mistakes, eliminates certain index errors in the vertical circle of the theodolite, and gives a better value for the horizontal distance and difference in elevation, each of these distances being taken as the mean of two values.

In the stadia traverse of Fig. 14-6(a), a traverse begins and closes on station *A*. The azimuth of *AD* has been previously established as 218° 00′. The elevation of *A* is 104.56 m. The HI at *A* is 1.52 m, having been measured by holding the rod alongside the transit, and is recorded as shown in the notes. The value of the azimuth of *AD* is set on the clockwise circle, and a backsight is taken on *D*, using the lower motion. This orients the theodolite circle at *A*. The middle cross hair is set on 1.52 on the rod held at *D*, and a vertical angle of +0° 21′ is read and recorded. The stadia interval of 1.770 m is now read and recorded (see Section 14-6).

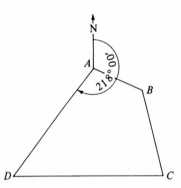

STA.	INT.	HOR. DIST.	AZIMUTH	VERT Δ	D.E.
	⋏ @ A;	H.I.=1.52			
D	1.770	176.99	218°00′	+0°21′	+1.08
B	0.709	70.52	114°16′	−4°12′	−5.18
	⋏ @ B;	H.I.=1.55			
A	0.706	70.22	294°16′	+4°12′	+5.16
C	1.148	114.73	166°53′	+1°25′	+2.84
	⋏ @ C;	H.I.=1.55			
B	1.147	114.63	346°53′	−1°24′	−2.80
D	1.989	198.85	270°14′	+0°55′	+3.18
	⋏ @ D;	H.I.=1.53			
C	1.988	198.77	90°14′	−0°44′ on 2.12	−3.13
A	1.770	176.99	38°08′	−0°22′	−1.13
			e = +8′		

FIGURE 14-6 Stadia traverse with corresponding notes.

Station B is now sighted using the upper motion, and the clockwise circle reading of 114° 16′ is recorded as the azimuth of AB. The middle cross hair is set on 1.52 m, and a vertical angle of −4° 12′ is read and recorded. A stadia interval of 0.709 m is read and recorded. This completes the stadia observations at A.

The instrument is next set up at station B, the HI above the ground is measured, and the back azimuth of the line AB is set on the horizontal circle. This is 294° 16′ and is recorded as shown. The theodolite is oriented by backsighting on Station A using the lower motion. The stadia interval is read with the rod held on A and is recorded as 0.706 m. When the middle cross hair is set at the HI, the vertical angle is +4° 12′. Station C is next sighted by using an upper motion, and the azimuth of BC, which is read on the horizontal circle, is recorded as 166° 53′. A stadia interval and a vertical angle to the HI are read and recorded.

The procedure just described is repeated at each station, and a final sight is taken on A from D to close the traverse. Notice that when the instrument was set up at station D and the rod was held at C, the HI could not be sighted with the middle cross hair. Therefore the rod reading of 2.12 m is recorded, together with the corresponding vertical angle of −0° 44′.

All the distances and differences in elevation (DEs) are computed using Eqs. (14-3) and (14-5) and are entered in the notes as shown.

The azimuth of line DA should be 218° 00′− 180° = 38° 00′, but the field results show 38° 08′. Thus the azimuth error is +8′. This is distributed throughout the traverse by reducing each *forward* azimuth in proportion to the number of instrument setups. Thus the measured azimuth of AB receives −02′ correction, BC receives −04′, CD receives −06′, and DA receives −08′ correction. The average distances, the adjusted azimuths, and the average differences in elevation for each line are shown in Table 14-1.

The sum of the plus DEs is less than the sum of the minus DEs, as shown in the table. These are adjusted in proportion to the length of each line. Line AB gets a correction of 0.29 × 70.37/560.85 = 0.04 m; line BC gets corrected 0.29 × 114.68/560.85 = 0.06 m, and so on. The adjusted DEs shown in the table are then used to compute the elevations of B, C, and D.

Latitudes and departures are computed for each line, and the traverse is adjusted by the method described in Chapter 8.

TABLE 14-1   Results of Measurements in Stadia Traversing

Line	Distance	Adjusted Azimuths	Average DEs +	Average DEs −	Adjusted DEs +	Adjusted DEs −
AB	70.37	114° 16′ − 02′ = 114° 14′		5.17		5.13
BC	114.68	166° 53′ − 04′ = 166° 49′	2.82		2.88	
CD	198.81	270° 14′ − 06′ = 270° 08′	3.17		3.27	
DA	176.99	38° 08′ − 08′ = 38° 00′		1.11		1.02
	560.85		5.99	6.28	6.15	6.15
				−5.99		
				$e = -0.29$		

14-9   Details About a Point

When the transit stadia method is used to locate details about a point for the purpose of plotting these details, the theodolite is set up over the point and the HI above the ground is determined by the method discussed in the first situation of Section 14-7. The horizontal circle is oriented by setting the known azimuth of a line on the circle and backsighting along this line by using the lower motion. The upper motion is then used for all subsequent pointings from the station, so that the azimuth of each line directed to a detail point may be determined. As stadia readings are taken by the instrumentman, the note keeper should keep a detailed sketch up to date in the field notebook, numbering the detail points to correspond with the numbers assigned in the notes. Word descriptions of detail points—such as "fence corner," "3′ oak," and "road int's'n,"—are used to advantage in documenting the stadia observations.

A form of notes for gathering details about a point is shown in Fig. 14-7. The instrument is set up on station G and oriented by backsighting on station F with the azimuth of the line GF on the horizontal circle. The value of the HI is the height of the telescope axis above the ground, and all differences in elevation are applied to the elevation of the ground point. For example, on the sight to point 1, the value of the vertical distance by Eq. (14-5) = −18.9 ft. The difference in elevation is then +5.2 − 18.9 − 7.4 = −21.1 ft, and the elevation of point 1 is 491.0 − 21.1 = 469.9 ft. On the sight taken to point 3, the telescope bubble was centered when the rod reading was made, and since there was no vertical angle, the difference in elevation is +5.2 − 8.8 = −3.6 ft. On the sights taken to points 4, 5, and 7, the middle cross hair was sighted on the HI on the rod, and the difference in elevation in each instance is simply the vertical distance determined by Eq. (14-5).

An alternative method of determining elevations is discussed in the second situation of Section 14-7. The HI above the vertical datum at the instrument setup is determined, and then the difference in elevation between the telescope axis and the ground point is obtained as described in Section 14-7.

14-10   Stadia Leveling

The operations in stadia leveling are quite similar to those in differential leveling, but in stadia leveling the line of sight does not have to be horizontal. This feature is advantageous in hilly country where a relatively low order of accuracy is acceptable because the lines of sight can be quite long.

In Fig. 14-8 a line of stadia levels is run from BM43 to BM44. The instrument is set up, and backsight readings are taken on BM43. These consist of an interval, a rod reading, and a corresponding vertical or zenith angle. Foresight readings consisting of an interval, a rod reading, and a corresponding vertical or zenith angle are then taken on TP1. This procedure is repeated at the second and third setups.

The following procedure is used to detect mistakes in reading the vertical circle and the interval, and in computing the values for the backsights and the foresights. The line of sight is directed at the rod, and the lower stadia hair is set on some foot mark (or decimeter mark) as shown in Fig. 14-9(a). The readings of the

POINT	INTERVAL	HOR. DIST.	AZIMUTH	VERT. ∠	ROD READ.	D.E.	ELEV.
	⅄ @ STA. G	BS. 43°22' on STA.F			H.I.= 5.2 ft.		491.0
1	3.28	328	37°10'	-3°18'	7.4	-21.1	469.9
2	3.32	333	39°15'	-3°16'	8.1	-21.8	469.2
3	4.06	407	152°50'	—	8.8	-3.6	487.4
4	4.51	452	158°05'	-1°06'	5.2	-8.7	482.3
5	1.29	129	292°22'	+3°45'	5.2	+8.4	499.4
6	1.20	111	316°30'	+16°20'	8.2	+29.4	520.4
7	1.31	124	316°40'	+14°10'	5.2	+31.1	522.1

FIGURE 14-7 Stadia detail notes.

FIGURE 14-8 Stadia leveling.

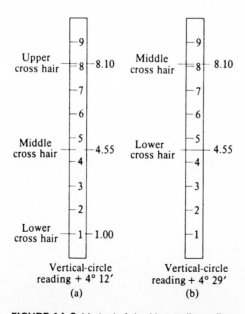

FIGURE 14-9 Method of checking stadia readings.

lower, middle, and upper cross hairs are observed and recorded. The vertical circle is then read, and the angle is recorded opposite the middle cross-hair reading. Now the line of sight is either raised or lowered by means of the vertical tangent screw until the middle cross hair is set on the original reading of the upper or lower cross hair. In Fig. 14-9(b) the middle cross hair is set on the original reading of the upper cross hair. The vertical circle is again read, and the angle is recorded opposite this cross-hair reading.

If the value of K for the transit is 100, or nearly so, then the angle between the upper and middle cross hairs or between the middle and lower cross hairs is $\tan^{-1}$ $(1/200) = 0°\ 17'$, approximately. Therefore the two values of the vertical angle should differ by about $0°\ 17'$. If they do not, then a mistake has been made in either a cross-hair or vertical-circle reading and the readings can be checked before leaving the setup. The computation of the backsight and foresight values can be checked, because there will be two values for each distance.

A form of notes for stadia leveling is shown in Fig. 14-10. The readings shown are those for the three setups of Fig. 14-8. The stadia interval for the first backsight taken on BM43 is $9.80 - 3.00 = 6.80$. Assuming the stadia constant to be zero, the value of V for $2°\ 13'$ is 26.3 ft by Eq. (14-5). The value of the backsight is $+6.40 - 26.3 = -19.9$ ft. The value of V for $2°\ 30'$ is 29.6. The value of the backsight using this angle is then $+9.80 - 29.6 = -19.8$ ft. The mean, rounded to 0.1 ft, is -19.9 ft, as shown in the notes.

The stadia interval for the foresight taken on TP1 is $8.25 - 1.00 = 7.25$. The value of V for $1°\ 41'$ is 21.3. The value of the foresight is then $-21.3 - 4.60 = -25.9$ ft. Using $1°\ 23'$, $V = 17.5$. the value of the foresight is then $-17.5 - 8.25 = -25.75$. The mean, rounded to 0.1 ft, is -25.8 ft, as shown in the notes.

After all the values of the backsights and foresights between benchmarks have been computed, the difference in elevation between the points is then the alge-

STATION	ROD	VERT.∠	B.S.	ROD	VERT.∠	F.S.
	9.80	+2°30′				
B.M. 43	6.40	+2°13′	−19.9			
	3.00					
	7.70	−3°08′		8.25	−1°23′	
T.P. 1	5.85	−3°25′	+27.9	4.60	−1°41′	−25.8
	4.00			1.00		
	6.20	−2°21′		8.00	+3°20′	
T.P. 2	4.10	−2°38′	+23.4	4.50	+3°03′	+32.6
	2.00			1.00		
				5.00	+0°27′	
B.M. 44				3.50	+0°10′	−2.6
				2.00		

FIGURE 14-10 Stadia leveling notes.

braic sum of the backsights and the foresights. In the example given here, DE = −19.9 − 25.8 + 27.9 + 32.6 + 23.4 − 2.6 = +35.6 ft.

14-11   Errors in Stadia Measurements

The accuracy of stadia measurements is largely dependent on the instrument and the rod used, and on atmospheric conditions. For sights up to 400 ft, it is usually possible to obtain the intercepted distance on the rod to 0.01 ft. At 100 m, the intercept can be measured with an accuracy of 3 mm. Consequently, if the stadia interval factor and the rod used are correct, the resulting horizontal distance should be within 1 ft or 3 dm of the correct distance. For longer sights the error depends on the magnifying power of the telescope, the coarseness of the stadia hairs, the type of rod used, and the weather conditions. Under reasonably favorable conditions, the error should not exceed 10 ft in a 1000-ft sight or 3 m in a 300-m sight.

Errors due to imperfections in the graduation of the rod can be kept to a negligible quantity by standardizing the rod. For most stadia work some type of pattern rod, with coarser graduations than are found on leveling rods, is used. Unless the painting of the rod has been carelessly done, no correction should be necessary for work of moderate precision.

If the rod is not held plumb, the resulting error in horizontal distance will be small when the vertical angle is small, but may be considerable when large vertical angles are encountered. The effect on an observed distance of 1000 ft, determined by reading a 12-ft rod the top of which is 0.5 ft out of plumb, for vertical angles of 0° 30′, 10°, and 25°, is given in Table 14-2.

Errors from this source can be eliminated by using a rod equipped with a plumbing level. In computing the values just given, the constant C has been assumed to be zero.

When a cross hair appears to cover an appreciable space on the rod, the observations can be made best by using the tops of the hairs, rather than by attempting to use their centers.

Errors in observing the intercepted distance on the rod will be random in character. On important sights all three hairs can be read. If the hairs are equally spaced, the middle-hair reading should be the mean of the upper and lower readings. When sights are taken between instrument stations, observations should be made from both ends of the line.

TABLE 14-2   Error in Horizontal Distance Due to Rod Being Out of Plumb

Vertical Angle	Observed Distance	Corrected Horizontal Distances		
		Rod Plumb	Rod Leaning Toward	Rod Leaning Away
+ 0° 30′	1000	999.9	999.4	998.7
+ 10° 00′	1000	969.8	976.1	961.9
+ 25° 00′	1000	821.4	836.7	804.7

14-12   Errors in Stadia Elevations

The errors in elevations obtained from stadia measurements are caused by errors in measuring the vertical angles and in reading the rod, and by not keeping the rod plumb.

The effect of an error of 1′ in a vertical angle is practically independent of the size of the angle, but is proportional to the length of the sight. For a distance of 100 ft, the error in elevation is about 0.03 ft.

The effect of an error of 1 ft in an observed inclined distance increases with the size of the vertical angle. The error in elevation is 0.05 ft for a 3° angle, about 0.1 ft for a 5° angle, about 0.2 ft for a 10° angle, and about 0.4 ft for a 25° angle. It is necessary therefore that intervals be observed more carefully when the vertical angles are large.

The error in elevation caused by the rod not being plumb is considerable on a long sight when the vertical angle is large. The effect of the top of a 12-ft rod being 0.5 ft out of plumb when the observed distance is 1000 ft is shown in Table 14-3.

14-13   Self-Reducing Stadia Instruments

When a major portion of the surveying to be performed consists of stadia measurements, the self-reducing stadia instrument is of particular value. The theodolite shown in Fig. 14-11 is a self-reducing tacheometer in which the stadia hairs are replaced with three curved lines, as shown in Fig. 14-12. The lower curve is referred to as the zero curve. It is placed on a convenient full graduation, which is 1.000 m in Fig. 14-12. The upper curve then determines the horizontal distance interval. In (a) this interval is 1.572 − 1.000 = 0.572 m. Multiplying by 100 gives the horizontal distance as 57.2 m. In (b) the interval is 0.485, giving a horizontal distance of 100 × 0.485 = 48.5 m. The middle curve determines the vertical-distance interval, together with the factor for the part of the curve that is being used. In (a) the vertical-distance interval is 0.401 m, or 40.1 cm. The factor +0.2 is to be applied to the interval in centimeters, or the factor to be applied to the interval in meters is +20. The vertical distance is therefore 0.401 × 20 = 40.1 × 0.2 = +8.02 m. In (b) the vertical-distance interval is 21.7 cm and the factor is − 1. Hence the vertical distance is 21.7 × − 1 = −21.7 m.

As the line of sight is inclined, the curves approach or recede from the zero curve to give a continuous solution of Eqs. (14-3) and (14-5). The vertical-distance curve is interrupted to give a new curve as the line of sight becomes more and more inclined. Each new curve has a different multiplying factor, which is analogous to a different stadia interval factor in the transit. There is only one horizontal-distance curve, however, and its factor is always 100.

TABLE 14-3   Error in Vertical Distance Due to Rod Being Out of Plumb

| | | Differences of Elevation | | |
Vertical Angle	Observed Distance	Rod Plumb	Rod Leaning Toward	Rod Leaning Away
+ 0° 30′	1000	8.7	8.7	8.7
+ 10° 00′	1000	171.0	172.1	169.6
+ 25° 00′	1000	383.0	390.1	375.2

FIGURE 14-11 Automatic stadia-reduction instrument.
Courtesy of Leica, Inc.

(a)
Distance : 57.2 m
Difference of elevation: $+0.2 \cdot 40.1 = 8.02$ m

(b)
Distance: 48.5 m
Difference of elevation : $-1 \cdot 21.7 = -21.7$ m

FIGURE 14-12 Stadia intercept curves in field of view of reduction tacheometer.

14-14   Subtense Bar

The subtense bar, shown in Fig. 14-13, establishes a short but very precise baseline at one end of a line to be measured. It contains a target at each end, together with a sighting target at its middle point. The separation of the end targets is controlled by invar wires under a slight but firm spring tension. Although temperature fluctuations cause the bar to expand and contract, the low thermal expansion of the invar holds the targets at practically a fixed distance apart. Slack is increased or taken up by the springs.

The subtense bar is leveled by means of a bull's-eye level and leveling screws, just as is a transit or theodolite. The bar is thus brought into a horizontal plane. Also, the bar is brought perpendicular to the line to be measured by taking a sight to the far end of the line through a small low-power telescope located at the midpoint of the bar.

Figure 14-14 shows a plan view of a subtense measurement. The instrument used to measure the horizontal angle γ between the two targets T and T' is located at A. The horizontal distance AB is given by the relationship

$$AB = \frac{TT'}{2} \cot \frac{\gamma}{2}$$

or

$$D = \frac{b}{2} \cot \frac{\gamma}{2} \tag{14-6}$$

in which D is the horizontal distance between the ends of the line; b is the length of the subtense bar, or the distance between the targets; and γ is the subtended horizontal angle. Both D and b are, of course, expressed in the same units. The subtense bar is

FIGURE 14-13 Subtense bar. Courtesy of Leica, Inc.

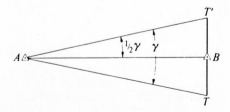

FIGURE 14-14 Subtense bar principles.

usually 2 m long. The length of the line in meters is then cot $\gamma/2$. A table supplied with the subtense bar gives the distances in meters corresponding to values of γ.

The instrument used to measure the subtended angle must be capable of measuring to 1″ of angle or less. This accuracy can be obtained by measuring the angle with a 1″ theodolite in several positions of the circle. There is no need to reverse the telescope in these measurements because both targets are at the same vertical angle and at the same distance from the theodolite. Also, the angle is obtained by the difference between the two directions to the targets. Thus an instrumental error for one pointing equals that for the other pointing. These errors are eliminated by the subtraction of one direction from the other.

The accuracy of subtense measurement is a function of the subtended angle and the length of the bar. Differentiating Eq. (14-6) with respect to γ, and with the approximation cot $\dfrac{\gamma}{2} \cong \csc \dfrac{\gamma}{2}$, gives

$$dD_\gamma = -\frac{D^2}{b}d\gamma$$

and differentiation with respect to b gives

$$dD_b = -\frac{D}{b}db$$

If it is assumed that the uncertainty in $d\gamma$ is 1″, or 0.00000485, and the uncertainty in db is 0.2 mm, then the uncertainty dD in length for various distances is given in Table 14-4, together with the ratio of dD to the distance D.

The precision of a distance measured by use of the subtense bar is seen to fall off quite rapidly as the distance increases. This change is caused by the effect of the angular error. As would be expected, an error in the length of the bar produces an error of constant percent in the distance because the ratio of dD_b to D is constant. If the accuracy of the angular measurement is increased by using multiple pointings, so that the uncertainty is reduced to ± 0.5″, each ratio in the second column is halved. Thus up to a distance of about 150 m or about 500 ft, the precision can be held to about 1 part in 5000.

TABLE 14-4   Accuracy of Subtense Measurements

D (m)	dD_γ (m)	Ratio $dD_\gamma:D$	dD_b (m)	Ratio $dD_b:D$
50	0.006	1 : 8330	0.0050	1 : 10,000
75	0.014	1 : 5350	0.0075	1 : 10,000
100	0.024	1 : 4170	0.010	1 : 10,000
150	0.055	1 : 2720	0.015	1 : 10,000
200	0.097	1 : 2060	0.020	1 : 10,000
250	0.152	1 : 1640	0.025	1 : 10,000
300	0.218	1 : 1380	0.030	1 : 10,000
400	0.388	1 : 1030	0.040	1 : 10,000
500	0.606	1 : 820	0.050	1 : 10,000
600	0.874	1 : 690	0.060	1 : 10,000

PROBLEMS   **14-1.** Determine the stadia interval factor K of an external-focusing transit from the accompanying data. The taped distances are from the point of focus of the telescope, which is 1.3 ft in front of the center of the instrument.

Taped Distance (ft)	Stadia Interval (ft)	Taped Distance (ft)	Stadia Interval (ft)
50.0	0.501	300.0	2.997
100.0	1.001	400.0	3.988
150.0	1.502	500.0	4.985
200.0	2.000	600.0	5.982

14-2. The following stadia notes were recorded for a closed traverse:

Station	Interval	Horizontal Distance	Azimuth	Vertical Angle	DE	Elevation	Adjusted Elevation
1						250.0	250.0
	3.69		120° 32′	+0° 56′			
2							
	5.70		138° 33′	−0° 16′			
3							
	4.39		180° 19′	−7° 06′			
4							
	4.87		141° 12′	−1° 05′			
5							
	4.63		209° 45′	−7° 29′			
6							
	5.63		231° 03′	−3° 26′			
7							
	4.92		316° 32′	+0° 07′			
8							
	4.39		320° 23′	+0° 14′			
9							
	5.02		285° 06′	−0° 18′			
10							
	5.78		277° 16′	−0° 32′			
11							
	4.35		62° 38′	+1° 06′			
12							
	4.53		83° 38′	+0° 50′			
13							
	5.92		46° 46′	+2° 33′			
14							
	6.50		7° 05′	+10° 50′			
1							

It is to be noted that the observations were made in the forward direction only, and that the middle cross hair was set on the HI to obtain the vertical angle. The stadia interval factor K is 100, and the constant C is zero.

(a) Compute the horizontal distance and the difference in elevation for each line to the nearest tenth of a foot.

(**b**) Compute the elevations of the traverse stations. Adjust the elevations in accordance with the principles of Section 14-8.

(**c**) Plot the stadia traverse to a scale of 1 in. = 200 ft by the method shown in Fig. 8-32. Using a sheet of drafting paper 18 by 24 in., lay off a $1\frac{1}{2}$-in. border on all sides. With the long dimension in the east-west direction, locate station 1 in the center of that dimension and 1 in. down from the top border. This position will allow the traverse to be conveniently centered on the map sheet.

(**d**) Distribute the error of closure in the position of station 1 by proportion.

14-3. The following stadia observations were made in a closed traverse, using an internal-focusing transit in which $K = 100$:

Occupied Station	Observed Station	Stadia Interval	Azimuth	Vertical Angle
A	G	4.71	180° 00′	+5° 22′
	B	7.68	38° 13′	+1° 49′
B	A	7.66	218° 13′	−1° 49′
	C	6.00	126° 54′	+0° 50′
C	B	6.00	306° 54′	−0° 50′
	D	6.38	67° 51′	−1° 57′
D	C	6.40	247° 51′	+1° 57′
	E	11.30	192° 32′	+9° 29′
E	D	11.33	2° 32′	−9° 31′
	F	9.70	263° 07′	−4° 52′
F	E	9.71	83° 07′	+4° 52′
	G	4.39	302° 44′	−9° 02′
G	F	4.39	122° 44′	+9° 02′
	A	4.70	0° 07′	−5° 22′

The center cross hair was set on the HI in each instance when the vertical angle was read.

(**a**) Compute the average horizontal distance and the average difference in elevation for each line.

(**b**) The given elevation of station A is 763.2 ft. Compute the adjusted elevations of the remaining stations in accordance with the principles of Section 14-8.

(**c**) After having adjusted the azimuths of the lines, plot the traverse to a scale of 1 in. = 100 ft by the method shown in Fig. 8-32. Using a sheet of drafting paper 18 by 24 in., lay off a 1-in. border on all sides. With the long dimension in the east-west direction, locate station A $3\frac{1}{2}$ in. in from the west border and $7\frac{1}{2}$ in. down from the north border of the sheet. This position will allow the traverse to be conveniently centered on the map sheet.

(**d**) Distribute the error of closure in the position of station A by proportion.

14-4. With a theodolite set 1.58 m above station A, a sight is taken on a rod held at station B. The interval is 1.420 m; the rod reading of the middle cross hair is 3.24 m; the vertical angle is −4° 08′. With the instrument set 1.55 m above station B, a sight is taken on the rod held at station A. The interval is 1.430 m, the rod reading of the middle cross hair is 4.00 m, the vertical angle is +5° 55′. The stadia constant is zero, and the stadia interval factor $K = 99.2$. What is the average length of the line AB? What is the difference in elevation from A to B?

14-5. With an instrument set 5.0 ft above station R, a sight is taken on a rod held at station S. The interval is 5.80 ft, the rod reading of the middle cross hair is 9.9 ft, the vertical angle is +5° 02′. With the instrument set 5.1 ft above station S, a sight is taken on the rod held at R. The interval is 5.78 ft, the rod reading of the middle cross hair is 12.6 ft, the vertical angle is – 3° 52′. The instrument is internally focusing and $K = 100$. What is the average length of RS and what is the difference in elevation from R to S?

14-6. With the instrument set 5.0 ft above station C, a sight is taken on a rod held at station D. The interval is 3.80 ft, the rod reading of the middle cross hair is 12.8 ft, the vertical angle is +0° 18′. With the instrument set 5.1 ft over station D, a sight is taken on the rod held at C. The interval is 3.83 ft, the middle cross hair reading is 11.8 ft, the vertical angle is +1° 56′. The value of C is 0 and $K = 100$. What is the average length of the line CD? What is the average difference in elevation from C to D?

14-7. The following notes were recorded when carrying stadia levels between two temporary benchmarks. The value of C is 0 and the value of $K = 100$. The elevation of TBM20 is 868.3 ft. Compute the elevation of TBM21.

Station	Backsight			Foresight		
	Interval	Rod	Vertical Angle	Interval	Rod	Vertical Angle
TBM20	3.46	5.8	+0° 12′			
TP1	5.45	8.6	+2° 45′	3.96	6.2	+0° 45′
TP2	3.24	8.2	−1° 15′	5.40	5.7	+4° 08′
TP3	5.16	10.0	−0° 20′	3.85	8.2	−0° 25′
TP4	3.02	8.1	−4° 52′	4.10	3.0	−3° 02′
TP5	2.12	6.2	−1° 02′	4.22	5.0	+4° 15′
TBM21				3.27	4.9	+0° 08′

14-8. The following notes were recorded when carrying stadia levels between two temporary benchmarks. The value of $C = 0$ and $K = 100$. The elevation of TBM32 is 386.16 m. Compute the elevation of TBM33.

Station	Backsight			Foresight		
	Interval	Rod	Vertical Angle	Interval	Rod	Vertical Angle
TBM32	1.015	2.41	+2° 02′			
TP1	1.543	1.84	+1° 15′	1.426	2.16	+3° 15′
TP2	1.433	1.75	−0° 05′	0.895	1.80	Level
TP3	2.162	2.00	+1° 18′	0.758	1.54	−0° 08′
TP4	1.005	1.80	+3° 45′	1.816	0.56	−3° 13′
TP5	2.050	2.81	−0° 15′	1.505	2.02	−2° 54′
TBM33				1.024	1.81	+1° 12′

14-9. The mean angle measured between the two targets of a 2-m subtense bar is 0° 44′ 18.35″. The standard error of the angle is ±0.68″. Assume that the standard error of the distance between the targets is ±0.30 mm. Compute the distance, in meters, between the theodolite and the subtense bar. Compute the standard error of this distance, in meters.

14-10. In Problem 14-9, if the subtended angle is 1° 02′ 18.62″ ±0.45″, what is the distance, in feet, between the theodolite and the subtense bar? What is the standard error of this distance, in feet?

14-11. In Problem 14-9, if the subtended angle is 0° 20′ 05.54″ ± 0.54″, what is the distance, in meters, between the theodolite and the subtense bar? What is the standard error of this distance, in meters?

BIBLIOGRAPHY   Easa, S. M. 1991. Maximizing accuracy in stadia surveying. *Surveying and Land Information Systems* 51:3.

Miller, C. H., and Odum, J. K. 1983. Calculator program for reducing alidade or transit stadia traverse data. *Surveying and Mapping* 43:393.

Woodworth, B. 1986. Twopence topo—confessions of a stadia topographer. *Tech. Papers Ann. Conv. Amer. Cong. Surv. & Mapp.* 2:10.

15 Topographic Surveys

15-1 Introduction

Topographic surveying is the process of determining the positions, on the earth's surface, of the natural and man-made features of a given locality and of determining the configuration of the terrain. The location of the features is referred to as *planimetry,* and the configuration of the ground is referred to as *topography* or *hypsography.* The purpose of the survey is to gather data necessary for the construction of a graphical portrayal of planimetric and topographic features. This graphical portrayal is a topographic map. Such a map shows both the horizontal distances between the features and their elevations above a given datum. On some maps the character of the vegetation is shown by means of conventional symbols.

Topographic surveying or mapping is accomplished by ground methods requiring the use of the transit or theodolite, total station, level, hand level, tape, and leveling rod in various combinations. Total station instruments are used to advantage in topographic surveying. The vast majority of topographic mapping is accomplished by aerial photogrammetric methods, as described in Chapter 16. In the photogrammetric methods, however, a certain amount of field surveying and field editing must be done by ground methods described in this chapter.

The preparation of a topographic map, including the necessary control surveys, is usually the first step in planning and designing an engineering project. Such a map is essential in the layout of an industrial plant, the location of a railway or highway, the design of an irrigation or drainage system, the development of hydroelectric power, city planning and engineering, and landscape architecture. In time of war, topographic maps are essential to persons directing military operations.

15-2   Scales and Accuracy

Since a topographic map is a representation, on a comparatively small plane area, of a portion of the surface of the earth, the distance between any two points shown on the map must have a known definite ratio to the distance between the corresponding two points on the ground. This ratio is known as the scale of the map. As stated in Section 8-34, this scale can be expressed in terms of the distance on the map, in inches, corresponding to a certain distance on the ground, in feet. For example, a scale may be expressed as 1 in. = 200 ft. The scale can be expressed also as a ratio, such as 1 : 6000, or as a fraction, such as 1/6000. In either of these two cases, 1 unit on the map corresponds to 6000 units on the ground. A fraction indicating a scale is referred to as the *representative fraction.* It gives the ratio of a unit of measurement on the map to the corresponding number of the same units on the ground.

The scale to which a map is plotted depends primarily on the purpose of the map, that is, the necessary accuracy with which distances must be measured or scaled on the map. The scale of the map must be known before the field work is begun, since the field methods to be employed are determined largely by the scale to which the map is to be drawn. When the scale is to be 1/600 (1 in. = 50 ft) distances can be plotted to the nearest $\frac{1}{2}$ or 1 ft, whereas if the scale is 1/12,000 (1 in. = 1000 ft) the plotting will be to the nearest 10 or 20 ft and the field measurements can be correspondingly less precise.

15-3   Methods of Representing Topography

Topography may be represented on a map by hachures, contour lines, form lines, or tinting. *Hachures,* or hill shading, are a series of short lines drawn in the direction of the slope. For a steep slope the lines are heavy and closely spaced. For a gentle slope they are fine and widely spaced. Hachures are used to give a general impression of the configuration of the ground, but they do not give the actual elevations of the ground surface.

A *contour line,* or *contour,* is a line that passes through points having the same elevation. It is the line formed by the intersection of a level surface with the surface of the ground. A contour is represented in nature by the shoreline of a body of still water. The *contour interval* for a series of contour lines is the constant vertical distance between adjacent contour lines. Since the contour lines on a map are drawn in their true horizontal positions with respect to the ground surface, a topographic map containing contour lines shows not only the elevations of points on the ground, but also the shapes of the various topographic features, such as hills, valleys, escarpments, and ridges.

Figure 15-1 is the classical illustration used to show the relationship between the configuration of the ground and the corresponding contour lines. This illustration used to be printed on the backs of the U.S. Geological Survey quadrangle maps along with an explanation of the topographic map and how it is interpreted and used. Unfortunately the U.S. Geological Survey discontinued this feature of its quadrangle series in the early 1950s. The upper part of the illustration shows a stream lying in a valley between a cliff on the left and a rounded hill on the right.

FIGURE 15-1 Contour line representation of terrain. By permission of U.S. Geological Survey.

The stream is seen to empty into the ocean in a small bay protected by a sand hook. Other features such as terraces, gullies, and a gentle slope behind the cliff can be identified. An abrupt cliff to the right of the sand hook plunges almost vertically to the ocean. The lower part of the figure is the contour line or topographic map representation of this terrain. The contour interval of this map is 20 units and could represent either 20 ft or 20 m.

On maps intended for purposes of navigation, peaks and hilltops along the coast are sometimes shown by means of *form lines*. Such lines resemble contours but are not drawn with the same degree of accuracy. All points on a form line are supposed to have the same elevation, but not enough points are actually located to conform to the standard of accuracy required for contour lines.

On aeronautical charts and on maps intended for special purposes, such as those that may accompany reports on some engineering projects, elevations may be indicated by tinting. The area lying between two selected contours is colored one tint, the area between two other contours another tint, and so on. The areas to be flooded by the construction of dams of different heights, for example, might be shown in different tints.

15-4 Contour Lines

The configuration of the ground and the elevations of points are most commonly represented by means of contour lines, because contours give a maximum amount of information without obscuring other essential details on the map. Some of the principle of contours are represented in Fig. 15-2. Four different contour intervals are shown in views (a), (b), (c), and (d). The steepness of the slopes can be determined from the contour interval and the horizontal spacing of the contours. If all four of these sketches are drawn to the same scale, the ground slopes are the steepest in (d), where the contour interval is 20 ft, and are the flattest in (c), where the interval is 1 ft.

The elevation of any point not falling on a contour line can be determined by interpolating between the two contour lines that bracket the point. Quite often, when the scale of the map is large and the terrain is flat, the successive contours are spaced so far apart horizontally that interpolation between adjacent contours does not have much significance. Therefore, in such an instance, the accuracy and utility of the map is greatly increased by showing the elevations of points at regular intervals in some form of a grid pattern. Elevations between these points are then determined by interpolation. Spot elevations are shown on Fig. 15-3.

A contour cannot have an end within the map. It must either close on itself or commence and end at the edges of the map. A series of closed contours represents either a hill or a depression. From the elevations of the contour lines shown, a hill is represented in Fig. 15-2(a) and a depression in view (b). As indcted, a depression contour is identified by short hachures on the downhill side of the contour. A ravine is indicated by the contours in Fig. 15-2(c). If the elevations were reversed, the same contours would represent a ridge. View (e) is incorrect, as two contours

FIGURE 15-2 Contour lines.

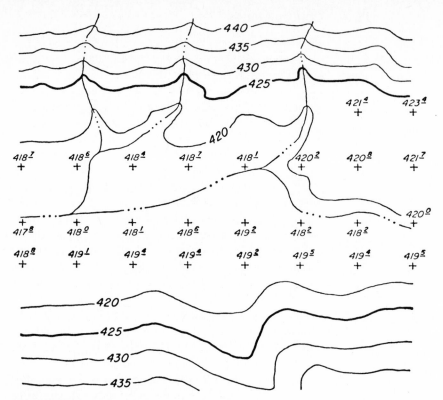

FIGURE 15-3 Spot elevations to supplement contour lines.

are shown meeting and continuing as a single line; this would represent a knife-edged ridge or ravine, something not found in nature. View (f), if not incorrect, is at least unusual. Several contours are shown merging and continuing as a single line. This would be correct only in the case of a vertical slope or a retaining wall. Also, one contour is shown to cross two others. Thus each point marked x has two elevations, a condition found only at a cave or an overhanging cliff.

A series of equally spaced contour lines represents a constant slope along a line normal to the contours. A series of straight, parallel, equally spaced contours represents man-made excavations or embankments. The steepest direction from any point on a topographic map is that which runs normal to successive contour lines near the point.

The drainage of the terrain is the primary agent in shaping the topography. Its influence on the shape of the contour lines can be seen in Figs. 15-1, 15-2(d), and 15-3. Note that as contour lines cross gullies or streams or other drainage features, the contour lines form modified Vs pointing upstream. The form of the Vs is determined by the type of underlying soil or rock. In general, if the underlying material is fine grained like a clay soil, the V will be smooth and rounded, and if the material is coarse and granular, the V will be quite sharp.

As a convenience in scaling elevations from a topographic map, each fifth contour is drawn as a heavier line. This is called an *index contour*. When the interval is 1 ft, contours whose elevations are multiples of 5 ft are shown heavy. When the interval is 10 ft, the heavy contours have elevations that are multiples of 50 ft. Enough contours should be numbered to prevent any uncertainty regarding the elevation of a particular contour. Where the contours are fairly regular and closely spaced, only the heavy contours need be numbered.

15-5 Field Methods

Among the factors that influence the field method to be employed in the compilation of a topographic map are the scale of the map, the contour interval, the type of terrain, the nature of the project, the equipment available, the required accuracy, the type of existing control, and the extent of the area to be mapped. The area to be mapped for highway or railroad location and design takes the form of a strip with a width varying from 100 ft to perhaps more than 1000 ft (30 to 300 m). The control lines are the sides of a traverse that have been established by a preliminary survey and that have been stationed and profiled as outlined in Chapter 3. The method of locating topography most commonly employed for this purpose is the *cross-section* method.

To make an engineering study involving drainage, irrigation, or water impounding or to prepare an accurate map of an area having little relief, each contour line must be carefully located in its correct horizontal position on the map by following it along the ground. This is the *trace contour* method.

When an area of limited extent is moderately rolling and has many constant slopes, points forming a grid are located on the ground and the elevations of the grid points are determined. This is the *grid* method of obtaining topography.

If the area to be mapped is rather extensive, the contour lines are located by determining the elevations of well-chosen break points from which the positions of points on the contours are determined by computation. This is known as the *controlling-point* method.

15-6 Cross-Section Method

The cross-section method of obtaining topography can be performed by transit and tape, by transit stadia, by level and tape, by a combination EDM-theodolite, or by a total station instrument. Horizontal control is established by a theodolite-tape traverse, by an EDM-theodolite traverse, or by a traverse run using a total station instrument. This traverse is run between fixed control points. Stakes are set every 50 or 100 ft or at other pertinent intervals, the spacing depending on the terrain. Vertical control is obtained by profile leveling which may be performed either before the topography is taken or concurrently with the cross sectioning. When using the total station instrument, vertical control is obtained as the traverse progresses.

When the theodolite and tape are used, the instrumentman occupies each station or plus station on the line. He determines the HI by holding the leveling rod alongside the instrument. A right angle is turned off the line, and the rodman, holding one end of the tape, proceeds along this cross line until a break in the slope occurs. If possible, the instrumentman takes a level sight on the rod and records the reading and the distance to the point. If the rod cannot be sighted with the telescope level, a vertical angle is read and the slope distance is recorded. The rise or fall of the line of sight equals the slope distance times the sine of the vertical angle, as discussed in Section 3-3. The horizontal distance to the point is obtained by Eq. (2-1). The rod reading must be taken into account in determining the elevation of the point, as discussed in Section 14-7.

The rodman proceeds to the next break in the topography, and the process is repeated. If the width of strip on each side of the line is greater than the length of the tape, a third man is necessary to hold the tape. As the distance out becomes greater, the slope distance cannot be measured directly in case vertical angles must be read, and the true rise or fall of the line of sight will be somewhere between the measured distance times the sine of the vertical angle and the measured distance times the tangent of the angle.

On a large project in which the width of the project is greater than the length of a tape, the EDM attached to the theodolite or else the total station instrument is employed to advantage. Either of these systems measures the horizontal distance to the point and the difference in elevation between the instrument setup and the point on the cross section. The reflector is held on a staff or rod that contains a level bubble to bring the rod vertical. The height of the reflector above ground together with the HI of the instrument are then used in determining the difference in elevation to the point.

Table 15-1 shows the data recorded for nine regular cross sections spaced at 50-ft intervals, and additional ones at station 22 + 28 and station 23 + 70. The center line elevations are obtained either from prior profile leveling or else by leveling carried along as the cross-section measurements are made. If by the latter method, then profile-level notes would be recorded on the left side of the field book in combination with the cross-section notes on the right side. The cross-section notes consist of the station number, the horizontal distance to the right or left of the center line to the point as the denominator, and the elevation of the point as the numerator. Note that the stationing increases from the bottom upward. A glance at the right side shows that when someone views the notes, he can imagine looking in the forward direction of the line, with the values to the left of him entered as such as those to the right of him entered as such.

If the instrument used in measuring cross sections incorporates an electronic field book or a data collector, the instrumentman then needs only to key in the station number and the direction left or right from the centerline. The distance to and the elevation of the point are automatically stored in the data collector, and are available in the office for plotting. This eliminates the recording of cross-section notes in the field.

The total station instrument can be used in an altogether different manner than that in which each station is occupied, provided that the center line is placed on a coordinate system. This implies that the azimuth of the center line is known and that the coordinates of the stations are also known. The instrument is set up on a station, and the azimuth of the center line together with the coordinates and elevation of the

TABLE 15-1   Recorded Data for Nine Cross Sections

Station	Center Line Elevation	Cross Sections L			C	R		
26 + 00	550.0			550.2 / 93	550.0 / 0	547.6 / 23	539.0 / 61	532.4 / 106
25 + 50	534.5			545.0 / 107	534.5 / 0	532.1 / 59	527.7 / 109	
25 + 00	539.3		551.3 / 103	549.1 / 31	539.3 / 0	537.3 / 10	521.7 / 100	
24 + 50	544.2		551.0 / 94	555.1 / 52	544.2 / 0	532.3 / 52	524.5 / 106	
24 + 00	551.0		550.0 / 101	556.3 / 65	551.0 / 0	529.1 / 100		
23 + 70	545.0	548.0 / 100	557.3 / 67	553.9 / 45	545.0 / 0	539.0 / 27	528.6 / 100	
23 + 50	546.6		547.4 / 102	557.8 / 61	546.6 / 0	525.8 / 94		
23 + 00	550.0		552.0 / 108	559.8 / 69	550.0 / 0	547.6 / 20	532.0 / 109	
22 + 50	542.7		554.8 / 89	549.2 / 47	542.7 / 0	538.0 / 46	535.0 / 102	
22 + 28	539.5			550.0 / 92	539.5 / 0	530.0 / 62	530.7 / 96	
22 + 00	539.3	546.6 / 92	545.5 / 61	541.4 / 19	539.3 / 0	539.7 / 32	528.6 / 102	

station are keyed in. The reflector is set to the HI of the instrument. A sight is taken along the center line to orient the circle. The person on the reflector then must remain on or near the perpendicular to the center line when selecting breaks in the topography. When a sight is taken to the reflector, the data collector computes the coordinates and elevation of the point and stores it for future plotting. In this manner several cross sections can be measured from a single instrument setup.

In any of the foregoing methods the located points are plotted in the office. The positions of the contour lines are obtained by interpolating between the elevations of the plotted points as discussed below in Section 15-7.

In compiling topography by the cross-section method, the positions of all planimetric features, such as buildings, fences, streams, and property lines, must be located with respect to the control line and plotted on the topographic map. Methods of locating the positions of points with respect to a traverse line are discussed in Section 6-13.

15-7 Methods of Interpolating

In locating contours on a map by interpolation, the positions of points on the contours can be determined either mathematically, mechanically or with software. In Fig. 15-4 it is required to locate the 5-ft contours on the line connecting points a and b, whose elevations are 873.4 and 896.2, respectively. It is evident that contours at elevations 875, 880, 885, 890, and 895 will cross this line.

FIGURE 15-4 Interpolating with triangle and scale.

The distances on the map from the point a to the contours can be calculated by proportion. The horizontal distance between a and b on the map is scaled and found to be 2.78 in. The corresponding vertical distance on the ground is 896.2 – 873.4 = 22.8 ft. The vertical distance from a to the 875-ft contour is 875.0 – 873.4 = 1.6 ft. The horizontal distance from a to the 875-ft contour is (2.78/22.8) 1.6 = 0.20 in. The distance between two adjacent 5-ft contours is (2.78/22.8) 5 = 0.61 in. The horizontal distances can also be expressed in terms of the distances on the ground.

Figure 15-4 illustrates a method by which the points on the contours can be located mechanically by using a triangular engineer's scale and a small celluloid triangle. The method is an application of the geometric method of dividing a line into any number of equal parts. The 7.34-in. mark on the scale is pivoted on a, whose elevation is 873.4 ft; the corner of the triangle is placed at the 9.62-in. mark; and both the scale and the triangle are turned until the edge of the triangle passes through b, whose elevation is 896.2 ft. The scale is then held in place while the corner of the triangle is moved successively to the 9.50-, 9.00-, 8.50-, 8.00-, and 7.50-in. points on the scale. Where the edge of the triangle crosses the line ab in the various positions are the corresponding contour points.

This method is very rapid and accurate and entirely eliminates mathematical computations. Any edge of the triangular scale can be used, provided the length on the scale corresponding to the difference in elevation between the two plotted points is shorter than the length of the straight line between the points on the map. When the difference in elevation is considerable and the map distance is short, it may be necessary to let one division on the scale represent several feet in elevation. Thus the smallest division on the scale may correspond to a difference in elevation of 10 ft, instead of 1 ft as in Fig. 15-4. By changing the value of a division, some side of the scale can always be used, regardless of the difference in elevation of the length of the line on the map.

Instead of the triangle and scale, a piece of tracing cloth, on which equally spaced horizontal lines have been ruled, can be used in exactly the same manner. The tracing cloth is turned until lines corresponding to the given elevations pass through the two points on the map. The contour points are then pricked through the tracing cloth to the map beneath.

In most cases the position of the contour crossings can be estimated with sufficient precision, and exact interpolation is unnecessary. When the positions of the contour lines have been plotted, the contour lines are sketched in freehand by joining points at the same elevation.

The cross-section notes shown in Table 15-1 are used to plot the contour lines shown in Fig. 15-5. The center line is first laid out in the proper position on the

FIGURE 15-5 Contour lines interpolated from cross-section notes.

map sheet, usually by the coordinates of its points of intersection. The center line tangent is then stationed, and cross-section lines are drawn through the station marks. Distances are laid out to the points where elevations were determined by reference to the cross-section notes. These points are represented by small dots in the illustration. Elevations are entered at the points. The positions of the contour-line crossings are then determined; and, finally, the contour lines are sketched in and numbered. Notice than an extra cross section was needed at station 22 + 28 to pick up the stream bed, and at station 23 + 70 to locate the swale in the side of the hill.

If a total station instrument is used to obtain cross sections, the data held in the data collector are downloaded to the control unit of an automatic drafting machine. Then the points are plotted according to their coordinates. Contour lines can then be drawn as discussed previously, or they are located by any of several computer programs used to generate contour lines.

15-8   Trace Contour Method

Under certain critical conditions the surveyor is required to locate contours with a higher degree of accuracy than can be obtained by other field methods. This requires that the positions of a string of points at constant elevation must be determined for each contour line. A network of control points is established in the area to be mapped. The elevations of the control points are determined by leveling or as part of the control survey.

Two methods are discussed here for tracing out contour lines. In the first method a theodolite and leveling rod are employed. In Fig. 15-6 the theodolite is set up at control station S, whose elevation is 232.6 ft. A backsight is taken on control station R, and the horizontal circle is read (or stored). The height of the instrument from the control station up to the horizontal axis of the theodolite is measured (5.2 ft in Fig. 15-6), and added to the elevation of the station to give the elevation of the line of sight of 237.8 ft.

To locate the distance and direction to a point on the 236-ft contour line, the rodman is so positioned as to give a horizontal line of sight reading of 1.8 ft (237.8 − 1.8 = 236). A stadia interval is read to determine the distance to the point. The horizontal circle is then read (or stored). In a similar manner the rodman moves along the contour line with the rod reading of 1.8 ft always maintained. The stadia interval and horizontal circle reading are read and stored for each point. This results in a string of points on the 236-ft contour line.

The 234-ft contour line is located in a similar manner with the rod reading of 3.8 ft (237.8 − 3.8 = 234). This procedure is continued for each subsequent contour line until the instrumentman can no longer read the rod. A new setup at a different elevation is then required.

In the second method a total station instrument with continuous tracking capability is used to advantage in that a larger difference in elevation is no impediment that exists with the leveling rod. In Fig. 15-7 the instrument is set up at the same station S as before. A backsight is taken on station R with the azimuth of SR

FIGURE 15-6 Trace contouring using theodolite and leveling rod.

entered. The EDM reflector is set to the height of the horizontal axis (5.2 ft in this example). The elevation of station S is entered into the instrument before measurements begin. With the instrument set to the tracking mode, the reflector is moved until the elevation scale reads 236 ft. This puts the foot of the reflector rod on the 236-ft contour line. The horizontal circle (the azimuth), the horizontal distance, and the elevation for this point are all stored. This procedure is repeated for the string of points on each contour line. Since the line of sight can be raised or lowered, there is no such thing as "running out of rod."

FIGURE 15-7 Trace contouring using a total station instrument.

Plotting of the contour lines can be performed using a 360° protractor and scale or else, in the case of the total station instrument, its data can be downloaded to an automatic plotter.

15-9 Grid Method

The grid method of obtaining topography may be used in areas of limited extent where the topography is fairly regular. A level is usually used for determining elevations of the grid points, although a theodolite can be used by bringing the telescope horizontal for each sighting.

If the boundary of the area to be mapped has not been previously surveyed, the first step is to run a traverse around the area and establish the corners. Then, to determine the topography, the area is usually divided, as far as possible, into squares or rectangles of uniform size. The dimensions of these divisions depend on the required accuracy and the regularity of the topography but are usually between 25 and 100 ft. The form chosen for the divisions will depend somewhat on the shape of the area. The size of the divisions should be such that, for the most part, the ground slopes can be considered uniform between the grid points at the corners of the divisions. The grid points either are defined by stakes or are located by ranging out and measuring from stakes already set.

Figure 15-8 represents a tract of land of which a topographic map is to be prepared. It is assumed that a traverse survey locating the boundaries has already been made. The tract is to be divided by means of lines running in two perpendicular directions and spaced 100 ft apart. When a tract is divided in this manner, it is customary to designate by letters the lines that extend in one direction, and by figures the lines at right angles to that direction. The point of intersection of any two lines is then designated by the letter and figure of the respective intersecting lines. The intersections of the dividing lines with the exterior boundary lines on the far

FIGURE 15-8 Topography by grid method.

sides of the tract are designated by the proper letters or numbers affected with an accent or a subscript, as A', B', $2'$, $3'$.

Different methods may be followed in laying out the tract. Any method is satisfactory if it accurately defines the positions of the points of intersection so that they can be readily located when the levels are being taken. For this purpose it is not usually necessary to mark all points of intersection, but enough points should be marked by stakes to permit the remaining points to be located easily and quickly by merely ranging them in from the points that are marked. The rougher and more irregular the surface of the tract, the more stakes must be set. Also, a tract of irregular form usually requires a comparatively greater number of stakes than a tract of rectangular form.

After the required stakes have been set, levels are taken over the tract to determine the elevations at all points of intersection and also at any intermediate points where the slope changes abruptly. Such an intermediate point is generally located in a direct line between two intersections by its distance from the intersection having the lower letter or number. This distance is measured with a tape, approximated by pacing, or merely estimated by the eye, according to the conditions and to the degree of accuracy required, and is recorded as a plus. Thus on line CC' there is a low point 80 ft beyond stake $C5$ and its elevation is 730.8; this point would be designated by $C5 + 80$. The high point, whose elevation is 746.5 and which is situated between lines 4 and 5 and between lines A and B, would be designated in the notes as $A + 80$, $4 + 45$. The levels should be taken in the order that is most advantageous for the nature of the ground. The object is to take rod readings at each of the intersections and at other points with as few settings of the level as possible. To be sure that rod readings are taken at all the intersections, those taken from each setting are checked off on the sketch, or sketches, in which they are all shown.

An EDM mounted on a theodolite or a total station instrument can be used in place of the level. A point that commands a view of a substantial portion of the area to be mapped is located with reference to the control that is used to establish the grid intersections. Its elevation is also determined. After the grid has been staked out, the point is occupied by the EDM instrument, and the reflector is held at each grid intersection. The slope distance to the grid point is measured along with the vertical angle from which the elevation of the grid point can be determined. To locate planimetry, the theodolite is oriented in azimuth by backsighting on a control point, and then the upper motion is used to determine the direction to each planimetric feature. The EDM is used to measure the distance to the point.

After the field work has been completed, the control points, tract boundary, and grid are plotted to the desired scale. The values of the elevations of the grid intersections are then written at the corresponding map positions of the intersections. The positions of the contour lines are located and sketched by interpolation between the grid intersections. Finally, the planimetric features are plotted using a 360° protractor, as discussed in Section 15-8. If a total station instrument is used, the data from the data collector are downloaded into a drafting machine (see Section 8-33) which then plots the points and enters the elevations automatically. Programs are available which automatically generate contour plots from the data.

15-10   Controlling-Point Method

The compilation of a topographic map by determining the positions and elevations of carefully selected controlling points is applicable to nearly every condition encountered in mapping. It is the method used most extensively in mapping a large area to a relatively small scale because of the economy realized. This method can be applied, instead of the cross-section method, to the mapping of a strip of terrain for route-location studies.

The controlling-point method has in the past employed the plane table and alidade. However, the capability of the total station instruments for storing data and downloading into automatic drafting instruments for the generation of contour lines and planimetry has all but displaced the use of the plane table.

A system of control, both horizontally and vertically, is established in the area to be mapped. This presumes that the coordinates and elevations of the control points have now been determined. The total station instrument occupies a control point, the HI is determined, the coordinates of the point have been entered, and a backsight is taken along a line of known azimuth. The person on the reflector now chooses critical points on the terrain that will most closely reflect the true shape of the topography. These points are along ridges, streams, gullies, draws, the top and bottom of constant slopes in the terrain, peaks and depressions—that is, those points that would permit interpolation to determine the positions of contour lines. As the coordinates and elevations of these points are located, the instrumentman must designate the type of point observed to be understood by the program that operates the drafting machine. Ridges, valleys, streams, and gullies are referred to as *break lines* because the terrain changes slope abruptly along these lines.

All planimetric features are located and identified by the unique codes associated with the instrument system being used. Some of the more advanced instruments allow the operator to designate how points are to be joined together just as a draftsman would logically join points.

After all of the control points in the area have been occupied, the data collector is downloaded into the control unit of the plotter. The planimetric features are automatically drawn to the desired scale. The contours are generated by various computer programs, the details of which can be found in the manufacturer's literature. Figure 15-9 illustrates a topographic map generated by an automatic plotter from data downloaded from a data collector. The tick marks are rectangular coordinate grid intersections plotted by the machine.

The controlling-point method depends to a great extent on the reflector person's knowledge of land shapes, slopes, and stream gradients, and his ability to decide where to select points so that he selects neither too many nor too few points. A distinct advantage of the "one-person" instrument described in Section 4-24 is that the instrumentman has total control over the selection of points to be located.

FIGURE 15-9 Topographic map generated from output of data collector.
By permission of Leitz, Inc.

PROBLEMS

15-1. A 100-ft grid is located over an area that measures 800 ft in the north-south direction and 800 ft in the east-west direction. The southwesterly corner is designated *A*-1 and the southeasterly corner is designated *I*-1. The northwesterly corner is designated *A*-9 and the northeasterly corner is designated *I*-9. Elevations are obtained at the grid points using a total station instrument, rounded off to the nearest foot. These are tabulated as follows:

9	375	372	366	360	354	350	340	322	302
8	372	372	370	368	365	362	361	339	312
7	366	365	365	366	364	362	361	347	319
6	363	353	350	360	362	359	350	347	328
5	361	340	335	343	352	350	338	334	328
4	360	344	329	328	337	339	325	321	321
3	360	343	320	317	318	318	313	312	312
2	358	343	323	316	310	308	303	302	303
1	353	340	326	318	310	303	297	292	291
	A	B	C	D	E	F	G	H	I

Plot a grid to a scale of 1 in. = 200 ft and draw in 10-ft contour lines.

15-2. Repeat Problem 15-1 with the following values:

9	493	490	488	491	492	493	491	486	483
8	480	474	473	477	484	490	487	480	477
7	479	467	457	460	476	482	479	474	468
6	478	468	459	450	462	470	470	468	461
5	478	470	459	440	440	452	451	460	450
4	480	467	451	438	434	430	427	420	430
3	476	465	459	447	440	433	425	415	409
2	475	472	466	460	454	450	440	422	402
1	472	472	470	468	465	462	461	439	412
	A	B	C	D	E	F	G	H	I

15-3. Repeat Problem 15-1 with the following values:

9	586	583	579	570	560	550	528	520	509
8	580	577	577	574	570	563	558	550	542
7	574	568	568	570	570	565	562	557	551
6	568	561	556	561	564	559	553	550	552
5	560	550	537	546	561	550	539	540	547
4	520	530	522	534	544	537	523	520	536
3	515	509	514	525	532	531	520	510	518
2	522	502	500	507	520	525	520	508	507
1	539	512	494	483	490	510	517	506	498
	A	B	C	D	E	F	G	H	I

15-4. Repeat Problem 15-1 with the following values:

9	722	702	700	707	720	725	720	708	707
8	739	712	694	683	690	710	717	706	698
7	747	719	700	685	676	690	707	700	689
6	747	728	703	688	670	673	690	687	679
5	734	728	707	690	673	659	664	670	669
4	721	721	710	692	674	660	650	653	658
3	712	712	707	693	671	657	647	639	642
2	702	703	700	683	664	651	641	634	636
1	692	691	682	668	655	641	633	626	630
	A	B	C	D	E	F	G	H	I

15-5. The cross-section notes on page 546 were taken for the purpose of plotting a strip of topography for road location. The positions of stream crossings are marked (*). Plot the tangent between stations 18 and 27 to a scale of 1 in. = 100 ft, and sketch in each 2-ft contour line.

15-6. The cross-section notes on page 547 were taken for the purpose of plotting contour lines of a strip of topography. Plot the tangent from station 0 to station 3, using a scale of 30 ft to the inch. Plot the notes on a sheet of $8\frac{1}{2} \times 11$-in. paper, and sketch the 5-ft contour lines. Show the location of the stream (*).

15-7. The elevations for the gridded area shown in Fig. 15-10 are as follows:

F	690	709	726	732	735	740	743	747
E	696	715	729	736	740	737	739	740
D	702	723	733	742	737	737	745	737
C	707	726	732	732	729	740	740	730
B	709	725	723	722	730	738	731	727
A	711	721	715	715	732	730	719	722
	1	2	3	4	5	6	7	8

$$P = 741 \qquad Q = 751$$

Construct a 1-in. grid on a sheet of $8\frac{1}{2} \times 11$-in. paper and sketch in the 5-ft contour lines. Label and broaden every fifth contour line.

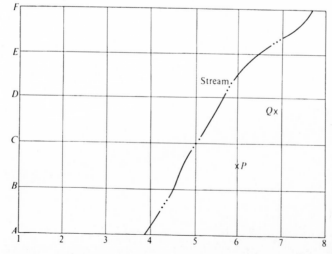

FIGURE 15-10 Grid for solution to Problem 15-7.

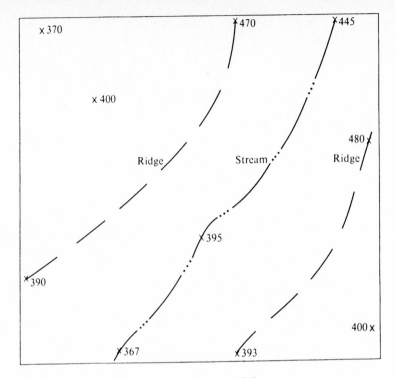

FIGURE 15-11 Diagram to accompany Problem 15-8.

15-8. The controlling points shown in Fig. 15-11 were obtained by a total station instrument. Trace the figure and sketch the 10-ft contour lines. Number and broaden every fifth contour line, including the 400-ft line.

Cross-Section Notes

Station	Left			Center Line		Right		
27 + 00	$\frac{473.1}{200}$	$\frac{461.4}{107}$	$\frac{450.08}{53}$	$\frac{450.4}{0}$	$\frac{439.4}{200}$			
26 + 50	$\frac{458.0}{200}$	$\frac{449.8}{97}$	$\frac{429.3}{17}$	$\frac{427.7}{0}$	$\frac{425.9}{116}$	$\frac{421.3^*}{200}$		
26 + 00		$\frac{445.2}{200}$	$\frac{428.2}{65}$	$\frac{425.7}{0}$	$\frac{425.3^*}{8}$	$\frac{423.6}{94}$	$\frac{421.9}{200}$	
25 + 93				$\frac{425.5^*}{0}$				
25 + 50		$\frac{430.8}{176}$	$\frac{429.0}{140}$	$\frac{428.6}{0}$	$\frac{424.3}{154}$	$\frac{422.0}{200}$		
25 + 00		$\frac{436.1}{200}$	$\frac{430.8}{91}$	$\frac{431.0}{0}$	$\frac{430.3}{65}$	$\frac{423.2}{200}$		
24 + 90				$\frac{431.6}{0}$	$\frac{422.5^*}{200}$			
24 + 50			$\frac{439.3}{200}$	$\frac{432.4}{0}$	$\frac{425.0^*}{136}$	$\frac{424.8}{200}$		

Cross-Section Notes—*continued*

Station	Left			Center Line	Right		
24 + 00		443.2/200	435.8/93	432.1/0	427.7*/94	428.1/115	425.8/200
23 + 76		444.9/200	433.8/73	431.4*/0	426.9/200		
23 + 15		438.4*/182	438.0/150	437.5/0	437.5/80	431.2/200	
23 + 00		440.0/200	438.1/87	439.2/0	438.8/113	432.1/200	
22 + 00		448.9/200	448.2/164	450.0/0	439.8/139	433.7/200	
21 + 00	459.9/200	460.0/153	457.2/88	446.8/0	433.4/102	424.6/200	
20 + 00			467.0/200	438.6/0	420.7/143	416.1/200	
19 + 00			479.2/200	443.5/0	422.4/111	414.0/200	
18 + 00	499.1/200	475.4/118	461.2/63	447.0/0	433.2/80	419.2/200	

Cross Section Notes

Station	Left							Center Line	Right					
3 + 00			405/50	400/43	395/32	390/17	385/3	384/0	380/18	376*/35				
2 + 50				400/46	395/35	390/25	385/9	382/0	380*/8	385/39	390/50			
2 + 00			400/47	395/35	390/27	385/20	384*/10	385/0	390/5	395/11	400/20	400/33	395/50	
1 + 89								390/0						
1 + 78								395/0						
1 + 69								400/0						
1 + 62								405/0						
1 + 54								410/0						
1 + 50	400/50	395/42	390*/32	395/19	400/13	405/9	410/4	412/0	415/6	415/12	410/18	405/27	400/33	395/50
1 + 42								415/0						
1 + 25								417/0						
1 + 00				400/50	405/44	410/36	415/24	416/0	415/4	410/10	405/21	400/30	395/48	

Cross Section Notes—*continued*

Station	Left				Center Line	Right		
0 + 95					$\frac{415}{0}$			
0 + 84					$\frac{410}{0}$			
0 + 68					$\frac{405}{0}$			
0 + 50	$\frac{415}{50}$	$\frac{410}{33}$	$\frac{405}{14}$		$\frac{403}{0}$	$\frac{400}{11}$	$\frac{395}{27}$	$\frac{390}{50}$
0 + 32					$\frac{400}{0}$			
0 + 00	$\frac{415}{52}$	$\frac{410}{44}$	$\frac{405}{33}$	$\frac{400}{14}$	$\frac{397}{0}$	$\frac{395}{7}$	$\frac{390}{28}$	$\frac{385}{48}$

15-9. Reduce the following transit-stadia notes, and plot the positions of the points to a scale of 1 in. = 20 ft. Sketch 1-ft contour lines. Assume $K = 100$ and $C = 0$. Line AB is due north.

		Transit at B, Elevation = 459.5 ft BS *0° 00′ on A*				
Point	Interval	Horizontal Distance	Azimuth	Vertical Angle	DE	Elevation
1	0.56		0° 00′	−1° 20′		
2	0.99		1° 00′	−0° 28′		
3	0.25		10° 15′	+4° 36′		
4	0.85		12° 15′	+1° 58′		
5	1.06		16° 45′	−0° 03′		
6	0.58		17° 45′	−2° 22′		
7	0.76		30° 00′	−1° 49′		
8	1.08		40° 45′	−2° 14′		
9	1.43		45° 00′	−2° 48′		
10	0.83		50° 00′	−3° 15′		
11	0.59		61° 45′	−2° 32′		
12	1.12		61° 45′	−4° 41′		
13	0.86		68° 45′	−4° 05′		
14	1.00		68° 45′	−4° 40′		
15	1.04		75° 15′	−6° 07′		
16	0.91		77° 00′	−6° 21′		
17	0.45		80° 30′	−3° 04′		
18	0.68		83° 30′	−3° 38′		
19	0.80		86° 30′	−6° 12′		
20	1.00		88° 45′	−9° 05′		

BIBLIOGRAPHY

Cheves, M. S. 1988. Recent developments in data collection and automated contour mapping. *Tech. Papers Fall Conv. Amer. Cong. Surv. & Mapp.* p. 102.

Christenson, L. L. 1988. Digital terrain modeling for Michigan highways. *Tech. Papers Ann. Conv. Amer. Cong. Surv. & Mapp.* 1:206.

Hogan, R. E. 1988. New tools to make the photogrammetrist a key member of the engineering/graphics information system. *Tech. Papers Ann. Meet. Amer. Cong. Surv. & Mapp.* 1:87.

Kellie, A. C. 1979. An evaluation of field techniques for topographic mapping. *Proc. Ann. Meet. Amer. Cong. Surv. & Mapp.* p. 471.

Lee, J. 1991. Assessing the existing methods for TIN extraction. *Tech. Papers Ann. Conv. Amer. Cong. Surv. & Mapp.* 2:194.

Marth, R. 1991. Topographic surveyor-transition into the 1990's. *Tech. Papers Ann. Conv. Amer. Cong. Surv. & Mapp.* 1:173.

Nicholson, J. 1986. Automated earthwork design using the HASP digital system. *Tech. Papers Ann. Conv. Amer. Cong. Surv. & Mapp.* 2:80.

Paiva, J. V. R. 1991. An intelligent solution for kinematic GPS mapping. *Tech. Papers Ann. Conv. Amer. Cong. Surv. & Mapp.* 1:202.

16 Photogrammetry

16-1 Scope

Photogrammetry is the science of making measurements on photographs. Terrestrial photogrammetry applies to the measurement of photographs that are taken from a ground station, the position of which usually is known or can be readily determined. Aerial photogrammetry applies to the measurement of photographs taken from the air. As the science of aerial photogrammetry has developed, it has come to include all operations, processes, and products involving the use of aerial photographs. Among these are the measurement of horizontal distances, the determination of elevations, the compilation of planimetric and topographic maps, the measurement of profiles and cross sections, the preparation of orthophotos, and the interpretation and analysis of aerial photographs for geological, agricultural, and engineering investigation, and for evaluating timber stands and other natural resources.

This chapter discusses the elementary principles of aerial photogrammetry as they apply to the work of the survey engineer. Some of the operations to be discussed are only approximations to more rigorous operations involved in photogrammetry but which are most helpful to the surveyor. These operations involve the use of paper print aerial photographs that can be used for determining approximate distances and elevations, for planning a survey, and for the construction of mosaics.

More rigorous operations are involved in the compilation of very accurate topographic maps, in the production of orthophotos, and in the digitizing of points on the terrain to obtain digital terrain models.

The entire field of photogrammetry cannot be incorporated into a single chapter in a textbook. This chapter is intended as an introduction to photogrammetry in order that the student may gain some appreciation of the potential applications of the science to surveying. References at the end of the chapter should be studied by the student who is interested in pursuing this field in depth.

16-2 Aerial Camera

The aerial camera is comparable to a surveying instrument in that it gathers data in the form of light rays with certain directions and records the data on a photographic negative. Figure 16-1 shows a schematic diagram of an aerial camera with its component parts. The optical axis usually is supposed to be essentially vertical when a photograph is taken.

The *lens assembly,* which includes the shutter and the diaphragm, forms the photographic image with the proper amount of light admitted. The lens itself is composed of several elements, as shown in Fig. 16-2. It must be highly corrected to satisfy the requirements of a rather wide angular coverage, a high resolution, and a very minor amount of lens distortion. Because of atmospheric haze that contains an overabundance of blue light, the lens is invariably fitted with a yellow, orange, or red filter to absorb some of the blue light. The *cone* and the *camera body* hold the lens assembly in its proper position with respect to the *focal plane.*

The camera body houses the drive mechanism for the shutter assembly and the *magazine.* The upper surface of the camera body, in general, defines the focal plane of the camera. This surface contains four or eight *fiducial marks,* located either with one in the middle of each side of the focal plane opening, as seen in the aerial photograph of Fig. 16-4; in each corner, as shown in Fig. 16-10; or both. Aerial cameras used for mapping have focal-plane openings with dimensions of 9 × 9 in. (230 × 230 mm), and produces aerial photographs of corresponding sizes. The purpose of the fiducial marks is to define the point at which the optical axis of the lens intersects the focal plane. This point is called the *principal point* of the photograph.

The magazine is a light-tight container for the film. It contains a feed spool, a take-up spool, a mechanism for advancing the film after each exposure, and a device for holding the film flat in the focal plane at the instant the exposure is made.

FIGURE 16-1 Component parts of aerial camera.

FIGURE 16-2 Aus Jena Lamegon P1 camera.

The two points on the optical axis marked N and N' in Fig. 16-1 are the *front* and *rear nodal points* of the lens system. On striking the front nodal point, any ray of light is so refracted by the lens that it appears to emerge from the rear nodal point parallel with its original direction.

The focal length of the lens is the distance between the rear nodal point of the lens and the focal plane. It is designated by the symbol f. The focal length is fixed for a given camera, and its value is precisely determined by calibration. The nominal values of focal lengths used in mapping are generally 3.5, 6, 8.25, and 12 in. (88, 152, 210, and 305 mm).

The camera is suspended in a camera mount by means of a pair of trunions on either side of the camera. It is free to rotate in the mount about all three axes. The mount is secured over an opening in the bottom of the airplane. Figure 16-3 shows an aerial camera installed in the mount. Some of the features just described can be seen in the picture.

16-3 Types of Aerial Photographs

An aerial photograph taken with the optical axis held essentially vertical is called a *vertical photograph.* Because of movement of the airplane at the instant of exposure, virtually all vertical photographs contain a certain amount of tilt. The optical axis may be accidentally inclined as much as 5° from the vertical. Tilt, however, does not present serious difficulties in mapping. In some of the discussions in this chapter, a vertical photograph is one that is assumed to be truly vertical unless otherwise stated. A vertical photograph is shown in Fig. 16-4.

FIGURE 16-3 RC20 aerial camera.
Courtesy of Leica, Inc.

An aerial photograph taken with the optical axis purposely tilted by a sizable angle from the vertical is called an *oblique photograph*. A high oblique is a photograph on which the apparent horizon appears. A low oblique is a photograph taken with the optical axis purposely tilted from the vertical but not enough to include the horizon. Oblique photographs are shown in Fig. 16-5(a) and (b).

Most planimetric and topographic mapping and orthophoto production are done by using vertical photographs, although high obliques are sometimes used in the preparation of small-scale planimetric maps and charts because of their greater ground coverage.

16-4   Photographic Scale

Photographic scale is the ratio between a distance measured on a photograph and the corresponding ground distance. A vertical photograph resembles a planimetric map in that it shows the planimetric and cultural features of a portion of the ground in their relative positions. It is different from a planimetric map, however, in two respects: (1) The vertical photograph does not contain standard map symbols, which are essential to a map; (2) the planimetric map has a uniform scale throughout, whereas the scale of a vertical photograph varies in different portions of the photograph because of ground relief (see Section 16-5). The photograph is a perspective projection of the ground onto the focal plane of the camera. Consequently points lying in a plane closer to the camera at the time of exposure will have larger images than those points lying in a plane farther from the camera. The scale will vary across the area of the photograph also because of tilt of the optical axis at the time of exposure.

FIGURE 16-4 Vertical photograph showing fiducial marks at the middle of the four edges. Courtesy Pacific Aerial Surveys, Oakland, California.

FIGURE 16-5 Oblique photographs. (a) High oblique. (b) Low oblique. Courtesy of Pacific Aerial Surveys, Oakland, California.

In Fig. 16-6(a) points A, O, and B all lie at the same elevation. The horizontal distances AO and OB are equal to $A'O'$ and $O'B'$ on a reference datum. A truly vertical photograph taken with the camera at L_1 would show the positions of A, O, and B at a, o, and b. The ratio ao/AO equals the ratio ob/OB, and the scale of the photograph is uniform across the photograph.

In Fig. 16-6(b) points A, O, and B are at different elevations, and the horizontal distances between A and O and between O and B are $A'O'$ and $O'B'$, respectively. If a vertical photograph is taken with the camera at L_2, the points A, O, and B would appear at a, o, and b on the photograph. The ratio $ao/A'O'$ does not equal the ratio $ob/O'B'$, and the scale of the photograph is seen to vary from point to point because of the variation in the elevations of the ground points.

In Fig. 16-6(c) points A, O, and B all lie at the same elevation. A tilted photograph taken with the camera at L_3 shows the positions of the ground points at a, o, and b. The ratio ao/AO does not equal the ratio ob/OB, and consequently the scale varies across the photograph. In Fig. 16-6(d) the combined effect of both relief and tilt on the scale of a photograph is shown.

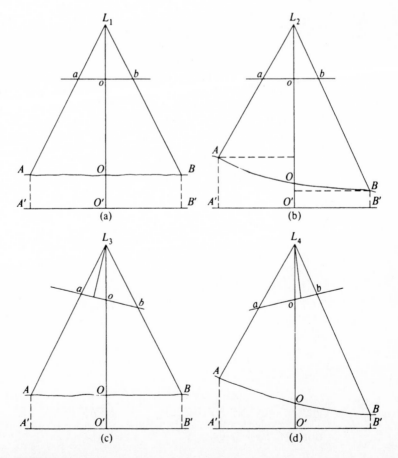

FIGURE 16-6 Effect of relief and tilt on photographic scale.

The scale of a vertical photograph at a point, along a line or in an area, can be determined from the relationship between the focal length of the camera, the flying height of the aircraft at the time of exposure, and the elevation of the point, the line, or the area. In Fig. 16-7 points O and A are at the same elevation, and so $AA' = OO'$ or $h_A = h_O$. Points B and C lie at elevation $h_B = h_C$, and D lies at elevation h_D. By similar triangles,

$$\frac{ao}{AO} = \frac{Lo}{LO}$$

or

$$\frac{ao}{AO} = \frac{f}{H - h_A} = \frac{f}{H - h_0}$$

Similarly,

$$\frac{bc}{BC} = \frac{f}{H - h_B} = \frac{f}{H - h_C}$$

But the ratios ao/AO and bc/BC are the scales of the photograph along lines ao and bc, respectively. Therefore, in general,

$$S_E = \frac{f}{H - h} \tag{16-1}$$

in which S_E is the scale of a vertical photograph for a given elevation; f is the focal length, either in inches or in millimeters; H is the flying height above the datum, in feet or meters; and h is the elevation of the point, line, or area above the datum, in feet or meters. The focal length given for most modern aerial camera lenses is expressed in millimeters. Although a scale can be expressed in terms of the num-

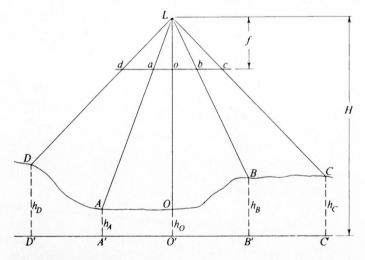

FIGURE 16-7 Photographic scale.

ber of feet per millimeter, this is quite uncommon. When working with photo scales, it is more convenient to convert millimeters to either inches or feet by dividing by 25.4 or 304.8, respectively.

EXAMPLE 16-1   In Fig. 16-7 the elevation of points O and A is 267 ft, that of points B and C is 524 ft, and that of D is 820 ft. The flying height above sea level is 1769 ft, and the focal length of the camera is 8.23 in., or 0.6858 ft. Determine the scale of the photograph along the line ao, along the line bc, and at point d, expressing each scale as a proportion.

Solution: The value of S_E for line ao is $0.6858/(1769 - 267)$. Reducing the numerator to unity gives a scale of 1:2190. The value of S_E for line bc is $0.6858/(1769 - 524)$, and the scale is 1:1815.

The value of S_E at point d is $0.6858/(1769 - 820)$, and the scale is 1:1384. ◆

Quite often the *average* scale of a single photograph or of a set of photographs is desired to be able to measure distances in any area of the photograph or photographs. If the average scale is known, it can be applied to a scaled distance to give a reasonable value of the corresponding ground length, provided that the relief is not extremely variable. The average scale is given by the relationship.

$$S_A = \frac{f}{H - h_{av}} \tag{16-2}$$

in which S_A is the average scale of the photograph, which may be reduced to either an engineer's scale or a representative fraction; f is the focal length, in millimeters or inches; H is the flying height above the datum, usually sea level, in feet or meters; and h_{av} is the average elevation of the area covered by the photography, in meters or feet.

The scale of a photograph can be determined by comparing a distance measured on the photograph with the corresponding known ground distance. The ground distance may have been measured directly, or it may be a distance of common knowledge as, for example, the length of a section line (see Chapter 18), a city block, or a stretch of a highway.

The scale of a photograph can be determined by comparing a distance measured on the photograph with the corresponding distance measured on a map of known scale. The photograph scale is then found by the following relationship:

$$\frac{\text{photo scale}}{\text{map scale}} = \frac{\text{photo distance}}{\text{map distance}} \tag{16-3}$$

Both the photo scale and the map scale must be in the same units and must be expressed as a fraction. For example, if the distance between two road intersections is 4.34 in. on a photograph and 1.55 in. on a map drawn to a scale of 1 in. = 800 ft,

$$\frac{1 \text{ in.}/X \text{ ft}}{1 \text{ in.}/800 \text{ ft}} = \frac{4.34 \text{ in.}}{1.55 \text{ in.}}$$

Hence $X = 285$ ft, and the photo scale is 1 in. = 285 ft. Note that the quantity X is the denominator of the photograph scale. Equation (16-3) can be expressed also as

$$D_P = \frac{M}{P} D_M \tag{16-4}$$

in which D_P is the denominator of photograph scale; D_M is the denominator of map scale, in the same units as D_P; P is the photograph distance; and M is the map distance, in the same units as P.

EXAMPLE 16-2   The distance between two points measures 24.62 mm on a map whose scale is 1/24,000. The distance between the same two points appearing on a vertical aerial photograph measures 32.05 mm. What is the scale of the photograph?

Solution: By Eq. (16-4),

$$D_P = \frac{24.62}{32.05} \times 24{,}000 = 18{,}436 \text{ or } 18{,}450 \text{ (approx.)}$$

The photo scale is thus 1/18,450. ◆

EXAMPLE 16-3   If the focal length of the camera lens of Example 16-2 is 152.4 mm, what is the flying height above the ground in meters?

Solution: Assuming a datum to lie at the average elevation of the area, then by Eq. (16-2),

$$\frac{1}{18{,}450} = \frac{152.4 \text{ mm}}{(H-0) \text{ m}} = \frac{0.1524 \text{ m}}{(H-0) \text{ m}}$$

giving $H = 2812$ m above the ground (approx.). ◆

16-5   Relief Displacement

The term *relief displacement* is applied to the displacement of the image of a ground point on a photograph from the position the image would have if the point were on the datum. This displacement is due to the elevation of the ground point above or below the datum. In Fig. 16-8 ground points A and B lie at elevations h_A and h_B, above the datum. Their images on a vertical photograph are at points a and b, respectively. The datum positions A' and B' would have images at a' and b' on the photograph. Point o is the principal point of the photograph. It is located by joining opposite fiducial marks at the corners or the edges of the photograph. Because of the elevations of A and B above datum, both a and b have been displaced outward along the radial lines oa and ob, respectively.

The amount of relief displacement for a point from proportional triangles is given by

$$d = \frac{rh}{H} \tag{16-5}$$

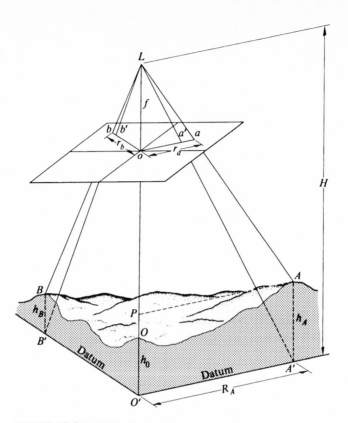

FIGURE 16-8 Relief displacement.

in which d is the relief displacement ($a'a$ for point A and $b'b$ for point B in Fig. 16-8), expressed in inches or millimeters; r is the radial distance from the principal point to the image of the point, in the same units as d; h is the elevation of the point above the datum, in feet or meters; and H is the flying height above the datum, in the same units as h.

From Eq. (16-5) it is seen that the relief displacement of a point depends on the position of the point on the photograph. At the principal point, the displacement is zero. It increases outward toward the edges of the photograph. Also, the relief displacement increases as the elevation of the ground point increases, and decreases with an increase in flying height.

Assume that a photograph contains all the corners of a tract of land and that the elevations of these corners are known. Unless all the corners are at the same elevation, the sides of the tract cannot be scaled directly from the photograph because of variation in scale discussed in Section 16-4. The relief displacement of each point can be computed, provided the flying height above sea level has been determined, by measuring the radial distance to each point and applying Eq. (16-5). The positions of the points are then corrected by the amount of relief displacement, and the new positions are therefore at a common datum. In Fig. 16-9 points a, b, c, d, and e are the photographic images of the corners of a tract

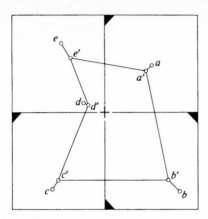

FIGURE 16-9 Points brought to common datum.

of land. The elevations of the points and the flying height above sea level are known. The positions of the five points, corrected to a common datum, are a', b', c', d', and e'. These are the positions of the five points on a map that is at the datum scale of f/H, in which H is the flying height used to compute the relief displacements. The corrected lines may be scaled; and the lengths of the lines, the angles between the lines, and the area of the tract can be determined to a fair approximation depending on how much the photograph was tilted at the instant of exposure.

If the flying height above the datum for a given photograph is not known, it can be determined with varying accuracy by measuring the length of a ground line, the end points of which lie at about the same elevation, and then scaling the corresponding photograph distance. The scale at the elevation of the line being known, H may be determined by Eq. (16-1). The photographic distance may also be compared with a corresponding map distance to give the scale by Eq. (16-3), and then H may be determined.

Figure 16-10 shows a vertical photograph taken from an altitude of approximately 2000 ft above sea level and covering an area at an average elevation of about 500 ft. The scale is, by Eq. (16-2), about 1/3000 or 1 in. = 250 ft. Lines ab and ac are actually straight fence lines, and the angle at a from b to c is approximately 90° if measured at ground point A from B to C. Because of relief displacement, however, it is obvious that the fence lines are displaced outward from the principal point as the lines go over ridges, and are displaced inward as the lines cross the gullies. This relief displacement, which is nothing more than a manifestation of perspective in an aerial photograph, is of rather minor consequence in a photograph taken at a great altitude over ground with very little relief. However, if a photograph is to be used as a map substitute, then the effect of relief displacement and scale variation must be recognized, especially where the scale of the photograph is large and there is relatively large terrain relief.

FIGURE 16-10 Effect of relief displacement on straight lines.

16-6   Photograph Overlap

When aerial photography is used for mapping, mosaic construction, or orthophotography, flight lines are laid out on a flight map with a spacing that will cause photographs to cover an overlapping strip of ground. This overlap between flight strips amounts to about 25% of the width of the photograph. The actual spacings on the flight map may be determined from the scale at which the photographs are to be made by solving Eq. (16-3). The photographic distance is 75% of the width of the photograph, provided that a 25% overlap between flight strips is to be maintained. This overlap is shown in Fig. 16-11.

Each photograph in the line of flight covers an area that overlaps the area covered by the previous photograph by about 60%. This is illustrated in Figs. 16-11 and 16-12. The large overlap between successive photographs serves three primary purposes. First, it provides coverage of the entire ground area from two viewpoints, such coverage being necessary for stereoscopic viewing and measuring.

FIGURE 16-11 Photograph overlap.

FIGURE 16-12 Overlap along flight line.

Second, it allows only the central portion of each photograph to be used in mosaic construction, eliminating to a great extent the effect of relief displacement. Third, the small overlap area between alternate photographs allows control to be extended along the strip by photogrammetric methods.

The 60% forward overlap provides 20% overlap of alternate photographs, which allows control to be extended along the strip by photogrammetric methods.

16-7 Ground Control for Photogrammetry

For aerial photographs to be used for making simple measurements of distances and elevations, for constructing planimetric and topographic maps, for measuring profiles and cross sections, and for constructing orthophotos, a certain amount of ground control is necessary to fix the scale of the photographs or the map, and to establish a vertical datum with which to establish contour lines on the map. This control may be obtained simply by measuring the ground distance between two identifiable points that appear on a photograph. Or an elaborate system of horizontal and vertical control nets may be required to control the horizontal and vertical positions of identifiable points that appear on the photographs. In general, triangulation and traverse stations and benchmarks cannot be identified on the photographs unless they have been marked in the field before the pictures were taken.

As a result, shorter traverse lines and level lines, usually of a lower order of accuracy than the main system, are run to points that can be identified. These points are called *picture points,* and the network of picture points in a group of photographs is called *photo control.*

The horizontal photo control may be obtained by running a traverse from a fixed point to the picture points, or it may be obtained by GPS observations. A large amount of photo control work is conducted by a combination of a theodolite and an electronic distance measuring system or by a total station instrument. Vertical photo control is obtained by running differential or stadia levels from benchmarks to the points, or, for large contour-interval mapping by using GPS.

The choice of the field method to be used in establishing photo control depends on the desired accuracy, which in turn depends on the scale of the map or mosaic and the contour interval to be used in compiling the map.

A picture point must be positively identifiable on the photograph and its conjugate point on the ground. It must be sharp and well defined as seen on the photograph under magnification. The picture point must fall in the correct position on the photograph. A picture point falling, for example, near the very edge of the photograph may be entirely useless for the purpose for which it is intended. It should be reasonably accessible on the ground so that the expense of the photo-control survey can be kept to a minimum. Finally, the picture point must be well described and documented directly on the photograph used in the field when the photo-control survey is performed.

When a very high degree of accuracy is required in the photogrammetric process, natural features are generally not used as picture points. Instead, the field survey crew sets out panels, referred to as *premarked points* or panel points, just prior to the photography. The most common panel configuration is shown in Fig. 16-13. The size of the inner square is a function of the proposed scale of the photography and the size of the measuring mark in the instrument to be used for measuring the photographs (see Section 16-9). The outer four legs of the panel are used to find and identify the panel point on the photograph. The control survey is run to the center of the inner square to determine its position and elevation.

FIGURE 16-13 Panel configuration.

When surveys are executed for the purpose of obtaining photo control for mapping, the surveyor must consult with the photogrammetrist as to the accuracy required in the field survey and also as to the proper location of the photo control point or the premarked point. If the project does not require premarked points, the photo control survey can be performed after the aerial photographs have been processed and printed. The photogrammetrist then selects the areas on each photograph where he wants the surveyor to select suitable picture points; or the photogrammetrist may himself choose the picture points. This coordination between the surveyor and the photogrammetrist is essential for the proper orientation of the photographs when the map is being compiled (see Section 16-9). If the ground is to be premarked, then a flight map prepared by the photogrammetrist must be available to show the surveyor where the panels are to be set.

16-8   Stereoscopy and Parallax

The word *stereoscopy* is defined as the viewing of an object or image in three dimensions and necessarily implies binocular, or two-eyed, vision. The optical axes of a person's eyes will converge when looking at an object, and the angle measured at the object between the two optical axes is called the *parallactic angle.* This angle increases as the object is brought closer to the eyes and decreases as the object recedes from the eyes. Two objects at different distances is interpreted by the eyes as forming two different parallactic angles, and the difference in the two angles is interpreted as depth.

In Fig. 16-14(a) two marks of the same size and shape are plotted on a sheet of paper separated by a distance of about $2\frac{1}{4}$ in. If each eye concentrates on but

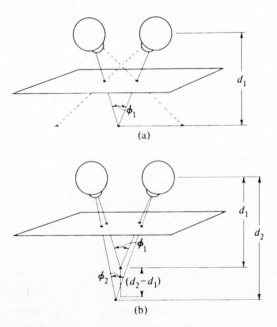

FIGURE 16-14 Stereoscopic vision.

one mark, the image of the two marks will appear as one mark at a distance d_1, and the parallactic angle formed at this image is ϕ_1. In Fig. 16-14(b) an additional pair of marks is plotted with a slightly larger separation. The angle ϕ_2 is formed at the single image of this pair, the image appearing to lie at a distance d_2 from the eyes. Thus the four marks on a plane surface viewed stereoscopically will appear as two marks separated in the third dimension of depth by the amount $d_2 - d_1$.

In Fig. 16-15(a) a tower is photographed from two consecutive camera stations L and L' and forms images as shown. Notice the relief displacement of the top of the tower with respect to the bottom. The two photographs lined up side by side with the flight line parallel with the eye base will form a three-dimensional image of the tower, as shown in Fig. 16-15(b). Furthermore all points in the area of overlap between the two photographs will be seen stereoscopically, that is, in three dimensions. Because of the dimensions of the photograph and because of the eyes' resistance to focus closely and at the same time diverge abnormally, a stereoscope is used to form the stereoscopic image. The stereoscope can be of a simple lens type shown in Fig. 16-16(a) or of a mirror type shown in Fig. 16-16(b). When the photographs are laid down, the flight line of one photograph must line up with the same flight line of the other photograph before the stereoscope is placed over the pair. This adjustment is shown in Fig. 16-17.

In Fig. 16-15(a) it is noted that the image of the top of the tower has moved through a total algebraic distance of $x_t - x'_t$ between two successive exposures, and that the image of the bottom of the tower has moved through a total distance of $x_b - x'_b$ between exposures. This movement of an image over the focal plane between exposures is called the *parallax* of the point. Parallax increases as the point lies closer to the camera, that is, as the elevation of the point increases. It is

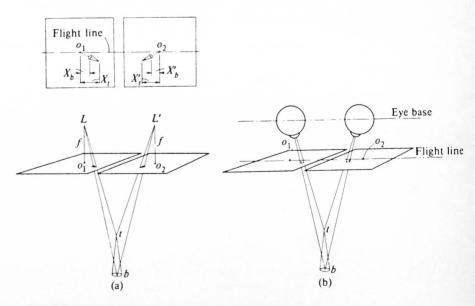

FIGURE 16-15 Images of tower on pair of aerial photographs.

(a)

(b)

FIGURE 16-16 Stereoscope. (a) Lens stereoscope. (b) Mirror stereoscope.

the gradual difference in parallax between successive points as they are photographed from two successive camera stations that gives a gradual difference in the parallactic angles formed between the optical axes of the eyes. Thus a continuous three-dimensional image is formed.

Parallax is a direct indication of elevation and can be measured on a pair of overlapping photographs by means of a *parallax bar*. Such a measuring device is shown in use under a mirror stereoscope in Fig. 16-18. A schematic view of the

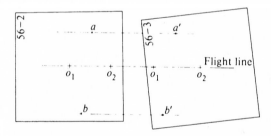

FIGURE 16-17 Adjusting photographs for stereoscopic viewing.

parallax bar is shown in Fig. 16-19(a). Two measuring marks are located on the bottom surfaces of two pieces of glass attached to the bar. The left mark is fixed to the bar while the right mark is moved left or right by turning the micrometer screw. A movement to the left is an increase in parallax and to the right is a decrease. If the bar is placed under a mirror stereoscope, the two measuring marks will appear as a single mark (see Fig. 16-14). This is called the floating mark. By turning the micrometer screw one way or the other, this single mark appears to move up or down according to the increase or decrease in the parallax of the marks.

In Fig. 16-19(b) two photographs are oriented on a large sheet of paper such that their flight lines are carefully aligned as shown in Fig. 16-17. Prior to this alignment, the position of the left principal point o_1 must be transferred to its correct position on the right-hand photograph, and the right principal point o_2 must be

FIGURE 16-18 Parallax bar used under a mirror stereoscope.

FIGURE 16-19 Diagram of parallax bar.

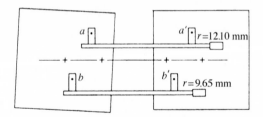

FIGURE 16-20 Micrometer readings of parallax bar.

transferred to the left-hand photograph. The separation S between any two *conjugate* points such as o_2 and o_2 is set to be equal to the spacing between the measuring marks when the movable mark is in its midposition. The photographs are now taped to the paper. The parallax bar is placed on the photographs as shown, and the mirror stereoscope is then placed over the photographs as shown in Fig. 16-18.

Figure 16-20 shows how the *difference* in parallax between two points can be measured. The parallax bar is moved over the photographs to point a, and the micrometer screw is turned until the floating mark appears to touch the stereoscopic image at a. The micrometer is read as 12.10 mm. The bar is moved to point b, the floating mark is again brought to the surface of the stereoscopic image at b. The micrometer is read as 9.65 mm. The difference in parallax, dp, from a to b is thus $9.65 - 12.10 = -2.45$ mm. The difference in parallax between any two points, such as the top and bottom of a tree or building or between the top and bottom of a hill, can be measured in this manner. The distance o_1-o_2 is measured with a scale on each photograph to give an average photograph base b. The difference in elevation dh between two points is then given by

$$dh = \frac{dp \, H}{dp + b} \tag{16-6}$$

in which dh is in feet or meters, dp and b are in millimeters, and H is the flying height *above the ground,* expressed in feet or meters. The value of H is obtained from Eq. (16-2) in which $h_{av} = 0$.

EXAMPLE 16-4   A pair of overlapping vertical photographs were taken with a lens having a 6-in. focal length. The average scale of the photography is 1 : 10,000. A vertical control point, whose elevation is 680 ft, appears in the overlap area. The average distance between principal points is 3.660 in., or 92.96 mm. Parallax bar measurements give the measured differences of parallax between the known control point and five unknown points. These differences are shown in the table below. Compute the elevations of the five points.

Point	dp (mm)	$dp + b$ (mm)	dh (ft)	Elevation (ft)
1	+ 4.62	97.58	+ 237	917
2	−0.88	92.08	− 48	632
3	+ 1.90	94.86	+ 100	780
4	+ 5.76	98.72	+ 292	972
5	−3.35	89.61	−187	493

Solution: We must assume that the elevation of the known point is nearly equal to the average elevation of the area shown on the photographs. Then, by Eq. (16-2), in which the average elevation is assumed to be zero,

$$S_A = \frac{6/12}{H} = \frac{1}{10,000} \text{ or } H = 5000 \text{ ft}$$

With b given as 92.96 mm, the differences of elevation are computed by Eq. (16-6), as shown in the table. Finally, these are used with the elevation of the known point to give the values in the last column. ◆

The parallax bar is useful for obtaining rough profiles along any selected line drawn on one of the two photographs used under the stereoscope. Since the photos may be tilted, the vertical accuracy obtained by Eq. (16-6) will not give absolute elevations any better than about 1/300 times the flying height. However differences in elevation between two points located fairly close together are reliable up to about 1/1000 times the flying height.

16-9   Stereoscopic Plotting Instruments

So far in this chapter, discussion has been confined to the use of paper prints in which the photographs were assumed to be vertical. These conditions would not be sufficient to obtain measurements necessary for the compilation of accurate maps. To take into account the tilts and difference of flying heights of consecutively exposed photographs, and to provide dimensional stability, photographs are measured in different kinds of stereoscopic plotting instruments. A stereoscopic plotting instrument, or plotter, is used to plot planimetric and topographic maps to a predetermined scale by using photographs that have been taken with a precision aerial camera, and to measure profiles and cross sections. The elevations of selected ground points measured in a plotter can be determined with varying degrees of accuracy, the error depending on the precision of the plotter. This ranges from 1/1000 times the flying height to as little as 1/10,000 times the flying height. The error in contour lines drawn by means of the plotter will vary from 1/500 to 1/2500 times the flying height, the amount again depending on the precision of the plotter and also on the type of ground cover existing at the time the photographs were made.

To maintain accuracy, the photographs used in a plotting instrument are sometimes printed on glass plates. These glass plates are either the same size as the negatives or they are reduced in size by a carefully controlled ratio. For use in some plotters, the glass plates are made by projection through a lens that compensates within tolerance the distortions due to the aerial camera lens. In other plotters the camera-lens distortion is compensated for directly in the projection systems.

Every plotting instrument has four main features: a projection system, a viewing system, a measuring system, and a drawing system. These features will be described with reference to Fig. 16-21, which shows a plotting instrument of simple design. This is the Multiplex plotter. Although the Multiplex instrument has not been used for mapping for many years, the simplicity of its design lends itself admirably to the explanation of the principles of a plotting instrument. In this instrument each of the two projectors contains a projector lens with a principal distance of 30 mm, a light source, and a condenser lens that concentrates the light so that it passes through the projector lens. A small reproduction on glass, called a *diapositive* and measuring 64 × 64 mm, is placed in each of the two projectors, corresponding to two overlapping photographs. A blue-green filter is placed below the light source in one projector, and a red filter is put in the other projector. When the lamps of the projectors are turned on, two cones of rays are projected downward toward the map table, each cone of rays having a different color.

Because of tilt in the photographs and because of unequal flying heights, a ray of light through a ground point and coming from one projector may fail to intersect the ray through the same point coming from the other projector; and similarly for other pairs of rays through other points. Each projector is free to rotate about each of the three axes and also to move along the three axes. By a systematic manipulation of these motions of each projector, the discrepancies existing between the rays are eliminated, and the two overlapping cones of rays form a stereoscopic image in the area above the map table. This process is referred to as *relative orientation*. It sets into the two projectors the proper amount of relative tilts that existed in the aerial camera at the time the two aerial photographs were taken along the flight line.

The stereoscopic image is viewed by the operator through a pair of spectacles, one lens of which is blue-green and the other red. In this way the cone of rays projected downward from the left projector can be seen with the left eye only, whereas that coming from the right projector is seen with the right eye only. The

FIGURE 16-21 Stereoscopic plotting instrument. Courtesy of U.S. Geological Survey.

operator thus is able to see a three-dimensional image of the overlap area covered by the two successive photographs. This is called the *anaglyphic* method of viewing. The stereoscopic image seen by the operator is called a *stereoscopic model*.

The small movable table that sits on the map table is called the *tracing table*. The tracing table contains a white circular platen that moves up and down by means of a knurled knob, the up and down movement being recorded on a vernier or a counter. In some types of tracing tables, this movement is recorded in millimeters; in other types it is given directly in feet or meters of elevation. At the center of the platen is a small hole, underneath which is a small light bulb. When the operator views the model formed on the platen, he also sees a small pinpoint of light at the center of the platen. This is the measuring mark, and its up-and-down motion is recorded as just mentioned.

Directly beneath the measuring mark is the plotting pencil, which can be raised from or lowered to the map sheet. The position of the plotting pencil can be adjusted in both directions so that the pencil and the measuring mark form a line that is perpendicular to the surface of the map table.

Before the stereoscopic model can be measured and plotted in the form of a planimetric or topographic map, or for profiling and cross sectioning, it must be brought to the correct scale and datum. The scale is controlled by the horizontal positions of at least two photo-control points appearing in the area of overlap, their positions having been plotted on the map sheet. By varying the spacing between the two projectors in the direction of the flight line, the scale is varied until the images of the control points received by the measuring mark are directly over the corresponding plotted positions of the control points. This operation is called *scaling* the model.

Three, four, or five vertical control points appearing in the area of overlap are used to bring the model to the correct datum. The operator raises or lowers the measuring mark until it appears to be in contact with a vertical control point as seen in the model. He reads the elevation counter and determines the discrepancy between the known elevation and the measured elevation. Repeating this at each vertical control point, the operator then analyzes the discrepancy at each point to decide the amount by which the projectors as a unit must be tipped and tilted to make the measured elevations in the model agree with the known elevations. After the tips and tilts have been introduced, the elevations are again read to determine any minor amount of adjustment that must be made. This process is referred to as *leveling* the model.

Taken together, the process of scaling and leveling the stereoscopic image is referred to as *absolute orientation.* Thus the tilts and flying heights of the original photographs have been recreated in the instrument. The plotter thus creates a spatial model of a portion of the earth's surface. For the instrument shown in Fig. 16-21, the distance between the two projector lenses is the air base reduced to the scale of the map. The height of the projectors above the map table represents, to the scale of the map, the flying height above a level datum, although not necessarily the sea-level datum. Vertical distances measured above the map sheet to points on the stereoscopic model are elevations, to scale, of the points above the level datum. The stereoscopic model can be measured in all three dimensions by means of the measuring mark.

When plotting planimetric features on the map sheet, the operator follows the features with the measuring mark, raising it when the terrain rises and lowering it when the terrain is falling. The plotting pencil traces these features in their correct orthographic positions on the map sheet.

When plotting a contour line, the operator sets the measuring mark to correspond to the elevation of the contour line to be plotted by reference to the tracing table elevation counter. With the pencil raised from the map sheet, he moves the measuring mark until it appears to come in contact with the spatial model. The measuring mark is thus on the contour line. The operator lowers the pencil to the map sheet and moves the tracing table in such a manner that the measuring mark is at all times in apparent contact with the model. The pencil traces this movement as a contour line orthographically on the map sheet. To verify that he has not missed a contour line at high points, the operator usually measures the elevations of these points and records their values on the map sheet.

FIGURE 16-22 Zeiss D2 Planimat. Courtesy of Carl Zeiss, Oberkochen.

The instrument shown in Fig. 16-22 performs the same function as the multiplex plotter previously discussed, but the design and operating principles are quite different. The projection takes place by means of two steel space rods machined to remarkable precision. These rods are analogous to two rays of light coming from two conjugate points on the diapositives and intersecting in space below the projectors. The two space rods can be seen in Fig. 16-22. The two diapositive holders can be rotated about three axes and can also be translated in a direction parallel with the flight line. Viewing of the three-dimensional image formed by two diapositives takes place by means of a pair of optical trains commencing on the undersides of the diapositives, continuing through a series of prisms and lenses and ending at a pair of binocular eyepieces, which are shown in Fig. 16-22.

Two measuring marks, one in each optical train, are superimposed onto the images of the diapositives, and they appear as one floating mark when the viewer observes the stereoscopic image. This feature is similar to the parallax bar principle mentioned in Section 16-8. The image of the mark is caused to move through the model in X and Y directions by moving the intersection of the space rods using a pair of handwheels, which are shown in Fig. 16-22. This movement of the floating mark enables the details seen in the model to be plotted on the map sheet directly in front of the observer. The floating mark is made to appear to move up or down by rotating the large foot disk seen in the illustration. The up-and-down motion is registered on a height display, and the mark can be set to a selected elevation for following a contour line.

There are no projector lenses in the instrument shown in Fig. 16-22. The perspective centers, replacing the lenses, are located at the pivot pints of the space rods and are not seen in the illustration. By physically varying the vertical distance between these pivot points and the plane of the diapositives, all of the focal lengths given in Sec. 16-2 can be accommodated.

FIGURE 16-23 Wild Aviomap stereoplotter. Courtesy of Leica, Inc.

The instrument shown in Fig. 16-23 also uses space rods in place of optical rays to produce the stereoscopic model. Its design and operating principles are similar to those of the instrument shown in Fig. 16-22. This instrument includes the coordinatograph shown to the right. The movement of the measuring mark actuated by the two handwheels is translated to the coordinatograph pencil by means of electric pulses. The map can be drawn at widely different scales by setting off the appropriate ratios between the model scale and the map scale.

The instrument shown in Fig. 16-24 is called an analytical plotter because all of the tilts and tips of the pair of photographs are, in effect, generated by a computer based on the principles of spatial analytical geometry. Thus the diapositives need not be rotated about the three axes to perform relative and absolute orientation. After the two photographs, or diapositives, have been placed in their respective carriers, the operator sets the measuring mark on the fiducial marks of each diapositive and executes a command to find the principal point of each photograph. He then enters the focal length of the camera. Following this, the measuring mark is set on a series of selected points in the photographs from which the relative orientation of the pair is computed. The measuring mark is then set on all of the ground control points visible in the stereo model. The operator then gives the command by which the computer scales and levels the model. The model is now ready for compilation as previously discussed.

The instruments briefly discussed above are representative of a wide variety of stereoscopic plotting instruments in use at the present time. More detailed descriptions are given in modern photogrammetry textbooks (see Kraus and Karl, 1993, in the Bibliography).

FIGURE 16-24 Wild BC3 Analytical plotter. Courtesy of Leica, Inc.

16-10 Advantages and Disadvantages of Photogrammetric Mapping

The main advantages of the compilation of topographic maps by using aerial photographs over ground methods are as follows: the speed of compilation, the reduction in the amount of control surveying required to control the mapping, the high accuracy of the locations of planimetric features, the faithful reproduction of the configuration of the ground by continuously traced contour lines, and the freedom from interference by adverse weather and inaccessible terrain. Also, by the proper selections of flying heights, focal lengths, and plotting instruments, and by proper placement of ground control, photogrammetric mapping can be designed for any map scale ranging from 1 in. = 20 ft down to 1 in. = 20,000 ft and smaller (1/250 to 1/250,000), and for a contour interval as small as $\frac{1}{2}$ ft (1 dm). Because of the wealth of detail that can be seen in a spatial model, the resultant photogrammetric map will be more complete than will a comparable map produced by ground methods.

Among the disadvantages of mapping by using aerial photographs are the following: the difficulty of plotting in areas containing heavy ground cover, such as high grass, timber, and underbrush; the high cost per acre of mapping areas 5 acres or less in extent; the difficulty of locating positions of contour lines in flat terrain; and the necessity for field editing and field completion. Field completion is required where the ground cannot be seen in the spatial model because of ground cover, where spot elevations must be measured in flat terrain, and where such planimetric features as overhead and underground utility lines must be located on the map. Editing is necessary to include road classification, boundary lines not showing on the photography, drainage classification, and names of places, roads, and other map features.

16-11 Digitized Stereoscopic Model

Most modern stereoscopic plotting instruments can be made digital in all three directions, with the three-dimensional position of a point being read out to about the nearest 0.001 mm. The Zeiss Planimat shown in Fig. 16-22 is typical of such an instrument. The X and Y spindles actuated by the two handwheels can be shaft encoded to give the X- and Y-model coordinates of any point in digital form. The foot disk turns the Z spindle, which can also be shaft encoded to give the Z coordinate of the point in digital form. The X, Y, and Z coordinates are then recorded on cards, magnetic tape, or disks for direct input into a data processor.

All desired points in a model are digitized in the above fashion. If a property or cadastral survey is being executed by photogrammetric methods, then all property corners, which have been prepaneled, are measured, along with the horizontal and vertical control points. The entire stereoscopic model can then be transformed mathematically in three dimensions to fit the ground coordinates and elevations of the control points. This process is just like absolute orientation. Since the points in the cadastral survey have also been measured, they are also transformed into the ground coordinate system.

A stereoscopic model can be completely digitized in three dimensions on a regular grid array in the X and Y directions, the grid spacing depending on the roughness of the terrain and the purpose to which the *digitized terrain model* is to be put. Here again the grid points are all transformed mathematically into the ground system by means of the control. The digitized terrain can then be analyzed in a data processor to determine optimum highway alignment and grade, high-water lines, earthwork quantities, and many other engineering studies limited only by the imagination of the engineer. Since much of the process of digitizing can be made automatic, this method lends itself to rapid computer location of contour lines by interpolation between grid points. This is analogous to the grid method of contouring discussed in Section 15-9. The contour lines are then plotted from the digital data by means of a coordinatograph similar to that shown in Fig. 8-34.

Figure 16-25(a) shows a portion of a contour map as it was plotted manually in a stereoscopic plotting instrument. The same area was digitized on a regular grid array in the plotter. The resulting contour map generated in a computer and plotted on a coordinatograph drive by the computer output is shown in Fig. 16-25(b). The main difference between the two drawings is the smoothness of the contour lines generated by the computer.

The use of the stereoscopic model to digitize highway profiles and cross sections is discussed in Section 17-14. In this process the points to be digitized lie along predetermined lines in the model. These lines have been defined with reference to a completed topographic map at the same scale as the stereoscopic model.

16-12 Stereo Plotter with Computer System Interface

Modern stereoscopic plotting instruments, including the analytical plotters, can be interfaced with powerful computer systems to allow a large range of operations to be performed by the plotter-computer system. Using the proper computer commands, the computer will select and drive the plotting pens of the coordinato-

FIGURE 16-25 Contour lines plotted. (a) Manually. (b) By computer control.

graph to square up buildings, draw straight lines between points, pass circles through three measured points in the model, interpolate between points to draw irregular curves, generate contour lines based on digitized models, plot profiles and cross sections, draw standard symbols on the map, match the map features at the edges of contiguous map sheets, draw parallel lines at a given separation, and a host of other graphical operations. In addition, based on data measured in the stereoscopic model, the computer can calculate plane areas and volumes, inverse lines to obtain lengths and azimuths, compute settings for performing absolute orientation of the model, and a multitude of computations.

Since all data measured in the stereoscopic model can be coded and stored in digital form in the computer system, a multilevel digital database can be created and stored. For example, the storage of contour line data can be recalled to generate a plot of contours only. Planimetric features such as transportation networks, street networks, utilities, property lines, vegetation types, waterways and drainage, and other features can be stored separately and then recalled to generate separate maps of the various themes.

16-13 Orthophotos

The effect of relief and tilt displacements on the positions of points on photographs and mosaics was discussed in previous sections. These effects are caused by the perspective nature of a photograph. A pair of photographs, however, if properly oriented in a stereoscopic plotting instrument, will present a theoretically perfect model of the terrain for measurement. The measurements are made orthogonally or orthographically to form a map underneath the model. This is imagined to be accomplished first by setting in the proper tilts in the projectors by the orientation process, and second by raising or lowering the tracing table when tracing the map to compensate exactly for scale changes due to relief. The features compiled on the map sheet are represented by lines and symbols, between which accurate scaling is possible.

To combine the accuracy of scaling a map with the pictorial representation of a mosaic, the images on the aerial photograph can be manipulated optically mechanically or optically electronically to remove the perspective aspect and the tilt effect. The resulting photo is called an *orthophoto,* and the process is referred to as *orthophotography.* There are at present several different ingenious methods for preparing orthophotos. The easiest to comprehend is the *fixed-line-element* rectification of a photograph and will be presented here. Various methods of this type, called analog methods, are being replaced by digital methods described in Section 16-14.

Referring to Fig. 16-26, a vertical-moving table containing a large film stage is located in the model area of an optical-projection type of stereoplotter (see Fig. 16-21). During operation, which takes place in a darkened room with a red safelight, the film is covered by a large curtain on rollers to protect it from the light. This is shown in Fig. 16-27. The curtain contains an opening channel across its width, which is in turn covered by a ribbon curtain that moves at 90° with the larger curtain. The ribbon curtain contains a small slit surrounded by a white area, which constitutes the viewing platen. This platen is a strict analogy with the tracing table of the plotting instrument, and the small slit constitutes the floating mark. The slit lies perpendicular to the motion of the ribbon curtain.

FIGURE 16-26 Fixed-line-element rectification.

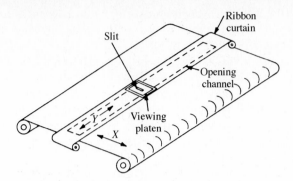

FIGURE 16-27 Curtain used to cover film during exposure in fixed-line-element rectification.

After the stereoscopic model has been completely oriented, orthochromatic film, which is not sensitive to red light, is taped in place on the film stage and covered with the compound curtain, and is then placed in the model area. The projector lamps, which are filtered for anaglyphic viewing, are turned on to allow the operator to view a small portion of the model on the viewing platen. He can raise or lower the entire film support to keep the small slit in contact with the surface of the model. As the operator moves along a line in the model from front to back (or side to side), keeping the slit in constant contact with the model, an exposure of the image is "painted" onto the film in true orthographic projection. The film is exposed only by the blue light. When the model has been traversed completely along the strip, the slit is automatically stepped over a distance equal to the slit length, carrying the large curtain with it. The model is then traversed in the opposite direction. This back and forth motion exposes the film along contiguous strips over the entire model, producing a negative in orthographic projection. After the negative is developed, prints are made to produce the orthophoto at a known scale. The instrument shown in Fig. 16-28 produces orthophotos by the method just described.

The different types of a rectification employed to obtain orthophotos in a variety of instruments are shown in Fig. 16-29. The terrain and its corresponding topographic map is shown in view (a). The fixed-line element method described above is shown in view (b). In view (c) the line elements are rotated to give better match between the adjacent strips. This is referred to as the *rotating-line* method. Some orthophoto rectifiers employ *plane area* rectification, diagrammed in Fig. 16-29(d), while others use the *curved area* method shown in view (e).

A series of orthophotos can be pieced together in the form of an ortho mosaic to cover a large area. Also, contour lines obtained in the stereoplotting process can be combined with the imagery, as shown in Fig. 16-30.

The orthophoto on the left in Fig. 16-30 is the equivalent of a planimetric map in all respects except that it shows features photographically rather than symbolically. Careful examination of the orthophoto will reveal slight and unavoidable mismatching between scan lines, particularly in the building on the waterfront along the right edge of the illustration. This mismatching has virtually no effect on the overall accuracy with which the orthophoto can be scaled.

FIGURE 16-28 SFOM 693 Orthophotograph.
Scanning is done in the X direction.
Courtesy of Matra Division Optique.

The utility of the orthophoto can be considerably enhanced by adding the contour lines, as shown on the right in Fig. 16-30. If the tone of the orthophoto is generally light, then black contour lines are used. On the other hand, white contour lines would be more visible if the tone is fairly dark.

The difference between an aerial photograph and the corresponding orthophoto can be appreciated by a study of Fig. 16-31. A square grid has been ruled on a negative used to produce a diapositive from which the orthophoto was prepared. The left picture is the aerial photograph showing the grid. The right picture is the processed orthophoto; notice how irregular the grid lines appear. This irregularity is a manifestation of relief displacement having been corrected by the orthophoto process. The instrument that produced this orthophoto uses the curved area method of rectification diagrammed in Fig. 16-29(e).

16-14   Digital Photogrammetry

Photogrammetry is a fast-changing field in which the impact of high-speed computers with large storage capacity has been enormous. Many of the operations described in previous sections are today accomplished automatically or semiautomatically. This change is a result of what is referred to as *digital photogrammetry*. In previous sections the use of optical-mechanical instruments represented analog photogrammetry. The introduction of computerized equipment in which the instrument is under computer control is a form of analytical photogrammetry embodied in various analytical plotters.

FIGURE 16-29 Methods of differential rectification used to produce orthophotos. (a) Terrain and corresponding contour map. (b) Fixed-line-element strip rectification. (c) Rotating-line-element strip rectification. (d) Plane area element rectification. (e) Curved area element rectification. After Edmond, *Bendix Technical Journal,* Vol. 1, No. 2, 1968.

FIGURE 16-30 Orthophoto of southern approach to the Golden Gate Bridge. Courtesy of Pugh-Nolte & Associates.

FIGURE 16-31 Comparison between aerial photograph and orthophotograph. Aerial photograph on the left. Courtesy of Hobrough, Ltd.

FIGURE 16-32 Photographic scanner. Courtesy of Vexcal Imaging Corp.

In digital photogrammetry, after film development, the original aerial photograph is digitized in a scanner in the form of a microdensitometer. Figs. 16-32 and 16-33 are representative of scanners currently in use to digitize aerial photographs. To ensure that no information in the photograph is lost, the digitizing interval must be adapted to the resolution of the photograph.

The digitizing interval and the size of the sensor element must not be confused. In an ideal process the digitizing interval and the size of the sensor element are the same. Nonetheless, the digitizing interval is very small in some

FIGURE 16-33 Photographic scanner. Courtesy of Carl Zeiss, Inc.

digitizing instruments, but the sensor element is substantially larger. In such a case the photograph has overlapping pixels with an unwanted smearing of the original information.

The essence of digital photogrammetry can be gleaned by studying a digital photogrammetric image, as shown in Fig. 16-34. A digital image is a two-dimensional matrix $\mathbf{I}$, with elements $\mathbf{I}_{nm}$. Each element is called a pixel (picture element). The row index n runs from 1 to n in steps of 1, and the column index m goes from 1 to m. Since each matrix element represents an area, the word *pixel* is used rather than *point*. The dimensions of the pixel are Δx and Δy.

The pixels $\mathbf{I}_{nm}$ are carriers of information. The information is the value of the density-gray level of the pixel. The most widely used range of values start at zero and end at 255. The information contained in the 256 different values can be stored in 8 bits (2^8) and, as in standard practice, the 8 bits are treated as a single element, the byte.

Pixel values usually with black as zero and white as 255, represent the gray values or densities of black-and-white pictures. Color images can be represented as three black and white images, where each image is assigned a primary color. Color pictures play an important role in remote sensing.

For photogrammetric purposes, if a digital image is to be used, a relationship between the pixel position and an X, Y coordinate system is required. Fig. 16-34 shows this image coordinate system with its origin half a pixel outside the image matrix. Multiplying the index n by Δy gives the image coordinate y of the center of pixel $\mathbf{I}_{nm}$. In the same way, multiplying the index m by Δx, produces the image coordinate x. In digital photogrammetry the traditional method for measurement of image coordinates is replaced by the identification of pixels. Furthermore the ultimate goal of digital photogrammetry is to automate this process.

Once the data are in digital form, a softcopy photogrammetric workstation can be used to create automatically many of the products described in the previous sections of this chapter. Figure 16-35 shows one such workstation.

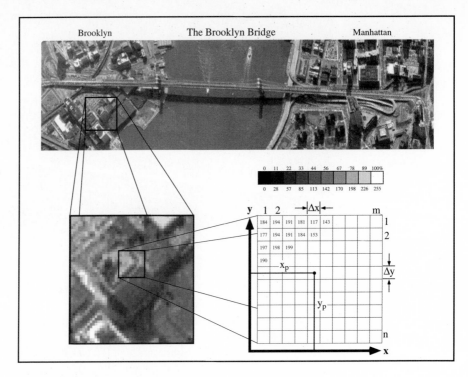

FIGURE 16-34 Digital image configuration.

FIGURE 16-35 Photogrammetric workstation. Courtesy of Leitz, Inc.

In compiling a topographic map, the operator can view two digital images stereoscopically to measure discrete points or to trace out planimetry and contour lines. The images are under the control of the computer just as in the analytical plotter. All discrete points, all planimetry, and all contour lines are stored in the computer in digital form for input into an automatic plotter.

Generating a digital orthophoto requires a digital description of the terrain, usually in the form of a digital elevation model (DEM). Referring to Fig. 16-29(a), suppose that the position and elevation of each pixel in a digitized photograph has been obtained by measuring profiles along parallel lines across the area. The computer then rectifies each pixel as suggested graphically in Fig. 16-29(d), and places the pixel in its correct orthographic position. The positions of the rectified pixels are stored in the computer and then used to generate the orthophoto electronically. Digital orthophotos are currently replacing the analog-analytical methods described in Section 16-13.

The advantage of digital photogrammetry is that it generates files compatible with the input formats to numerous Geographic Information Systems (GIS), as described in Chapter 20. Also, various governmental agencies require digital files for map making, such as the digital line graph (DLG) of the U.S. Geological Survey, and for data transfer that meets the spatial data transfer standard (SDTS) of the Federal Geographic Data Committee (FGDC). Most modern photogrammetric workstations are capable of producing these formats. Obviously digital capability increases the productivity of these procedures.

Future developments in digital photogrammetry will be based on the use of CCD cameras to acquire digital images directly in the picture plane and then use the computer techniques to emulate the eye and vision system.

If high accuracy is to be reached, the pixel size in the plane of the measuring camera must be as small as a few micrometers (10^{-6}m.). If high accuracy is not a primary need, but rather one of the other advantages of digital photogrammetry is sought, a significantly larger pixel size will be adequate. A digital orthophoto is a very important product of digital photogrammetry that can be produced with significantly larger pixel sizes. Digital photogrammetry is essential if there is an extremely short period of time between making the image and the creation of the end product. Processing digital photogrammetric data with large numbers of pixels and huge volumes of data is possible today, but the resolving power of present digital cameras falls short of the resolving power of ordinary photographic film cameras.

PROBLEMS

16-1. A camera with a focal length of 8.25 in. And a picture size of 9 in. by 9 in. is used to photograph an area from an altitude of 9500 ft above sea level. The average ground elevation is 725 ft above sea level. What is the average scale of the photography: (**a**) as an engineer's scale 1 in. = so many feet; (**b**) as a representative fraction.

16-2. What is the ground area covered by a single photograph of Problem 16-1?

16-3. Overlap and sidelap of the photography in Problem 16-1 are 65% and 20%, respectively. What is the ground spacing, in feet, between exposures? What is the ground spacing, in ft, between flight lines?

16-4. The flight lines in Problem 16-3 are to be plotted on a map that is drawn to a scale of 1 : 20,000. What is the flight line spacing on the map, in inches?

16-5. If the aircraft flies at a ground speed of 180 mph along the flight line of Problem 16-3, what is the time interval, to the nearest 0.1 sec between exposures along the flight line?

16-6. Two points lying at an elevation of 650 ft appear on a photograph taken as described in Problem 16-1. The distance between these points scales 31.85 mm on the photograph. What is the ground distance, in feet, between the two points?

16-7. A surveyor uses an aerial photograph in the field for general reconnaissance. To determine an approximate scale of the photograph, he scales 61.25 mm between two road intersections. The distance between these two points measures 11.80 mm on a USGS quadrangle map which is at a scale of 1 : 62,500. What is the approximate scale of the photograph?

16-8. If the focal length of the aerial camera used to take the photograph described in Problem 16-7 is 152.4 mm, and if the points lie at an elevation of 855 ft, from what height above sea level was the photograph taken?

16-9. The image of a tall building with vertical sides appears on an aerial photograph taken with a camera in which the focal length is 210.4 mm. The photo scale is 1 : 6000. The distance from the principal point to the top of the building is measured as 78.65 mm, and to the bottom of the building measures 69.20 mm. Compute the height of the building, to the nearest meter.

16-10. The photograph diagrammed in Fig. 16-9 was taken with a 6-in. focal length lens from a flying height of 5500 ft above sea level. The following data are given as an aid in solving this problem:

Point	Photographic Azimuth	Radial Distance (in.)	Ground Elevation (ft)
a	45°	3.255	1460
b	138°	5.510	1585
c	212°	4.788	1120
d	288°	1.020	1320
e	320°	3.472	1845

The bearing of *AB* on the ground is S 32° 30′ W. Plot a 9 × 9-in. photograph, and plot the position of the photograph axes and the positions of *a*, *b*, *c*, *d*, and *e*. Compute the relief displacement for each point, and plot the datum position *a′*, *b′*, *c′*, *d′*, and *e′*. Make a graphical survey of the reduced polygon, determining ground lengths and bearings of the lines, and the area of the figure in square feet and in acres.

16-11. A 10-mile strip of terrain is to be photographed for highway mapping. The aerial camera contains a 6-in. focal length lens and takes 9 × 9-in. photographs. The ground elevation varies from a low of 500 ft to a high of 1100 ft, and the average elevation is 750 ft. The average width of coverage is to be 1500 ft.

(**a**) What flying height above sea level should be used for this flight?

(**b**) If the overlap between photos is 60% and the centers of the first and last photographs are to fall outside the limit of the strip, how many photographs will be required?

(**c**) What will be the largest and smallest photographic scales? Express each as an engineer's scale.

(**d**) What percent of the flying height above the 750-ft level is represented by the total variation in the elevation of the terrain?

16-12. Panels are to be set out in an area to be photographed with a 210-mm focal length lens for the purpose of making a control survey by photogrammetric methods. What are the theoretical dimensions, in centimeters, of the inner square shown in Fig. 16-13 if the photo image of this square is to be 40×40 μm and if the flying height is to be 1800 m above the ground? What are the dimensions in inches?

16-13. Photographs are taken at a height of 1260 m above ground with a 152.50-mm focal length lens. The photo base is 90.55 mm. The difference in parallax between a control point A whose elevation is 245.5 m and five other points are measured on an overlapping pair of these photographs using a parallax bar. These differences are as follows:

Point	Difference in Parallax (mm)
B	− 1.46
C	+ 2.82
D	+ 4.95
E	− 5.84
F	+ 0.45

Determine the elevations of the five points, to the nearest 0.1 m.

BIBLIOGRAPHY   American Society of Photogrammetry. *Photogrammetric Engineering and Remote Sensing.* 1996. Vol. LXII, No. 6.

American Society of Photogrammetry. 1968. *Canadian Surveyor* 22. Entire issue.

—— 1981. *Manual of Photogrammetry,* 4th ed. Bethesda, MD.

Hallert, B. 1960. *Photogrammetry.* New York: McGraw-Hill.

Kraus, Karl. 1993. *Photogrammetry,* 4th ed. Bonn, Germany: Ferd. Dümnlers Verlag.

Moffitt, F. H., and Mikhail, E. M. 1980. *Photogrammetry,* 3rd ed. New York: Harper & Row.

Scher, M. B. 1969. Orthophoto maps for urban areas. *Surveying and Mapping* 29:431.

Schwidefsky, K. 1959. *An Outline of Photogrammetry.* New York: Pitman.

Wolf, P. R. 1983. *Elements of Photogrammetry.* New York: McGraw-Hill.

Ventura, S. J. 1991. Evaluation of prototype 1:12,000 orthophoto quarter quadrangles. *Surveying and Land Information Systems* 51:2.

Zeller, M. 1952. *Text Book of Photogrammetry.* London: H. K. Lewis.

17 Earthwork

17-1 Introduction

Earthwork operations involve the determination of the volumes of materials that must be excavated or embanked on an engineering project to bring the ground surface to a predetermined grade, and the setting of stakes to aid in carrying out the construction work according to the plans. Although the term *earthwork* is used, the principles involved in determining volumes apply equally well to volumes of concrete structures, to volumes of stock piles of crushed stone, gravel, sand, coal, and ore, and to volumes of reservoirs. The field work includes the measurements of the dimensions of the various geometrical solids that make up the volumes, the setting of grade stakes, and the keeping of the field notes. The office work involves the computations of the measured volumes and the determination of the most economical manner of performing the work.

The units used for measurement in earthwork computations are the foot, the square foot, and the cubic yard. In the metric system the corresponding units are the meter, the square meter, and the cubic meter. This chapter employs the foot-yard system. However, metric units can be substituted with equal validity.

Because of the inherent accuracy of modern topographic maps of large scale produced by photogrammetric methods, much of the field work to be discussed in this chapter is eliminated, except for earthwork of limited extent. The measurements for the determination of volumes can be made directly from stereoscopic models or on topographic maps prepared by photogrammetric methods (see Sections 16-9, 17-13, and 17-14).

17-2 Cross Sections

A cross section is a vertical section taken normal to the direction of the proposed center line of an engineering project, such as a highway, railroad, trench, earth dam, or canal. A simple cross section for a railroad embankment is shown in Fig. 17-1. The cross section for a highway or an earth dam would have similar characteristics. It is bounded by a base b, side slopes, and the natural terrain. The inclination of a side slope is defined by the horizontal distance s on the slope corresponding to a unit vertical distance. The slope may be a rise (in excavation) or a fall (in embankment). A side slope of $3\frac{1}{2}$: 1, for example, means that for each $3\frac{1}{2}$ ft of horizontal distance the side slope rises or falls 1 ft. This can be designated as $3\frac{1}{2}$: 1 or 1 on $3\frac{1}{2}$.

17-3 Preliminary Cross Sections

In making a preliminary estimate and in determining the location of a facility, such as a highway or railroad, a preliminary line is located in the field as close to the final location of the facility as can be determined from a study of the terrain supplemented by maps or aerial photographs of the area. The preliminary line is stationed, and profile levels are taken. The configuration of the ground normal to the line is obtained by determining the elevations of points along sections at right angles to the line. This is identical to the process of obtaining elevations for topographic mapping described in Section 15-6.

The values of the elevations and the corresponding distances out to the right or left of the preliminary line can be plotted on specially printed cross-section paper, at a relatively large scale of from 1 in. = 5 ft to 1 in. = 20 ft (however, see Sections 17-13 and 17-14). When the location and grade of a trial line representing a tentative location of the center line of the facility have been established, the offset distance from the preliminary line to the trial line is plotted, and the grade elevation of this trial line is plotted in relation to the terrain cross section. In Fig. 17-2 the elevations of, and distances to, the points plotted on the ground line were determined with reference to the preliminary center line. These are shown as

FIGURE 17-1 Cross section in fill for railroad embankment.

FIGURE 17-2 Preliminary cross section.

fractions, with elevations as the numerators and distances from the preliminary line as the denominators. The offset distance of 15 ft from the preliminary line to the trial line is plotted, and the base of the roadbed is plotted at grade elevation 672.00 ft. At the edges of the roadbed, the side slopes of 2 : 1 are laid off and drawn to intersect the terrain line at the scaled distances and elevations shown as fractions lying under the terrain line. These points of intersection are called *catch points.*

The cross-sectional area bounded by the base, the side slopes, and the ground line of each trial cross section along the trial line is determined from the plotted cross section by using a planimeter, by computation based on the formulas for areas given in section 8-21, or by surveying software. This procedure is discussed in Section 17-8. The volumes of excavation and embankment for this trial line are computed from the successive areas and the distances between the areas by the methods described in Sections 17-9 to 17-11. The volumes for various trial lines are compared. The necessary changes in line and grade are then made to locate the final line and establish the final grade. This location will require a minimum of earthwork costs and, in the case of a highway project, for example, it will at the same time meet the criteria designed for curvature, maximum grade, and safe sight distances.

17-4   Final Cross Sections

The line representing the adopted center line of a facility is staked out in the field and stationed. This line is located by computing and running tie lines from the preliminary line as discussed in Section 8-24. Deflection angles are measured between successive tangents, and horizontal curves are computed and staked out. Reference stakes are sometimes set opposite each station on both sides of the center line at distances of 25, 50, or 100 ft from the center line. These stakes are used to relocate the center line after grading operations are begun. Stakes at a distance on either side equal to half the base width are sometimes driven to facilitate taking final cross sections and setting construction or slope stakes. The center line and the reference lines are then profiled.

When the final line has been located and profiled, a cross section is taken at each station to determine the area of the cross section and at the same time to locate the limits of excavation of embankment. These limits are defined on the ground by stakes. The process of setting these stakes is called *slope staking.*

In Fig. 17-3 the level is set up and a backsight is taken on the leveling rod held at a station on the center line whose elevation has been determined from the profile levels. The HI is established as 880.2 ft. The base width given on the plans is 24 ft, the side slopes are $1\frac{1}{2}$: 1, and the grade elevation (the design elevation of the road surface) at the station is 878.4 ft. If the leveling rod were held so that the foot of the rod were at grade, the reading of the rod would be 1.8 ft. This is equal to the HI minus the grade elevation and is called the *grade rod*. The grade rod can be plus or minus. The vertical distance from a point on the ground line to the grade line at any section is called the *fill* at the point if the section is in embankment, or the *cut* if the section is in excavation. Figure 17-3 shows that, at the left edge of the cross section, the fill is 8.6 ft, at the center line the fill is 6.2 ft, and at the right edge the fill is 4.2 ft.

The grade rod may be determined from the center-line cut or fill and the rod reading at the center line by the following relationship:

$$\text{grade rod} = \text{ground rod} + \text{center cut}$$

or   (17-1)

$$\text{grade rod} = \text{ground rod} - \text{center fill}$$

in which the ground rod is the rod reading at the center line. The ground rod in Fig. 17-3 is 8.0 ft. So the grade rod is 8.0 − 6.2 = 1.8 ft, as determined before.

With the grade rod established for the cross section, the amount of cut or fill at any point in the section can be determined by reading the rod held at the point and applying the following relationship:

$$\text{cut or fill} = \text{grade rod} - \text{ground rod} \tag{17-2}$$

If the result is plus, the point is above grade indicating cut (+); if the result is minus, the point is below grade, indicating fill (−).

The location, on the ground, of the slope stake is determined by trial. When the ground surface is horizontal, the position of the slope stake is at a distance from the center line equal to one-half base width plus the product of the side-slope ratio and the center cut or fill. This is shown in the dashed-line portion of Fig. 17-1.

FIGURE 17-3 Slope staking. Side slopes are 1.5 : 1.

If the ground in Fig. 17-3 were horizontal, each slope stake would be located at a distance from the center line equal to

$$12 + 1.5 \times 6.2 = 21.3 \text{ ft}$$

However, since the ground slopes transversely to the center line, it can be seen that the right slope stake will be less than 21.3 ft from the center line while the left slope stake will be greater than 21.3 ft from the center line. By trial, a point is found where 12 ft plus 1.5 times the fill computed by Eq. (17-2) equals the actual distance from the center line. With some experience the point can be found by one or two trials. If it is assumed that the ground appears to rise about 1.5 ft between the center line and the right-hand edge, the fill at the right-hand edge will be 6.2 − 1.5 = 4.7 ft and the distance out should be

$$12.0 + 1.5 \times 4.7 = 19.0 \text{ ft}$$

On the basis of this estimate, the rod is held 19 ft from the center line and a rod reading is taken. If the actual reading at this point is found to be 5.8 ft, then by Eq. (17-2) the depth below the grade is

$$1.8 - 5.8 = -4.0 \text{ ft}$$

To position the point at the intersection of the side slope with the ground surface, the distance from the center should be

$$12.0 + 1.5 \times 4.0 = 18.0 \text{ ft}$$

Since the actual distance was 19.0 ft, the trial point is incorrect. The rod is therefore moved nearer the center. Had the ground surface at the right edge been horizontal, the slope stake could have been set at the 18.0-ft point. Since the ground is sloping downward toward the center line, the fill will be somewhat more than 4.0 ft, and the distance out will be greater than 18.0 ft. Consequently the next trial is taken at 18.2 ft, where a rod reading of 6.0 ft is obtained. The depth below the grade is

$$1.8 - 6.0 = -4.2 \text{ ft}$$

and the distance from the center line should be 18.3 ft. This is within 0.1 ft of where the rod is being held, so the stake is set at 18.3 ft from the center and the fill is recorded as 4.2 ft.

In the field notes these two dimensions are recorded in fractional form, the numerator representing the fill and the denominator representing the distance out from the center. To distinguish between cut and fill, either the letters C and F or the signs + and − are used to designate them. As the point is located with respect to the finished grade, a point below grade indicates a fill and is designated by a − sign. The amount of cut or fill is marked on the side of the stake toward the center stake, and the distance out is marked on the opposite side. The stake is usually driven slantingly to distinguish it from a center-line stake and to prevent it from being disturbed during the grading operations.

In Fig. 17-3 there is a decided break in the ground surface between the center line and the right slope stake. This break is located by taking a rod reading there and measuring the distance from the center line. In a similar manner the break on the left side of the center and the left slope stake are located. The field notes for this particular station could be recorded as shown in the following table:

Station	Grade Elevation	Ground Elevation	Cross Section				
			L		C		R
15 + 00	878.4	872.2	−8.6	−8.4	−6.2	−5.8	−4.2
			24.9	21.0	0	10.5	18.3

17-5 Cross Sectioning by Slope Measurement

The position of a slope stake can be determined by measuring the distance out from the center line along the slope and determining the difference in elevation between the center-line stake and the tentative position of the slope stake. Figure 17-4 shows a cross section in cut with a roadbed width b and side slope $s : 1$. The cut at the center line is designated c. A plus slope measurement

$$M'P' = t$$

is made from a point M' vertically above the center-line stake at M to a point P' vertically above the tentative position of the slope stake at P. The slope of $M'P'$ is the vertical angle α. In the figure

$$MM' = PP'$$

Then

$$V = t \sin \alpha \text{ and } H = t \cos \alpha$$

The vertical distance V is the difference in elevation between M and P.

If P is the correct position of the slope stake, then the total rise of the side slope is $c + V$ (or $c - V$ on the downhill side). The side slope is in the ratio

$$\frac{1}{s} = \frac{c + V}{D} = \frac{c + V}{H - b/2}$$

FIGURE 17-4 Slope staking by slope measurements.

Thus considering both an uphill and a downhill sight,

$$H = \frac{b}{2} + s(c \pm V) \qquad (17\text{-}3)$$

If the computed value of $H = t \cos \alpha$ is the same as the value computed by Eq. (17-3), the slope stake is in the correct position. If

$$H = t \cos \alpha$$

is larger than that obtained by Eq. (17-3), the slope stake is too far out and must be moved in. Thus the final position of the slope stake is determined by trial.

If the cross section is a fill section, then

$$H = \frac{b}{2} + s(f \mp V) \qquad (17\text{-}4)$$

in which f is the center fill, considered a positive quantity in this equation. If the sight is to the uphill side, the negative sign is used for V, and if downhill the plus sign is used.

Various methods of making the measurements can be used. If a theodolite is set up at M and MM' is measured, the vertical angle α is measured to a point P' on the leveling rod or staff such that

$$PP' = MM'$$

The slope distance is then measured with a steel or metallic tape.

The same procedure as above can be performed using the principle of transit stadia. The values of H and V are those discussed in Section 14-5. The leveling rod held at the slope-stake position should be carefully plumbed using a rod level. This method eliminates the use of the tape.

Several of the EDMs described in Chapter 2 are mounted directly onto transits or theodolites. Others are equipped with vertical circles. These instruments can be used for slope staking in a manner similar to the use of the transit and tape. The reflector is set on a staff at the same height as the horizontal axis of the transit or theodolite or of the instrument itself. The vertical angle α and the slope distance t are measured with the instrument combination. After a trial position of the slope stake has been made, the reflector is moved inward or outward along the cross section according to the difference between the two computed values of H, and a second determination is made. Thus the slope-stake position is fixed by trial and error.

Some EDMs have provisions for computing the values of H and V internally, as discussed in Chapter 2. This eliminates having to do some of the computations in the field as the slope staking progresses. Any of the total station instruments discussed in Chapter 4 can be employed for slope staking since they compute the values of H and V.

17-6 Location of Slope Stakes by Inversing

In modern transportation engineering, all of the surveys are placed on a coordinate system. The preliminary line from which preliminary cross sections are measured (however, see Sections 17-13 and 17-14) is surveyed on the coordinate system. The final alignment is laid out with respect to the control points as discussed in Section 8-24. The preliminary cross sections as shown in Fig; 17-2 are used to compute the positions of the catch points. The base width, side slopes, and grade elevation for each cross section are entered into a computer program. The computer then intersects the side slope lines with the appropriate lines on the preliminary cross section, using Eqs. (8-15) and (8-16). Output from the computer gives the distance to the left (d_L) and to the right (d_R) where slope stakes are to be located, together with the cuts and fills at the catch points and the areas of the cross section. This is described in detail in Section 17-13.

If each station on the final located line is occupied with a theodolite, theodolite-EDM combination, or a total station instrument, a right angle is laid off from the line, and the calculated distance left or right is measured off and the slope stake is set. This, however, requires an instrument setup at each station. If a control point with a commanding view of a substantial portion of the project is convenient, the instrument can be set up at this control point from which angles and distances are measured off to locate several slope stakes.

In Fig. 17-5 the location center line is on a coordinate system. Thus the coordinates of each center-line station are known, as are the coordinates of a control

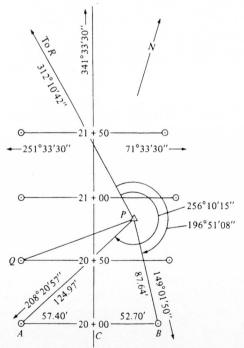

FIGURE 17-5 Location of slope stakes by inversing.

point *P*. The azimuth of the center line is shown as 341° 33′ 30″. The azimuth of each of the crosslines going to the left of the center line is 90° less, or 251° 33′ 30″. The azimuth of these lines going to the right is 90° more, or 71° 33′ 30″. The latitudes and departures can be computed for each crossline going to the left and to the right since the distances (the d_L's and d_R's) and the azimuths are known. The coordinates of the catch points where the slope stakes are to be set can then be computed based on the coordinates of the center-line stations.

The length and azimuth of any line from the control point *P* to a catch point such as point *Q* to the left of station 20 + 50 are computed by Eqs. (8-12) and (8-13). The angles to be laid off at *P* from a backsight on another control station to each catch point can be computed from the known azimuths. Some of the total station instruments can be programmed to compute the angle and distance to be laid off in order to set the slope stake. The program generally follows the computations shown in Example 17-1.

EXAMPLE 17-1   In Fig. 17-5, the coordinates of the center line at station 20 + 00 (point *C*) are *X* = 1,508,697.85 and *Y* = 457,715.20. The coordinates of control point *P* are *X* = 1,508,702.74 and *Y* = 457,807.02. The azimuth from *P* to control point *R* is 312° 10′ 42″ as shown. The distance *CA* to the slope stake to the left of station 20 + 00 is 57.40 ft; the distance *CB* to the slope stake to the right of station 20 + 00 is 52.7 ft. Compute the clockwise angle at *P* from *R* to *A* and from *R* to *B*, and compute the distance *PA* and *PB*.

Solution: The latitudes and departures of *CA* and *CB* and the coordinates of *A* and *B* are computed as follows:

Point	Length	Azimuth	Latitude	Departure	Y	X
C					457,715.20	1,508,697.85
	57.40	251° 33′ 30″	−18.16	−54.45		
A					457,697.04	1,508,643.40
C					457,715.20	1,508,697.85
	52.70	71° 33′ 30″	+16.67	+49.99		
B					457,731.87	1,508,747.84

Inversing line *PA*:

$X_A = 1{,}508{,}643.40 \quad Y_A = 457{,}697.04$
$X_P = 1{,}508{,}702.74 \quad Y_P = 457{,}807.02$
$\overline{\quad -59.34 \qquad\qquad -109.98\quad}$

azimuth *PA* = 208° 20′ 57″
length *PA* = 124.97 ft

Clockwise angle from *R* to *A*:

azimuth *PA* = 208° 20′ 57″
+ 360°
568° 20′ 57″
azimuth *PR* = 312° 10′ 42″
angle = 256° 10′ 15″

Inversing line *PB*:

$X_B = 1{,}508{,}747.84 \quad Y_B = 457{,}731.87$
$X_P = 1{,}508{,}702.74 \quad Y_P = 457{,}807.02$
$\overline{\quad +45.10 \qquad\qquad -75.15\quad}$

azimuth *PB* = 149° 01′ 50″
length *PB* = 87.64 ft

Clockwise angle from *R* to *B*:

azimuth *PB* = 149° 01′ 50″
+ 360°
509° 01′ 50″
azimuth *PR* = 312° 10′ 42″
angle = 196° 51′ 08″   ◆

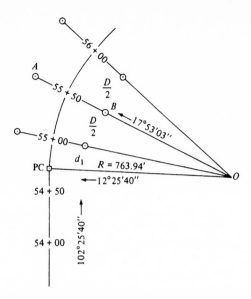

FIGURE 17-6 Location of slope stakes on a curve by inversing.

If the center line is a curve, the cross sections lie on radii from the center of the curve. In Fig. 17-6, if the center line is on a coordinate system, the coordinates of the PC are known. The coordinates of the center of the curve at O can be computed since the radius is known and the azimuth of the radius from the PC to O is (in this example) the azimuth of the center line plus (or minus) 90°. The azimuths of the radii to the curve points can then be determined by adding (or subtracting) the central angles d_1, $d_1 + D/2$, $d_1 + D/2 + D/2$, and so on to (or from) the azimuth of $O - $ PC. The distance from O to any left slope stake is then R plus (or minus) the distance from the center line to the left slope stake; the distance from O to any right slope stake is then R minus (or plus) the distance from the center line to the right slope stake. (The terms in parentheses apply to a curve that deflects to the left.)

EXAMPLE 17-2 In Fig. 17-6 the azimuth of the center line approaching the curve is 102° 25′ 40″. The coordinates of the PC are $X = 1,418,065.24$ and $Y = 434,312.12$. The degree of curve is 7° 30′ (arc definition). The radius is 763.94 ft. The PC is at station 54 + 77.25. then d_1 is 1° 42′ 23″ and $D/2$ is 3° 45′. The distance to the left slope stake A at station 55 + 50 is 73.60 ft from the center line and to the right slope stake at B is 83.54 ft. Compute the coordinates of A and B.

Solution: The azimuth from the PC to O is 102° 25′ 40″ + 90° = 192° 25′ 40″. The azimuth from O to the PC is then 12° 25′ 40″. The azimuth from O to station 55 + 50 is 12° 25′ 40″ + 1° 42′ 23″ + 3° 45′ = 17° 53′ 03″. Computation of latitudes, departures, and coordinates are as follows:

Point	Length	Azimuth	Latitude	Departure	Y	X
PC					434,312.12	1,418,065.24
	763.94	192° 25′ 40″	−746.04	−164.41		
O					433,566.08	1,417,900.83
O					433,566.08	1,417,900.83
	837.54	17° 53′ 03″	+797.07	+257.20		
A					434,363.15	1,418,158.03
O					433,566.08	1,417,900.83
	680.40	17° 53′ 03″	+647.52	+208.95		
B					434,213.60	1,418,109.78

The calculation of the coordinates of slope stakes given above for simple circular curves can also be applied to compound and reversed curves and to spirals, taking into account the geometry of these curves.

All of the above computations are carried out for each slope stake in one program in a data processor, and the output is given to the field engineer. The control point is occupied with a combination theodolite/EDM or a total station instrument, all of the catch points that can be seen from the control point are located, and the slope stakes are set.   ◆

17-7   Distance Between Cross Sections

The horizontal distance between cross sections is dependent on the precision required, which in turn is dependent on the price per cubic yard paid for excavation. As the unit cost for highway or railroad grading is usually small, a station distance of 100 ft is generally sufficiently precise. For rock excavation and for work done under water, the cost per cubic yard mounts very rapidly, and the distance between sections is often reduced to 10 ft.

In addition to the cross sections taken at regular intervals, other sections are taken at the point of curvature (PC) and the point of tangency (PT) of each curve, at all breaks in the ground surface, and at all grade points. A *grade point* is a point where the ground elevation coincides with the grade elevation. In passing from cut to fill or from fill to cut, as many as five sections may be needed in computing the volume when the change occurs on a side hill. In Fig. 17-7 these sections are located at *A, B, C, D,* and *E*. Unless extreme precision is required, the sections at *A* and *D* are usually omitted. Although stakes are set only at the grade points *B, C,* and *E*, full cross sections are taken at the stations and the measurements are recorded in the field notes.

It will be noted in Fig. 17-7 that a wider base of cross section is used when excavation is encountered. The additional width is needed to provide ditches for draining the cut.

17-8   Calculation of Areas

The purpose of cross sectioning is to obtain the measurements necessary to compute the area of the plane of the cross section bounded by the ground, the side slopes, and the roadbed. The cross-section areas are then used to compute

18 + 00

E · 17 + 81

D

C

17 + 63

A B

17 + 49

17 + 00

FIGURE 17-7 Transition from fill to cut.

earthwork volumes, either by the average end-area method discussed in Section 17-9 or by the prismoidal formula given in Section 17-10.

Four classes of cross-section configurations are shown in Fig. 17-8. The first is the level section shown in view (a). The area of the cross section is given by

$$A = (b + fs)f \tag{17-5}$$

in which b is the width of the roadbed, f is the center fill, and s is the side-slope ratio. If the cross section is in excavation, then c is the center cut and is substituted for f in Eq. (17-5).

Figure 17-8(b) shows a three-level section. The cross-section notes contain information only at the center line and at the two slope stakes. The area of the cross section is the sum of the areas of four triangles labeled I, II, III, and IV. The areas are

$$A_{\mathrm{I}} = \frac{1}{2} \times \frac{b}{2} \times f_l; \quad A_{\mathrm{II}} = \frac{1}{2} f \times d_l; \quad A_{\mathrm{III}} = \frac{1}{2} f \times d_r; \quad A_{\mathrm{IV}} = \frac{1}{2} \times \frac{b}{2} \times f_r$$

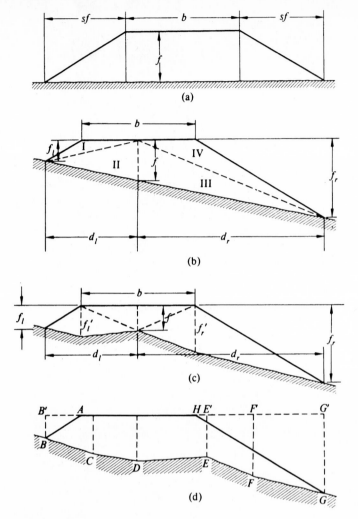

FIGURE 17-8 Types of cross sections. (a) Level section. (b) Three-level section. (c) Five-level section. (d) Irregular section.

Combining the areas of the four triangles gives

$$A = \frac{b}{4}(f_l + f_r) + \frac{f}{2}(d_l + d_r) \qquad (17\text{-}6)$$

in which f_l and f_r are the fill at left and right slope stakes, respectively; d_l and d_r are the distances to left and right slope stakes, respectively; and f is the center fill. If the section is in cut, the symbol c is substituted for f.

A five-level section is shown in view (c). The fill is determined at points measured out a distance equal to one-half roadbed width on both sides of the center line. The area of this class of section is

$$A = \frac{1}{2}(f_l'd_l + fb + f_r'd_r) \qquad (17\text{-}7)$$

in which f'_l and f'_r are the fill at the left and right edge of the roadbed, respectively; and d_l and d_r are the distances to left and right slope stakes, respectively.

The cross section shown in Fig. 17-8(d) is called an *irregular section* in which the ground surface is so irregular that a three- or five-level section would not give enough information to obtain an accurate determination of the area. The area of the section may be obtained by computing the areas of the trapezoids forming the figure $B'BCDEFGG'$ and subtracting from their sum the areas of the two triangles ABB' and GHG'.

Another method, which can also be applied to any of the preceding figures, is to consider the cross section as a traverse. The field notes provide the coordinates of the corners with respect to the finished grade and the center line as coordinate axes, the horizontal distances from the center line being the X coordinates with due regard for algebraic sign, and the vertical cuts or fills being the Y coordinates. For an eight-sided traverse for example, the area can be expressed by one of the following equations, which are like Eq. (8-25) or (8-26) in Section 8-21:

$$\text{area} = \tfrac{1}{2}[X_1(Y_2 - Y_8) + X_2(Y_3 - Y_1) + X_3(Y_4 - Y_2) + ...]$$
$$\text{area} = \tfrac{1}{2}[Y_1(X_2 - X_8) + Y_2(X_3 - X_1) + Y_3(X_4 - X_2) + ...]$$

Since for a cross section in earthwork the Y coordinates of two of the points are zero, the computations will be shortened if the second equation is used. An application of this equation to the area in Fig. 17-8(d) is given in Table 17-1, in which the roadbed width is 24 ft.

17-9 Volume by Average End Areas

According to the end-area formula, the volume, in cubic feet, between two cross sections having areas A_0 and A_1 is

$$V_e = \tfrac{1}{2}(A_0 + A_1)L \tag{17-8}$$

TABLE 17-1 Computations for Area of Cross Section

A	B	C	D	E	F	G	H
F0.0	F5.2	F6.8	F7.2	F6.1	F7.4	F9.6	F0.0
−12.0	−19.8	−10.0	0	+15.0	+20.0	+26.4	+12.0
(1)	(2)	(3)	(4)	(5)	(6)	(7)	(8)

$Y_i(X_{i+1} - X_{i-1})$ = double area

0.0(− 19.8 − 12.0)	= 0.0
5.2(− 10.0 + 12.0)	= + 10.4
6.8(0.0 + 19.8)	= + 134.6
7.2(+ 15.0 + 10.0)	= + 180.0
6.1(+ 20.0 − 0.0)	= + 122.0
7.4(+ 26.4 − 15.0)	= + 84.4
9.6(+ 12.0 − 20.0)	= − 76.8
0.0(− 12.0 − 26.4)	= + 0.0
	2)454.6
Area =	227.3 ft²

in which L is the distance between the sections. The volume in cubic yards, is

$$V_e = \frac{1}{2}(A_0 + A_1)\frac{L}{27} = \frac{L}{54}(A_0 + A_1) \tag{17-9}$$

Although this relationship is not an exact one when applied to many earth-work sections, it is the one most commonly used because of the ease of its application and because of the fact that the computed volumes are generally too great and thus the error is in the favor of the contractor.

An example of the calculation of earthwork volumes by the average end-area formula for cut sections is given in Table 17-2. Note that the section at station 36 + 00 is three-level, and at station 37 + 00 is five-level. The roadbed is 36 ft, and side slopes are 2 : 1.

17-10 Volume by Prismoidal Formula

When the more exact volume must be known, it can be calculated by means of the prismoidal formula

$$V_p = \frac{L}{6}(A_0 + 4M + A_1) \tag{17-10}$$

in which M is the area of the middle section and V_p is the volume, in cubic feet. In general, M will *not* be the mean of the two end areas. It can be shown that this formula is correct for determining the volumes of prisms, pyramids, wedges, and prismoids that have triangular end sections and sides that are warped surfaces. Since the earth-work solids are included in this group, except for slight irregularities of the ground, the prismoidal formula gives very nearly the correct volume of earthwork. The error in the use of the end-area formula arises chiefly from the fact that in its application the volume of a pyramid is considered to be one-half the product of the base and the altitude, whereas the actual volume is one-third the product of those quantities.

The area of the middle section can be obtained by taking intermediate cross sections on the ground or when the same number of points have been taken on adjacent sections, by computing the area of a section that has dimensions equal to

TABLE 17-2 Computation of Volume of Earthwork by Average End-Area Formula

Station	Cross-Section Notes	Area
36 + 00	$\dfrac{C11.7}{41.4} \quad \dfrac{C7.4}{0.0} \quad \dfrac{C4.1}{26.2}$	$\dfrac{36}{4}(11.7 + 4.1)$ $+\dfrac{7.4}{2}(41.4 + 26.2) = 392.3 \text{ ft}^2$
37 + 00	$\dfrac{C12.4}{42.8} \quad \dfrac{C9.1}{18.0} \quad \dfrac{C7.6}{0.0} \quad \dfrac{C6.3}{18.0} \quad \dfrac{C2.9}{23.8}$	$\dfrac{1}{2}(9.1 \times 42.8 + 7.6 \times 36$ $+ 6.3 \times 23.8) = 406.5 \text{ ft}^2$

$$V = \frac{100}{54}(392.3 + 406.5) = 1479 \text{ yd}^3$$

the means of the corresponding dimensions of the two end sections. An example of this method for three-level sections having bases of 24 ft and side slopes of $1\frac{1}{2}$: 1 is given in Table 17-3.

By the end-area formula,

$$V_e = (100/54)(215.04 + 82.20) = 550.4 \text{ yd}^3$$

Thus the error in applying the end-area formula to this solid is 11.3 yd³, or about 2%.

17-11   Volume Through Transition

In passing through a transition from cut to fill, at least one of the cross sections will contain both cut and fill areas (the section at C in Fig. 17-7). Thus there will be both excavation volumes and embankment volumes between some of the cross sections. Let the cross sections of Fig. 17-7 be represented as shown in Fig. 17-9. The cross section at station 17 + 00 is a complete three-level section in fill; that at station 17 + 49 is a partial three-level section in fill in which the fill goes to zero at the left edge of the roadbed. The section at station 17 + 63 is a side-hill section with the grade point at the center line. The left side of the section is in cut and the right side in fill.

The cross section at station 17 + 81 is a partial three-level section in cut in which the cut begins from zero at the right edge of the roadbed. Finally, at station 18 + 00 the section is a full three-level section in cut.

The volume between station 17 + 00 and station 17 + 49 is all embankment. Between station 17 + 49 and station 17 + 63, the left part of the volume is a prism in excavation and the right side is embankment. The same is true between station 17 + 63 and station 17 + 81. Finally, the volume lying between station 17 + 81 and station 18 + 00 is all excavation.

The volume of the prism of excavation between station 17 + 49 and station 17 + 63 and the prism of embankment between station 17 + 63 and station 17 + 81 should be computed as pyramids—that is, one-third the base times the altitude—since they are pyramidal in shape. The area of the partial cross section corresponds to the base of the pyramid, and the difference in stationing corresponds to the altitude of the pyramid.

TABLE 17-3   Computation of Volume of Earthwork by Prismoidal Formula

Station	L	C	R	Area
15 + 00	$\dfrac{-8.6}{24.9}$	$\dfrac{-6.4}{0.0}$	$\dfrac{-4.2}{18.3}$	$\frac{1}{2} \times 12 \times 12.8 + \frac{1}{2} \times 6.4 \times 43.2 = 215.04 \text{ ft}^2$
16 + 00	$\dfrac{-4.6}{18.9}$	$\dfrac{-2.8}{0.0}$	$\dfrac{-1.4}{14.1}$	$\frac{1}{2} \times 12 \times 6.0 + \frac{1}{2} \times 2.8 \times 33.0 = 82.20 \text{ ft}^2$
M	$\dfrac{-6.6}{21.9}$	$\dfrac{-4.6}{0.0}$	$\dfrac{-2.8}{16.2}$	$\frac{1}{2} \times 12 \times 9.4 + \frac{1}{2} \times 4.6 \times 38.1 = 144.03 \text{ ft}^2$

$$V_p = \frac{100}{6} \times (215.04 + 4 \times 144.03 + 82.2) = 14{,}556.0 \text{ ft}^3 = 539.1 \text{ yd}^3$$

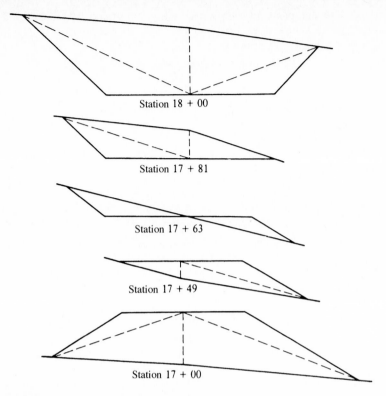

Station 18 + 00

Station 17 + 81

Station 17 + 63

Station 17 + 49

Station 17 + 00

FIGURE 17-9 Cross sections in transition from fill to cut.

17-12 Volume by Truncated Prisms

The volume of a borrow pit (see Section 17-19) or of the excavation for the foundation of a building is frequently found by dividing the surface area into squares or rectangles, according to the methods of Section 15-9. The original ground-surface elevations at the corners of these figures are usually determined by direct leveling. The final elevations are obtained either by repeating the leveling after the excavation has been completed or, in the case of building foundations, from the plans of the structure. The vertical depth at each corner is calculated by subtracting the final elevation from the original ground-surface elevation.

The volume of any prism is taken as the horizontal area of the prism multiplied by the average of the corner depths. Thus if a, b, c, and d represent the depths at the corners of a truncated square or rectangular prism, the volume is

$$V = A \times \frac{a + b + c + d}{4} \qquad (17\text{-}11)$$

in which A is the horizontal area of the prism.

The computations can be shortened by noting that some corner heights are common to more than one prism. The volume of the assembled prisms is

$$V = A \times \frac{\Sigma h_1 + 2\Sigma h_2 + 3\Sigma h_3 + 4\Sigma h_4}{4} \qquad (17\text{-}12)$$

in which A is the area of one of the equal squares or rectangles, Σh_1 is the sum of the vertical heights common to one prism, Σh_2 is the sum of those heights common to two prisms, Σh_3 is the sum of those common to three prisms, and Σh_4 is the sum of those common to four prisms.

As an example of the application of Eq. (17-12), consider that the area of Fig. 15-8 bounded by points $B1$, $B4$, $F4$, $F2$, $D2$, and $D1$ is to be brought to a finished grade elevation of 700.00 ft. Assuming that the sides of the excavation are vertical, the values of the corner heights are as follows:

h_1	h_2	h_3	h_4
B1 = 22.5	B-2 = 27.3	D-2 = 13.2	C-3 = 27.3
B4 = 45.6	B-3 = 34.8	Σh_3 = 13.2	D-3 = 18.3
F4 = 19.9	C-4 = 28.2		E-3 = 11.3
F2 = 4.2	D-4 = 22.7		C-2 = 17.4
D1 = 11.7	E-4 = 19.8		Σh_4 = 74.3
Σh_1 = 103.9	F-3 = 12.9		
	E-2 = 10.0		
	C-1 = 14.3		
	Σh_2 = 170.0		

If each square in Fig. 15-8 is, say, 25 ft on a side, then by Eq. (17-12) the volume in cubic feet is

$$V = 625 \times \frac{103.9 + 2 + 170.0 + 3 \times 13.2 + 4 \times 74.3}{4} = 122{,}000 \text{ ft}^3$$

17-13 Volumes from Topographic Maps

In modern highway-location practice, accurate topographic maps with a large scale and a small contour interval are prepared by photogrammetric methods supplemented by field completion surveys. These maps are known as design maps. They are compiled typically at a scale of 1 in. = 50 ft and with a 2-ft contour interval. The positions of all horizontal-control monuments are accurately plotted by means of their rectangular coordinates with respect to a grid that shows on the map.

The design map is used to study various possible positions for the location of the highway, to aid in establishing the various geometrical properties of the roadbed, to establish limits of the right of way, and to determine the quantities of earthwork for the various possible locations. These quantities are used for comparison of the different lines, for estimates in bidding, and in some instances for determining the actual pay quantities the contractor is held to.

The design map covers a strip of terrain that usually varies from about 600 to 2000 ft in width. When it has been compiled and verified in the field, the design engineer can then project trial lines onto the map with different grade lines. A study of the earthwork involved, together with other controlling factors, will establish a final line on the design map. The coordinates of the points of intersection are scaled from the grid lines, as described in Section 8-33, and tie lines from the control monuments are computed. Thus the final line may be located and staked out in the field by running these tie lines as outlined in Section 8-24.

The earthwork quantities for a given trial line are determined from the design map by the method that will now be described. The trial line is laid out on the map, and the curves joining the tangents at their intersections are drawn in. The line is marked from beginning to end at every 50 or 100 ft, the interval depending on the regularity of the terrain and the accuracy desired. These marks are stations and plusses.

A crossline is plotted normal to the trial line at each station and plus, and the elevations of, and distances right or left from, the center line to points representing breaks in the terrain are determined by interpolation between the contour lines and by scaling on the map. These elevations and distances are recorded opposite each station number on a set of notes, with the elevations as the numerators and the distances out from the center line as the denominators. The notes representing these cross sections are called *terrain notes*. The cross sections are taken far enough out to allow for deep cuts and fills, and to allow for line changes on successive trial lines. A cross section, with the terrain notes shown, is represented in Fig. 17-10. The actual terrain cross sections, however, are not plotted.

A grade line that is established by the design engineer is superimposed on a profile of the line plotted from elevations obtained from the design map. The lengths of the vertical curves are established, and the elevations of the stations on the vertical curves are computed. The grade elevation at each center-line station and plus is thus obtained.

Roadbed notes are prepared for each station and plus, taking into account the roadbed width, ditches in cuts, side slopes, roadbed crown, and superelevation on curves. The roadbed notes for a section coinciding with the section of Fig. 17-10 are given in Fig. 17-11, the roadbed itself being shown graphically. The actual roadbed cross sections, however, are not plotted. Point a is at the center line. Points b and c are known as hinge points, because the roadbed cross section breaks at these points when the section is in cut. Points e and f are at the edges of the shoulder in cut or in fill.

The configuration of the roadbed will remain constant as long as the highway is on a tangent. Since the notes will be used in subsequent computations, the

$$\frac{429^0}{69} \qquad \frac{426^0}{41} \qquad \frac{429^0}{9} \quad \frac{426^2}{0} \qquad \frac{420^0}{20} \qquad \frac{408^0}{48} \qquad \frac{407^0}{76}$$

FIGURE 17-10 Terrain notes.

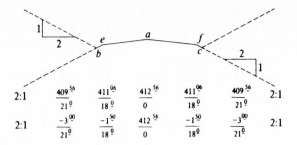

FIGURE 17-11 Roadbed notes.

values of the elevations at points *b, e, f,* and *c* can be tabulated as vertical distances from the center-line elevation, as shown in the lower notes of Fig. 17-11. Thus as long as the configuration of the roadbed does not change, the numerators of the lower set of fractions will not change, except for the numerator representing the grade elevation at the center line. This elevation is shown as 412.56 in Fig. 17-11. In the process of computation these vertical offset distances will be converted to elevations. The roadbed notes will vary on curves because of superelevation, and will vary in different materials because of changes in the side slopes.

The notes for the terrain cross section and the roadbed notes are input into a high-speed computer and are brought together by a prepared computer program, the details of which can be found from the vendor. The computer determines the distances to, and the elevations of, the catch points at each station and the area of each cross section bounded by the roadbed, the side slopes, and the terrain. The distance to the catch point (on a final line) is used in the field to locate the position of each slope stake. A graphical portrayal of the solution of Fig. 17-10 and Fig. 17-11 in the computer is shown in Fig. 17-12. The distances to the left and right slope stakes are, respectively, 58.6 and 33.0 ft. The area of the cross section is computed as that of a closed traverse in the computer, and the result is 949.3 ft^2.

Using the average end-area method, the computer determines the volumes between successive cross sections from the computer areas and the spacing of the cross sections in stations and plusses. These values are part of the data output by the computer and furnished to the design engineer.

When computations are to be made for a second trial line, the terrain notes need not be changed, provided they were taken from the map at sufficient distances from the center line. The roadbed notes need not be changed if the grade line has not been changed. However, the amount of line shift at each station must

FIGURE 17-12 Terrain and roadbed notes combined.

be introduced into the computer. For example, if the section of Fig. 17-12 is shifted to the right by 20 ft, this is portrayed in graphical form in Fig. 17-13. The resulting distances to the left and right slope stakes determined by the computer are, respectively, 55.6 and 23.0 ft. The area of the cross section is 667.6 ft^2.

If a grade change is to be made, the elevation of the roadbed center line at each section must be changed correspondingly before the data are introduced into the computer.

When the final line has been established by analysis of various trial lines, the computer output will give, in addition to the cross-sectional areas and earthwork volumes, the distances out from the center line to the left and right slope stakes. These listings are used by the field party to locate the slope stakes for construction.

A device used to scale distances automatically and record elevations along terrain cross sections is shown in Fig. 17-14. The round base plate is oriented on the topographic map in such fashion that the cylindrical movable arm is caused to move along a line perpendicular to the center line of a trial line. With the pointing device, located at the end of the arm and set at the center line, the elevation counter is set to read the ground elevation at the center line as interpolated between the contour lines on the map. The vertical counter is then set for the contour interval. As the arm is moved, say, to the left of the center line, the pointer is set on the first contour line encountered, and the operator depresses the register button. This operation records the distance to the left of the center line and the contour elevation of the point. When the movement of the arm is continued to the next contour line, the vertical counter automatically adds or subtracts the contour

FIGURE **17-13** Cross section after line shift.

FIGURE **17-14** Automatic recording scaler. Courtesy of Autotrol Corp.

interval while the movement of the arm measures the distance, and the values are again registered. When the terrain cross section to the left has been completed, the pointer is again indexed on the center line. The measurements are then taken from the center line to the right. Information pertinent to the terrain cross section, such as the station number, is set into the control console shown in Fig. 17-14 to be registered automatically with the cross section data.

The operator must depress a plus-minus button when going across ascending or descending contour lines in order that the elevation counter will register the correct value of the contour line. If the elevation of a point between contour lines is to be recorded, the operator interpolates between the contour lines to obtain the value of the intermediate elevation. By depressing a button on the small control unit seen in Fig. 17-14, the small increments of elevation are added to or subtracted from the elevation contour until the intermediate elevation is registered on the console display. This value can be obtained by the operator, without having to observe the console display, simply by counting audible signals emitted by the contour. The register is then depressed to record this intermediate elevation together with the corresponding distance from the center line.

The automatic scaling device is accurate to about 0.001 in. over the travel of the arm. It is certainly far more accurate than are the locations of the map features and the contour lines. The scaler eliminates mistakes and scaling errors. The scaling is much easier and far more efficient than that done by means of an ordinary engineer's scale. By automatically recording the data necessary for machine computation, the scaler eliminates not only the data-recording time but also the keyboard entry operation, which is always subject to mistakes.

The foregoing method of determining earthwork quantities in highway location entirely eliminates the field surveys for determining preliminary and final cross sections. Furthermore, the planning can be conducted well in advance of all but the initial control survey, and time is provided for purchasing the necessary right of way without too much danger of unscrupulous land speculation. The success of the entire method, of course, depends on the accuracy and reliability of the design map.

17-14   Earthwork Data from Photogrammetric Models

The compilation of a design map by photogrammetric methods requires that a pair of overlapping aerial photographs be oriented in a stereoscopic plotting instrument and then that the resulting spatial model be oriented to control that has been plotted on the map sheet. When these conditions are satisfied, the operator of the stereoplotter then proceeds to compile the planimetric and topographic features. This process is discussed in Section 16-9. The resulting topographic map is a graphical representation of the more complete and more detailed spatial model from which it was derived. To increase the accuracy of earthwork quantities, it would be logical to measure the spatial model directly, thus eliminating the inaccuracy and the incompleteness of the map.

Figure 17-15 shows the scaling device discussed in Section 17-13 in which the pointer has been replaced by the pencil holder of the tracing table of a stereoplotter.

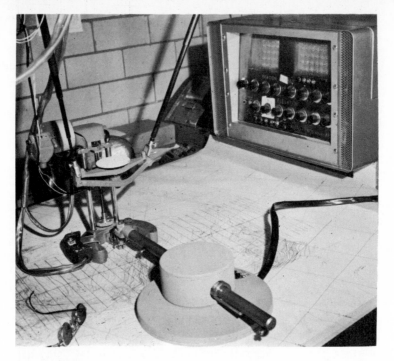

FIGURE 17-15 Automatic scaler used to measure cross sections directly in stereoscopic model.

The topographic map that was compiled from the aerial photography has been used to study possible trial center-line locations for a highway facility. After a trial center line has been drawn on the map, the map in turn is reoriented under the stereoscopic model (see Fig. 16-21).

The base of the scaler is now oriented so that the tracing table is constrained to move perpendicular to the center line drawn on the map. This condition is readily apparent from a study of Fig. 17-15. The stereoplotter operator then sets the tracing-table pencil on the center line at a desired station, sets the measuring mark on the surface of the stereo model, and indexes the elevation movement of the tracing table. An encoder senses and displays the elevation of the tracing table at all times.

After registering the center-line elevation, the operator then moves the tracing table, say, to the left of the center line until a significant break in the terrain appears in the stereoscopic model. After setting the measuring mark in apparent contact with the surface of the model at this point, he then depresses the register button, causing the elevation of the point and its distance from the center line to be automatically recorded. A repetition of this procedure thus provides terrain cross-section notes that can be used to compute earthwork quantities along this trial line.

The analytical plotter discussed briefly in Section 16-9 can be used to obtain profiles and cross sections through the interaction between the plotter and the computer system discussed in Section 16-12. If the location has been established on a coordinate system, the coordinates of the PT station of one curve and those of the PC of the next successive curve are entered into computer memory, along with the interval at which profile elevations are to be taken. The computer then drives the measuring mark along the line to the appropriate station or fractional station, and the operator then sets the measuring mark on the point and records the elevation. The computer drives the measuring mark to the next point at which the operator again measures and records the elevation, and so on. With the necessary programming, the computer will drive the measuring mark along lines perpendicular to the alignment so the operator can measure cross-section elevations at selected points. The output consists of profile elevations at each station or plus, together with the distance from the center line and the elevations of the points on the cross sections at which the operator measured terrain elevations.

If the radius, the PC, and the PT coordinates for a curve, together with the curve direction, are entered, the computer can be programmed to drive the measuring mark to each curve station and along the radius at each station to allow the operator to measure profile and cross-section elevations.

The use of the plotter-computer interactively eliminates much manual effort on the part of the operator. It also eliminates the use of a map to define the center line and the cross-section lines, since all measurements are with reference to the stereoscopic model, which itself has been established on the coordinate system. Furthermore, since all of the cross-section data are stored in memory, they can be recalled to be used to compute cross-sectional areas and earthwork volumes.

17-15   Earthwork Quantities by Grading Contours

A grading contour is a line of constant elevation that is plotted on a topographic map to represent a true contour line after the proposed grading has been performed. Because grading operations produce smooth surfaces with regular slopes, grading contours are either a series of straight, equally spaced lines or a series of curved lines that are equally spaced. Simple landscape grading is shown in Fig. 17-16. The original contour lines are shown as solid lines with a contour interval of 1 ft. The grading contours are drawn on the map as straight dashed lines. The irregular dashed lines have been drawn through the grade points to show the areas of cut and fill. The area marked I is enclosed by the original 879-ft contour line and the 879-ft grading contour. It is a horizontal surface at elevation 879 ft. Similarly, the area marked II is a horizontal surface at elevation 878 ft. The volume of fill necessary between these two surfaces is obtained by planimetering each area, determining the average of the two areas, and multiplying the average by the vertical separation or the contour interval which, in this instance, is 1 ft. The volume between areas II and III and the volume between areas III and IV are similarly obtained. The solid defined by the embankment between point F and area I is pyramidal in shape, as is the solid defined by the embankment between point A and area IV. Each of these two volumes should be taken as $\frac{1}{3}bh$, in which b is the shaded area representing the base of the solid and h is the contour interval, or 1 ft. The volume of excavation between K and P may be determined in the same manner.

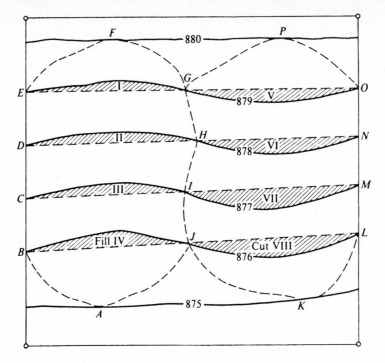

FIGURE 17-16 Landscape grading.

In Fig. 17-17 the location of a highway roadbed is plotted to scale on a design map on which the contour interval is 5 ft. The line is stationed by scaling 100-ft distances. The grade elevation at station 55 + 00 is to be 465.00 ft and that at station 58 + 33 is to be 462.00 ft. The side slopes are 2 : 1 in cut and are 3 : 1 in fill. The grade line along the center line intersects the ground surface at stations 53 + 62 and 58 + 64. These grade points are located by finding where the grade elevations coincide with the ground elevations determined by interpolation between contour lines.

The road is in cut between stations 53 + 62 and 58 + 64. Since the side slopes in cut are 2 : 1, a distance 10 ft out from either edge at station 55 + 00 will be at elevation 470 ft after the slopes have been formed. Similarly a point 16 ft out from an edge at station 58 + 33 will also be at elevation 470 after the slopes have been formed.

A straight line shown dashed in the figure on either side of the road as a line joining two points at elevation 470 is a grading contour line. The 470-ft grading contours intersect the 470-ft contour lines on the ground at points a and b. Grading contours at elevation 475 will lie 10 ft farther out; those at 480, another 10 ft farther out, and so on. In Fig. 17-17 these points are plotted along lines normal to the center line at stations 55 + 00 and 58 + 33, and the elevations of the grading contours are shown at these stations. The top of the cut runs out at elevation 513 ft on one side of the road and at 497 ft on the other side.

FIGURE 17-17 Design map showing grading contours for roadbed in cut.

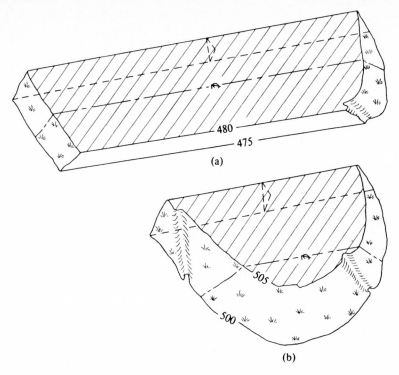

FIGURE 17-18 Volumes between successive contour surfaces.

A line joining the intersections of the grading contours with the corresponding ground contours represents the edge of the cut or the intersection of the ground by the side slopes. Any point on this line represents a slope-stake position. For example, the slope stakes at station 57 + 00 would be located at points c and d.

Figure 17-18(a) shows the prismoid lying between the 475- and 480-ft contour lines in Fig. 17-17. To determine its volume, the area A_{475}, enclosed by the 475-ft ground contour and the 475-ft grading contour on the map is measured by using a planimeter. The area A_{480}, enclosed by the 480-ft contour lines, is also measured. The volume found by the average end-area method is

$$V_{475-480} = 5 \times \frac{A_{475} + A_{480}}{2}$$

in which the volume is in cubic feet and the areas are in square feet, since the vertical distance between the two areas is in feet.

Figure 17-18(b) shows the prismoid between the 500- and 505-ft contour lines in Fig. 17-17, and its volume is determined in the same manner.

The volume in Fig. 17-17 lying between the 510-ft contour and the point at elevation 513 ft is computed by treating the earthwork as a pyramid. This volume is

$$V_{510-513} = \tfrac{1}{3} \times 3 \times A_{510}$$

Similarly, the volume lying between the 495-ft contour and the point at elevation 497 ft is

$$V_{495-497} = \tfrac{1}{3} \times 2 \times A_{495}$$

The method of procedure to be used in determining earthwork volumes from a topographic map by means of grading contours is summarized as follows:

1. Plot the proposed facility, such as a building, highway, or dam, to scale in the desired position. This can be done by referencing the facility to the horizontal control shown on the map, or by using coordinates. (A design map must always show coordinate grid lines or grid ticks.)

2. Establish all grade points from the grade elevations shown on the plans of the facility by interpolating between existing contour lines. Locating these points allows a visualization of areas of cut and fill.

3. From known grade elevations at certain points and the side-slope ratios given on the plans, plot the positions of points that will lie on the grading contours.

4. Draw the grading contours in their proper positions, and terminate them where they join the existing ground-contour lines.

5. Measure the areas enclosed by successive contours, and record the areas in a tabular form.

6. Compute the volumes lying between the successive contour surfaces by using the average end-area method. Apply the pyramidal formula where the cut or fill runs out. In some instances the prismoidal formula may be applied. In these instances the vertical separation between surfaces will be twice the contour interval, since the area enclosed by the middle one of three consecutive contour lines will be the area M of the middle section.

The method of determining earthwork quantities by the grading-contour technique presupposes a reliable topographic map of reasonably large scale. The accuracy obtainable is equivalent to that obtained by field cross-sectioning methods and the work is performed more rapidly. The method is advantageously applied where the facility involves complicated warped grading surfaces, the dimensions of which are difficult to compute. Such would be the case in computing earthwork volumes at proposed highway interchanges.

17-16 Reservoir Volumes from Contour Maps

Two general methods are used in determining reservoir volumes from contour maps. One method is to planimeter the area enclosed by each contour. The volume is then computed by the average end-area method, A_0 and A_1 being the planimetered areas of two adjacent contours and L the contour interval. The prismoidal formula can also be used where there is an odd number of contours, alternate contours being considered as middle sections and L as twice the contour interval.

A second method, which can be used when the reservoir is regular in shape, is to scale the dimensions of vertical cross sections from the contour map. The volume is then calculated from the cross-sectional areas, as in the case of a route survey.

17-17   Mass Diagram

Many of the problems connected with the handling of earthwork on highway and railroad construction can be solved by means of a mass diagram. It is usually plotted directly below the profile of the route, the ordinate at any station representing the algebraic sum of the volumes of cut and fill up to that station. From the diagram the most economical distribution of materials can be determined.

There are two common types of grading contracts. In one, the contractor bids a lump sum for the work, handling the materials as he sees fit. This method of payment is becoming more preferred, particularly on large projects. The mass diagram is of great value to him in determining the most economical manner of doing the work. In the other form of contract, payment is on the basis of the number of cubic yards of material handled. In this contract the successful bidder is paid a specified price per cubic yard for excavation. Embankment is taken care of by a clause stating that the excavation shall be deposited as directed by the engineer, who naturally directs that the excavated material be placed in embankments.

Since it would be unfair to expect the contractor to haul the materials taken from the excavations unlimited distances in forming the embankments, the contract will state the distance—500, 1000, or 1500 ft—that the excavation must be hauled without increased compensation. As soon as the contract limit has been exceeded, an additional sum per cubic yard is paid for each 100 ft of excess distance. The specified distance is the *limit of free haul,* and any excess is *overhaul.* The volumes and lengths involved in the calculation of overhaul payments can be taken from the mass diagram.

17-18   Shrinkage

Material taken from an excavation may occupy a greater or less volume when deposited in an embankment. As the material is first excavated and placed in a truck or other conveyance, it usually will occupy a greater volume than in its original position. Solid rock that must be broken up for handling may occupy about twice as much space as it did before being excavated. When earth is placed in an embankment, it will be compacted and may occupy less space, especially if it has been removed from the surface of a cultivated field and deposited at the bottom of a 15- or 20-ft embankment. The weight of a high fill may cause the original ground surface to settle or subside. When the fill is across marshy ground, this subsidence may be considerable. As the material is transported from excavation to embankment, some of it is generally lost in transit. Hence, for embankments composed of anything but rock, a volume somewhat greater than the calculated volume of the embankment will be required to make the actual embankment. An allowance is usually made for this shrinkage. The amount allowed may vary from 5% to 15%, the actual allowance depending on the character of the material handled and the condition of the ground on which the embankment is placed.

17-19   Computation of Mass Diagram

In studies that are made to decide on the best line location and grade line, earthwork quantities are determined from preliminary cross sections taken in the field, from topographic maps, or from stereoscopic models. If the mass diagram is to be used in computing overhaul payments, the volumes used are usually taken from the final cross sections unless the contract specifies the use of large-scale maps for this purpose. In the latter event the quantities computed for the final line discussed in Section 17-13 are used to compute the mass diagram.

Points beyond which it is not feasible to haul material define the limits of a mass diagram. A limit point may be the beginning of a project, the end of a project, a bank of a river, or an edge of a deep ravine where a bridge will be constructed.

The ordinates to the mass diagram are computed by adding the volumes between successive stations to the previous volumes, thus obtaining the accumulated volume up to any station. The volumes, and thus the ordinates, are in cubic yards. The abscissas are in stations. Excavation volumes are considered plus, and embankment volumes are minus. Before embankment volumes are used to compute ordinates, they must be increased to allow for shrinkage. This allowance is usually expressed as a percent shrinkage or as a shrinkage factor. In the tabulation in Table 17-4, the shrinkage factor is 10%.

TABLE 17-4   Calculation of Ordinates to Mass Diagram

Station	Cut Volume (yd³)	Fill Volume (+10%)	Mass Diagram	Station	Cut Volume (yd³)	Fill Volume (+10%)	Mass Diagram
0			10,000.0	28			6,040.4
		183.5		29		1,611.2	
1			9,816.5	29			4,429.2
		622.2				1,338.2	
2			9,194.3	30			3,091.0
		1,034.7				1,002.9	
3			8,159.6	31			2,088.1
		1,268.2				652.7	
4			6,891.4	32			1,435.4
		1,231.5				357.2	
5			5,659.9	33			1,078.2
		919.1			39.2	150.0	
6			4,740.8	34			967.4
		502.8			235.8	51.8	
7			4,238.0	35			1,151.4
	20.6	163.6			465.2		
8			4,095.0	36			1,616.6
	190.2	12.0			711.5		
9			4,273.2	37			2,328.1
	615.7				904.4		
10			4,888.9	38			3,232.5
	941.5				904.4		
11			5,830.4	39			4,136.9
	1,150.0				757.0		
12			6,980.4	40			4,893.9

TABLE 17-4 Calculation of Ordinates to Mass Diagram (Continued)

Station	Cut Volume (yd³)	Fill Volume (+10%)	Mass Diagram	Station	Cut Volume (yd³)	Fill Volume (+10%)	Mass Diagram
	1,500.0				516.5		
13			8,480.4	41			5,410.4
	1,772.7				280.2	90.4	
14			10,253.1	42			5,600.2
	1,754.6				126.7	316.1	
15			12,007.7	43			5,410.8
	1,540.4				98.4	640.0	
16			13,548.1	44			4,869.2
	1,262.3				20.0	770.6	
17			14,810.4	45			4,118.6
	931.8					789.0	
18			15,742.2	46			3,329.6
	546.5					728.0	
19			16,288.7	47			2,601.6
	203.1	30.6				577.2	
20			16,461.2	48			2,024.4
	101.0	278.5				355.7	
21			16,283.7	49			1,668.7
	18.0	586.2				115.9	
22			15,715.5	50			1,552.8
		1,005.3			407.0	100.0	
23			14,710.2	51			1,859.8
		1,377.4			735.5	26.6	
24			13,332.8	52			2,568.7
		1,676.0			983.3		
25			11,656.8	53			3,552.0
		1,860.3			1,285.9		
26			9,796.5	54			4,837.9
		1,917.1			1,522.4		
27			7,879.4	55			6,360.3
		1,839.0					

The initial ordinate in the tabulation is arbitrarily set at some large number so that all ordinates are positive. In Table 17-4 the initial ordinate is 10,000 yd³. Between stations 7 and 9, stations 19 and 22, stations 33 and 35, stations 41 and 45, and stations 50 and 52, both excavation and embankment occur. These are portions of the route similar to that shown in Fig. 17-9. Since excavation will be carried from one side of the roadway to the other during grading, only the net amount of cut or fill is used to compute the mass-diagram ordinate. In Fig. 17-19 the mass diagram is shown plotted below the center-line profile.

An inspection of the mass diagram will show that a rising curve indicates excavation and a descending curve indicates embankment. Maximum and minimum points on the mass diagram occur at grade points on the profile. If a horizontal line is drawn to intersect the diagram at two points, excavation and embankment (adjusted for shrinkage) will be equal between the two stations represented by the points of intersection. Such a horizontal line is called a *balance*

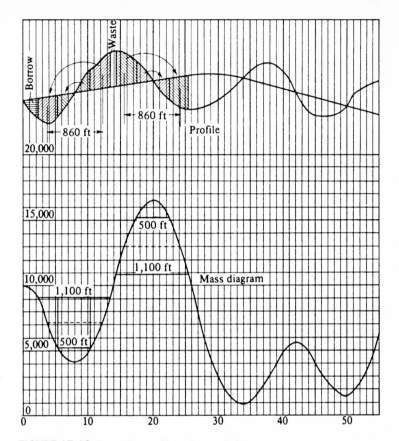

FIGURE 17-19 Center-line profile and mass design.

line because the excavation balances the embankment between the two points at its ends.

Since the ordinates to the diagram represent the algebraic sums of the volumes of excavation and embankment referred to the initial ordinate, the total volumes of excavation and embankment will be equal where the final ordinate equals the initial ordinate. If the final ordinate is greater than the initial ordinate, there is an excess of excavation; if it is less than the initial ordinate, the volume of embankment is the greater, and additional material must be obtained to complete the embankments.

In deciding on the grade line to be adopted, reasonable rates of grade should not be exceeded, and an attempt should be made to balance the volumes of cut and fill over moderately short stretches of the line. If this is not done, some of the material may have to be hauled excessively long distances. Where it is impossible to avoid long hauls, it is often cheaper to waste material in one place and obtain the volumes necessary to complete the embankments from borrow pits located along the right of way.

17-20   Calculation of Overhaul

When the contract prices are known, the economical limit of haul can be determined. This limit is reached when the cost of haul equals the cost of excavation. For longer distances it will be cheaper to waste in one place and borrow in another. As an example, the economical limit of haul will be 1100 ft if the cost of excavation and hauling 500 ft or less is 24 cents per cubic yard, and if the cost of hauling each additional 100 ft is 4 cents. At this distance the cost for 1 yd^3 will be 24 cents for excavation and hauling 500 ft, plus $6 \times 4 = 24$ cents for overhaul. For longer distances the cost of excavating and hauling will be greater than 48 cents, or the cost of excavating twice, and it will be cheaper to borrow and waste. For distances under 1100 ft the cost of excavating and hauling will be less than 48 cents, and hauling will be the cheaper method. These figures are on the assumption that fill material is available within the limits of the right of way. If additional land for borrow pits must be purchased, the economical limit of haul would be greater.

To determine the cost of the grading shown in Fig 17-19, two horizontal lines are drawn. The length of one is 500 ft, the limit of free haul; the length of the other is 1100 ft, the assumed economical limit of haul. Although excavation and embankment balance between station 0 + 00 and station 14 + 00, it would not be economical to haul material that distance. It will be cheaper to borrow 900 yd^3 needed to make the fill from station 0 + 00 to station 2 + 30. The cost of this fill will be

$$900 \times \$0.24 = \$216.00$$

The 500-ft line intersects the mass diagram at station 5 + 40 and at station 10 + 40. When the excavation between the grade point at station 8 + 00 and station 10 + 40 is moved backward along the line, it will complete the fill from station 5 + 40 to station 8 + 00. The volume, as scaled from the diagram, is 1200 yd^3. Since no part of it is hauled more than 500 ft, the cost of this grading will be

$$1200 \times \$0.24 = \$288.00$$

A horizontal line 1100 ft long extends from station 2 + 30 to station 13 + 30. The volume from station 10 + 40 to station 13 + 30 is 3800 yd^3. This is the volume needed to complete the fill between station 2 + 30 and station 5 + 40. All of this material is hauled more than 500 ft. The positions of the centers of gravity of the masses of excavation and embankment can be found by drawing a horizontal line midway—vertically—between the 1100- and 500-ft horizontal lines. This line intersects the mass diagram at station 3 + 70, the center of gravity of the embankment, and at station 12 + 30, the center of gravity of the excavation. The average haul is the difference between these two stations, or 860 ft. Since the charge for overhaul is only for the distance in excess of 500 ft, the average overhaul distance is 860 – 500 = 360 ft. The cost of excavating and moving the material between station 10 + 40 and station 13 + 30 will be excavation and 500-ft haul $\qquad 3800 \times \$0.24 = \$\ 912.00$

overhaul—3.6 stations $\qquad 3800 \times 3.6 \times \ 0.04 = \underline{\quad 547.20}$

$$\text{total cost} = \$1459.20$$

17-21 Economical Handling of Material

When there are frequent grade points on the profile, some study may be necessary to decide whether a given excavation should be moved forward or back along the line, or perhaps wasted. The mass diagram facilitates the choice of the plan requiring the least amount of work. Since the areas between horizontal balance lines and the mass diagram are the graphical products of volumes and distances, they are indicative of the amount of work involved in the grading. The smaller the areas, the less will be the work, and the cheaper the cost. Where the work may be done in two or more ways, the most economical way will be the one for which the areas between the mass diagram and the horizontal lines are a minimum.

These facts will be apparent from an investigation of the profile and mass diagram shown in Fig. 17-20. Although excavation and embankment balance between station 40 + 00 and station 70 + 00, there must be some waste and borrow if the economical limit of haul is not to be exceeded. If the horizontal line ab is lowered, some material would be hauled more than 2500 ft. Consequently the line cannot be below ab. With the line in this position, there will be a volume of waste represented by the ordinate Aa. However, the total amount of waste will not be increased if the line ab is raised since there is to be additional waste beyond b and any increase to the left of a will be compensated for by an equal decrease beyond b. The most economical position for the line will be efg, where $ef = fg$. With the line in this position, there is no increase in the total amount of waste, but the labor involved is a minimum. The area $ehf + fgb$ is a measure of the work required to make the corresponding excavations and embankments. Should the line be lowered to the position jkl, the corresponding area is $jhk + klb$. This area is greater than the former since the approximate trapezoid that has been added, or $efkj$, is greater than the one taken away, or $fglk$.

FIGURE 17-20 Economic hauling of materials.

Similarly, if the line *efg* is raised, the area taken away from *ehf* will be less than the area added to *fgb,* and the resulting area will be greater. Consequently the labor involved will be a minimum when *ef = fg.*

Beyond station 50 + 00, the horizontal line cannot be below *cd* without exceeding the economical limit of haul. For this line the most economical position is *mnop,* where *mn + op − no* equals the economical limit of haul. By projecting these points onto the profile, it will be evident which parts of the excavations are to be moved forward and which backward in making the embankments.

PROBLEMS

17-1. Plot the following preliminary cross-section notes to a scale of 1 in. = 20 ft.

$\frac{474.8}{86.5}$	$\frac{480.2}{44.1}$	$\frac{486.6}{19.6}$	$\frac{479.8}{0}$	$\frac{480.8}{13.2}$	$\frac{480.0}{32.2}$	$\frac{486.2}{46.6}$	$\frac{487.3}{65.0}$	$\frac{481.8}{91.0}$

Assume a 50-ft roadbed at grade elevation of 460.00 and side slopes of 2 : 1. The road-bed center line coincides with the preliminary center line at this station. Plot the road-bed cross section, and scale the distances from the center line to the catch points and their elevations. Compute the cross-sectional area by the method of coordinates.

17-2. Change the grade elevation in Problem 17-1 to 463.80 ft and shift the roadbed 10 ft to the right. Then compute the cross-sectional area.

17-3. Change the grade elevation in Problem 17-1 to 456.00 ft and shift the roadbed 15 ft to the left. Then compute the cross sectional area.

17-4. Plot the following preliminary cross-section notes to a scale of 1 in. = 20 ft.

$\frac{217.2}{70.5}$	$\frac{210.4}{55.2}$	$\frac{211.0}{33.8}$	$\frac{215.3}{17.4}$	$\frac{212.8}{0}$	$\frac{208.8}{20.0}$	$\frac{211.6}{34.6}$	$\frac{214.9}{60.3}$

Assume a 40-ft roadbed at grade elevation of 225.00 ft and side slopes of $1\frac{1}{2}$: 1. The roadbed center line coincides with the preliminary center line at this station. Plot the roadbed cross section, and scale the distances from the center line to the catch points and their elevations. Compute the cross-sectional area by the method of coordinates.

17-5. Change the grade elevation in Problem 17-4 to 226.80 ft and shift the roadbed 15 ft to the left. Then compute the cross-sectional area.

17-6. Change the grade elevation in Problem 17-4 to 222.00 ft and shift the roadbed 20 ft to the right. Then compute the cross-sectional area.

17-7. The center cut at a cross section to be slope-staked is 4.8 ft. The roadbed width is 50 ft and the side slopes are 2 : 1. A vertical angle (refer to Fig. 17-4) of +7° 12′ is measured to a tentative position of the slope stake, and a slope distance of 42.5 is measured. Is the tentative position too far out or too far in, and approximately by how much?

17-8. Change the center cut in Problem 17-7 to 7.2 ft and the vertical angle to +6° 10′, with a tentative slope distance of 54.7 ft.

17-9. Change the center cut in Problem 17-7 to 12.2 ft and the vertical angle to +4° 00′, with a tentative slope distance of 50.0 ft.

17-10. From the accompanying final cross-section notes, compute the total volume of cut and the total volume of fill between station 26 + 00 and station 31 + 00 by the average end-area method. The roadbed width is 30 ft in cut and 26 ft in fill, and the side slopes are $1\frac{1}{2}$: 1 in cut and 2 : 1 in fill.

Station	Ground Elevation	Grade Elevation	Cross Section		
			L	C	R
31 + 00	431.2	437.32	$\dfrac{F8.9}{30.8}$	$\dfrac{F6.1}{0}$	$\dfrac{F4.6}{22.2}$
30 + 00	435.9	438.32	$\dfrac{F5.1}{23.2}$	$\dfrac{F2.4}{0}$	$\dfrac{F2.1}{17.2}$
29 + 62	437.7	438.64	$\dfrac{F2.7}{18.4}$	$\dfrac{F0.9}{0}$	$\dfrac{F0.0}{13.0}$
29 + 00	439.3	439.32	$\dfrac{F1.5}{16.0}$	$\dfrac{C0.0}{0}$	$\dfrac{C2.5}{18.8}$
28 + 70	441.5	439.50	$\dfrac{C0.0}{15.0}$	$\dfrac{C2.0}{0}$	$\dfrac{C4.2}{21.3}$
28 + 00	443.8	440.32	$\dfrac{C2.1}{18.2}$	$\dfrac{C3.5}{0}$	$\dfrac{C6.6}{24.9}$
27 + 50	446.0	440.82	$\dfrac{C4.0}{21.0}$	$\dfrac{C5.5}{0}$	$\dfrac{C7.9}{26.8}$
27 + 00	446.6	441.32	$\dfrac{C4.0}{21.0}$	$\dfrac{C5.3}{0}$	$\dfrac{C7.9}{26.8}$
26 + 30	449.7	441.92	$\dfrac{C5.1}{22.6}$	$\dfrac{C7.8}{0}$	$\dfrac{C9.3}{29.0}$
26 + 00	450.4	442.32	$\dfrac{C5.9}{23.9}$	$\dfrac{C8.1}{0}$	$\dfrac{C11.4}{32.1}$

17-11. Compute the volume in cubic yards between stations 30 and 31 in Problem 17-10 by the prismoidal formula.

17-12. Compute the volume in cubic yards between stations 27 and 28 in Problem 17-10 by the prismoidal formula, using the cross-sectional area at station 27 + 50 as the area of the midsection.

17-13. Assuming that the freehaul distance is 500 ft, the cost of excavation is 30 cents per cubic yard, and the overhaul cost is 5 cents per hundred feet, compute the cost of grading in Fig. 17-19.

17-14. Given the following five-level sections, compute the volume of earthwork lying between them by (**a**) the average-end-area method, and (**b**) by the prismoidal formula. Compute the side slopes.

$$\frac{C0.8}{27.4} \quad \frac{C1.2}{25.0} \quad \frac{C4.6}{0} \quad \frac{C7.8}{25.0} \quad \frac{C9.1}{52.3}$$

$$\frac{C3.8}{36.4} \quad \frac{C3.2}{25.0} \quad \frac{C5.6}{0} \quad \frac{C6.2}{25.0} \quad \frac{C6.1}{43.3}$$

17-15. Given the following notes for three irregular sections, compute the volume of earthwork lying between station 37 and station 38 by (**a**) the average end-area method and (**b**) by the prismoidal formula. The roadbed width is 60 ft, and the side slopes are $1\frac{1}{2}$: 1.

Station 37 + 00 $\quad \dfrac{C4.6}{36.9} \quad \dfrac{C6.2}{28.1} \quad \dfrac{C5.9}{17.9} \quad \dfrac{C8.8}{7.1} \quad \dfrac{C7.1}{0} \quad \dfrac{C7.9}{14.1} \quad \dfrac{C12.7}{42.2} \quad \dfrac{C20.0}{60.0}$

Station 37 + 50 $\quad \dfrac{C5.6}{38.4} \quad \dfrac{C6.2}{31.2} \quad \dfrac{C14.4}{21.8} \quad \dfrac{C9.9}{0} \quad \dfrac{C9.7}{12.0} \quad \dfrac{C11.2}{29.0} \quad \dfrac{C5.4}{38.1}$

Station 38 + 00 $\quad \dfrac{C1.8}{32.7} \quad \dfrac{C3.2}{27.8} \quad \dfrac{C2.0}{21.0} \quad \dfrac{C4.1}{5.9} \quad \dfrac{C4.0}{0} \quad \dfrac{C8.8}{11.7} \quad \dfrac{C6.8}{31.2} \quad \dfrac{C10.2}{45.3}$

17-16. In Fig. 15-8 the area bounded by points $B2$, $B6$, $F6$, $F1$, $D1$, and $D2$ is to be brought to a level grade of 690.00 ft. The sides of the squares are 50 ft. Assuming vertical faces for all sides of the excavations, compute the volume in cubic yards to be removed, using Eq. (17-12).

17-17. Change the grade elevation of Problem 17-16 to 685 ft, and compute the earthwork volume in cubic yards.

17-18. In Fig. 15-8, the area bounded by points $C1$, $C3$, $E3$, $E5$, $F5$, $F4$, $G4$, and $G1$ are to be brought to a level grade of 690.00 ft. The sides of the squares are 30 ft. Assuming vertical faces for all sides of the excavation, compute the volume of cubic yards to be removed, using Eq. (17-12).

17-19. Change the grade elevation of Problem 17-18 to 685 and compute the earthwork volume in cubic yards.

BIBLIOGRAPHY

El-Taklawi, M. R., Rashad, M. Z., El Rakim, A. A., and El-Din, S. S. 1989. Earthwork calculations from digital terrain models. *Tech. Papers Fall Conv. Amer. Cong. Surv. & Mapp.* p. 7.

Meyer, C. F., and Gibson, D. W. 1980. *Route Surveying and Design.* New York: Harper & Row.

Nicholson, J. 1986. Automated earthwork design using HASP digital system. *Tech. Papers Am. Conv. Amer. Cong. Surv. & Mapp.* V. 2, p. 80.

Rashad, M. Z. 1986. Accuracy of earthwork calculations from multiquadric technique data. *Tech. Papers Ann. Conv. Amer. Cong. Surv. & Mapp.* V. 2, p. 27.

Skelton, R. R. 1949. *Route Surveys.* New York: McGraw-Hill.

Stipp, D. W. 1968. Trigonometric construction staking. *Surveying and Mapping* 28:437.

18 United States Public Land Surveys

18-1 Historical Background

The United States rectangular surveying system was devised with the object of marking on the ground and fixing for all time legal subdivisions for the purposes of description and disposal of the public domain under the general land laws of the United States. This system has been used in 30 states including Alaska. It has not been used in the older states along the Atlantic seaboard and in a few others where the lands were in private hands before the federal government came into being.

The rectangular system of survey of the public lands was inaugurated by a committee appointed by the Continental Congress. In 1784 this committee reported "An ordinance for ascertaining the mode of locating and disposing of lands in the western territory, and for other purposes therein mentioned." The ordinance, as finally passed on May 20, 1785, provided for townships 6 miles square, containing 36 sections 1 mile square. The first public surveys were made under the direction of the Geographer of the United States. The area surveyed now forms a part of the state of Ohio. In these initial surveys only the exterior lines of the townships were surveyed, but the plats were marked by subdivisions into sections 1 mile square, and mile corners were established on the township lines. The sections were numbered from 1 to 36, commencing with number 1 in the southeast corner of the township and closing with number 36 on the northwest corner thereof. By this method number 6 was in the northeast corner, and number 7 was west of, and adjacent to, number 1.

The act of congress approved May 18, 1796, provided for the appointment of a surveyor general and directed the survey of the lands northwest of the Ohio River and above the mouth of the Kentucky River. Under this law it was provided that "The sections shall be numbered, respectively, beginning with number one in the northeast section and proceeding west and east alternately through the township, with progressive numbers, till the thirty-sixth be completed." This method of numbering sections, shown in Fig. 18-1, is still in use.

6	5	4	3	2	1
7	8	9	10	11	12
18	17	16	15	14	13
19	20	21	22	23	24
30	29	28	27	26	25
31	32	33	34	35	36

FIGURE 18-1 Method of numbering sections.

Since that time the laws relating to the survey of the public lands have been amended several times. Until quite recently it was the practice of the government to award contracts for the survey of certain portions of the public lands. These awards were frequently made, in payment of political obligations, to persons wholly unfit to be entrusted with such work. As a result, much of the work has been carelessly done. Fraudulent returns have been made to the advantage of speculators in timber and mineral lands. In more than one instance the government has been furnished with the field notes of a survey executed in the comparative comfort of a tent rather than on the ground. In 1910, after most of the damage had been done, Congress abolished this contract system and authorized the interior department to employ a permanent body of surveyors, known as United States surveyors.

The interest of the present-day engineer or surveyor in the rectangular system is in the retracement of old lines and in the subdivision of the section into smaller units. For this reason it is imperative that he be familiar with the methods used in the original survey.

18-2 Manual of Surveying Instructions

Various regions of the United States have been surveyed under different sets of instructions issued at periods ranging from 1785 to the present time. The earliest instructions were issued to surveyors in manuscript or in printed circulars. More detailed regulations that improved the system for greater accuracy, permanency, and uniformity were issued in book form in editions of 1855, 1881, 1890, 1894, 1902, 1919, 1930, and 1947. The latest edition is the *Manual of Surveying Instructions, 1973*. The methods outlined in the following sections are taken from the edition of the *Manual of Instructions for the Survey of the Public Lands of the United States, 1947*, prepared and published under the direction of the Director of the Bureau of Land Management and printed by the U.S. Government Printing Office. These methods are the same as those outlined in the 1973 manual.

Although the methods have, in general, been the same on all surveys, there have been important variations in detail. The local engineer or surveyor, before beginning the retracement of old land lines, should familiarize himself with the exact methods in use at the time the original survey was made.

18-3   General Procedure

The first step in the survey of an area is to divide it into tracts approximately 24 miles square by means of meridians and parallels of latitude. These 24-mile tracts are then divided into 16 townships, which are approximately 6 miles on a side. The last step, so far as the federal government is concerned, is to divide the township into 36 sections, each approximately 1 mile square. The subdivision of the sections into smaller units is the task of the local engineer and surveyor.

18-4   Initial Point

The place of beginning of the survey of any given region is called the *initial point*. Through this point is run a meridian, called the *principal meridian,* and a parallel of latitude, called the *baseline.* The initial point is selected with a view to its control of extensive agricultural areas within reasonable geographical limitations. On the establishment of an initial point, the position of the point in latitude and longitude is determined by accurate field astronomical methods.

Since surveys in widely separated sections of the country have been in progress simultaneously, a large number of initial points have been established. The positions of these points and the areas governed by them are given in Table 18-1. Since many of the points were established before present-day instrumentation for accurate field astronomical determinations were available, some of the values shown are only approximately correct. Present instructions call for the establishment of the initial point in such a manner as to make it as nearly permanent as possible.

18-5   Principal Meridian

The principal meridian is an astronomical meridian that is extended from the initial point, either north or south, or in both directions, as the conditions may require, to the limits of the area being subdivided. Monuments are placed on this line at intervals of 40 chains ($\frac{1}{2}$ mile), and at its intersection with navigable bodies of water, streams 3 chains or more in width, and lakes of an area of 25 acres or more.

Two independent sets of measurements of this line are made. Should the difference between the two sets exceed 20 links (13.2 ft) per 80 chains, it is required that the line be remeasured to reduce the difference. If tests for alignment show the line to have deviated more than 3′ from the true cardinal course, the alignment must be corrected.

TABLE 18-1 Meridians and Baselines of the United States Rectangular Surveys

Meridian	Governing Surveys (Wholly or in Part) in the States of—	Longitude of Principal Meridian West From Greenwich	Latitude of Baseline North from Equator
Black Hills	South Dakota	104° 03′ 16″	43° 59′ 44″
Boise	Idaho	116° 23′ 35″	43° 22′ 21″
Chicasaw	Mississippi	89° 14′ 47″	35° 01′ 58″
Choctaw	Mississippi	90° 14′ 41″	31° 52′ 32″
Cimarron	Oklahoma	103° 00′ 07″	36° 30′ 05″
Copper River	Alaska	145° 18′ 37″	61° 49′ 04″
Fairbanks	Alaska	147° 38′ 26″	64° 51′ 50″
Fifth Principal	Arkansas, Iowa, Minnesota, Missouri, North Dakota, and South Dakota	91° 03′ 07″	34° 38′ 45″
First Principal	Ohio and Indiana	84° 48′ 11″	40° 59′ 22″
Fourth Principal	Illinois	90° 27′ 11″	40° 00′ 50″
Fourth Principal	Minnesota and Wisconsin	90° 25′ 37″	42° 30′ 27″
Gila and Salt River	Arizona	112° 18′ 19″	33° 22′ 38″
Humboldt	California	124° 07′ 10″	40° 25′ 02″
Huntsville	Alabama and Mississippi	86° 34′ 16″	34° 59′ 27″
Indian	Oklahoma	97° 14′ 49″	34° 29′ 32″
Kateel River	Alaska	158° 45′ 31″	65° 26′ 17″
Louisiana	Louisiana	92° 24′ 55″	31° 00′ 31″
Michigan	Michigan and Ohio	84° 21′ 53″	42° 25′ 28″
Mount Diablo	California and Nevada	121° 54′ 47″	37° 52′ 54″
Navajo	Arizona	108° 31′ 59″	35° 44′ 56″
New Mexico Principal	Colorado and New Mexico	106° 53′ 12″	34° 15′ 35″
Principal	Montana	111° 39′ 33″	45° 47′ 13″
Salt Lake	Utah	111° 53′ 27″	40° 46′ 11″
San Bernardino	California	116° 55′ 48″	34° 07′ 13″
Second Principal	Illinois and Indiana	86° 27′ 21″	38° 28′ 14″
Seward	Alaska	149° 21′ 26″	60° 07′ 37″
Sixth Principal	Colorado, Kansas, Nebraska, South Dakota, and Wyoming	97° 22′ 08″	40° 00′ 07″
St. Helena	Louisiana	91° 09′ 36″	30° 59′ 56″
St. Stephens	Alabama and Mississippi	88° 01′ 20″	30° 59′ 51″
Tallahassee	Florida and Alabama	84° 16′ 38″	30° 26′ 03″
Third Principal	Illinois	89° 08′ 54″	38° 28′ 27″
Uintah	Utah	109° 56′ 06″	40° 25′ 59″
Umiat	Alaska	152° 00′ 05″	69° 23′ 30″
Ute	Colorado	108° 31′ 59″	39° 06′ 23″
Washington	Mississippi	91° 09′ 36″	30° 59′ 56″
Willamette	Oregon and Washington	122° 44′ 34″	45° 31′ 11″
Wind River	Wyoming	108° 48′ 49″	43° 00′ 41″

It should be borne in mind that the only equipment available for most of the early surveys was a compass and a chain. Therefore discrepancies far in excess of these limits will be found in rerunning many old lines. In other cases the agreement will be surprisingly close, when one considers that much of the country was often heavily wooded at the time the surveys were made.

18-6   Baseline

From the initial point the baseline is extended east and west on a true parallel of latitude to the limits of the area being surveyed. Monuments are placed on this line at intervals of 40 chains and at its intersection with all meanderable bodies of water. The manner of making the measurement of the baseline and the accuracy of both the alignment and the measurement must be the same as that required in the survey of the principal meridian.

Because of the convergence of meridians, a line making an angle of 90° with the principal meridian will, theoretically, be an east-and-west line for only an infinitesimal distance from the meridian. If extended, the resulting line is an arc of a great circle and gradually departs southerly from the true parallel, which is a small circle. On this account the baseline, being a parallel of latitude, is run as a curve with chords 40 chains in length. Three methods are employed in running the baseline: the solar method, the tangent method, and the secant method.

The solar method is based on the fact that a line making a right angle with the meridian does not differ appreciably from the theoretical parallel of latitude in a distance of 40 chains. If a transit, equipped with a solar attachment and in good adjustment, is employed, the true meridian may be determined by observation at the end of each half-mile interval. At each point an angle of 90° is turned from the new meridian, and this line is extended 40 chains to a point at which another meridian is determined. Thus the direction of the line beyond each half-mile monument is determined by turning off a right angle from the meridian at that monument.

18-7   Baseline by Tangent Method

In the tangent method of locating a parallel of latitude, the direction of the tangent is determined by turning off a horizontal angle of 90° to the east or the west from the meridian. Points on the true parallel are established at half-mile intervals by offsets to the north from this tangent. The establishment of a baseline in latitude 42° 30′ N is illustrated in Fig. 18-2.

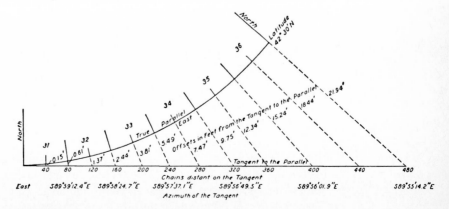

FIGURE 18-2 Baseline by offsets from tangent.

TABLE 18-2   Offsets (In Feet) from Tangent to Parallel

Latitude	1 Mile	2 Miles	3 Miles	4 Miles	5 Miles	6 Miles
30°	0.38	1.54	3.46	6.15	9.61	13.83
31°	0.40	1.60	3.60	6.40	10.00	14.40
32°	0.42	1.66	3.74	6.65	10.40	14.97
33°	0.43	1.73	3.89	6.91	10.80	15.56
34°	0.45	1.80	4.04	7.18	11.22	16.16
35°	0.47	1.86	4.19	7.45	11.65	16.77
36°	0.48	1.93	4.35	7.73	12.09	17.40
37°	0.50	2.01	4.51	8.02	12.53	18.05
38°	0.52	2.08	4.68	8.32	12.99	18.71
39°	0.54	2.15	4.85	8.62	13.47	19.39
40°	0.56	2.23	5.02	8.93	13.95	20.09
41°	0.58	2.31	5.20	9.25	14.46	20.82
42°	0.60	2.40	5.39	9.58	14.97	21.56
43°	0.62	2.48	5.58	9.92	15.50	22.33
44°	0.64	2.57	5.78	10.28	16.06	23.12
45°	0.66	2.66	5.98	10.64	16.62	23.94
46°	0.69	2.75	6.20	11.02	17.21	24.79
47°	0.71	2.85	6.42	11.41	17.83	25.67
48°	0.74	2.95	6.65	11.81	18.46	26.58
49°	0.76	3.06	6.88	12.24	19.12	27.53
50°	0.79	3.17	7.13	12.68	19.81	28.52

The values of the offsets, which depend on the latitude and are proportional to the squares of the distances from the starting meridian, are shown in Table 18-2. Also the values of the azimuth of the tangent from the south point at the end of each mile are shown in Table 18-3.

The field instructions require the location of all landmarks with respect to measurements along the true line and also require the blazing of trees along the true line, therefore this method is not so convenient as the secant method, particularly when the country is so heavily wooded as to require the clearing of two lines, one for the tangent and another for the parallel.

18-8   Baseline by Secant Method

As shown in Fig. 18-3 the secant used in locating a parallel of latitude is the one passing through the 1- and 5-mile points on the parallel. The secant is located by offsetting to the south from the initial point and turning off a known angle from the meridian at the offset point. The secant is extended 6 miles, points on the parallel being located at half-mile intervals by offsets from the secant. The offsets at the $\frac{1}{2}$- and $5\frac{1}{2}$-mile points are to the north. The 1- and 5-mile points are on the secant, and the remaining points are south of the secant. The offsets and the values of the azimuth of the secant from the north at half-mile intervals up to 3 miles are shown in Table 18-4. For the other 3 miles the offsets are the same as for points at corresponding distances from the 3-mile point; and the azimuths are numerically equal to those at corresponding distances from the 3-mile point but are measured from the south.

TABLE 18-3 Azimuths of the Tangent to the Parallel

Latitude	1 Mile	2 Miles	3 Miles	4 Miles	5 Miles	6 Miles
30°	89° 59′ 30.0″	89° 59′ 00.0″	89° 58′ 29.9″	89° 57′ 59.9″	89° 57′ 29.9″	89° 56′ 59.9″
31°	59′ 28.8″	58′ 57.5″	58′ 26.3″	57′ 55.0″	57′ 23.8″	56′ 52.6″
32°	59′ 27.5″	58′ 55.0″	58′ 22.5″	57′ 50.0″	57′ 17.5″	56′ 45.1″
33°	59′ 26.2″	58′ 52.5″	58′ 18.7″	57′ 45.0″	57′ 11.2″	56′ 37.4″
34°	59′ 24.9″	58′ 49.9″	58′ 14.8″	57′ 39.7″	57′ 04.6″	56′ 29.6″
35°	59′ 23.6″	58′ 47.2″	58′ 10.8″	57′ 34.4″	56′ 58.0″	56′ 21.6″
36°	59′ 22.2″	58′ 44.5″	58′ 06.7″	57′ 28.9″	56′ 51.1″	56′ 13.4″
37°	59′ 20.8″	58′ 41.7″	58′ 02.5″	57′ 23.3″	56′ 44.1″	56′ 05.0″
38°	59′ 19.4″	58′ 38.8″	57′ 58.2″	57′ 17.6″	56′ 36.9″	55′ 56.3″
39°	59′ 17.9″	58′ 35.8″	57′ 53.7″	57′ 11.6″	56′ 29.5″	55′ 47.5″
40°	59′ 16.4″	58′ 32.8″	57′ 49.2″	57′ 05.6″	56′ 21.9″	55′ 38.3″
41°	59′ 14.8″	58′ 29.6″	57′ 44.5″	56′ 59.3″	56′ 14.1″	55′ 28.9″
42°	59′ 13.2″	58′ 26.4″	57′ 39.6″	56′ 52.8″	56′ 06.0″	55′ 19.3″
43°	59′ 11.5″	58′ 23.1″	57′ 34.6″	56′ 46.2″	55′ 57.7″	55′ 09.2″
44°	59′ 09.8″	58′ 19.6″	57′ 29.5″	56′ 39.3″	55′ 49.1″	54′ 58.9″
45°	59′ 08.0″	58′ 16.1″	57′ 24.1″	56′ 32.2″	55′ 40.2″	54′ 48.2″
46°	59′ 06.2″	58′ 12.4″	57′ 18.6″	56′ 24.7″	55′ 31.0″	54′ 37.2″
47°	59′ 04.3″	58′ 08.6″	57′ 12.9″	56′ 17.2″	55′ 21.4″	54′ 25.7″
48°	59′ 02.3″	58′ 04.6″	57′ 06.9″	56′ 09.2″	55′ 11.5″	54′ 13.9″
49°	59′ 00.2″	58′ 00.5″	57′ 00.7″	56′ 01.0″	55′ 01.2″	54′ 01.4″
50°	58′ 58.1″	57′ 56.2″	56′ 54.3″	55′ 52.4″	54′ 50.5″	53′ 48.6″

The advantage of the secant method lies in the fact that the offsets are much smaller than those for the tangent method. By clearing a line of moderate width, both the secant and the parallel will be contained in the same clearing. Measurements to landmarks will be substantially the same along both the secant and the parallel.

18-9 Standard Parallels

The next step in the subdivision of the district being surveyed is to run the *standard parallels* or *correction lines*. These lines are parallels of latitude that are established in exactly the same manner as the baseline. They are located at intervals of 24 miles north and south of the baseline and extend to the limits of the district being surveyed.

FIGURE 18-3 Baseline by offsets from secant.

TABLE 18-4 Azimuths of the Secant, and Offsets (in Feet) to the Parallel

Latitude	0 Miles	$\frac{1}{2}$ Mile	1 Mile	$1\frac{1}{2}$ Miles	2 Miles	$2\frac{1}{2}$ Miles	3 Miles
30°	89° 58′ 30″ 1.92 N	89° 58′ 45″ 0.86 N	89° 59′ 00″ 0.00	89° 59′ 15″ 0.68 S	89° 59′ 30″ 1.16 S	89° 59′ 45″ 1.44 S	90° (E or W) 1.54 S
31°	89° 58′ 26″ 2.00 N	89° 58′ 42″ 0.90 N	89° 58′ 58″ 0.00	89° 59′ 13″ 0.70 S	89° 59′ 29″ 1.20 S	89° 59′ 44″ 1.50 S	90° (E or W) 1.60 S
32°	89° 58′ 23″ 2.08 N	89° 58′ 39″ 0.94 N	89° 58′ 55″ 0.00	89° 59′ 11″ 0.72 S	89° 59′ 28″ 1.24 S	89° 59′ 44″ 1.56 S	90° (E or W) 1.66 S
33°	89° 58′ 19″ 2.16 N	89° 58′ 36″ 0.97 N	89° 58′ 53″ 0.00	89° 59′ 09″ 0.76 S	89° 59′ 26″ 1.30 S	89° 59′ 43″ 1.62 S	90° (E or W) 1.73 S
34°	89° 58′ 15″ 2.24 N	89° 58′ 32″ 1.01 N	89° 58′ 50″ 0.00	89° 59′ 07″ 0.79 S	89° 59′ 25″ 1.35 S	89° 59′ 42″ 1.69 S	90° (E or W) 1.80 S
35°	89° 58′ 11″ 2.32 N	89° 58′ 29″ 1.05 N	89° 58′ 47″ 0.00	89° 59′ 05″ 0.81 S	89° 59′ 24″ 1.39 S	89° 59′ 42″ 1.74 S	90° (E or W) 1.86 S
36°	89° 58′ 07″ 2.42 N	89° 58′ 26″ 1.09 N	89° 58′ 45″ 0.00	89° 59′ 03″ 0.84 S	89° 59′ 22″ 1.45 S	89° 59′ 41″ 1.81 S	90° (E or W) 1.93 S
37°	89° 58′ 03″ 2.50 N	89° 58′ 22″ 1.13 N	89° 58′ 42″ 0.00	89° 59′ 01″ 0.88 S	89° 59′ 21″ 1.51 S	89° 59′ 40″ 1.88 S	90° (E or W) 2.01 S
38°	89° 57′ 58″ 2.60 N	89° 58′ 19″ 1.17 N	89° 58′ 39″ 0.00	89° 58′ 59″ 0.91 S	89° 59′ 19″ 1.56 S	89° 59′ 40″ 1.95 S	90° (E or W) 2.08 S
39°	89° 57′ 54″ 2.70 N	89° 58′ 15″ 1.21 N	89° 58′ 36″ 0.00	89° 58′ 57″ 0.94 S	89° 59′ 18″ 1.61 S	89° 59′ 39″ 2.02 S	90° (E or W) 2.15 S
40°	89° 57′ 49″ 2.79 N	89° 58′ 11″ 1.26 N	89° 58′ 33″ 0.00	89° 58′ 55″ 0.97 S	89° 59′ 16″ 1.67 S	89° 59′ 38″ 2.09 S	90° (E or W) 2.23 S
41°	89° 57′ 45″ 2.89 N	89° 58′ 07″ 1.30 N	89° 58′ 30″ 0.00	89° 58′ 52″ 1.01 S	89° 59′ 15″ 1.73 S	89° 59′ 37″ 2.17 S	90° (E or W) 2.31 S
42°	89° 57′ 40″ 2.99 N	89° 58′ 03″ 1.35 N	89° 58′ 26″ 0.00	89° 58′ 50″ 1.05 S	89° 59′ 13″ 1.80 S	89° 59′ 37″ 2.25 S	90° (E or W) 2.40 S
43°	89° 57′ 35″ 3.10 N	89° 57′ 59″ 1.40 N	89° 58′ 23″ 0.00	89° 58′ 47″ 1.08 S	89° 59′ 12″ 1.86 S	89° 59′ 36″ 2.32 S	90° (E or W) 2.48 S
44°	89° 57′ 30 3.21 N	89° 57′ 55″ 1.45 N	89° 58′ 20″ 0.00	89° 58′ 45″ 1.12 S	89° 59′ 10″ 1.93 S	89° 59′ 35″ 2.41 S	90° (E or W) 2.57 S
45°	89° 57′ 24″ 3.32 N	89° 57′ 50″ 1.50 N	89° 58′ 16″ 0.00	89° 58′ 42″ 1.16 S	89° 59′ 08″ 2.00 S	89° 59′ 34″ 2.49 S	90° (E or W) 2.66 S
46°	89° 57′ 19″ 3.45 N	89° 57′ 46″ 1.55 N	89° 58′ 12″ 0.00	89° 58′ 39″ 1.20 S	89° 59′ 06″ 2.06 S	89° 59′ 33″ 2.58 S	90° (E or W) 2.75 S
47°	89° 57′ 13″ 3.57 N	89° 57′ 41″ 1.61 N	89° 58′ 09″ 0.00	89° 58′ 36″ 1.24 S	89° 59′ 04 2.14 S	89° 59′ 32″ 2.67 S	90° (E or W) 2.85 S
48°	89° 57′ 07″ 3.70 N	89° 57′ 36″ 1.66 N	89° 58′ 05″ 0.00	89° 58′ 33″ 1.29 S	89° 59′ 02″ 2.21 S	89° 59′ 31″ 2.77 S	90° (E or W) 2.95 S
49°	89° 57′ 01″ 3.82 N	89° 57′ 31″ 1.72 N	89° 58′ 01″ 0.00	89° 58′ 30″ 1.34 S	89° 59′ 00″ 2.30 S	89° 59′ 30″ 2.87 S	90° (E or W) 3.06 S
50°	89° 56′ 54″ 3.96 N	89° 57′ 25″ 1.79 N	89° 57′ 56″ 0.00	89° 58′ 27″ 1.38 S	89° 58′ 58″ 2.38 S	89° 59′ 29″ 2.97 S	90° (E or W) 3.17 S

These standard parallels are numbered, as shown in Fig. 18-4, as the first (second, third, and so on) standard parallel north (or south). In the earlier surveys standard parallels have been placed at intervals of 30 or 36 miles north and south of the baseline. In connecting new work with old work, it has been necessary to establish intermediate correction lines which have been given local names, as "Fifth Auxiliary Standard Parallel North" or "Cedar Creek Correction Line."

18-10 Guide Meridians

The survey district is next divided into tracts approximately 24 miles square by means of guide meridians. These lines are true meridians that start at points on the baseline or standard parallels at intervals of 24 miles east and west of the principal meridian and extend north to their intersection with the next standard parallel, as shown in Fig. 18-4. Because of the convergence of the meridians, the distance between these lines will be 24 miles only at the starting points. At all

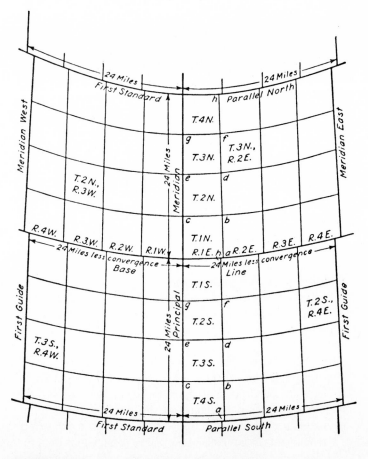

FIGURE 18-4 Standard parallels and guide meridians.

other points the distance between them will be less than 24 miles. Guide meridians are designated by number, as the first (second, third, and so on) guide meridian east (or west). As they extend from one standard parallel to the next, their lengths are limited to 24 miles (30 or 36 miles on older work). At their intersection with the standard parallel to the north, a new monument is established. On the standard parallels, two sets of monuments are thus to be found. The monuments that were established as the parallel was first located are called *standard corners* and apply to the area north of the parallel. The second set, found at the intersection of the parallel with the meridians from the south, are called *closing corners* and govern the area to the south of the parallel. The distance between the closing corner and the nearest standard corner should be measured and recorded in the field notes.

The guide meridians are located with the same care for securing accuracy of alignment and measurement as the principal meridian. The monuments are placed at half-mile intervals, except that all discrepancies in measurement are thrown into the last half mile. For this reason the distance from the closing corner to the first monument south of the standard parallel may be more or less than 40 chains.

18-11   Township Exteriors

The regular order of establishing the township exteriors can be followed by reference to Figure 18-4. Beginning at *a,* on the baseline (or standard parallel) and 6 miles east of the principal meridian (or guide meridian), a line is run due north for 6 miles, monuments being established at intervals of 40 chains. From the township corner *b,* a random line is run to the west to intersect the principal meridian (or a guide meridian) at *c.* As this line is run, temporary monuments are set at intervals of 40 chains. From the amount by which the random line fails to strike *c,* the direction of the true line between *c* and *b* can be calculated. The line is then run on this calculated course from *c* back to *b,* and the temporary monuments are replaced by permanent ones in proper position. The true line is blazed through timber, and distances to important items of topography are adjusted to the correct true-line measurements.

On account of the convergence of meridians and errors in field work, the northern boundary of the township will not be exactly 6 miles in length. *All discrepancies due to convergence and to errors in measurement or alignment are thrown into the most westerly half mile, all other distances being made exactly 40 chains.* Although every attempt is made to keep lines within 0° 14′ of the true courses, the boundaries of a township are not considered defective unless the error exceeds 0° 21′. If the random north boundary fails to close on the corner previously set by more than 3 chains for line or by more than 3 chains, minus the theoretical convergence, for distance, it is evident that the 0° 21′ limit has been exceeded, and the lines are retraced and the error is corrected.

The eastern boundary of the next township to the north is next run by extending the meridian from *b* to *d,* a distance of 6 miles, and locating permanent monuments at intervals of 40 chains. A random is then run westerly from *d* toward *e* and corrected back as the true line with monuments at intervals of 40 chains, measured from d, any discrepancy being thrown into the most westerly half mile. The next step is to run the line df, a random from f to g, and the true line from g to f. From f the meridian is extended to its intersection h with the baseline (or standard parallel), and a closing corner is established at the intersection. The distance from the closing corner to the nearest standard corner is measured and recorded in the field notes. Because of errors in measurement, the last half mile on the meridian may be more or less than 40 chains in length.

The two other meridians of the 24-mile block are run in a similar manner, random lines being run to the west at the township corners to connect with the corners previously set. From the township corners on the third meridian of the block, randoms are also run to the east to connect with the township corners on the guide meridian. As these lines are corrected back to the west, monuments are established at intervals of 40 chains, measured from the guide meridian, and the discrepancy is thus thrown into the most westerly half mile of the township, as before.

The instructions relative to the establishment of township exteriors in the case of irregular and partial surveys, as given in the *Manual of Surveying Instructions, 1947,* are as follows:

> As the remaining unsurveyed public lands are found to contain less and less extensive areas surveyable under the law, it becomes necessary to depart from the ideal procedure in order more directly to reach the areas authorized for survey. The many possible combinations are entirely too numerous to state in detail, but where an irregular order appears to be necessary such departure from the ideal order of survey will be specifically outlined in the written special instructions. Such departure should always be based on the principle of accomplishing, by whatever plan, the same relation of one township boundary to another as would have resulted from regular establishment under ideal conditions.
>
> In authorizing surveys to be executed, it will not usually be provided that exteriors are to be carried forward until the township is to be subdivided; thus, where causes operate to prevent the establishment of the boundaries in full, it is not imperative that the survey of the exterior lines be completed; under such conditions it may be found necessary to run section lines as offsets to township exteriors and such section lines will be run either on cardinal courses or parallel to the governing boundaries of such townships, or even established when subdividing, as existing conditions may require.

18-12   Numbering Townships

The townships of a survey district are numbered meridionally into ranges and latitudinally into tiers with respect to the principal meridian and the baseline of the district. As illustrated in Fig. 18-4, the third township south of the baseline is in tier 3 south. Instead of tier, the word *township* is more frequently used; thus any township in this particular tier is designated as "township 3 south." The fourth township west of the principal meridian is in range 4 west. By this method of numbering, any township is located if its tier, range, and principal meridian are given, as township 3 south, range 4 west, of the fifth principal meridian. This is abbreviated T3S, R4W, 5th PM.

18-13 Convergence of Meridians

The angular convergence between two meridians is dependent on the distance between the meridians, the latitude, and the dimensions of the earth. The convergence θ, in seconds of arc, can be expressed by the equation

$$\theta = \frac{s\sqrt{1 - e^2 \sin^2 \varphi}\, \tan \varphi}{a \text{ arc } 1''} \tag{18-1}$$

in which s is the distance between the meridians, e is the eccentricity of the spheroid, a is the semimajor axis of the spheroid, and φ is the latitude. When tables for computing geographic positions are available, the equation may be written as follows:

$$\theta = sA \tan \varphi \tag{18-2}$$

in which

$$A = \frac{\sqrt{1 - e^2 \sin^2 \varphi}}{a \text{ arc } 1''}$$

The value of A, for s in meters, may be taken from the tables.

The linear convergence on the parallel, for two meridians of length l and at a distance s apart, is given by the equation

$$c = slA \tan \varphi \sin 1'' \tag{18-3}$$

The values of the convergence for two meridians 6 miles long and 6 miles apart, for latitudes between 30° and 51°, are given in Table 18-5.

TABLE 18-5 Convergence of Meridians 6 Miles Long and 6 Miles Apart

	Convergence			Convergence	
Latitude	On the Parallel (ft)	Angle	Latitude	On the Parallel (ft)	Angle
30°	27.67	03′ 00.1″	41°	41.63	04′ 31.0″
31°	28.79	03′ 07.5″	42°	43.12	04′ 40.7″
32°	29.94	03′ 14.9″	43°	44.65	04′ 50.7″
33°	31.11	03′ 22.6″	44°	46.24	05′ 01.1″
34°	32.32	03′ 30.4″	45°	47.88	05′ 11.7″
35°	33.55	03′ 38.4″	46°	49.58	05′ 22.8″
36°	34.80	03′ 46.6″	47°	51.34	05′ 34.3″
37°	36.10	03′ 55.0″	48°	53.17	05′ 46.2″
38°	37.42	04′ 03.7″	49°	55.07	05′ 58.5″
39°	38.79	04′ 12.5″	50°	57.04	06′ 11.4″
40°	40.19	04′ 21.7″	51°	59.11	06′ 24.8″

18-14 Subdivision of Townships

In the work previously described, the exterior lines of the township are generally marked by monuments at half-mile intervals. However, because of errors in field work and the convergence of meridians, the most westerly and the most northerly half-miles of certain townships may be more or less than 40 chains in length. In subdividing the township into sections the aim is to secure as many sections as possible that will be 1 mile on a side. To accomplish this, the error due to convergence of meridians is thrown as far to the west as possible by running lines parallel to the east boundary of the township, rather than running them as true meridians. The earlier practice had been to run these lines as true meridians. Errors in linear measurements are thrown as far to the north as possible by locating monuments at intervals of 40 chains along the lines parallel to the east boundary of the township. Thus all the accumulated error falls in the most northerly half-mile, which may be more or less than 40 chains in length.

The order in which the lines are run to subdivide the township into sections is indicated by the numbers on the lines of the township shown in Fig. 18-5. The instructions for monumenting these lines, as given in the *Manual of Surveying Instructions, 1947,* are:

> The subdivisional survey will be commenced at the corner of sections 35 and 36, on the south boundary of the township, and the line between sections 35 and 36 will be run parallel to the east boundary of the township, or to the mean course thereof, if it is imperfect in alignment but within limits, establishing the quarter-section corner at 40 chains, and at 80 chains, the corner of sections 25, 26, 35, and 36. From the last-named corner, a random line will be run eastward, without blazing, parallel to the south

FIGURE 18-5 Subdivision of township.

boundary of section 36, to its intersection with the east boundary of the township, placing at 40 chains from the point of beginning, a post for temporary quarter-section corner. If the random line intersects said township boundary exactly at the corner of sections 25 and 36, it will be blazed back and established as the true line, the permanent quarter-section corner being established thereon, midway between the initial and terminal section corners. If the random intersects said township boundary to the north or south of said corner, the falling will be carefully measured, and from the data thus obtained, the true return course will be calculated, and the true line blazed and established, and the position of the quarter-section corner determined, as directed above. The meridional section line will be continued on the same plan, likewise the successive latitudinal sectional lines except that each random will be run parallel to the true south boundary of the section to which it belongs. After having established the west and north boundaries of section 12, the line between sections 1 and 2 will be projected northward, on a random line, parallel to the east boundary of the township, or to its mean course, as the case may be, setting a post for temporary quarter-section corner at 40 chains, to its intersection with the north boundary of the township. If the random intersects said north boundary exactly at the corner of sections 1 and 2, it will be blazed back and established as the true line, the quarter-section corner being established permanently in its original temporary position, and the fractional measurement thrown into that portion of the line between the permanent quarter-section corner and the north boundary of the township. If, however, said random intersects the north boundary of the township to the east or west of the corner of sections 1 and 2, the falling will be carefully measured, and from the data thus obtained the true course will be calculated and the true line established, the permanent quarter-section corner being placed upon the same at 40 chains from the initial corner of the random line, thereby throwing the fractional measurement in that portion lying between the quarter-section corner and the north boundary of the township. When the north boundary of a township is a baseline or standard parallel, the line between sections 1 and 2 will be run as a true line parallel to the east boundary of the township, or to its mean course, as the case may be; the quarter-section corner will be placed at 40 chains, and a closing corner will be established at the point of intersection with such base or standard line; and in such case, the distance from said closing corner, to the nearest standard corner off such base or standard line, will be carefully measured and noted.

The successive ranges of sections proceeding from east to west will be surveyed in the same manner; then after having established the west and north boundaries of section 32, a random line will be initiated at the corner of sections 29, 30, 31, and 32, which will be projected westward to the south boundary of the township, setting a temporary quarter-section corner at 40 chains to an intersection with the west boundary of the township, where the falling will be measured and the bearing of the true line calculated, whereupon the line between sections 30 and 31 will be permanently marked between the section corners, and the quarter-section corner thereon will be established at 40 chains from the east, thereby placing the fractional measurement in the west half mile as required by law. The survey of the west two ranges of sections will be continued on the same plan, and the random line between sections 6 and 7 will be run westward parallel to the true line between sections 7 and 18; the random will be corrected to a true line and the fractional measurement placed in the west half mile; finally the random line between sections 5 and 6 will be run northward parallel to the true line between sections 4 and 5; the random will be corrected to the true line and the fractional measurement placed in the north half mile.

In this manner the section lines are marked by monuments at half-mile intervals. The monuments on the east and west boundaries of a section are 40 chains apart, except for the fractional measurement placed in the most northerly half mile of the township. As the east and west boundaries are not exactly parallel and as there may be errors in chaining, the north and south boundaries of the section may not be exactly 80 chains. If the error does not exceed 50 links, the lines are allowed to stand, and the quarter-section corner is located midway between the two section corners except for the west range of sections, where the quarter-section corner is placed at a distance of 40 chains from the northeast corner of the section.

In addition to the section and quarter-section corners, meander corners are established at the intersection of the section lines with all navigable streams, with all streams 3 or more chains in width, and with lakes of the area of 25 acres and upward.

As noted in Section 18-2, some of the details of subdivision have been changed from time to time. According to the present instructions, double sets of corners are found only when the north boundary of the township is a baseline or a standard parallel. In some of the earlier surveys, closing corners are to be found on all four sides of a township. The first set of corners was established when the exterior boundaries of the township were located. Those monuments on the north boundary belong to sections lying to the north of the line. Those on the west boundary of the township belong to sections lying to the west of that line. The other set of corners was established at the time the township was subdivided. For instance, lines numbered 11, 22, 33, 44, and 60 in Fig. 18-5 were run to an intersection with the north boundary of the township, and closing corners were established at the intersections. In a similar manner lines numbered 47, 50, 53, 56, and 59 were extended to an intersection with the west line of the township, and a closing corner was established at each intersection. The local surveyor should acquaint himself with the regulations in effect at the time any particular survey was made.

18-15   Meandering

The purpose of meandering is to locate the mean high-water line of all important rivers and lakes to determine the areas of land that may be sold for agricultural purposes. The United States Supreme Court has given the principles governing the use and purpose of meandering shores in its decision in a noted case (R. R. Co. vs. Schurmeir, 7 Wallace, 286-287) as follows:

> Meander lines are run in surveying fractional portions of the public lands bordering on navigable rivers, not as boundaries of the tract, but for the purpose of defining the sinuosities of the banks of the stream, and as the means of ascertaining the quantity of land in the fraction subject to sale, which is to be paid for by the purchaser. In preparing the official plat from the field notes, the meander line is represented as the borderline of the stream, and shows to a demonstration that the watercourse, and not the meander line as actually run on the land, is the boundary.

Meander corners are set at every point where standard, township, or section lines intersect the bank of a navigable stream, any river 3 chains or more in width, and any lake with an area of 25 acres or more. The meander corners are connected by traverses, which follow the bank or shoreline. The true bearings and the lengths of the sides of the traverse are measured. Since the meander lines are not strict boundaries, the work is simplified by adjusting the locations of the stations on the meander line to give bearings to the exact quarter degree and lengths of whole chains, or multiples of 10 links, with odd links only in the final course.

The latitudes and departures of the meander courses should be computed before leaving the vicinity, and if misclosure is found, indicating an error in the field measurements, the lines should be rerun.

In the case of lakes that are entirely within the boundaries of a section, a quarter-section line is run, if one crosses the lake, to the margin of the lake, where a *special meander corner* will be established. If the lake is entirely within a quarter section, an *auxiliary meander corner* will be established at some suitable point on its margin and connected with a regular corner on the section boundary.

18-16   Marking Lines Between Corners

The marking of the survey lines on the ground was supposed to be done in such a manner as to perpetuate the lines. This was done by monumenting the regular corners, by recording in the field notes all natural topographic features, and, whenever living timber was encountered, by means of blazing and hack marks on the trees.

A *blaze* is an axe mark that is made on a tree trunk at about breast height, so that a flat scar is left on the tree surface. The bark and a small amount of the live wood tissue are removed, and the smooth surface that is exposed forever brands the tree. A sufficient number of trees standing within 50 links of the line, on either side of it, are blazed on two sides quartering toward the line, to render the line conspicuous and readily traceable in either direction. The blazes are placed opposite each other and so that they coincide in direction with the line where the trees stand very near it, but approach nearer to each other and toward the line the farther the line passes from the blazed trees.

A *hack* is also an axe mark that is made on a tree trunk at about breast height, but in this case a horizontal notch is cut into the surface of the tree. The notch is made V shaped. Two hacks are cut to distinguish those made in the survey from accidental marks resulting from other causes. All trees intersected by the line will have two hacks or notches resembling a double-V ($\lessgtr$) cut on each of the sides facing the line.

Unfortunately for the present-day surveyor, the monuments used on the early surveys were not permanent in character, and forest fires and settlers have destroyed most of the timber, so that the retracing of the old lines is often difficult and in many instances impossible.

18-17   Objects Noted in the Survey

The field notes and plat of a survey furnish not only a technical record of the procedure but also a report on the character of the land, soil, and timber traversed by the survey, and a detailed schedule of the topographical features along every line. It is necessary to include, in addition, accurate connections showing the relation of the rectangular surveys to other surveys, to natural objects, and to improvements. This information is often of considerable value in relocating obliterated lines.

According to the *Manual of Surveying Instructions, 1947,* the information that should appear in the field notes is as follows:

1. The precise course and length of every line run, noting all necessary offsets therefrom, with the reason for making them, and the method employed.
2. The kind and diameter of all bearing trees, with the course and distance of the same from their respective corners, and the markings; all bearing objects and marks thereon, if any; and the precise relative position of witness corners to the true corners.
3. The kind of material of which the corners are constructed, their dimensions and markings, depth set in the ground, and their accessories.
4. Trees on line. The name, diameter, and distance on line to all trees that intersect, and their markings.
5. Intersections by line of land objects. The distance at which the line intersects the boundary lines of every reservation, townsite, or private claim, noting the exact bearing of such boundary lines, and the precise distance to the nearest boundary corner; the center line of every railroad, canal, ditch, electric transmission line, or other right-of-way across public lands, noting the width of the right-of-way and the precise bearing of the center line; the change from one character of land to another, with the approximate bearing of the line of demarcation, and the estimated height in feet of the ascents and descents over the principal slopes typifying the topography of the country traversed, with the direction of said slopes; the distance to and the direction of the principal ridges, spurs, divides, rim rock, precipitous cliffs, and so on; the distance to where the line enters or leaves heavy or scattering timber, with the approximate bearing of the margin of all heavy timber, and the distance to where the line enters or leaves dense undergrowth.
6. Intersections by line of water objects. All unmeandered rivers, creeks, and smaller watercourses that the line crosses; the distance measured on the true line to the center of the same in the case of the smaller streams, and to both banks in the case of the larger streams, the course downstream at points of intersection, and their widths on line, if only the center is noted. All intermittent watercourses, such as ravines, gulches, arroyos, draws, dry drains, and so on.

7. The land's surface: whether level, rolling, broken, hilly or mountainous.

8. The soil: whether rocky, stony, gravelly, sandy, loam, clay, and so on, and also whether first, second, third, or fourth rate.

9. Timber: the several kinds of timber and undergrowth, in the order in which they predominate.

10. Bottom lands to be described as upland or swamp and overflowed, and the depth of overflow at seasonal periods to be noted. The segregation of lands fit for cultivation without artificial drainage, from the swamp and overflowed lands.

11. Springs of water, whether fresh, saline, or mineral, with the course of the stream flowing therefrom. The location of all streams, springs, or water-holes, which because of their environment may be deemed to be of value in connection with the utilization of public grazing lands.

12. Lakes and ponds, describing their banks, tributaries, and outlet, and whether the water is pure or stagnant, deep or shallow.

13. Improvements; towns and villages; post offices; Indian occupancy; houses or cabins, fields, or other improvements, with owner's name; mineral claims; mill sites; United States mineral monuments, and all other official monuments not belonging to the system of rectangular surveys will be located by bearing and distance or by intersecting bearings from given points.

14. Coal banks or beds, all ore bodies, with particular description of the same as to quality and extent; all mining surface improvements and underground workings; and salt licks.

15. Roads and trails, with their directions, whence and whither.

16. Rapids, cataracts, cascades, or falls of water, in their approximate position and estimated height of their fall in feet.

17. Stone quarries and ledges of rocks, with the kind of stone they afford.

18. Natural curiosities, petrifactions, fossils, organic remains, and so on, also all archaeological remains, such as cliff dwellings, mounds, fortifications, or objects of like nature.

19. The general average of the magnetic declination in the township, with the maximum known range of local attraction and other variations, will be stated in the general description, and the general average for the township, subject to local attraction, will be shown upon the plat.

In addition to the field notes, the surveyors are required to prepare, as the work progresses, an outline diagram showing the course and length of all established lines with connections, and a topographical sketch embracing all features usually shown on the completed official township plat. These maps will be made to scale and will be kept up with the progress of the field work. The interiors of the sections will be fully completed, and the topographical features will be sketched with care while in the view of the surveyor. These maps will then form the basis of the official plat, the ultimate purpose of which is a true and complete graphic representation of the public lands surveyed.

18-18   Specimen Field Notes

To give some ideas of the manner in which the field notes are recorded, the notes in Table 18-6 for 1 mile of the subdivision of T15N, R20E of the principal base and meridian of Montana have been taken from the manual of 1947. That portion of the notes covering the adjustment of the transit, and a comparison of the chain with the one used in the survey of the township exterior, has been omitted.

TABLE 18-6   T 15 N, R 20 E, Principal Base and Meridian of Montana[a]

Chains	
	Beginning the subdivisional survey at the cor. of secs. 1, 2, 35, and 36, on the S bdy. of the Tp., which is monumented with a sandstone 8 x 6 x 5 in. aboveground, firmly set, marked, and witnessed as described in the official record.
	N 0° 01′ W, bet. secs. 35 and 36.
	Over level bottom land.
20.00	Enter scattering timber.
29.30	SE cor. of field; leave scattering timber.
31.50	A cabin bears W, 6.00 chs. dist.
39.50	Enter State Highway No. 25, bears N along section line, and E.
40.00	Point for the $\frac{1}{4}$ sec. cor. of secs. 35 and 36.
	Bury a granite stone, 12 × 12 × 12 in. mkd. *X*, 2 ft underground, from which
	An iron post, 30 in. long, 2 in. diam., set 24 in. in the ground, for a reference monument, with brass cap mkd. with an arrow pointing to the cor. and $\frac{1}{4}$ S 36 RM, bears E 46 lks. dist.
	An iron post 30 in. long, 2 in. diam., set 24 in. in the ground, for a reference monument with brass cap mkd. with an arrow pointing to the cor. and $\frac{1}{4}$ S 35 RM, bears W 46 lks. dist.
50.50	NE cor. of field.
51.50	Highway turns to N 70° W.
57.50	Enter heavy timber and dense undergrowth, edge bears N 54° E and S 54° W.
72.00	Leave undergrowth.
80.00	Point for the cor. of secs. 25, 26, 35, and 36.
	Set an iron post, 30 in. long, 2 in. diam., 24 in. in the ground with brass cap mkd.

T 15 N	R 20 E
S 26	S 25
S 35	S 36

1945

from which

A green ash, 13 in. diam., bears N 22°E, 26 lks. dist. mkd. T 15 N R 20 E S 25 BT.
A green ash, 23 in. diam., bears S 71$\frac{1}{4}$°E, 37 lks. dist. mkd. T 15 N R 20 E S 36 BT.
A green ash, 17 in. diam., bears S 64°W, 41 lks. dist. mkd. T 15 N R 20 E S 35 BT.
A cottonwood, 13 in. diam., bears N 21$\frac{1}{4}$°W, 36 lks. dist. mkd. T 15 N R 20 E S 26 BT.
Land, level bottom; northern 20 chs. subject to overflow.
Soil, alluvial, silt, and loam.
Timber, green ash, and cottonwood; undergrowth, willow.

[a]*Manual of Instructions for the Survey of Public Lands of the United States of 1947.*

18-19   Corner Monuments

The law provides that the original corners established during the progress of the survey shall forever remain fixed in position, and that even evident errors in the execution of the survey must be disregarded where these errors were undetected prior to the sale of the lands. The original monuments thus assume extreme importance in the location of land boundaries. Unfortunately most of the public lands were surveyed before the present regulations relative to the character of monuments went into effect, and as a consequence most of the monuments used were of a very perishable nature. Their disappearance or destruction has rendered the relocation of old lines a very difficult task.

The present regulations call for the use of iron-pipe monuments, from 1 to 3 in. in diameter and about 3 ft long. One end of the pipe is split for a distance of about 4 or 5 in., and the two halves are spread to form flanges or foot plates. A brass cap is securely riveted to the opposite end of the pipe, and finally the pipe is filled with concrete. The 3-in. monuments are used for township corners, the 2-in. ones for section corners, and the 1-in. ones for quarter-section corners. The caps of the iron posts are suitably and plainly marked with steel dies at the time the posts are used. The posts are set with about three-fourths of their length in the ground, and earth and stone, if the latter is at hand, are tamped into the excavation to give the post a solid anchorage.

When the procedure has been duly authorized, the use of durable native stone may be substituted for the iron post. No stone can be used that is less than 20 in. in length, or less than 6 in. in either of its minor dimensions, or less than 1000 in.^{3} in volume. The stone is marked by letters chiseled on the faces. To aid in the identification of the sections to which it refers, grooves or notches are cut in the faces or edges. The number of these grooves or notches is a measure of the number of miles to the township line in that direction. Thus the corner common in sections 10, 11, 14, and 15 would have two notches on the east edge and four notches on the south edge.

Where a corner point falls upon solid surface rock and excavation is prevented, a cross (**X**) is cut at the exact corner point. If feasible, the monument is erected in the same position and is supported by a large stone mound of broad base, which is so well constructed that the monument will possess thorough stability.

Where the corner point falls exactly at the position occupied by a sound living tree, which is too large to be removed, the tree should be appropriately marked for the corner.

According to the earlier instructions, a corner monument can be a wooden post, a mound of earth under which was buried a charred stake or a quart of charcoal, or some other fairly permanent and distinguishable mark when nothing more permanent is available.

18-20   Witnesses

The corner monuments are witnessed by measuring the bearing and distance to natural or artificial objects in the immediate vicinity. Bearing trees, or other natural objects, are selected for marking when they are available within 5 chains of the corner monument. One tree or object is marked in each section cornering at the monument, when available, and the true course and horizontal distance from the exact corner point to the center of the tree is at its root crown, or to the cross (**X**) upon a marked object, are carefully determined and recorded with the description of the tree, or object, and its marks.

Trees that are used as witnesses are blazed, and the blazes are scribed with letters and figures to aid in identifying the location. Thus a tree used as a witness for a section corner might bear the inscription T20N R12E S24BT; this indicates that the tree stands in section 24, T20N, R12E, BT (standing for bearing tree). A tree used as a witness for a quarter-section corner is marked $\frac{1}{4}$ S 32BT when the tree is in section 32.

A cross (**X**) and the letters BO and the section number are chiseled into a bearing object if it is of rock formation, and the record is made sufficiently complete to enable another surveyor to determine where the marks will be found.

Where stone is available and the surface of the ground is favorable, a mound of stone is employed as a witness when a full quota of trees or other bearing objects are not available. The mound consists of not less than five stones and has a base not less than 2 ft wide and a minimum height of $1\frac{1}{2}$ ft. In stony ground the size of the mound is sufficiently increased to make it conspicuous. The nearest point on the base is about 6 in. distant from the monument, and the size and position of the mount are recorded in the notes.

On open prairies where no trees, rocks, or other natural objects are available, the corners may be witnessed by digging pits 18 in. square and 12 in. deep, with the nearest side 3 ft distant from the corner monument. The pits are oriented with a square side (and not a corner) toward the monument. The field notes should contain a description of the pits, including the size and position, as the regulations relative to this form of witness have been changed from those set forth in the earlier editions of the manual.

When it is impossible to utilize any of the types of witnesses just mentioned, a suitable memorial will be deposited at the base of the monument. A memorial may consist of any durable article that will serve to identify the location of the corner in case the monument is destroyed. Such articles as glassware, stoneware, a stone marked with a cross (**X**), a charred stake, a quart of charcoal, or pieces of metal constitute a suitable memorial. A full description of such articles is embodied in the field notes wherever they are so employed.

18-21   Filing of Field Notes and Plats

Copies of the field notes and plats of the public land surveys may be procured from the Eastern States Office, Bureau of Land Management, 7450 Boston Blvd., Springfield, VA 22153 (703) 440-1600, and from the Bureau of Land Management State offices in the following states (as of 1996):

Alaska State Office	Nevada State Office
Bureau of Land Management	Bureau of Land Management
222 W. 7th Ave. #13	850 Harvard Way
Anchorage, AK 99513	PO Box 12000
(907) 271-5960	Reno, NV 89502
	(702) 785-6500 or
	(702) 785-6505

Arizona State Office
Bureau of Land Management
3707 N. 7th St.
PO Box 16563
Phoenix, AZ 85011
(602) 650-0528

California State Office
Bureau of Land Management
2135 Butano Dr.
Sacramento, CA 95825
(916) 979-2800

Colorado State Office
Bureau of Land Management
2850 Youngfield St.
Lakewood, CO 80215
(303) 239-3600

Idaho State Office
Bureau of Land Management
3380 Americana Terrace
Boise, ID 83706
(208) 384-3192

Montana State Office
(N. Dakota, S. Dakota)
Bureau of Land Management
Granite Tower
PO Box 36800
222 N. 32nd St.
Billings, MT 59107
(406) 255-2885

New Mexico State Office
(Oklahoma, Kansas, Texas)
Bureau of Land Management
1474 Rodeo Rd.
PO Box 27115
Santa Fe, NM 87504-1449
(505) 438-7575

Oregon State Office
(Washington)
Bureau of Land Management
PO Box 2965
Portland, OR 97208
(503) 952-6001

Utah State Office
Bureau of Land Management
324 South State St., Suite 301
Salt Lake City, UT 84111-2303
(801) 539-4001

Wyoming State Office
Bureau of Land Management
2515 Warren Ave.
PO Box 1828
Cheyenne, WY 82003
(307) 775-6256

Nebraska
Newcastle RA
1101 Washington Blvd.
Newcastle, WY 82701-2972
(307) 746-4453

18-22   Subdivision of Sections

The function of the United States surveyor ends with the establishment of monuments at half-mile intervals on the external lines of the sections and the filing of the detailed field notes and a plat. No monuments are set by him within the section, except in the case of a meanderable lake entirely within a section. As the rectangular system provides for the disposal of the public lands in units of quarter-quarter sections of 40 acres, the function of the local surveyor begins when he is employed as an expert to identify the lands that have passed into private ownership. This may be a simple task or a most complicated one, the difficulty depending on the condition of the original monuments. The work of the local surveyor usually includes the subdivision of the section into the smaller units shown on the official plat. It is apparent that he cannot properly serve his client unless he is familiar with the legal requirements concerning the subdivision of sections. If any of the original monuments are missing, he needs to know not only the methods used in the original survey but also the principles that have been adopted by the courts before he can restore the missing corners legally.

After all the original monuments have been found or any missing ones have been replaced, the first step in the subdivision is the location of the center of the section. Regardless of the location of the section within the township, this point is always at the intersection of the line joining the east and west quarter-section corners, called the *east-and-west quarter line*, with the line joining the north and south quarter-section corners, called the *north-and-south quarter line*. By locating these lines on the ground, the section is divided into quarter sections containing approximately 160 acres each.

The method of dividing these quarter sections into the 40-acre parcels depends on the position of the section within the township. For any section except those along the north and west sides of the township, the subdivision is accomplished by bisecting each side of the quarter section and connecting the opposite points by straight lines. The intersection of these lines is the center of the quarter section.

The aim in subdividing the sections along the north and west sides of the township is to secure as many regular parcels as possible and to throw all irregularities caused by convergence or field work as far west and north as possible. The method of subdividing these sections is shown in Fig. 18-6. The corners on the north-and-south section lines are set at intervals of 20.00 chains, measured from the south, the discrepancy being thrown into the most northerly quarter mile. Similarly, the monuments on the east-and-west lines are set at intervals of 20.00 chains, measured from the east, the fractional measurement being thrown into the most westerly quarter mile. The fractional lots formed are numbered in a regular series progressively from east to west or from north to south in each section. As section 6 borders on both the north and the west boundaries of the township, the fractional lots are numbered commencing with number 1 in the northeast, thence progressively west to number 4 in the northwest, and south to number 7 in the southwest fractional quarter-quarter section.

It is the duty of the local surveyor to set these corners in the same positions they would have occupied had they been set in the original survey. For this reason distances shown as 20.00 chains indicate distances of 20 chains according to the chain used by the original surveyor, and not 1320.00 ft according to the tape of the local surveyor. It is therefore necessary to establish these points by proportionate measurement, the local surveyor determining the length of the chain, in terms of his tape, by measuring the distance between the two existing monuments. Thus if the measured distance from the north quarter-section corner of section 6 to the northwest corner is 2637.24 ft, and if the field notes show this distance to be 39.82 chains, the local surveyor will measure $20.00 \times 2637.24/39.82 = 1324.58$ ft westward from the quarter-section corner in establishing the quarter-quarter-section corner.

Where some corners fall within meanderable bodies of water, every effort is made to run the necessary division lines in the same positions they would occupy in a regular section. Thus, when a quarter-section corner for an interior

FIGURE 18-6 Subdivision of sections.

section could not be set, the quarter-section line is run from the opposite existing corner on a bearing that is the mean of the bearings of the two adjacent section lines. This same principle holds in the establishment of quarter-quarter-section lines.

18-23 Legal Descriptions

The rectangular system of subdivision provides a very convenient method of describing a piece of land that is to be conveyed by deed from one person to another. If the description is for a 40-acre parcel, the particular quarter of the quarter section is first given, then the quarter section in which the parcel is located, then the section number, followed by the township, range, and principal meridian. Thus the parcel of land adjacent to lots 3, 4, and 5 in section 6, Fig. 18-6, can be described as the SE $\frac{1}{4}$ NW $\frac{1}{4}$, Section 6, T8N, R12E of the Third Principal Meridian. Lot 1 in section 7 may also be described as the NW $\frac{1}{4}$ NW $\frac{1}{4}$, Section 7, T8N, R12E of the Third Principal Meridian. The legal descriptions of the larger parcels appear in Fig. 18-6.

18-24   Relocating Lost Corners

Nothing in the practice of surveying calls for more skill, persistence, and good judgment on the part of the surveyor than the relocation of a missing corner. In the original survey it was presumed that permanent monuments were being carefully established and witnessed, so that long lines of monuments would be perfect guides to the place of any one that chanced to be missing. Unfortunately the "monuments" were often nothing but green sticks driven into the ground, lines were carelessly run, monuments were inaccurately placed, witnesses were wanting in permanence, and recorded courses and lengths were incorrect. As the early settlers made little effort to perpetuate either the corner monuments or the witnesses, the task of the present-day surveyor in reconciling much conflicting evidence (or in even finding any kind of evidence) is often an extremely difficult task.

A *lost corner* is a point of a survey whose position cannot be determined, beyond reasonable doubt, either from original traces or from other reliable evidence relating to the position of the original monument, and whose restoration on the earth's surface can be accomplished only by means of a suitable surveying process with reference to interdependent existent corners.

A thorough search will often show that an apparently lost corner is really not lost. Where wooden stakes were used, a careful slicing of the earth may disclose a discoloration in the soil caused by the decaying of the stake. Roots or depressions in the soil may give an indication of the location of the original witness tress. Fence lines, the testimony of nearby residents, and the field notes of adjacent surveys may be of value in ascertaining the location of the corner. It must be remembered that no matter how erroneously the original work is done, lines and monuments that can be identified still govern the land boundaries. The surveyor has no power to "establish" corners. In any case of disputed lines, unless the parties concerned settle the controversy by agreement, the determination of the line is necessarily a judicial act, and it must proceed upon evidence. The surveyor serves as an expert witness in the court proceedings, presenting all evidence he has been able to discover relative to the proper location of the corner in question. As a disinterested surveyor of proper training should be much more competent than any court to decide on the proper location of any questionable line or corner, the surveyor should exert all his influence toward effecting a settlement by agreement and thus save his client the expense of court proceedings. If this is impossible, he should insist that proper monuments, which are consistent with the court decree, be established, and if possible, be made a part of the decree.

When it is certain that a corner is lost, it is replaced in a manner that is in accord with the methods used in its original location. Lost corners are established by either single or double proportionate measurement. The method of double proportionate measurement is generally applicable to the restoration of lost corners of four townships and of lost interior corners of four sections. Its application is illustrated in Fig. 18-7. In the lower enlarged sketch, E represents the location of the corner if only the existing corners A and B are considered, the point being

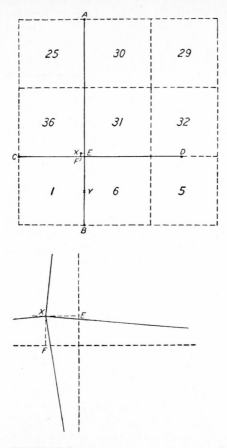

FIGURE 18-7 Restoring lost corner by double-proportionate measurement.

determined by proportionate measurement. If the known points C and D are used, a second location F is found. The restored corner is placed at X, the intersection of a west line through E with a north line through F. In this manner the distances from the replaced corner to the existing corners are consistent with the original field notes.

As a quarter-section corner was supposedly located on the straight line between the adjacent section corners, if there is no evidence on the ground to the contrary, a missing quarter-section corner is located on the straight line connecting the adjacent section corners. The distances from the section corners are determined by proportion, by comparing the present-day measurement between existing corners with the same distance as recorded in the field notes. Thus in Fig. 18-7 the quarter-section corner Y between sections 1 and 6 would be located on the line between B and X by a single proportionate measurement.

Before attempting work of this sort, the detailed rules and examples shown in the *Manual of Instructions for the Survey of the Public Lands of the United States,* from which much of the material in this chapter has been taken, should be carefully examined. Moreover, BLM has developed policy related to the positioning of public land survey system corners using GPS. These corners are now tied to the geodetic reference system and hence tied to the NAD 83.

BIBLIOGRAPHY

1991. *Surveying and Land Information Systems.* 51:4. 1991. Entire issue.

Bales, R. F. 1993. Center of section—center of controversy. *P.O.B.* 18:2.

Ball, W. E., Jr. 1986. Software for computing the geographic coordinates of Public Land Survey System corners. *Tech. Papers Ann. Conv. Amer. Cong. Surv. & Mapp.* V. 2, p. 50.

Brown, C. M. 1957. *Boundary Control and Legal Principles.* New York: Wiley.

——— and Eldridge, W. H. 1962 *Evidence and Procedures for Boundary Location.* New York: Wiley.

Buhler, D. A. 1987. GPS and the development of a national cadastre through cadastral surveys. Bureau of Land Management. *Tech. Papers Ann. Conv. Amer. Cong. Surv. & Mapp.* p. 162.

Bureau of Land Management, 1947 and 1973. *Manual of Instructions for the Survey of the Public Lands of the United States.* Washington, D.C.: U.S. Government Printing Office.

Cole, G. M. 1978. Florida's restoration of corners program. *Proc. Ann. Meet. Amer. Cong. Surv. & Mapp.* p. 28.

Crossfield, J. K. 1984. Evolution of the United States Public Land System. *Surveying and Mapping* 44:259.

Eickbush, F. D. 1993. The public land survey system—a timeless American asset. *Professional Surveyor* 13:6.

Eickbush, F. D. 1975. Alaskan surveys based on protraction diagrams. *Proc. Ann. Meet. Amer. Cong. Surv. & Mapp.* p. 59.

French, S. G. 1987. A plane survey method for dependent resurveys in the Missouri lead belt. *Tech. Papers Fall Conv. Amer. Cong. Surv. & Mapp.* p. 131.

Grimes, J. S. 1959. *Clark on Surveying and Boundaries.* Indianapolis: Bobbs-Merrill.

Keating, John B., Jr. 1993. The Geo-positioning selection guide for resource management. Bureau of Land Management, PO Box 1828, Cheyenne, WY 82003, BLM/SC/PT - 93/002 + 9160.

Knowles, D. R. 1982. Subdivision of the section as ruled by the courts (Missouri). *Surveying and Mapping* 42:41.

Lindsey, J. A. 1969. Relocating government corners on the state coordinate systems. *Surveying and Mapping* 29:401.

Mann, C. V. 1966. How shall we preserve the federal public land survey within Missouri? *Surveying and Mapping* 26:85.

McEntyre, J. G. 1981. Subdivision of a section. *Surveying and Mapping* 41:385.

—— 1982. *Modernization of the Public Land Survey System.* Washington, D.C.: National Academy Press.

Proportionate measurements—a panel presentation. *Surveying and Mapping* 28:261.

Restoration of Lost or Obliterated Corners and Subdivision of Sections. 1955. Washington, D.C.: U.S. Government Printing Office.

Schmidt, S. J. and Burns, T. B. 1992. Perpetuation of the U.S. Public Land Survey System in Illinois. *Surveying and Land Information Systems* 52:4.

Scruggs, R. 1991. Integrating GPS into BLM's cadastral survey program. *Tech. Papers Ann. Conv. Amer. Cong. Surv. & Mapp.* 1:225.

Von Meyer, N., and Mahun, G. 1988. Corner identification method for a public land survey system data base. *Tech. Papers Ann. Meet. Amer. Cong. Surv. & Mapp.* 1:105.

Wahl, J. W. 1991. Development of an electronic field book for cadastral retracement surveys. *Tech. Papers Ann. Conv. Amer. Cong. Surv. & Mapp.* 1:259.

Wattles, G. H. 1976. *Writing Legal Descriptions.* Orange, CA: Gurdon H. Wattles Publications.

Wattles, W. C. 1956. *Land Survey Descriptions.* Los Angeles: Title Insurance and Trust Company, revised by G. H. Wattles. 1974. P.O.B. 5702, Orange, CA 92667.

19 Municipal and Subdivision Surveys

19-1   Control Monuments and Maps

The value of a comprehensive city survey is hard to overestimate. All public agencies, utilities, land developers, engineers, architects, and others make constant use of the information furnished by city surveys. The money saved by the users of the information in planning, developing, engineering, and construction far exceeds the cost of the surveys. In fact, these are the primary reasons behind the growth of Geographic Information Systems, which are discussed in Chapter 20.

Information furnished by well-planned and well-executed city surveys consists of horizontal-control and vertical-control monuments, together with their coordinates or elevations or both, large-scale base maps, topographic maps, property maps, and overall city maps to smaller scale. Horizontal-control monuments are established by GPS or traverse of first- and second-order accuracy (see Tables 9-1 and 9-5). Vertical-control monuments, or benchmarks, must be established by first- or second-order lines of differential levels. Elevations derived from GPS observations are not sufficiently accurate for the basic vertical-control network of a city.

Base maps are compiled for a number of uses. In general they are used to form a common basis for topographic maps, property maps, subdivision maps, and maps showing the positions of utility lines. For the compilation of base maps, scales of 1/600, 1/1200, and 1/2400 are most satisfactory, the choice depending on the intensity of land use. Topographic maps should be published at a scale of 1/2400; property maps, at a scale of 1/600, and the overall city map at a scale no smaller than 1/24000.

19-2 Steps in a City Survey

Where it is possible to make a complete survey, the first step is to establish a network of GPS control. The network should be tied to existing National Geodetic Survey control stations, and should provide one control point for every 1 to 2 square miles throughout the metropolitan area. The next step is to run traverses connecting the GPS stations. These traverses should furnish the equivalent of second-order control for the base maps and property surveys, and of third-order control for the topographic work. Leveling of the same three orders should follow the lines of horizontal control.

From this basic control, base maps are prepared as described in Section 19-6, and topographic maps are compiled usually by photogrammetric methods supplemented by field methods, as discussed in Chapter 15. Much of the information obtained in the topographic survey facilitates the property surveys that are made to locate the monuments and boundaries of all public property, supplementary information being obtained by a search of all public records and by additional field work.

19-3 Triangulation for City Surveys

Although the principles governing the control survey for a city are the same as those given in Chapters 9 and 10, the work is modified somewhat because of the shorter lengths of the sides and because of the special conditions encountered. In planning the network, the tops of hills and of prominent buildings are used as stations, as far as possible, to keep the cost at a minimum. A line of sight that passes close to the ground along a hillside or bluff, or close to a building or chimney giving off heated air, is likely to produce uncertain results because of lateral refraction of the line of sight. Where such a sight is necessary, the observations should be made when the wind is blowing toward the object.

In the past, the control was developed by triangulation and was made up of either quadrilaterals or central-point figures; no single triangles were used. The diagram in Fig. 19-1 shows the main scheme first- and second-order triangulation covering the city of Oakland, California, and including neighboring smaller cities. This is an excellent example of a strong triangulation network that serves to control all subsequent municipal surveys in the area. The network is integrated directly into the triangulation network of the National Geodetic Survey, and the positions of all the stations are given in state plane coordinates.

Baselines for triangulation were measured with the EDMs discussed in Chapter 2. If EDM is to be used on a relatively short base typical of city surveys, the instrument must be capable of high resolution and small inherent error to maintain a higher relative accuracy. Because of the ease of measuring distances using EDM today, a mixture of traverse and GPS can help to build the network. Traverse angles are measured with 1-second direction instrument. To attain the required precision, four or eight positions with the direction instrument (the number depending on the accuracy of the instrument and the conditions surrounding the measurements) will generally be required. Because of the relatively short lengths of the sides, the signals and the instrument must be centered within 0.01 ft. The

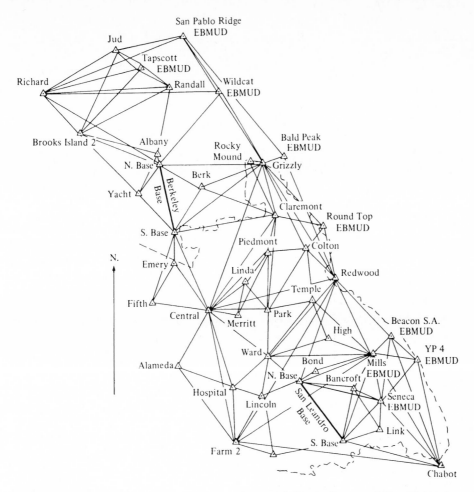

FIGURE 19-1 Triangulation diagram of the city of Oakland, California. Courtesy of City of Oakland.

instrument should be shielded from the wind during the observation period by means of a canvas shield when wind conditions are severe. Traverse observations and procedures are discussed in the following section.

The monuments should be of permanent character and carefully referenced. An iron pipe that is provided with a bronze cap and is set in a concrete post at least 12 in. in diameter makes a suitable monument if it extends below frost line and is so located as to minimize the chances of destruction by public or private construction. By using a bronze cap with a spherical top, the monument will serve also as a benchmark. The exact station mark can be designated by a small drill hole or an etched cross in the top of the cap. When the monument is set several inches below the surface of the ground, a small cast-iron box of the manhole type makes the monument more accessible (see Fig. 9-2).

To make the GPS stations usable to a maximum number of people, the computations should result in state plane coordinates.

For the use of local surveyors and engineers, an index map showing the locations of all control stations should be published. This map should be accompanied by a list of descriptions of the stations and their references, a tabulated list of elevations, coordinates, and the azimuths and lengths between stations.

19-4   Traverse for Control of City Surveys

The use of first-order traverse as the principal horizontal control in a city survey, independent of GPS, is advisable only for small areas. The principal use of first-order traverse is in establishing additional control points for originating second-order traverses, which control property surveys, and third-order traverses, which control the topographic surveys.

A first-order traverse begins at one triangulation or GPS station and closes on another such station. This practice tends to eliminate the effects of unknown systematic errors in measuring the traverse sides. The traverse lines are marked at intervals not greater than a mile by pairs of permanent intervisible monuments, 500 to 1000 ft apart. These marked lines provide positions for subsequent traversing of lower order. Set in pairs, the monuments provide beginning azimuths for subsequent traverses. Intermediate stations are semipermanent, are marked by iron pins or by lead or copper hubs in pavements or sidewalks, and are referenced for future recovery. The stations should be so located as to provide long clear sights free from horizontal refraction. Where a series of short sights is necessary, azimuths should be carried by means of cutoff lines, as discussed in Section 8-8, to maintain accuracy in the directions of the lines. Long narrow loop circuits should be avoided; connections subsequently made across the narrow dimension may show unallowable discrepancies, despite satisfactory closures obtained in the original circuit.

The average angular error, in seconds, should not exceed $2\sqrt{N}$, in which N is the number of sides or angles in the traverse, and the maximum error should not exceed $4\sqrt{N}$. The angles may be measured by four, six, or eight positions of a 1-second direction instrument, the number depending on the conditions surrounding the measurements. Triangulation lights or specially designed traverse lights can be used for signals in both day and night operations.

The traverse sides are measured by EDM, and the measured lengths are corrected for meteorological conditions and instrumental errors. Since the lines in the traverse are relatively short, slope reductions can be made by the methods outlined in Sections 2-11 and 2-12. The lines are then reduced to grid lengths for computing on the state coordinate system. After the azimuth errors are distributed, the average position closure should not exceed 1 : 50,000, and the maximum closure should not exceed 1 : 25,000.

The second-order traverse is used to furnish coordinates, bearings, and distances for street property monuments. The stations should coincide with the street monuments wherever possible. When a 1-second direction instrument is used to measure angles, two or four positions will suffice. The average angular closure, in seconds, should not exceed $6\sqrt{N}$, and the maximum closure should not exceed $8\sqrt{N}$. After the angular error is distributed, the average position closure should be 1 : 20,000 or better.

The diagram in Fig. 19-2 shows the traverse network covering the City of Alameda, California. The entire network is tied to five triangulation stations that appear in the lower-left part of Fig. 19-1. Nearly all the traverse lines are along the city streets (eliminated from the drawing for clarity), and the monuments are located at street corners. This is a flat-lying area of limited extent, and the traverse network is entirely adequate to control all subsequent surveys. The positions of the traverse stations are given by state plane coordinates.

The third-order traverse furnishes control for the topographic survey. Such a traverse should begin at a station of higher order and close on another station of higher order. Semipermanent monuments are used, since they can be replaced easily from the references or by retracing the survey. The average angular closure should not exceed 10″ per station. The angles can be measured with a 30-second engineer's transit or theodolite, each angle being measured once direct and once reversed. After the angular error is distributed, the position closure should not exceed 1 : 7500. The distances are most conveniently measured using EDM. This will assure a good linear closure at very little effort. If a tape is used, a spring balance and thermometer should be used to maintain the necessary closure accuracy.

Under favorable conditions, kinematic GPS surveys can be used to establish supplemental positions between higher-order triangulation and GPS stations.

The published data for the traverse stations should be the same as for the triangulation system, the order of the various stations being indicated. All rectangular coordinates should be referenced to the state plane coordinate system.

19-5   Leveling for City Surveys

One of the difficulties facing engineers, surveyors, and others who use city benchmarks to control their operations is the multiplicity of vertical datums in existence in one given city. This is troublesome when trying to tie a line of levels to two networks on different datums. A city should try to establish the elevations of all benchmarks with respect to a common level datum at the earliest possible time. This datum at present should preferably be the North American Vertical Datum of 1988 based on leveling operations of the National Geodetic Survey.

A first-order level net should be laid out along major streets and railroad lines if such a net has not already been established by the National Geodetic Survey. Traverse monuments should be incorporated into the net where possible. Additional benchmarks are set on bridge abutments or piers, masonry retaining walls, and masonry walls of buildings. These marks should be bronze tablets and should be carefully described and referenced for future use. The levels are run with a first-order level equipped with an optical micrometer, as discussed in Section 3-22. Only a single wire is read. Thus an auxiliary method must be used to determine the lengths of the sights to provide the proper weights in the least-squares adjustment of the level net (see Appendix A). These distances can be determined by stadia if the level is equipped with stadia hairs, or else they can be paced, scaled from a map, or obtained directly from values for the traverse that the lines happen to

FIGURE 19-2 Traverse diagram of the city of Alameda, California. Courtesy of City of Alameda.

follow. The levels are run in both directions over a section. The resulting differences of elevation, in feet, should agree within $0.017\sqrt{M}$, in which M is the length of the section in miles.

The second-order levels are run in closed circuits that begin and end on the same first-order benchmark, or are run from one such benchmark to another. These lines of levels follow the lines of second-order traverse, the traverse stations serving as benchmarks. Additional benchmarks are established in such places as on fire hydrants, curb catch basins, and crosses chiseled in concrete sidewalks. They are referenced by steel-tape measurements. The maximum closure, in feet, of any line or circuit should not exceed $0.035\sqrt{M}$, in which M is the length of the line or circuit in miles. Three-wire leveling or a tilting level equipped with an optical micrometer should be used.

The final elevations, adopted after a least-squares adjustment of the level net, should be published for the use of local engineers and surveyors. The order of each benchmark should be indicated, together with its complete description and references and the vertical datum with respect to which its elevation is established.

19-6   Base Maps

As previously stated, a base map is prepared to provide a common basis for other types of maps. It should show a maximum amount of basic information regarding horizontal and vertical control, city property, street center lines, boundary lines, and related features. Base maps are most useful at scales of 50, 100, and 200 ft to the inch (1/600, 1/1200, 1/2400). They should be indexed in series by means of the state plane coordinates of the lower-left corner of each sheet. A series at a scale of 1 in. = 50 ft with a working area of 20 in. in the north-south direction, and 30 in. in the east-west direction will have index X coordinates in multiples of 1500 ft and index Y coordinates in multiples of 1000 ft. A series at a scale of 1 in. = 200 ft on the same size sheet will have index coordinates in multiples of 6000 and 4000 ft in the X direction and Y direction, respectively.

The base-map manuscript should be prepared in pencil on dimensionally stable translucent material. The manuscript should contain grid lines for state plane coordinates at 10-in. intervals. GPS and traverse stations are plotted by coordinates, and their names and coordinate values are shown directly on the map. The positions of benchmarks, together with their elevations, should be indicated. The map should contain all construction-line notes, dimensions, and references; right-of-way lines; street center lines with bearings; all points of curvature, tangency, and intersections along the center lines; and curve radii and central angles. The corners of streets and angle points should be clearly defined. Any information that cannot otherwise be made clear at the map scale should be shown in enlarged detail on the margin of the sheet. Streets should be plotted and their widths noted on the map.

The information shown on the base map should be reliable and complete, so that it will not be necessary to consult original records. The manuscripts should be kept up-to-date as more survey information is obtained. Results of survey activities conducted by state, federal, and private agencies, as well as by public utilities, should be shown on the base maps when the reliability of such information is established.

Reproductions of the base maps should be made available to surveyors and engineers operating in the city, so that all activities are coordinated on one common data base.

19-7 Topographic Map of City

The topographic map should cover the built-up areas of the city and the development areas surrounding it. The map, to a scale of 1 in. = 200 ft and with a contour interval of 2 or 5 ft, should conform to the size of the base-map sheets. If the maps are to be published in color, the published sheets should show the topography by means of contours in brown. The drainage, comprising such features as streams, ponds, and lakes, is portrayed in blue. Black ink is used for the following features: structures, such as railroads, bridges, culverts, and curbs; benchmarks and other monuments; property lines, including street lines; boundaries of public property and political subdivisions; boundaries of recorded subdivisions; street names; names of parks, subdivisions, streets, and lakes; names of buildings; property dimensions; the height and character of buildings; grid lines or grid ticks; and lot numbers. Wooded areas and public property are shown in green.

The compilation of city topographic maps by means of aerial photogrammetry is recommended as superior to any other method of topographic mapping in accuracy, time, and cost. The control is obtained by second- and third-order traverse and levels. Photogrammetric compilation is supplemented by field mapping in areas where more information is necessary than can be obtained photogrammetrically.

Figure 19-3 shows a portion of Sheet G-18 of the Metropolitan Topographic Survey of the City of San Diego, California. The published scale is 1 in. = 200 ft, and the contour interval is 5 ft. Elevations are based on the City of San Diego Vertical Datum. Grid lines of the 1927 California Coordinate System, Zone VI, are shown at 5-in. intervals. This series of maps was produced by aerial photogrammetry supplemented by field-completion and field-editing surveys.

A final topographic map is judged to be sufficiently accurate if elevations determined from the contours on the map are correct within one-half the contour interval. The horizontal position of any well-defined point should be correct within 0.02 in. at the published scale. Errors in excess of these limits should not occur in more than 10% of the number of points tested. The mapping should be tested by running profile levels along traverse lines selected at random and then comparing these profiles with those obtained from the contours on the map.

Property lines located on the map are checked by assembling all available recorded plats of subdivisions. Distances between identified points and lines, such as monuments, fences, street lines, and block corners, are scaled from the manuscript sheets and compared with the corresponding distances shown on the recorded plats. Where there are many discrepancies, the manuscript sheet is checked in the field. When all identified lines have been checked, these lines are used as a basis for plotting the recorded information pertaining to the remaining lines and corners. The completed map should show street and alley lines, boundaries of recorded subdivisions, boundaries of public property, parks, playgrounds, schools, and other important areas.

FIGURE 19-3 Portion of sheet G18 of the topographic map series covering the city of San Diego, California. Original scale is 1 in. = 200 ft. Courtesy of City of San Diego.

19-8 City Property Survey

The city property survey marks by suitable monuments all corner, angle, and curve points of the street lines and establishes coordinate values for these and other critical points. When this is done, property monuments that become lost or obliterated can be conveniently and precisely reestablished from their known mathematical relations with other marked points. Thus each monument serves as a witness for every other monument.

The first aim of the property survey is the adequate location and recording of all street lines that constitute the boundaries of the public's property. The property survey tends to stabilize the boundaries of private property by furnishing an adequate control for the use of those engaged in property surveying. It is usually desirable to complete the work of establishing authoritative dimensions for public property as rapidly as time and opportunity permit, especially in the older sections of a city where original surveys are often very faulty. The work in the newer sections is usually of a better grade, having been controlled through plat and subdivision laws.

The work is begun by assembling all the available recorded information and the records of earlier surveys. When little or no information is available, this fact should furnish one of the strongest arguments for immediate action. The next step is the field location of all street intersection, angle, and curve points. Locating such points involves skill and judgment that is acquired only by years of experience with this class of work. The experienced surveyor is familiar with the more important court decisions and realizes the weight that must be given to occupation lines. As soon as the location of a point has been definitely determined, a monument should be set either directly at the point or at a standard offset distance from it. Because of present-day traffic conditions, the use of standard offset monuments is usually preferable. These monuments should be located from the second-order traverse with a precision of 1 : 20,000, or with such precision that any point whose location is dependent on that of the monument can be located within $\frac{1}{4}$ in. of its true position.

The property maps should be drawn to a scale of 1 in. = 50 ft, and the layout of the sheets should be of the best quality paper, and from them tracings should be made on the best quality translucent material. The maps should show the bearings and lengths of street lines, alley lines, and boundaries of public property; the street widths and intersections; the coordinates of all intersection, angle, and curve points; and the locations of traverse stations. In addition there should appear all existing monuments or other witness markers; public, industrial, and commercially important buildings; private buildings in sparsely settled areas; and such features as railroads and bridges. The lettering should include street names; names of parks, subdivisions, streams, and lakes; names of public or semipublic buildings; and the numbers of benchmarks and other monuments.

19-9 Underground Map

The purpose of the underground map is to show the locations of all sewers, water pipes, gas mains, electrical conduits, heating lines, subways, and tunnels, including their elevations at critical points. Much of the information to appear on such a map should be able to be taken from the original construction plans and the records of public-service companies, this information being supplemented by such field surveys as may be required to complete the map. The latter, called as-built surveys, are usually preferred since construction does not always follow the plans precisely.

The property maps can be used as a base for the underground map, many of the property dimensions being omitted to prevent overcrowding. The final maps, made on the best quality translucent material, should show street, alley, and easement lines, plotted to scale, with the various widths in figures; existing monuments and benchmarks; and surface structures, such as sidewalks, curbs, pavements, street railroads, transmission-line poles, and trees in parking areas. The underground structures should be differentiated by suitable symbols, and the sizes of the pipes, the percents of grade on gravity lines, and the elevations at critical points should be indicated. The leveling should be of second-order precision, or at least sufficiently accurate to permit the preparation of plans for new construction without fear of interference with existing structures. The horizontal locations should be shown with respect to the street and alley lines. As such a map soon becomes obsolete, some provision should be made for keeping it up-to-date as discussed in Sec. 20-9.

19-10   Digital Database

With the appropriate computer capabilities, digitizers, and computer programs, the information contained in the base maps, topographic maps, maps of city properties, subdivision maps, and underground maps can be coded and entered into computer files to form a database for a municipality for use in a GIS. Creation of a digital database to define all of the features of an area requires a tremendous amount of storage capacity in the computer. Several systems are available for accomplishing the task of establishing a digital database, and the interested student should consult the Bibliography at the end of Chapters 19 and 20 for information in greater depth.

Input to the database can take several different forms. Control information, for example, the names and coordinate values of triangulation and traverse stations, can be entered via the computer keyboard, or it can be transferred from files resulting from the computations for positions using separate computer programs. Data from maps that have already been compiled can be transferred to the database by means of various types of digitizers, in which each point or feature must be fully described by the code appropriate to the system's software. Modern methods of mapping employ the stereo plotter to measure aerial photographs. As discussed in Section 16-12, the output from the stereo plotter can be entered directly into the digital database, with keyboard entry made only to enter the appropriate feature identifiers as the compilation progresses.

Information from the database can be recalled to generate separate maps of interest to utilities, transportation managers, tax assessors, city and regional planners, and a multitude of other users of data contained in the database. A plotting table such as that shown in Fig. 8-35 is driven by commands from the computer to draw each separate map as needed.

Property ownership together with all of the information associated with a particular parcel of land can be stored and then recalled at a later time. Information concerning horizontal- and vertical-control points that has been stored can be recalled for use by the engineer and surveyor.

Several advantages to the creation of a digital database are cited as follows:

- As new subdivisions of land are completed, the new information is easily added to the database.
- If utility entities such as communications, gas and electric, water distribution, waste disposal, and sewerage, modify their distribution systems, the changes can be made to the database.
- The addition of streets, roads, and railroads to the transportation system, or the changes in routing of public transit, can be updated in the files.
- New horizontal and vertical control can be added to the database as it is completed and adjusted.
- Information from the database can be retrieved more rapidly than would be the case if the user had to resort to maps, reports, data filed in filing cabinets, and so on.
- Since the data are stored in digital form in ground units, a map of a particular theme can be generated by the plotter at any desired scale by a simple computer command.

In Chapter 20 a similar but broad-based list of actions is provided that are facilitated by a GIS.

The information stored in a digital database must, of course, be related to a single horizontal datum, such as a state plane coordinate system, and to a single vertical datum such as the North American Vertical Datum. If different portions of a city of municipality are on different datums, additional computations must be made to transform the desired data into a common system (see Section 8-23 for example) before being recalled. This additional computation can be quite simple as in converting from one vertical datum to another, but is somewhat more complicated when converting between two horizontal datums.

19-11 Town, City, and Village Plats

A building lot that is to be conveyed by deed can be described either with respect to existing land lines or by reference to a plat. In the first method the bearing and length of each side are given, and the description as a whole is tied to some section line or other well-defined property line. As the distance from the section line increases, the description becomes more and more complicated, particularly when curved boundaries are encountered and when the tie to the section line is a traverse of many sides. In the second method the description is greatly simplified, consisting merely of the lot number and the name of the recorded plat of which it is a part. If a database has been established for a municipality, information regarding the location of a lot can be described by a unique parcel identifier.

Most states have laws that govern the platting or subdividing of land into city or town lots. While some of the details may differ, the main requirements in the various states are essentially the same. The first requirement is usually a well-defined boundary of the tract to be platted, with the corners marked by permanent monuments, to which reference is made on the plat. The tract is divided into streets and blocks by placing permanent monuments at all intersections of two streets, of two alleys, of a street and an alley, and of a street or alley with a boundary of the plat. The lot corners are marked by semipermanent monuments, preferably iron pipe.

The survey is then platted on paper, the plat showing all blocks, lots, streets, alleys, parks, and monuments; the sizes of all lots, the lot numbers, and the names or numbers of all streets with their widths and courses; the directions and lengths of the boundaries of the tract, and all other information that will enable surveyors to retrace the lines accurately. The plat contains a written or printed description that should be so complete that from it, without reference to the plat, the starting point should be located with reference to some well-defined point, such as a certain point in the section when the land platted is in a state where the rectangular system of subdivision has been in effect.

The owners of the land sign a certificate on the plat, acknowledging that they have caused the plat to be made, and dedicating all streets, alleys, and parks to the use of the public. The surveyor certifies that the plat is a correct one and that the monuments described in it have been placed as shown on the plat.

Before a plat can be recorded, it must be approved by certain public officials or boards. In this manner the locations and widths of the streets, the sizes of the lots, and the proper grading of the streets can be controlled by the municipality or other interested public boards. When the plat has been finally approved, one copy is placed on record in the office of the register of deeds or similar official. Additional copies may be required for city and state records.

19-12   Survey for Plat

The first step in the subdivision of a tract into blocks and lots is an accurate survey of the boundaries of the parcel. This survey should be placed on a plane coordinate system. The next step is to prepare a topographic survey of the area. The blocks and lots together with coordinate grid ticks are then laid out on the topographic map.

Where zoning or planning boards are in existence, it is usually necessary to locate the principal streets in conformity with the general layout of the district adopted by such boards. In such cases the widths of the streets are also specified by the board.

Rectangular blocks with streets at right angles to one another should not be adopted when the topography is such that curved streets and irregular-shaped lots would be more advantageous. By fitting the locations of the streets to the topography of the ground, the rates of grade will be comparatively low. Less grading will then be required for the streets, and the location of storm and sanitary sewers will generally be simplified. Some of the older portions of San Francisco can be pointed out as horrible examples of the adoption of a rectangular layout to topography that is utterly unsuited to such a plan. The result has been that the grades on some streets are so excessive as to render them practically worthless as thoroughfares.

Minimum lot frontage and lot sizes are usually controlled by local ordinances. These and other constraints must be taken into account when laying out the lots in the tract on the map. The coordinates of all points on the street center lines, including PIs, PCs, and PTs of curves, as well as the coordinates of the centers of the curves, can be computed by scaling the coordinates of the beginning of the street center lines where they enter the tract (if these coordinates are not already known). Tangents and curves are laid out, and closed paper traverses are then computed along these lines between the fixed coordinates of the ends of the street center lines. Adjusting the closed traverse fixes the coordinates of the points on all of the streets. Alternatively, the coordinates of these points can be scaled directly from the paper location. Next the coordinates of the block and lot corners are computed on the coordinate system using the principles of analytical geometry discussed in Section 8-21.

After the paper layout has been completed, the field location is made. Street lines or block corners are first located from points on the boundaries of the tract by inversing the lines to these points. The individual lot corners are then established. As these lines are to serve as property boundaries, extreme care should be taken to ensure their accuracy. The work should be so conducted as to ensure a precision of about 1 : 10,000. The monuments marking important points should be as permanent as possible and referenced whenever possible. Some state plat laws now require that all important monuments be constructed of iron pipe that is at least 1 in. in diameter and at least 15 in. long, and be set in a concrete base at least 4 in. in diameter and 48 in. in depth.

19-13   Lot Survey

The survey of a lot that is described with references to a certain plat may be simple or complicated, the requirements depending on the shape of the block and on the monuments that may be available. Figure 19-4 illustrates an example of the simpler sort, where the monuments that were placed at the block corners can be found and identified. In locating the corners of Lot No. 40, the method of proportionate measurements must be used. Although the widths of the lots fronting on Ashland Avenue are shown on the plat as 50.00 ft, it is also evident that the total frontage has been equally divided among the seven lots. Consequently, if there is any excess or deficiency in the frontage, it should be equally distributed. For this reason the corners a and b are set at distances from the northwest corner of the block equal to, respectively, $\frac{2}{7}$ and $\frac{3}{7}$ times the total block frontage. Temporary marks are set at e and f by measuring the south side of the block and locating these points by proportionate measurements. The other two lot corners, c and d, are located at the midpoints of the lines ae and bf.

If corners set on the original survey can be found and identified, such points are usually held to be the true corners, even though they have been erroneously located. When land has been held by an owner for a certain period of time, ordinarily 20 years, and adjacent owners are unable to reach an agreement on the boundary lines and bring the case into court, the occupation lines are usually held by the court to be the property lines. This law is called "adverse possession." As

FIGURE 19-4 Lot survey.

pointed out in Section 18-24, the surveyor has no power to "establish" property lines. In the majority of cases, particularly where there is no serious disagreement, his findings will be accepted as final by the adjacent property owners. He should at all times so conduct his work as to be reasonably certain that his opinions will be supported if the question of location is brought before a court for final decision.

When many of the monuments are missing and when errors have been made in the original survey, the problem will be much more difficult than in the example shown in Fig. 19-4. Considerable skill and judgment will be required to weigh properly the conflicting evidence that may be disclosed as the survey progresses.

Some of the more important legal decisions relative to property surveys are cited in the books listed in the Bibliography at the end of Chapter 18.

19-14 Metes and Bounds

The survey of a piece of ground by metes and bounds is the oldest type of land surveying. It consists of determining the length and direction of each side of the parcel and planting permanent monuments at the corners of the property. The lands in the 13 original states were surveyed in this manner, but in many instances the descriptions are so vague and incomplete as to make the rerunning of the lines an impossibility. The following description, taken from the Hartford, Connecticut, Probate Court records for 1812, will illustrate what is often encountered:

> 147 acres, 3 rods, and 19 rods after deducting whatever swamp, water, rock and road areas there may be included therein and all other lands of little or no value, the same being part of said deceased's 1280 acre colony grant, and the portion hereby set off being known as near to and on the other side of Black Oak Ridge, bounded and

described more in particular as follows, to wit:—Commencing at a heap of stone, about a stone's throw from a certain small clump of alders, near a brook running down off from a rather high part of said ridge; thence, by a straight line to a certain marked birch tree, about two or three times as far from a jog in a fence going around a ledge nearby; thence, by another straight line in a different direction, around said ledge and the Great Swamp, so called; thence, in line of said lot in part and in part by another piece of fence which joins on to said line, and by an extension of the general run of said fence to a heap of stone near a surface rock; thence, as aforesaid, to the "Horn," so called, and passing around the same as aforesaid, as far as the "Great Bend," so called, and from thence to a squarish sort of a jog in another fence, and so on to a marked black oak tree with stones piled around it; thence, by another straight line in about a contrary direction and somewhere about parallel with the line around by the ledge and the Great Swamp, to a stake and stone bounds not far off from the old Indian trail; thence, by another straight line on a course diagonally parallel, or nearby so, with "Fox Hollow Run," so called, to a certain marked red cedar tree out on a sandy sort of a plain; thence, by another straight line, in a different direction, to a certain marked yellow oak tree on the off side of a knoll with a flat stone laid against it; thence, after turning around in anther direction, and by a sloping straight line to a certain heap of stone which is, by pacing, just 18 rods and about one half a rod more from the stump of the big hemlock tree where Philo Blake killed the bear; thence, to the corner begun at by two straight lines of about equal length, which are to be run by some skilled and competent surveyor, so as to include the area and acreage as herein before set forth.

For purposes of rerunning, a more favorable description would contain the length and bearing of each side and would include references to permanent objects that are still in existence. If the surveyor can find and identify the two ends of any one line, his work is greatly simplified, since that line affords him a comparison between his tape and that used in the original survey and make unnecessary any consideration of magnetic declinations. The interior angles of the tract can be calculated from the given bearings, and these computed angles can be turned off from the identified corners when searching for the remaining ones.

If only one of the original corners can be found, it will be necessary to know whether the bearings given in the descriptions are magnetic or true ones. If they are magnetic, the surveyor must ascertain the value of the declination at the time the survey was made and what it is at present, if he is to use a compass in determining the approximate directions of the lines from that corner. What has been said in Section 18-24 relative to the relocation of lost corners applies also to the resurvey of a tract described by metes and bounds.

BIBLIOGRAPHY   Burtch, R. 1987. Subdivision regulations: A look at the history and legal basis. *Tech Papers Ann. Meet. Amer. Cong. Surv. & Mapp.* p. 241.

Crane, E. 1990. Building the parcel map. *Amer. Cong. Surv. & Mapp. Bulletin No. 126.*

Donahue, J. G. 1988. Land base accuracy. Is it worth the cost? *Am. Cong. Surv. & Mapp. Bulletin 117.* p. 25.

Dracup, J. F. 1969. Control surveys for Akron, Ohio. *Surveying and Mapping* 29:243.

Gale, L. A. 1969. Control surveys for urban areas in Canada. *Surveying and Mapping* 29:669.

Gale, P. M. 1970. Control surveys for the City of Houston. *Surveying and Mapping* 30:95.

—— 1976. Cadastral surveys for small municipalities. *Proc. Fall Conv. Amer. Cong. Surv. & Mapp.* p. 387.

Hanigan, F. L. 1983. Houston's metropolitan common digital base: four years later. *Surveying and Mapping* 43:141.

Hebrank, A. J. 1981. Overlay drafting for surveyors. *Surveying and Mapping* 41:151.

Hokanson, J. B., and Brelsford, K. E. 1991. County government creates automated mapping system. *APWA Reporter* 58:4. Amer. Public Works Assn.

Kellie, A. C. 1983. Retracement by course and distance. *Surveying and Mapping* 43:35.

Pittman, R., Jr., and Tannen, G. 1971. Early, present, and future platting procedures. *Proc. Fall Conv. Amer. Cong. Surv. & Mapp.* p. 380.

Scher, M. B. 1969. Orthophoto maps for urban areas. *Surveying and Mapping* 29:413.

Urban, L. J. 1973. The use of geodetic control in a rapidly developing metropolitan area. *Proc. Fall Conv. Amer. Cong. Surv. & Mapp.* p. 48.

20 Geographic Information Systems

20-1 Introduction

One important function of the surveys described in Chapter 19 is the creation of a base map. The base map serves a variety of purposes. As the general population of our nation increases, and the number of regulations increases, data must be available to comply with those regulations and solve the problems of a more populated planet. For example, regulations related to the environment are critical today to sustain the earth in the long term. The decisions related to such things are complex; for example, where is the best site in the city or county to place landfill? While these decisions can be based on information assembled by using manual methods, the use of a *Geographic Information System (GIS)* will facilitate the task enormously. A GIS requires that a digital database be in place and that it contain (in the best case) accurate, current, and appropriate data. The data discussed in Chapter 19 comprise a subset of the total that are necessary. In general, for applications related to surveying, data about the land are necessary to appropriately populate the database for decision making. Data that are in demand includes environmental data, such as smog density; demographic data, such as population and average hourly wage; and data about the built environment, such as available commercial warehouse space, water availability, and quality. Since many of the required decisions are related to the use, maintenance, and economics of the land, the term *Land Information System* (LIS) is used in the literature somewhat synonymously with GIS. From the above discussion a general definition of a GIS emerges as hardware, software, data, and an environment for operations. A slightly more specific definition that has been widely accepted in the GIS community is a decision support system involving the integration of spatially referenced data in a problem-solving environment (Cowan, 1988). GISs are, by and large, a marriage of graphics software with database management software.

As control surveying becomes easier, even perfunctory, the surveyor will be drawn into making decisions about the land using GIS technology. The student therefore is advised to study this chapter closely and is referred to the Bibliography at the end of the chapter.

20-2 Spatial Data

Features shown on maps or those organized in a digital database that are tied to the surface of the earth by coordinates, addresses, or other means are called spatial or geographic data. Almost 80% of all data, and nearly all data collected by surveyors, are spatial in nature.

At small scales, spatial data are collected and organized by the federal government, including the U.S. Geological Survey (USGS), the National Oceanic and Atmospheric Administration (NOAA), the Environmental Protection Agency (EPA), and the Bureau of Land Management (BLM). These data are required by these agencies to perform their respective missions. For example, NOAA needs the depth of the water at numerous points (called soundings) in a body of water to create a nautical chart. These agencies have amassed a wealth of data that are increasingly important to the surveyor. Because of the need to cooperate both horizontally (among federal agencies) and vertically (among cities, counties, states, and the federal government) an initiative called the National Spatial Data Infrastructure (NSDI) has evolved from the federal government. The NSDI leadership resides in the office of the Secretary of the Interior in a coordinating organization known as the Federal Geographic Data Committee (FGDC). The FGDC attempts to develop catalogs of these data, sometimes referred to as *metadata,* and in general attempts to eliminate duplication and to organize the nation's spatial data.

The surveyor is well advised to become familiar with the NSDI initiative. It is important to know how to access these data (this can now be done on-line through the Internet) and to begin to use large-scale (such as 1/600) and small-scale (such as 1/24000) data together in a GIS to make decisions about the land. Moreover, large- and small-scale data will become integrated as use of the Internet and the World Wide Web (WWW) increases and as the NSDI becomes more robust.

In any event, accurate and current spatial data are essential to the use of GISs, and today there is almost a separate discipline associated with its acquisition, usage, storage, and dissemination. This discipline is central to the survival and growth of the surveying professional.

20-3 Computer-Readable Data

To be used in a GIS, data must be in computer-readable form. These data can be acquired in digital form from digital devices like CCDs, data collectors, and GPS receivers, or the data can be created by scanning or digitizing paper maps or plats. Data collected in digital form from various measuring devices or images has been discussed in the appropriate chapters in the text. These data can be loaded into a GIS by making a format transformation on the output file of the device. In many cases the output file of, say, a data collector is in one of several standard formats that allow ready transfer to a GIS.

The most common method of digitizing a paper map today is through the use of a digitizing workstation. This is a graphic workstation with a digitizing board or table and an associated mouse or puck. The mouse is connected to a light, flexible wire connected to a computer (the workstation data port). The system encodes the map features by entering an *XY* coordinate of the feature into the computer. This is possible because the digitizing table is connected to the same computer, and the position of the mouse on the table is known. Experience has shown that a good operator can digitize a line to within ±0.004″ at a rate of 1/10″ per second. Obviously the task of digitizing an entire map is formidable and quite expensive.

To speed up the digitizing process, a scanner is used. When a map or map layer is scanned on a device called a scanner, like that shown in Figure 20-1, the result is a digital file called a *raster image file*, shown schematically in Figure 20-2. The resolution or density of such a file is determined by how finely the scanner scans the document. The typical scanning resolution for scanning USGS 1/24,000 maps is 1000 dots per inch (dpi), which is a fairly dense (high) resolution. Most high-resolution scanners range from 500 dpi to 1000 dpi. The raster file can be plotted on paper, displayed on a cathode-ray tube (CRT) screen or used (and stored) in digital form. The smallest unit of consistent color or shading is called a pixel, or grid cell (see also Section 16-14). Each equal-sized cell is uniquely located by its row and column numbers and is coded with a numerical code that

FIGURE 20-1 High-resolution laser scanner. Courtesy of the Center for Mapping, Ohio State University.

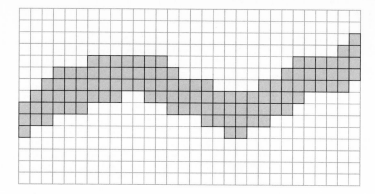

FIGURE 20-2 An enlargement of a scanned map with the raster lines shown by shading.

represents attributes of the area it covers. A point therefore is represented by a single cell, a line is depicted by a series of continuous cells having the same code, and an area having common properties would be shown as a group of identically coded contiguous cells.

Features on a map are usually classified into three categories: Points, lines, and areas. These data are generally represented by two data models: *raster* and *vector.* In the raster format, as previously discussed, points, lines, and areas are represented by pixels or cells. In the vector format, points are connected by lines, and areas are shown as polygons. These are discussed in detail later in this chapter.

Initially geographic information systems were designed to handle either raster or vector data models, but not both. Proponents and opponents for each data model argued in defense of or in opposition to each of the models. Raster data, it was argued, provided a rapid computational model: *Raster is faster.* The raster model is also better suited to representing continuously changing data, such as a pH measurement over a certain area. Vector data, in contrast, depicted geographical phenomena more realistically than raster: *Vector is correcter.* The vector model is better suited to representing objects, such as a type of land use over a certain area. After years of discourse and increasing levels of geographic data handling expertise, GIS developers now produce and market commercial GIS software products that facilitate the handling and processing of both data types. Provision of dual-use or dual-model GIS software is indicative of the increasing realization that both data models contain valuable tools for storing processing, and visualizing geographic data.

Both raster and vector data models strive to store and organize geographic information in efficient data formats. The entire set of geographic entities found in the real world can be represented using three graphic primitives: points, lines, and areas. Figure 20-3 shows a portion of a 1/24,000 (7.5-minute) USGS quadrangle of Millersport, Ohio, where houses are shown as points, roads as lines, and water bodies as polygons.

FIGURE 20-3 A portion of Millersport, Ohio, 1 : 24000-scale U. S. Geological Survey quadrangle map.

20-4 Topology in Spatial Data

Topology is a branch of mathematics dealing with the study of the properties of figures or solids that are not normally affected by changes in size or shape. In particular, changes in their relative relationships are studied. When dealing with spatial data under the control of a computer, we are at somewhat of a disadvantage compared to seeing the object with our eye. We have learned cognitive skills from the day we were born, therefore concepts such as near, far, next to, and behind are second nature to us. However, the computer needs to have these concepts detailed explicitly in its data and software or it will not realize that, for example, a set of lines forms a polygon. If this topology is known, then the area of the polygon can be computed easily.

The most important topological concepts required by GISs are

- Adjacency: On which side of the line is the swamp?
- Connectivity: Which lines are connected to which points?
- Orientation: From which point(s) do the lines begin and end?
- Nestedness: Are there other objects interior to the subject polygon?

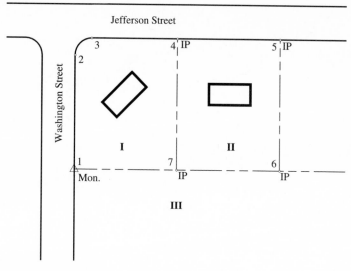

FIGURE 20-4 Map showing topological concepts.

Figure 20-4 illustrates these concepts. Adjacency is understood by noting that parcel I is adjacent to the two streets and parcels II and III. Furthermore each property line can be shown to be adjacent to the streets and/or other parcels. Connectivity is shown by noting that a line connects points 1 and 2, 2 and 3, and so on. Orientation is expressed by defining a line to go from point 4 to point 5. Finally, the houses are examples of nested features. Once these attributes and descriptions are placed in data files, the relative relationships, that is, the topology of these spatial data has been realized.

20-5 Raster Data Model

Raster data are frequently used to decompose and model geographic space. A raster space is referenced by numbered rows and columns, an $N \times N$ array or grid, that delimits an area located on the earth's surface. Generally the raster cell or minimum unit specifies the spatial resolution: smaller grid units, finer resolution; larger grid units, coarser resolution. Figure 20-5 illustrates the modeling of the Millersport, Ohio, 7.5-minute quadrangle in raster format. Examples of feature type are illustrated. Topology is not an explicit consideration in a raster-based GIS.

Over the years a robust set of cartographic modeling algorithms have been developed to facilitate spatial modeling using raster formats. Map-algebra algorithms based upon Boolean logic are used to analyze spatial phenomena and visualize geographic distributions. Simple map overlay operations are easily handled in raster formats. Raster data sets are plentiful because tremendous volumes of geographic data are recorded using instrumentation that collects spatial data in raster formats. Examples of raster data sets include satellite imagery (Landsat and SPOT), digital photography (Digital Ortho Photo Quadrangles), and digital raster output generated by scanning technology (Digital Raster Graphics).

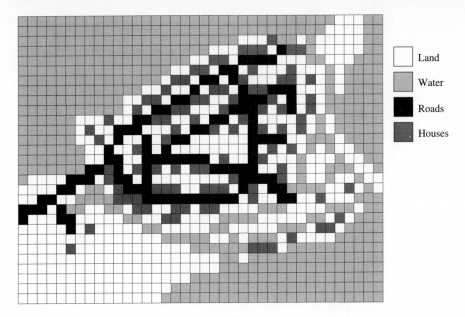

Land

Water

Roads

Houses

FIGURE 20-5 A course raster representation of the land cover of Liebs Island (Millersport, Ohio).

20-6   Vector Data Model

The use of a vector data format is quite common in digital mapping. A point location is represented by a single XY coordinate in the plane, a line by a linear and contiguous set of XY pairs in the plane, and areas by a series of contiguous XY pairs that enclose an areal unit in the plane. Points are used to specify locations of objects, such as corners of buildings, corners of boundaries, and features whose sizes dictate a point, such as telephone poles. Lines are used to represent road edges, railroad lines, boundaries, and others. Areas or polygons show regions having common attribute such as swamps, woodlands, property boundaries, and so on. The topology in a vector-based GIS helps ensure that the data are correct.

Figure 20-6 shows vectorized data obtained from the same location in the Millersport, Ohio, quadrangle as shown in Figure 20-5. In this figure, points are used to depict houses, lines to depict streets, and polygons to depict the areal extent of Liebs Island. Storage of descriptive attributes in the vector model is accomplished by constructing a set of database tables that are linked to the XY coordinates defining the spatial dimensions of the geographic object. A set of tables is constructed and linked so that descriptive attributes corresponding to each point, line, or area are searchable. The tables contain all descriptive information about the feature type encoded by the user.

Vector data sets are more common today than in years past. There are several reasons for this change. The costs of hardware and software designed for vector data collection have dropped significantly, and vector digitizing techniques are improving. Additionally, vector-based analytical models are becoming more

FIGURE 20-6 Vector representation of Liebs Island (Millersport, Ohio) including points (houses), lines (roads), and polygons (lake).

sophisticated. The vector model provides the geographic analyst with a large analytical tool set. Network analysis and routing are two types of analysis facilitated by a vector data model. The reason is that vector data uses topology. Topology (Section 20-4) provides points, lines, and areas with knowledge of their place in relation to other mapped features. For example, an area is coded with an inside and an outside, and a line can distinguish its left side and its right side.

Governments and private-sector organizations are actively providing access to high-quality vector data sets. Examples of high-quality vector data sets include the Digital Line Graph-3 data set (United States Geological Survey), TIGER (United States Census Bureau), and EtakMap from Etak, Inc.

20-7   Raster to Vector Conversion

The scanners like the one shown in Figure 20-1 are being used more frequently to convert paper graphic products (maps) to computer-readable or digital form. As stated previously, the result of the scanning process is a raster file or raster image. This raster product usually must be converted to vector form, as in Figure 20-7, because the jagged raster image confuses the points at which the line changes directions. Therefore it is important to define these points, that is, to determine in exactly which pixel the line changes direction. This vectorization process is complex and does not neatly lend itself to a unique solution. A number of commercial products exist, and generally speaking they work well for a specific conversion process, for example, a sewer line on an engineering drawing at a specific scale. However, the authors are unaware of a general product that cannot be significantly improved for specific applications.

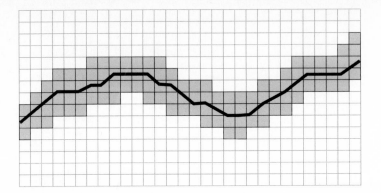

FIGURE 20-7 An enlargement of a scanned map with the raster line shown by shading and the vectorized line in bold.

20-8   Databases and Database Management Systems

In the 1970s it was becoming clear that a digital database could serve many masters. For example, data about the number of rooms in a condominium could be used by the assessor, the real estate board, the local engineer, and others. Therefore a common networked database made economic sense.

Clearly these data are spatial in nature, and it is easy to see that the foundation of such a database should be the coordinates or metes-and-bounds description of the property on which the condominium resides. The entire data set related to that parcel or property is sometimes referred to as a *multipurpose cadaster.* To use these data efficiently, the organization of such a database becomes very important, especially if it is large. There has been considerable time and energy spent on developing software systems capable of efficiently and effectively manipulating data residing in a database. This software is generally referred to as a database management system (DBMS). Digital databases, their structure, and DBMSs are topics in computer science, but there is a need for a brief explanation here.

Files of data that do not have DBMSs to help use them are usually contained in so called flat files, which are shown in Figure 20-8. They are called flat because each record in the file contains the same data items as all the others, for example, lists of coordinates. However, flat files cannot always be used efficiently, so other data structures have been developed to handle the access, retrieval, analysis, and storage of these data.

Three data structures that can be used in managing spatial data are hierarchical, network, and relational structures. In a hierarchical data structure, there are multiple records in the database and each record has a higher-level record associated with it. The advantage is that multiple sets of similar attributes are associated with any given record, without storing the repetitive data all in one file. This is generally referred to as a one-to-many relationship.

Figures 20-8 and 20-9 illustrate the difference between a flat file and a hierarchical file. In Figure 20-8 there is one record for each parcel, and all records are

Database systems for property records

Flat file system

Lot number	Lot address	Owner name	Owner address	Co-owner name	Co-owner address	Block	Section	Plat book	Building type	Square feet	% Hard surface	Assessed value	....

FIGURE 20-8 Flat file database system.

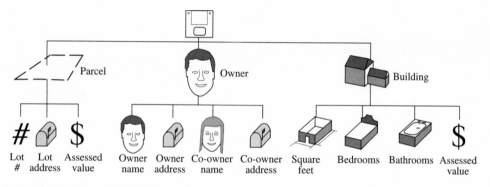

FIGURE 20-9 Hierarchial database system.

the same length. Access to the data in a record is accomplished by searching through all the records for a value in one field, usually denoted as a key field. The key field is the lot number. In Fig. 20-9 there are several records for each parcel with links that allow a one-to-many relationship, where each property file can have one or more owner detail records associated with it. Access to the data is limited to one type of record.

When detail records are associated with several records, the structure is called a network. Pointers link records between the files. This is called a many-to-many relationship, which allows more than one record in a file to be associated with more than one record in another file. Figure 20-10 illustrates the network structure.

One of the most useful data structures for manipulating spatial data is the relational database structure. If it is married with database management software, it is referred to as a relational database management system. Such a structure consists of a series of files called tables. Each table contains a data item that is the same as one or more other tables containing additional data.

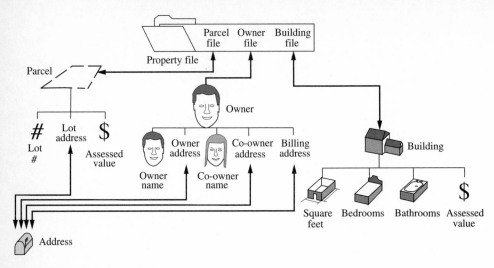

FIGURE 20-10 Network database system.

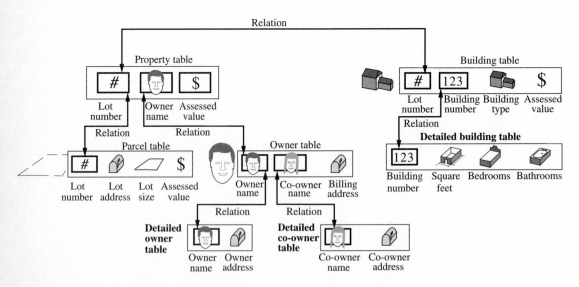

FIGURE 20-11 Relational database system.

Figure 20-11 illustrates a relational file structure for property data. Many commonly used GISs are combinations of graphic systems and relational database management systems.

Geographic information systems have unique and challenging data management requirements that are not efficiently handled by traditional database models, such as the hierarchical, network, or relational database. This is because the GIS manages both attribute (aspatial) and spatial information. The spatial information managed by the GIS can be stored easily in a traditional database, but the database does not know how to use these data. This means that the database cannot manipulate spatial information, such as determining if two areas overlap. This type of functionality is provided by various data structures. In some geographic information systems, when a query must be supported by the system, it is separated into spatial and aspatial parts. For example, if a GIS is used to store and manage information about soil types for the state of Ohio, then the query

Retrieve all regions in Clark County with a soil pH of less than 2.5

would be processed in several parts. The spatial portion of the query would retrieve soil polygons that contain, overlap, or are contained by the area of Clark County. The aspatial part of the query would search the attributes of the above-retrieved polygons to check the pH values.

Much progress has been made recently in providing the relational database with the ability to manipulate spatial information. Some state-of-the-art relational databases allow spatial data to be used directly by a relational database in answering queries. These tools have essentially moved the spatial data structures within the relational database, so that both spatial and aspatial can be understood by a single database.

20-9 Applications of GISs

The most common use of GISs today is in dealing with natural resources. However, it is predicted that GIS usage will become "enterprisewide" and, as time goes on, more use will be made of GISs in normal business decisions. This section will focus on the use of GISs in land planning and map updating since these functions are important and common to the surveyor.

Once a large section of land is in digital form in a computer and under the control of a GIS, updating information becomes easy. Although computer-aided drafting and design (CADD) systems will not be discussed here, many of the functions covered in this section are possible with a CADD system. For example, drawing parallel lines, buffering (centering two parallel lines on a given line), or shading can be performed. The following are typical examples of changes common to the task of the surveyor that can easily be made in a GIS environment:

- Easements or rights-of-way can be added or removed
- An entire section of a subdivision can be added
- Ownership and dimensional changes can be noted
- Lots can be combined
- Overlays containing additional information, such as taxes, can be seen on the screen
- Streets can be removed (vacated)
- The scale can be reduced[†]

In applications dealing with natural resources, overlaying is the central function of GISs. As an example, overlaying road networks with soil, aquifer, and

---

[†]This action brings us to a complicated topic called generalization, which is beyond the scope of this book. The interested student is referred to the Bibliography at the end of the chapter.

Roads

FIGURE 20-12 Overlaying themes.

topography will yield important information about where a land fill would be best placed within a county. These kinds of applications are the traditional uses of GISs. Figure 20-12 shows this situation. While this task could be accomplished manually using clear overlays, each overlay would have to be constructed manually with no flexibility to change its scale, attributes or other characteristics. Overlaying is one of the most powerful applications of GIS technology.

Table 20-1 lists a series of popular GISs. The capabilities listed in the table are explained below:

- Map analysis: This operation includes overlays, statistics on polygons, classification, and a series of logical operations.
- Buffer operations: See Fig. 20-13, which shows how areas around lines can be generated.
- COGO: Provides the suite of coordinate geometry operations and transforms the results into a GIS representation.
- Polygon operations: Includes polygon overlay, point and line in a polygon, and merge or dissolve capability.
- Vector editing: Includes raster conversion, line creation, edge matching, and topology checks.
- Map projections: Includes coordinate transformations and computations for a number of plane coordinate systems.
- Geocoding: Matching of street addresses to geographic locations

TABLE 20-1   Summary of Functionality of Major GIS Packages[†]
Source: GIS World Sourcebook 1996

GIS	Map Analysis	Buffer Operations	COGO	Polygon Operations	Vector Editing	Map Projections	Geocoding
Arc/Info	YES	YES	YES	YES	YES	YES	YES
ArcView	YES	YES	NO	YES	NO	YES	YES
AutoCAD/with Data Extension	YES	YES	YES	YES	YES	YES	NO
Grass	YES	YES	NO	YES	YES	YES	NO
MapInfo	YES	YES	NO	YES	NO	YES	NO
MGE/Microstation	YES	YES	YES	YES	YES	YES	NO
Smallworld	YES	YES	YES	YES	YES	YES	NO

[†]The functionalities described in the column headings refer mainly to functional categories. A package is indicated as supporting this category if it supports a substantial part of the operations that comprise that category, even if not every operation is supported.

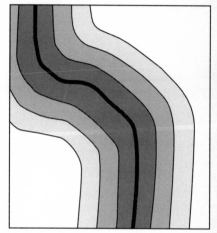

FIGURE 20-13 Raster (left) and vector (right) examples of a buffer or distance to/from a feature.

20-10   Conclusion

The field of Geographic Information Systems has seen active research in recent years, and some of the fruits of this work are appearing in commercial products. As mentioned in the discussion of databases, several well-supported commercial relational database products now support spatial queries directly. This will have a fundamental impact on the field of GIS, as it will move the GIS into the mainstream of corporate and commercial data processing operations from its current niche in natural resources and urban planning. The realization by corporations and governments that much of their data is spatial in nature will lead to the adoption of data management strategies that allow this decision-making power to be tapped. Furthermore the surveyor should benefit from this development since high-quality locational information will be required for accurate databases.

GIS is needed to support data of higher dimensionality. Currently the majority of GISs support two-dimensional spatial information. Higher dimensionality, including the ability to store and use time in an effective manner, will create a tool capable of modeling the physical environment of our world. Such a powerful GIS will have the potential to change the way businesses and governments make important decisions. In this environment the surveyor will be responsible for ensuring that the basic information used in such a powerful system is sound and accurate. This is a very important role and, in the authors' opinion, the future of the profession.

PROBLEMS

20-1. If a GIS were used to manage a cadaster for an urban region, what would be the best data model for the system, raster or vector? Explain.

20-2. If a GIS were used to manage information about insect infestations on a large commercial tree farm, with the base map created from a remotely sensed image, what would be the best data model for the system, raster or vector? Explain.

20-3. What is metadata, and why is it important?

20-4. Surveyors are often primary collectors of spatial information included in maps. Maps can then be digitized to create a digital database of the features described by the surveyor for a GIS. Would it be better if data collected by the surveyor were placed directly into the database rather than included in a map and digitized for the database?

20-5. Explain why topology is needed in a vector GIS? (Think of a polygon whose sides are not closed.)

20-6. Why is topology not an explicit consideration in a raster GIS? Does the raster have topology?

20-7. Why is the raster model ill-suited to managing information about the connections of a telephone network?

20-8. What is the many-to-many database model referred to as? The one-to-many?

20-9. Why does the traditional relational database not effectively handle the data in a GIS?

BIBLIOGRAPHY

Antennucci, J., et al. 1991 *Geographic Information Systems,* New York: Van Nostrand Reinhold.

Burrough, P. A. 1986. *Principles of Geographical Information Systems for Land Resource Assessment.* Oxford, U.K.: Clarendon Press.

Carver, S. J., and Brunsdon, C. F. 1994. Vector to raster conversion error and feature complexity: an empirical study using simulated data. *International Journal of Geographical Information Systems,* Vol. 8, No. 3, pp. 261–270.

Chrisman, N. R. 1996. *Exploring Geographic Information Systems.* New York: Wiley.

Committee on Geodesy, National Academy of Sciences/National Research Council. 1980. *Need for a Multipurpose Cadastre.* Washington, D.C.: National Academy Press.

Cowan, D. J. 1988. GIS versus CAD versus DBMs: What are the differences? *Photogrammetric Engineering and Remote Sensing.* 54: 1551.

Dale, P. F., and McLaughlin, J. D. 1988. *Land Information Management: An Introduction with Special Reference to Cadastral Problems in Third-World Countries.* Oxford, U.K.: Clarendon Press.

DeMers, M. N. 1997. *Fundamentals of GIS*. New York: Wiley.

Fotheringham, S., and Rogerson, P. 1994. *Spatial Analysis and GIS*. Bristol, PA: Taylor and Francis, Ltd.

Geo-Info Systems. Multiple issues.

GIS World. Multiple issues.

Goodchild, M., Gopal, S., et al. 1989. *Accuracy of Spatial Databases*. Bristol, PA: Taylor and Francis, Ltd.

Huxhold, W. E. 1991. *An Introduction to Urban Geographic Information Systems*. New York: Oxford University Press.

Mapping Science Committee, National Academy of Sciences/National Research Council. 1993. *Toward a Coordinated Spatial Data Infrastructure for the Nation*. Washington, D.C.: National Academy Press.

McMaster, R. B. and Shea, K. S. 1992. *Generalization in Digital Cartography*. Washington, D.C.: Association of American Cartographers.

Peuquet, D. J. 1990. A conceptual framework and comparison of spatial data models. *Introductory Readings in Geographic Information Systems*. Eds. Peuquet and Marble, London: Taylor and Francis.

Star, J. and Estes, J. 1990. *Geographic Information Systems: An Introduction*. New York: Prentice-Hall.

Stevenson, P. J. 1994. *Scanning for Automated Data Conversion of Cartographic Documents*. Report No. 426, Department of Geodetic Science and Surveying. Columbus, Ohio: Ohio State University.

Tomlin, C. D. 1990. *Geographic Information Systems & Cartographic Modeling Approach*. New York: Prentice Hall.

Adjustment of Elementary Surveying Measurements by the Method of Least Squares

A-1 Introduction

Whenever the engineer or surveyor conducts a field survey, no matter how simple or complex, he invariably makes more measurements than are absolutely necessary to locate the points in the survey. A line taped in two directions introduces one measurement more than is necessary to establish the length of the line. Measuring all three angles of a triangle also introduces one superfluous measurement. Closing a line of levels on a fixed benchmark introduces more measurements than are necessary to determine the elevations of the unknown benchmarks. A closed traverse has more than enough measurements necessary to fix the positions of the intermediate traverse stations. These extra measurements are termed *redundant measurements*. They immediately impose conditions that must be satisfied in order to resolve disagreements or inconsistencies in the measurements.

The redundant measurements are made for the purpose of checking other measurements, uncovering mistakes, adjusting the measurements, and estimating the magnitudes of the random errors. The mean of several measurements, for example, is the adjusted value of the measured quantity. The residuals resulting from subtracting the mean from the measured values furnish the data necessary to evaluate the random errors by determining the standard error of an individual measurement and of the mean value. The adjusted values of intermediate benchmarks in a level line are obtained by first determining the closure on a known benchmark and then distributing this closing error back through the line. This is the same procedure generally used for the adjustment of a traverse run between two fixed points.

In Chapter 5 procedures for adjusting some of the more elementary surveying measurements were discussed. These simple adjustments were based on one of the following principles of least squares: (1) In a set of measurements all of which are made with the same reliability, that is, all of which have equal weight, the most probable values that can be derived from the set are those which make the sum of the squares of the residuals or corrections a minimum. (2) In a set of measurements having unequal weights, the most probable values are those that make the sum of the products of the weights and the squares of the corrections a minimum. These principles may be stated briefly as follows: For equal-weight measurements,

$$\Sigma v^2 = \text{minimum} \tag{A-1}$$

or

$$[vv] = \text{minimum}$$

For unequal-weight measurements,

$$\Sigma pv^2 = \text{minimum} \tag{A-2}$$

or

$$[pvv] = \text{minimum}$$

The adjustment of measurements by the method of least squares accomplishes the double purpose of satisfying the above conditions and resolving discrepancies or closure errors of the measurements. The least-squares adjustment can be accomplished by either of two general methods. The first method makes use of the formation of *observation equations,* in which the corrections are stated as functions of indirectly determined values of parameters of the measurements. By way of example, the mean of the series of measurements in Section 5-3 was shown to be the most probable value by the formation of the following series of equations:

$$v_1 = M_1 - M$$
$$v_2 = M_2 - M$$
$$\cdot$$
$$\cdot$$
$$\cdot$$
$$v_n = M_n - M$$

These are observation equations. The corrections, or v's, appear on the left side. These are expressed as functions of the measurements themselves and the mean, which is an indirectly determined value. Then the condition of least squares was applied to the corrections by getting the sum of the squares of the corrections, differentiating with respect to the mean (the indirectly determined parameter), and setting the result equal to zero. The result of these operations is an equation of the form

$$\frac{d(\Sigma v^2)}{dM} = M_1 - M + M_2 - M + \cdots + M_n - M$$

$$= M_1 + M_2 + \cdots + M_n - nM = 0$$

Such an equation is known as a *normal equation*. The solution of this normal equation gives the value of the indirectly determined parameter M. The residuals are then found by solving the observation equations for the v's. This least-squares adjustment is so simple that we do not bother to form observation equations, differentiate to get the normal equation, and then solve the normal equation to arrive at the mean value. We simply add up all the measured values and divide by the number of measurements.

For a more complex series of measurements, the observation equations must be formed, and the normal equations are developed by differentiation of Σv^2 with respect to each parameter on the right side of the observation equations and setting equal to zero. These are solved simultaneously to arrive at the best set of parameters that will make Σv^2 a minimum. However, certain systematic procedures tend to simplify these operations, as will be shown later.

The second method for adjustment by least squares takes advantage of a priori conditions that must be imposed on the residuals. Such an a priori condition is the prior knowledge that the sum of three angles in a plane triangle must be 180°. Also, in a traverse between two fixed points, the sum of the latitudes must equal the difference between the Y coordinates of the two fixed points. This second method is the method of *condition equations*. For example, if three angles A_0, B_0, and C_0 of a triangle are observed and

$$A_0 + B_0 + C_0 = 179° \ 59' \ 56.7''$$

then we may state the condition by the equation

$$v_A + v_B + v_C = 3.3'' \quad \text{or} \quad v_A + v_B + v_C - 3.3'' = 0$$

in which v_A, v_B, and v_C are the corrections to be applied to the measured angles. The solution of this condition equation, after the condition of least squares is imposed, is

$$v_A = +1.1'' \qquad v_B = +1.1'' \qquad v_C = +1.1''$$

This simple adjustment was discussed in Section 5-9.

There are different approaches to the solution of the above condition equation. One approach is to express all the corrections as functions of only as many corrections as there are independent measurements. Thus, in the preceding example, there are only two independent angles. If A_0 and B_0 are considered independent, then C_0 must be considered dependent for the following reason: If there were no redundant measurements, then C_0 would equal $180° - (A_0 + B_0)$. Therefore, if the corrections v_A and v_B to A_0 and B_0 are independent, then

$$v_C = +3.3'' - (v_A + v_B)$$

In other words

$$v_A = v_A$$

$$v_B = v_B$$

$$v_C = +3.3'' - (v_A + v_B)$$

These are connecting equations, relating all the v's in terms of only the number of v's that are independent. On squaring we get

$$v_A^2 = v_A^2$$

$$v_B^2 = v_B^2$$

$$v_C^2 = [3.3'' - (v_A + v_B)]^2$$

Then

$$\Sigma v^2 = 2v_A^2 + 2v_B^2 + 3.3^2 - 6.6v_A - 6.6v_B + 2v_A v_B$$

To satisfy the condition of least squares, the partial derivative of Σv^2 with respect to each of the independent quantities v_A and v_B must be made zero. Thus we get the following normal equations:

$$\frac{\partial(\Sigma v^2)}{\partial v_A} = 0$$

$$\frac{\partial(\Sigma v^2)}{\partial v_B} = 0$$

When the differentiations are performed, the results are

$$4v_A - 6.6 + 2v_B = 0$$
$$4v_B - 6.6 + 2v_A = 0$$

By solving these normal equations, we get

$$v_A = +1.1'' \text{ and } v_B = +1.1''$$

Substitution of these values of v_A and v_B in the connecting equation

$$v_C = +3.3'' - (v_A + v_B)$$

gives

$$v_C = +1.1''$$

Another approach for solving condition equations is by the method of *correlates* or *Lagrange multipliers*. This approach, which is the more common, is as follows. Take the geometric condition equation and the condition of least squares

$$v_A + v_B + v_C - 3.3'' = 0 \qquad \text{(A-3)}$$

$$v_A^2 + v_B^2 + v_C^2 = \text{minimum} \qquad \text{(A-4)}$$

An infinite number of sets of values of the unknown v's will satisfy Eq. (A-3), but only one set will satisfy both Eqs. (A-3) and (A-4). To determine that set, differentiate each equation, giving

$$dv_A + dv_B + dv_C = 0 \qquad \text{(A-5)}$$

$$v_A dv_A + v_B dv_B + v_C dv_C = 0 \qquad \text{(A-6)}$$

Obviously an infinite number of sets of differentials given in Eq. (A-5) can satisfy that equation. If Eq. (A-5) is multiplied by an arbitrary multiplier C, the results are

$$C\, dv_A + C\, dv_B + C\, dv_C = 0 \qquad \text{(A-7)}$$

If Eq. (A-7) is now compared with Eq. (A-6), term for term, the following equalities result:

$$v_A = C$$

$$v_B = C \qquad \text{(A-8)}$$

$$v_C = C$$

Equations (A-8) are called *correlate* equations, and C is the Lagrange multiplier. If the values of v_A, v_B, and v_C given by Eq. (A-8) are substituted into the condition equation (A-3), the result is

$$C + C + C - 3.3'' = 0$$

or

$$3C - 3.3'' = 0 \qquad \text{(A-9)}$$

Eq. (A-9) is a normal equation whose solution gives $C = +1.1''$. If this value of the Lagrange multiplier is used in Eq. (A-8), then

$$v_A = +1.1'' \qquad v_B = +1.1'' \qquad v_C = +1.1''$$

as before.

When adjusting the three angles of a triangle, we do not bother to go through either of the elaborate procedures given above. We simply subtract 180° from the sum of the three measured angles, divide the discrepancy by three, and apply a correction of opposite sign to each angle. However, if a quadrilateral or some other net is to be adjusted, many conditions must be satisfied simultaneously. Therefore the equations must be formed in an orderly fashion so they may be solved simultaneously.

If the observations to be adjusted have different weights, then the weights must be incorporated into the adjustment procedure. Let us consider three measurements of the length of a line designated as M_1, M_2, and M_3, with corresponding weights p_1, p_2, and p_3. Also, let the most probable value be designated as $\bar{M}$. Then, by forming observation equations, squaring, and multiplying by the weights, we get

$$v_1 = M_1 - \overline{M} \qquad v_1^2 = (M_1 - \overline{M})^2 \qquad p_1 v_1^2 = p_1 (M_1 - \overline{M})^2$$

$$v_2 = M_2 - \overline{M} \qquad v_2^2 = (M_2 - \overline{M})^2 \qquad p_2 v_2^2 = p_2 (M_2 - \overline{M})^2$$

$$v_3 = M_3 - \overline{M} \qquad v_3^2 = (M_3 - \overline{M})^2 \qquad p_3 v_3^2 = p_3 (M_3 - \overline{M})^2$$

In order to satisfy the condition that $\Sigma\,(pv^2)$ is a minimum, it is necessary to differentiate $\Sigma\,(pv^2)$ with respect to $\overline{M}$ and to set the derivative equal to zero. Thus

$$\frac{d\Sigma(pv^2)}{d\overline{M}} = -2p_1(M_1 - \overline{M}) - 2p_2(M_2 - \overline{M}) - 2p_3(M_3 - \overline{M}) = 0$$

and

$$\overline{M} = \frac{p_1 M_1 + p_2 M_2 + p_3 M_3}{p_1 + p_2 + p_3} = \frac{\Sigma(pM)}{\Sigma p}$$

This result is the weighted mean given by Eq. (5-27).

In case of three angles of a triangle having different weights, the adjustment by the method of correlates is as follows: Let us assume that

$$A_0 + B_0 + C_0 = 179°\ 59'\ 30''$$

and that

$$p_A = 6 \qquad\qquad p_B = 3 \qquad\qquad p_C = 1$$

The condition is then

$$v_A + v_B + v_C - 30'' = 0 \qquad\qquad\qquad \text{(A-10)}$$

and the least-squares condition is

$$6v_A^2 + 3v_B^2 + 1v_C^2 = \text{minimum} \qquad\qquad\qquad \text{(A-11)}$$

Differentiation of these two equations gives

$$dv_A + dv_B + dv_C = 0 \qquad\qquad\qquad \text{(A-12)}$$

$$6v_A dv_A + 3v_B dv_B + 1v_C dv_C = 0 \qquad\qquad\qquad \text{(A-13)}$$

Multiplying Eq. (A-12) by the Lagrange multiplier C gives

$$C\,dv_A + C\,dv_B + C\,dv_C = 0 \qquad\qquad\qquad \text{(A-14)}$$

Equating coefficients between Eqs. (A-13) and (A-14) gives the correlate equations

$$6v_A = C \qquad \text{or} \qquad v_A = \frac{C}{6}$$

$$3v_B = C \qquad \text{or} \qquad v_B = \frac{C}{3} \qquad\qquad \text{(A-15)}$$

$$1v_C = C \qquad \text{or} \qquad v_C = \frac{C}{1}$$

Substituting the values of the v's from Eqs. (A-15) into the condition equation (A-10) gives the following normal equation:

$$\frac{C}{6} + \frac{C}{3} + \frac{C}{1} - 30'' = 0 \qquad \text{(A-16)}$$

whose solution gives

$$C = 20''$$

This value of C is substituted into the correlate equations (A-15) to give the solution for the v's. Thus

$$v_A = +3.3'' \qquad v_B = +6.7'' \qquad v_C = +20.0''$$

Compare these results with those of the example given in Section 5-10.

The concept of least-squares adjustment by the methods of observation equations and condition equations given here is very rudimentary. In fact observation and condition equations can be combined to form one system that is solved in such fashion as to make the weighted sums of squares of the corrections to the observations a minimum. To further investigate these methods, refer to the readings listed in the Bibliography at the end of this appendix.

A-2   Least-Squares Adjustment by Matrix Methods

The examples to follow in this appendix will be solved using either the method of observation equations or the method of condition equations as developed up to this point. These solutions can be simplified considerably by employing matrix algebra. Consequently the examples to follow will also be solved by matrix methods.

The matrix expression for the condition of least squares given in Eq. (A-1), in which the observation equations are all of equal weight, is

$$\mathbf{V}^T \mathbf{V} = \text{minimum} \qquad \text{(A-17)}$$

For observations of different weights, the least-squares condition given by Eq. (A-2) is expressed in matrix form as follows:

$$\mathbf{V}^T \mathbf{P} \mathbf{V} = \text{minimum} \qquad \text{(A-18)}$$

in which $\mathbf{V}$ is a column matrix of n residuals with dimensions $n \times 1$, $\mathbf{V}^T$ is the transpose of $\mathbf{V}$ with dimensions $1 \times n$, and $\mathbf{P}$ is a square diagonal matrix of weights with dimensions $n \times n$ or

$$\mathbf{P} = \begin{bmatrix} p_1 & 0 & \cdots & 0 \\ 0 & p_2 & & \vdots \\ \vdots & & & 0 \\ 0 & 0 & \cdots & p_n \end{bmatrix}$$

n being the number of observations or measurements. This weight matrix assumes that the measurements to be adjusted are not correlated.

The matrix method of satisfying the condition of least squares using *observation equations* will be developed in Section A-3(b).

A-3(a)   Adjustment of a Level Net by Observation Equations

The problem of adjusting a level net involves the tying of the net to previously established benchmarks, and the adjustment of the differences in elevations (DEs) between various junction points in the net so that no matter what route is taken to a point, the same elevation will be obtained. Furthermore, to satisfy the condition of least squares, the corrections to the observed DEs must be such that the sum of the weighted squares of the corrections will be a minimum. The weights of the DEs between points are generally considered to be inversely proportional to the lengths of the routes along which the DEs are obtained. For instance, if the DE between A and B is determined along two routes, the first one being 3 miles long and the second being 4 miles long, the weight of the first DE is $\frac{1}{3}$ and that of the second is $\frac{1}{4}$, or the first has weight 1, whereas the second has weight 0.75. Weights may also be assumed to be inversely proportional to the number of setups of the instrument between two points.

In the level net shown in Fig. A-1, there are 11 independent lines, along which the differences in elevation have been obtained as a result of three-wire leveling. The values shown in Table A-1 have been corrected for systematic errors discussed in Section 3-39. The subscripts for the v's correspond to the numbers assigned to the lines in Fig. A-1. The elevations of the three benchmarks A, D, and K are fixed as follows:

$$A = 856.425 \text{ ft} \qquad D = 878.337 \text{ ft} \qquad K = 867.271 \text{ ft}$$

The most common procedure for adjusting a set of measurements by the method of observation equations is first to compute an approximate set of values of the quantities sought, and then to compute small corrections to these initial values. In the case of a level net, an approximate elevation is computed for each

FIGURE A-1 Level net.

TABLE A-1

Line	Length (miles)	From To	Observed DE	Correction to DE
1	1.49	A–B	– 3.278	v_1
2	1.12	B–G	– 1.467	v_2
3	1.32	G–H	+ 9.756	v_3
4	1.15	H–A	– 4.975	v_4
5	1.27	B–C	+ 8.429	v_5
6	1.03	C–F	– 6.528	v_6
7	1.38	F–G	– 3.352	v_7
8	1.32	C–D	+16.724	v_8
9	0.96	D–E	– 12.563	v_9
10	1.29	E–F	– 10.721	v_{10}
11	2.36	F–K	+12.249	v_{11}

junction point by adding the DE of a line from a known elevation to the junction point. Thus

$$\text{approx. elevation } B = \text{elevation } A + \text{DE}_{AB}$$

$$\text{approx. elevation } B = 856.425 - 3.278 = 853.147 \text{ ft}$$

$$\text{approx. elevation } C = \text{approx. elevation } B + \text{DE}_{BC}$$

$$\text{approx. elevation } C = 853.147 + 8.429 = 861.576 \text{ ft}$$

Similarly the approximate elevations of the other junction points are obtained as follows:

Point	From Line	Approximate Elevation
B	AB	853.147
C	BC	861.576
E	CF and FE	865.769
F	CF	855.048
G	BG	851.680
H	GH	861.436

These approximate elevations must be adjusted by adding small corrections, which will be designated by x's with subscripts that correspond to the numbers of the junction points shown in parentheses in Fig. A-1. Thus the adjusted elevations may be represented as follows:

Point	Adjusted Elevation
B	$853.147 + x_1$
C	$861.576 + x_2$
E	$865.769 + x_3$
F	$855.048 + x_4$
G	$851.680 + x_5$
H	$861.436 + x_6$

We can now derive observation equations by applying the following general relationship:

$$(\text{observed DE}_{1-2}) + (\text{correction to DE}_{1-2})$$
$$= (\text{adjusted elevation of 2}) - (\text{adjusted elevation of 1})$$

For example, between A and B,

$$-3.278 + v_1 = (853.147 + x_1) - 856.425$$

Between B and G,

$$-1.467 + v_2 = (851.680 + x_5) - (853.147 + x_1)$$

Between G and H,

$$+9.756 + v_3 = (861.436 + x_6) - (851.680 + x_5)$$

Between H and A,

$$-4.975 + v_4 = 856.425 - (861.436 + x_6)$$

Between B and C,

$$+8.429 + v_5 = (861.576 + x_2) - (853.147 + x_1)$$

By writing an observation equation for each DE and combining numbers, we obtain 11 equations that express the v's, or the corrections to the DEs, in terms of the x's, or the corrections to the approximate elevations of the points. The results are

$$v_1 = x_1 + 0$$
$$v_2 = -x_1 + x_5 + 0$$
$$v_3 = -x_5 + x_6 + 0$$
$$v_4 = -x_6 - 0.036$$
$$v_5 = -x_1 + x_2 + 0$$
$$v_6 = -x_2 + x_4 + 0$$
$$v_7 = -x_4 + x_5 - 0.016$$
$$v_8 = -x_2 + 0.037$$
$$v_9 = x_3 - 0.005$$
$$v_{10} = -x_3 + x_4 + 0$$
$$v_{11} = -x_4 - 0.026$$

If each DE had equal or unit weight, the condition to be satisfied would be that $\Sigma v^2 = \text{minimum}$, or

$$v_1^2 + v_2^2 + v_3^2 + \cdots + v_{11}^2 = \text{minimum}$$

However, since each DE is obtained either through a different distance or by a different number of setups (the distance is used in this example), each DE has a weight that is inversely proportional to the distance involved. Consequently the condition to be satisfied is $\Sigma(pv^2) = \text{minimum}$, or

$$p_1 v_1^2 + p_2 v_2^2 + \cdots + p_{11} v_{11}^2 = \text{minimum}$$

The weights determined by taking the reciprocals of the lengths of the lines are shown in Table A-2. It is usually desirable to adjust each weight by a factor to make the average weight equal to unity. In this example it would be necessary to have the weights add up to 11. Since the sum of the tabulated weights is 8.676, each such weight would have to be multiplied by 11/8.676. However, it would be an unnecessary refinement to make the average weight equal to unity in the present example. When the square of each value of v is multiplied by the tabulated weight, the results are as follows:

$$0.672v_1^2 = 0.672(+x_1 + 0)^2$$

$$0.893v_2^2 = 0.893(-x_1 + x_5 + 0)^2$$

$$0.758v_3^2 = 0.758(-x_5 + x_6 + 0)^2$$

$$0.870v_4^2 = 0.870(-x_6 - 0.036)^2$$

$$0.787v_5^2 = 0.787(-x_1 + x_2 + 0)^2$$

$$0.971v_6^2 = 0.971(-x_2 + x_4 + 0)^2$$

$$0.725v_7^2 = 0.725(-x_4 + x_5 - 0.016)^2$$

$$0.758v_8^2 = 0.758(-x_2 + 0.037)^2$$

$$1.043v_9^2 = 1.043(+x_3 - 0.005)^2$$

$$0.775v_{10}^2 = 0.775(-x_3 + x_4 + 0)^2$$

$$0.424v_{11}^2 = 0.424(-x_4 - 0.026)^2$$

TABLE A-2

Line	From To	Length (miles)	Weight
1	A–B	1.49	0.672
2	B–G	1.12	0.893
3	G–H	1.32	0.758
4	H–A	1.15	0.870
5	B–C	1.27	0.787
6	C–F	1.03	0.971
7	F–G	1.38	0.725
8	C–D	1.32	0.758
9	D–E	0.96	1.043
10	E–F	1.29	0.775
11	F–K	2.36	0.424
		$\Sigma =$	8.676

Six normal equations can be formed by adding the right sides of the above equations to get $\Sigma\,(pv^2)$ and then taking the partial derivative of $\Sigma\,(pv^2)$ with respect to each of the x's and setting each partial derivative equal to zero. Thus

$$\frac{\partial\Sigma(pv^2)}{\partial x_1} = 0\,;\quad \frac{\partial\Sigma(pv^2)}{\partial x_2} = 0\,;\quad \dots\,;\quad \frac{\partial\Sigma(pv^2)}{\partial x_6} = 0$$

The solution of these equations would give the values of the six parameters (the x's) that would satisfy the condition of least squares.

The formation of the normal equations can be made quite mechanical by realizing that they take the following form:

$$[paa]x_1 + [pab]x_2 + [pac]x_3 + [pad]x_4 + [pae]x_5 + [paf]x_6 + [pak] = 0$$
$$[pba]x_1 + [pbb]x_2 + [pbc]x_3 + [pbd]x_4 + [pbe]x_5 + [pbf]x_6 + [pbk] = 0$$
$$[pca]x_1 + [pcb]x_2 + [pcc]x_3 + [pcd]x_4 + [pce]x_5 + [pcf]x_6 + [pck] = 0$$
$$[pda]x_1 + [pdb]x_2 + [pdc]x_3 + [pdd]x_4 + [pde]x_5 + [pdf]x_6 + [pdk] = 0$$
$$[pea]x_1 + [peb]x_2 + [pec]x_3 + [ped]x_4 + [pee]x_5 + [pef]x_6 + [pek] = 0$$
$$[pfa]\,x_1 + [pfb]\,x_2 + [pfc]\,x_3 + [pfd]\,x_4 + [pfe]\,x_5 + [pff]x_6 + [pfk] = 0$$

In these equations the p's are the weights, the a's are the coefficients of the x_1's in the observation equations, the b's are the coefficients of the x_2's, and so on, and the k's are the constant terms of the observation equations. The quantity $[pbd]$, for example, is the sum of 11 products, or $p_1b_1d_1 + p_2b_2d_2 \dots p_{11}b_{11}d_{11}$. The student should verify this formulation.

Arranging the observation equations in tabular form as shown in Table A-3 simplifies the formation of the normal equations. The values under x_1 are the a's, those under x_2 are the b's, and so on. The coefficients all happen to be plus or minus unity in this adjustment. The normal equations, which are formed as indicated, can also be tabulated as shown in Table A-4. Because $[pab] = [pba]$, $[pac] = [pca]$, and so on, the coefficients of the x's in the table are symmetrical about the

TABLE A-3   Tabulated Observation Equations

v	p	x_1	x_2	x_3	x_4	x_5	x_6	k
1	0.672	+1						0
2	0.893	−1				+1		0
3	0.758					−1	+1	0
4	0.870						−1	−0.036
5	0.787	−1	+1					0
6	0.971		−1	+1				0
7	0.725			−1	+1			−0.016
8	0.758		−1					+0.037
9	1.043			+1				−0.005
10	0.775			−1	+1			0
11	0.424				−1			−0.026

TABLE A-4   Tabulated Normal Equations

x_1	x_2	x_3	x_4	x_5	x_6	k	
+2.352	−0.787			−0.893		0	= 0
−0.787	+2.516		−0.971			−0.02805	= 0
		+1.818	−0.775			−0.00521	= 0
	−0.971	−0.775	+2.895	−0.725		+0.02262	= 0
−0.893			−0.725	+2.376	−0.758	−0.01160	= 0
				−0.758	+1.628	+0.03132	= 0

diagonal running through the coefficients [paa], [pbb], . . . , [pff]. Thus the coefficients in row i are equal to the coefficients in column i. Being familiar with this symmetry obviously permits a great saving of time in forming the normal equations.

The solution of the preceding set of six normal equations gives the values of the corrections to be added to the approximate elevations of the junction points to get their adjusted values. Solving the original observation equations for the v's gives the quantities needed to determine the standard errors of the measured values. The standard error of unit weight is

$$\sigma_0 = \sqrt{\frac{\Sigma p v^2}{n - u}} \qquad \text{(A-19)}$$

in which n is the number of lines and u is the number of junction points. The quantity $(n - u)$ is the number of redundant measurements or the number of degrees of freedom (see Section 5-4). In the case at hand, $n = 11$ and $u = 6$. Thus

$$\sigma_0 = \sqrt{\frac{\Sigma p v^2}{5}}$$

The standard error of each line is then obtained by Eq. (5-30).

After the approximate elevations of the junction points have been adjusted, the corrected values are considered as fixed elevations. The elevations of any intermediate benchmarks established along the various lines are then adjusted between two fixed junction points by the method discussed in Section 5-11.

A-3(b)   Adjustment of a Level Net Using Matrix Notation

The solution of the adjustment of a level net by the method of observation equations will now be performed using matrix notation. First, the matrix algebra involved in the solution must be developed. The matrix expression for a set of observation equations is given as follows: If there are n observations and u unknown parameters, then

$$\mathbf{V} = \mathbf{BX} + \mathbf{L} \qquad \text{(A-20)}$$

in which $\mathbf{V}$ is a column vector of n residuals with dimensions $n \times 1$, $\mathbf{X}$ is a column vector of unknown parameters with dimensions $u \times 1$, $\mathbf{B}$ is a coefficient matrix of dimensions $n \times u$, and $\mathbf{L}$ is a column vector of constant terms of dimensions $n \times 1$.

In Eq. (A-20) the coefficients of the residuals are all assumed to be unity. This is the usual case in the adjustment of elementary surveying measurements by observation equations.

Taking the general case of observations of different weights, the value of $\mathbf{V}$ from Eq. (A-20) is substituted into Eq. (A-18), giving

$$\mathbf{V}^T\mathbf{PV} = (\mathbf{BX} + \mathbf{L})^T\mathbf{P}(\mathbf{BX} + \mathbf{L}) = \text{minimum}$$

or

$$\mathbf{V}^T\mathbf{PV} = (\mathbf{X}^T\mathbf{B}^T + \mathbf{L}^T)\mathbf{P}(\mathbf{BX} + \mathbf{L})$$

Expanding gives

$$\mathbf{V}^T\mathbf{PV} = \mathbf{X}^T\mathbf{B}^T\mathbf{PBX} + \mathbf{X}^T\mathbf{B}^T\mathbf{PL} + \mathbf{L}^T\mathbf{PBX} + \mathbf{L}^T\mathbf{PL}$$

or, since

$$\mathbf{L}^T\mathbf{PBX} = \mathbf{X}^T\mathbf{B}^T\mathbf{P}^T\mathbf{L} \quad \text{and} \quad \mathbf{P}^T = \mathbf{P}$$

then

$$\mathbf{L}^T\mathbf{PBX} = \mathbf{X}^T\mathbf{B}^T\mathbf{PL}$$

Finally,

$$\mathbf{V}^T\mathbf{PV} = \mathbf{X}^T\mathbf{B}^T\mathbf{PBX} + 2\mathbf{X}^T\mathbf{B}^T\mathbf{PL} + \mathbf{L}^T\mathbf{PL} \tag{A-21}$$

To make the weighted sums of squares of the residuals a minimum, we take the partial derivative of $\mathbf{V}^T\mathbf{PV}$ with respect to the vector of unknown parameters $\mathbf{X}$ and set it equal to zero. Thus from Eq. (A-21)

$$\frac{\partial \mathbf{V}^T\mathbf{PV}}{\partial \mathbf{X}} = 2\mathbf{B}^T\mathbf{PBX} + 2\mathbf{B}^T\mathbf{PL} = 0$$

giving the set of normal equations in matrix form as

$$\mathbf{B}^T\mathbf{PBX} + \mathbf{B}^T\mathbf{PL} = 0 \tag{A-22}$$

in which $\mathbf{X}$, as before, is a column vector of unknown parameters with dimensions $u \times 1$, $\mathbf{B}^T\mathbf{PB}$ is a coefficient matrix of dimensions $u \times u$, and $\mathbf{B}^T\mathbf{PL}$ is a column matrix of constant terms of dimensions $u \times 1$.

The normal equations of Eq. (A-22) are solved for the $\mathbf{X}$ vector by matrix algebra, giving

$$\mathbf{X} = -(\mathbf{B}^T\mathbf{PB})^{-1}\mathbf{B}^T\mathbf{PL} \tag{A-23}$$

in which $(\mathbf{B}^T\mathbf{PB})^{-1}$ is the inverse of $\mathbf{B}^T\mathbf{PB}$

The vector $\mathbf{V}$ of residuals can then be obtained by substitution of the value of $\mathbf{X}$ from Eq. (A-23) into the original observation equations of Eq. (A-20).

In the level net adjustment, there are 11 observations, therefore $n = 11$. Since there are six junction points, there are six parameters to be computed, $x_1, x_2, \ldots, x_6$. Thus $u = 6$. The matrices making up the observation equation given by Eq. (A-20) are then as follows:

$$
\mathbf{V} = \begin{bmatrix} v_1 \\ v_2 \\ v_3 \\ v_4 \\ v_5 \\ v_6 \\ v_7 \\ v_8 \\ v_9 \\ v_{10} \\ v_{11} \end{bmatrix} \quad \mathbf{B} = \begin{bmatrix} +1 & 0 & 0 & 0 & 0 & 0 \\ -1 & 0 & 0 & 0 & +1 & 0 \\ 0 & 0 & 0 & 0 & -1 & +1 \\ 0 & 0 & 0 & 0 & 0 & -1 \\ -1 & +1 & 0 & 0 & 0 & 0 \\ 0 & -1 & 0 & +1 & 0 & 0 \\ 0 & 0 & 0 & -1 & +1 & 0 \\ 0 & -1 & 0 & 0 & 0 & 0 \\ 0 & 0 & +1 & 0 & 0 & 0 \\ 0 & 0 & -1 & +1 & 0 & 0 \\ 0 & 0 & 0 & -1 & 0 & 0 \end{bmatrix} \quad \mathbf{X} = \begin{bmatrix} x_1 \\ x_2 \\ x_3 \\ x_4 \\ x_5 \\ x_6 \end{bmatrix} \quad \mathbf{L} = \begin{bmatrix} 0 \\ 0 \\ 0 \\ -0.036 \\ 0 \\ 0 \\ -0.016 \\ +0.037 \\ -0.005 \\ 0 \\ -0.026 \end{bmatrix}
$$

$$11 \times 1 \qquad\qquad\qquad 11 \times 6 \qquad\qquad\qquad 6 \times 1 \qquad\qquad 11 \times 1$$

These matrices can be identified in the tabulated observation equations in Table A-3.

The weight matrix for the level-net adjustment is

$$
\mathbf{P} = \begin{bmatrix}
0.672 & 0 & . & . & . & . & . & . & . & . & 0 \\
0 & 0.893 & & & & & & & & & . \\
. & & 0.758 & & & & & & & & . \\
. & & & 0.870 & & & & & & & . \\
. & & & & 0.787 & & & & & & . \\
. & & & & & 0.971 & & & & & . \\
. & & & & & & 0.725 & & & & . \\
. & & & & & & & 0.758 & & & . \\
. & & & & & & & & 1.043 & & . \\
. & & & & & & & & & 0.775 & 0 \\
0 & . & . & . & . & . & . & . & . & 0 & 0.424
\end{bmatrix}
$$

in which all the off-diagonal terms are zeros, since no correlation is assumed to exist between the measured DEs. If the matrix multiplications indicated in Eq. (A-22) are performed, the resulting normal equations are then given in matrix form as follows:

$$
\underbrace{\begin{bmatrix}
+2.352 & -0.787 & 0 & 0 & -0.893 & 0 \\
-0.787 & +2.516 & 0 & -0.971 & 0 & 0 \\
0 & 0 & +1.818 & -0.775 & 0 & 0 \\
0 & -0.971 & -0.775 & +2.895 & -0.725 & 0 \\
-0.893 & 0 & 0 & -0.725 & +2.376 & -0.758 \\
0 & 0 & 0 & 0 & -0.758 & +1.628
\end{bmatrix}}_{\mathbf{B}^T \mathbf{PB}}
\underbrace{\begin{bmatrix} x_1 \\ x_2 \\ x_3 \\ x_4 \\ x_5 \\ x_6 \end{bmatrix}}_{\mathbf{X}}
+ \underbrace{\begin{bmatrix} 0 \\ -0.02805 \\ -0.00521 \\ +0.02262 \\ -0.01160 \\ +0.03132 \end{bmatrix}}_{\mathbf{B}^T \mathbf{PL}} = 0
$$

This is the matrix form of the tabulated normal equations shown in Table A-4.

The solution for the vector of parameters given by Eq. (A-23) and the adjusted elevations are as follows:

Solution (ft)	Adjusted Elevation (ft)
$\mathbf{X} = \begin{bmatrix} x_1 \\ x_2 \\ x_3 \\ x_4 \\ x_5 \\ x_6 \end{bmatrix} = \begin{bmatrix} +0.003 \\ +0.010 \\ +0.001 \\ -0.005 \\ -0.002 \\ -0.020 \end{bmatrix}$	$B = 853.147 + 0.003 = 853.150$ $C = 861.576 + 0.010 = 861.586$ $E = 865.769 + 0.001 = 865.770$ $F = 855.048 - 0.005 = 855.043$ $G = 851.680 - 0.002 = 851.678$ $H = 861.436 - 0.020 = 861.416$

The corrections to the measured DEs can now be obtained by substituting the value of $\mathbf{X}$ into Eq. (A-20), giving

$$v_1 = +0.003 \text{ ft} \qquad v_5 = +0.008 \text{ ft} \qquad v_9 = -0.004 \text{ ft}$$
$$v_2 = -0.005 \text{ ft} \qquad v_6 = -0.015 \text{ ft} \qquad v_{10} = -0.006 \text{ ft}$$
$$v_3 = -0.018 \text{ ft} \qquad v_7 = -0.013 \text{ ft} \qquad v_{11} = -0.021 \text{ ft}$$
$$v_4 = -0.016 \text{ ft} \qquad v_8 = +0.027 \text{ ft}$$

The standard error of unit weight for the network is given as

$$\sigma_0 = \sqrt{\frac{\mathbf{V}^T \mathbf{P} \mathbf{V}}{n - u}} \qquad \text{(A-24)}$$

For this level network, the standard error of unit weight is ±0.018 ft. The standard error of the measured DEs can then be found by Eq. (5-30), resulting in the following

$$\sigma_1 = \pm 0.022 \text{ ft} \qquad \sigma_5 = \pm 0.020 \text{ ft} \qquad \sigma_9 = \pm 0.018 \text{ ft}$$
$$\sigma_2 = \pm 0.019 \text{ ft} \qquad \sigma_6 = \pm 0.018 \text{ ft} \qquad \sigma_{10} = \pm 0.020 \text{ ft}$$
$$\sigma_3 = \pm 0.021 \text{ ft} \qquad \sigma_7 = \pm 0.021 \text{ ft} \qquad \sigma_{11} = \pm 0.028 \text{ ft}$$
$$\sigma_4 = \pm 0.019 \text{ ft} \qquad \sigma_8 = \pm 0.021 \text{ ft}$$

A-4   Adjustment of Horizontal Control by Variation of Coordinates

A control net of any degree of complexity can be adjusted by the method of observation equations commonly referred to as the *variation of coordinates*. If the work is being done on a plane coordinate system, the variations of the coordinates are designated ΔX and ΔY; and if on a geodetic system, the variation of coordinates are designated $\Delta \varphi$ and $\Delta \lambda$. This section will deal with a plane coordinate system.

In the adjustment of the level net of Section A-3(a) by observation equations, a set of approximate elevations was determined using selected measured DEs, as shown on page 696. Then a set of observation equations was developed that expressed the corrections to the DEs (the v's) as functions of corrections to the approximate elevations. In solving the problem, the elevation of each point i was allowed to vary by an amount x_i in such a way that the values of the x's made the weighted sum of squares of the corrections to the measured DEs a minimum. This condition was assured by the method of formation of the normal equations; that is, the partial derivative of the weighted sum of squares of the residuals with respect to each of the x's was set equal to zero.

In the adjustment of triangulation, intersection, or resection by observation equations, a set of approximate X and Y coordinates is first computed for each new station. These computations are made using unadjusted angles, together with preliminary azimuths. These approximate coordinates are then allowed to vary by small amounts ΔX and ΔY in such a way that the computed values of the ΔX's and ΔY's make the weighted sum of squares of the corrections to the measured angles a minimum.

Each measured angle in a control net involves three points in the net. These are the *at* station designated A, the *from* station designated F, and the *to* station designated T (see Section 4-16). If any of these are fixed stations, then the corrections to the coordinates of these stations will, of course, be zero.

The form of the observation equation that expresses the correction v to a measured angle as a function of the variations ΔX and ΔY of the control stations can be developed in a number of different ways. The development presented here is geometric in nature. The angle FAT in Fig. A-2 is equal to the azimuth of the line AT minus the azimuth of the line AF. Stated as an observation equation

$$\angle FAT + v_A = \alpha_{AT} - \alpha_{AF} \qquad \text{(A-25)}$$

in which $\angle FAT$ is the measured angle at A from F to T, v_A is the correction to the measured angle at A, α_{AF} is the adjusted azimuth of AF, and α_{AT} is the adjusted azimuth of AT.

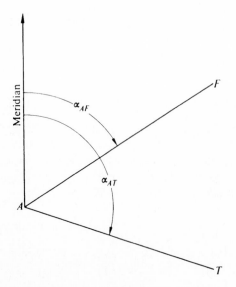

FIGURE A-2 Angle as difference in azimuths.

The adjusted azimuths of the two lines are obtained by applying corrections $\Delta\alpha_{AF}$ and $\Delta\alpha_{AT}$ to preliminary azimuths obtained from approximate coordinates of the stations by inversing according to Eq. (8-13). Thus

$$\alpha_{AF} = \alpha'_{AF} + \Delta\alpha_{AF}$$

$$\alpha_{AT} = \alpha'_{AT} + \Delta\alpha_{AT}$$

(A-26)

in which the primes (′) denote preliminary azimuths. The observation equation given by Eq. (A-25) can then be written as follows:

$$v_A = \Delta\alpha_{AT} - \Delta\alpha_{AF} + [\,\alpha'_{AT} - \alpha'_{AF} - \angle FAT\,]$$

(A-27)

The terms inside the brackets make up the constant term of the observation equation. Equation (A-27) can then be expressed as

$$v_A = \Delta\alpha_{AT} - \Delta\alpha_{AF} + K_A$$

(A-28a)

in which

$$K_A = \alpha'_{AT} - \alpha'_{AF} - \angle FAT$$

(A-28b)

Expressions for $\Delta\alpha_{AT}$ and $\Delta\alpha_{AF}$ will now be developed geometrically. In Fig. A-3, points A' and F' are approximate positions of, for example, two triangulation stations obtained by preliminary position computations, (see Section 9-8). The azimuth of this line α'_{AF} is obtained by Eq. (8-13), unless it has already been obtained for the position computation. The positions of the two points are allowed to vary by small amounts, ΔX_A, ΔY_A and ΔX_F, ΔY_F, to move them into the adjusted positions at A and F. The variation of the coordinates at each end of the line will change the preliminary azimuth of the line by a small amount $\Delta\alpha_{AF}$. The small distance e_A between the preliminary line $A'F'$ and the final line AF is expressed as

$$e_A = \Delta Y_A \sin\,\alpha'_{AF} - \Delta X_A \cos\,\alpha'_{AF}$$

(A-29a)

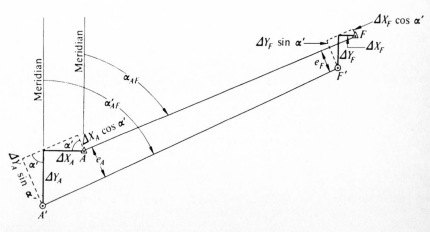

FIGURE A-3 Change in azimuth as function of change in coordinates.

FIGURE A-4 Change in preliminary azimuth.

and the small distance e_F at the other end of the line is

$$e_F = \Delta Y_F \sin \alpha'_{AF} - \Delta X_F \cos \alpha'_{AF} \tag{A-29b}$$

If the line AF is imagined to be moved parallel with itself to make A coincide with A', the results are shown in Fig. A-4. The small angle between the two lines is $\Delta\alpha_{AF}$ and can be expressed as

$$\Delta\alpha_{AF} = \frac{e_A - e_F}{AF} \tag{A-30}$$

in which the distance AF is obtained with sufficient accuracy from the approximate X and Y coordinates of A and F using Eq. (8-12). Since $\Delta\alpha_{AF}$ is a small angle, Eq. (A-30) can be expressed as

$$\Delta\alpha''_{AF} = \frac{e_A - e_F}{AF \text{ arc } 1''} \tag{A-31}$$

Substituting the value of e_A from Eq. (A-29a) and e_F from Eq. (A-29b) into Eq. (A-31) gives the expression for $\Delta\alpha_{AF}$ in seconds as a function of the variation in the coordinates of A and F, thus

$$\Delta\alpha''_{AF} = -\frac{\cos \alpha'_{AF}}{AF \text{ arc } 1''} \Delta X_A + \frac{\sin \alpha'_{AF}}{AF \text{ arc } 1''} \Delta Y_A$$

$$+ \frac{\cos \alpha'_{AF}}{AF \text{ arc } 1''} \Delta X_F - \frac{\sin \alpha'_{AF}}{AF \text{ arc } 1''} \Delta Y_F \tag{A-32}$$

By the same geometrical analysis an expression for $\Delta\alpha_{AT}$ can be developed in terms of the variation of the coordinates of A and T, giving

$$\Delta\alpha''_{AT} = -\frac{\cos \alpha'_{AT}}{AT \text{ arc } 1''} \Delta X_A + \frac{\sin \alpha'_{AT}}{AT \text{ arc } 1''} \Delta Y_A$$

$$+ \frac{\cos \alpha'_{AT}}{AT \text{ arc } 1''} \Delta X_T - \frac{\sin \alpha'_{AT}}{AT \text{ arc } 1''} \Delta Y_T \tag{A-33}$$

The observation equation for angle FAT is obtained in its final form by substituting the value of $\Delta\alpha_{AF}$ from Eq. (A-32) and $\Delta\alpha_{AT}$ from Eq. (A-28a), giving

$$v_A = \frac{\cos \alpha'_{AF}}{AF \text{ arc } 1''} \Delta X_A - \frac{\sin \alpha'_{AF}}{AF \text{ arc } 1''} \Delta Y_A - \frac{\cos \alpha'_{AF}}{AF \text{ arc } 1''} \Delta X_F + \frac{\sin \alpha'_{AF}}{AF \text{ arc } 1''} \Delta Y_F$$

$$- \frac{\cos \alpha'_{AT}}{AT \text{ arc } 1''} \Delta X_A + \frac{\sin \alpha'_{AT}}{AT \text{ arc } 1''} \Delta Y_A \qquad \text{(A-34)}$$

$$+ \frac{\cos \alpha'_{AT}}{AT \text{ arc } 1''} \Delta X_T - \frac{\sin \alpha'_{AT}}{AT \text{ arc } 1''} \Delta Y_T + K_A$$

The coefficients of the ΔX's and ΔY's in Eq. (A-34) are needed to only three or four significant figures at the most. However, the values of the preliminary azimuths α'_{AF} and α'_{AT} contained in the constant term must be determined with a precision consistent with the precision of the measured angles. The value of the residual v_A and the constant term K_A are both expressed in seconds of angle.

Each independent measured angle in a control net will introduce an observation equation of the form given by Eq. (A-34). These are then reduced by the methods given in Section A-3 in order to make the weighted sum of squares of the residuals a minimum.

An entire triangulation network such as that shown in Fig. 19-1 can be adjusted simultaneously by the method of variation of coordinates. The student is referred to the *Manual of Plane Coordinate Computation, Special Publication No. 193* of the U.S. Coast and Geodetic Survey for a complete discussion of this method. The calculation of a triangulation network adjustment is presented in detail.

Section A-6 shows the adjustment of an intersection survey, and Section A-7 presents the adjustment of a resection observation by this method.

A-5   Adjustment of a Traverse Net by Observation Equations

In Section A-3 a level net was adjusted by assuming an initial set of junction-point elevations and then letting these values be changed by small amounts to cause the sum of the products of the weights of the lines of levels and the squares of the corrections to the measured DEs of these lines to be a minimum. A traverse net consisting of several lines joined at junction points can be adjusted in a similar manner. In theory, however, there is one important difference. The corrections that we refer to as v's are supposed to be applied to quantities that are measured directly. In the level-net adjustment, we consider the DEs to be measured directly since they represent a series of additions and subtractions of directly observed readings of the leveling rod. The corrections that will be referred to in this section as v_x's and v_y's are applied to derived quantities and not to direct observations.

The basis for the adjustment of a traverse net by the method of observation equations as set forth in this section is the compass rule discussed in Section 8-15. This rule states that the correction to the latitude or departure of a line is directly

FIGURE A-5 Traverse section.

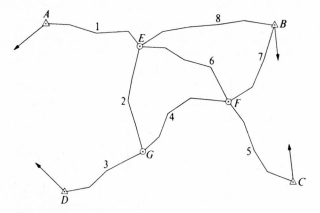

FIGURE A-6 Traverse net.

proportional to the length of that line.[†] Let us consider a series of traverse lines joining two points A and B, as shown in Fig. A-5. Such a series of lines is referred to as a section. A traverse net is composed of several sections, each of which joins either two junction points or a junction point and a fixed point, as indicated in Fig. A-6. For each section we compute the sum of the latitudes, called ΣL, and the sum of the departures, called ΣD. Each sum is analogous to the DE of a level line. The corrections to these two derived quantities are the v_y's and the v_x's, respectively. The quantities ΣL and ΣD are derived, since they are products of measured distances and functions of angles that have been obtained from measured angles, as explained in Chapter 8 (see also Section 5-7).

In Fig. A-6 traverse sections have been run from each of the four control points A, B, C, and D to each of the other three through the junction points E, F, and G. The fixed coordinates of the four control points in feet are as follows:

[†]This procedure is not rigorous since the angles are not included in the adjustment. However, this example should permit the student to understand the application of least squares to traverse net adjustments.

Point	Y	X
A	28,696.58	51,582.20
B	28,878.62	56,454.81
C	25,674.32	56,890.80
D	25,256.32	52,064.40

The lines with the arrowheads at the four control points indicate azimuth backsights. A traverse can be carried, for example, by backsighting at A, running in succession to E, F, and C, and then closing in azimuth at C. When all the traversing shown in Fig. A-6 has been executed, either the angular closures can be adjusted to give adjusted azimuths or else the computed azimuths can be used as unadjusted azimuths. The latter procedure is more valid. The latitudes and departures of all the lines are next computed by applying Eqs. (8-1) and (8-2) or Eqs. (8-3) and (8-4). The values of ΣL and ΣD for the sections are then determined and listed in tabular form. The measured and computed values for the traverse in Fig. A-6 are shown in Table A-5. The weights of the sections are assumed to be inversely proportional to their lengths, which is the basis of the compass rule. These weights have been modified so that their average is unity.

An approximate set of values for the coordinates of the junction points are now computed by adding the proper values of ΣL to known Y coordinates and adding the proper values of ΣD to known X coordinates. Thus for point E

$$Y_E = Y_A + \Sigma L_{AE} = 28{,}696.58 - 357.30 = 28{,}339.28$$
$$X_E = X_A + \Sigma D_{AE} = 51{,}582.20 + 2076.02 = 53{,}658.22$$

The values of the approximate coordinates of E, F, and G are as follows:

Point	From Section	Approx. Y	Approx. X
E	AE	28,339.28	53,658.22
F	BF	27,256.14	55,480.21
G	DG	26,142.34	53,754.08

TABLE A-5

Section	Length 1000 ft	From To	ΣL	ΣD	Correction to ΣL	Correction to ΣD	p
1	2.24	A–E	− 357.30	+ 2076.02	v_{y_1}	v_{x_1}	1.04
2	2.20	E–G	− 2196.62	+ 96.08	v_{y_2}	v_{x_2}	1.06
3	1.88	G–D	− 886.02	− 1689.68	v_{y_3}	v_{x_3}	1.24
4	2.74	G–F	+ 1113.50	+ 1725.88	v_{y_4}	v_{x_4}	0.85
5	2.44	F–C	− 1582.23	+ 1410.42	v_{y_5}	v_{x_5}	0.96
6	2.51	E–F	− 1082.82	+ 1822.00	v_{y_6}	v_{x_6}	0.93
7	2.00	F–B	+ 1622.48	+ 974.60	v_{y_7}	v_{x_7}	1.17
8	3.02	B–E	− 539.59	− 2796.74	v_{y_8}	v_{x_8}	0.77

We obtain the adjusted coordinates of each junction point by adding an as-yet unknown correction ΔY to the approximate Y coordinate and a correction ΔX to the approximate X coordinate. The adjusted coordinates may therefore be represented as follows:

Point	Adjusted Y	Adjusted X
E	$28{,}339.28 + \Delta Y_E$	$53{,}658.22 + \Delta X_E$
F	$27{,}256.14 + \Delta Y_F$	$55{,}480.21 + \Delta X_F$
G	$26{,}142.34 + \Delta Y_G$	$53{,}754.08 + \Delta X_G$

We can now derive an observation equation in Y and in X for each section. For example, for the section between A and E,

$$\Sigma L_{AE} + v_{y_1} = \text{adjusted } Y_E - Y_A$$

or

$$-357.30 + v_{y_1} = 28{,}339.28 + \Delta Y_E - 28{,}696.58$$

Hence,

$$v_{y_1} = + \Delta Y_E + 0$$

Also,

$$\Sigma D_{AE} + v_{x_1} = \text{adjusted } X_E - X_A$$

or

$$+2076.02 + v_{x_1} = 53{,}658.22 + \Delta X_E - 51{,}582.20$$

Hence,

$$v_{x_1} = + \Delta X_E + 0$$

The complete set of observation equations is as follows:

$$v_{y_1} = +\Delta Y_E + 0 \qquad\qquad v_{x_1} = +\Delta X_E + 0$$
$$v_{y_2} = -\Delta Y_E + \Delta Y_G - 0.32 \qquad\qquad v_{x_2} = -\Delta X_E + \Delta X_G - 0.22$$
$$v_{y_3} = -\Delta Y_G + 0 \qquad\qquad v_{x_3} = -\Delta X_G + 0$$
$$v_{y_4} = +\Delta Y_F - \Delta Y_G + 0.30 \qquad\qquad v_{x_4} = +\Delta X_F - \Delta X_G + 0.25$$
$$v_{y_5} = -\Delta Y_F + 0.41 \qquad\qquad v_{x_5} = -\Delta X_F + 0.17$$
$$v_{y_6} = -\Delta Y_E + \Delta Y_F - 0.32 \qquad\qquad v_{x_6} = -\Delta X_E + \Delta X_F - 0.01$$
$$v_{y_7} = -\Delta Y_F + 0 \qquad\qquad v_{x_7} = -\Delta X_F + 0$$
$$v_{y_8} = +\Delta Y_E + 0.25 \qquad\qquad v_{x_8} = +\Delta X_E + 0.15$$

The matrix form of these equations is

$$
\begin{bmatrix} v_{y_1} \\ v_{y_2} \\ v_{y_3} \\ v_{y_4} \\ v_{y_5} \\ v_{y_6} \\ v_{y_7} \\ v_{y_8} \end{bmatrix} = \begin{bmatrix} +1 & 0 & 0 \\ -1 & 0 & +1 \\ 0 & 0 & -1 \\ 0 & +1 & -1 \\ 0 & -1 & 0 \\ -1 & +1 & 0 \\ 0 & -1 & 0 \\ +1 & 0 & 0 \end{bmatrix} \begin{bmatrix} \Delta Y_E \\ \Delta Y_F \\ \Delta Y_G \end{bmatrix} + \begin{bmatrix} 0 \\ -0.32 \\ 0 \\ +0.30 \\ +0.41 \\ -0.32 \\ 0 \\ +0.25 \end{bmatrix}
$$

$$
\mathbf{V}_y \qquad\qquad \mathbf{B} \qquad\qquad \mathbf{X}_y \qquad\qquad \mathbf{L}_y
$$

$$
\mathbf{P}_y = \begin{bmatrix} 1.04 & \cdot & \cdot & \cdot & \cdot & \cdot & \cdot & 0 \\ 0 & 1.06 & & & & & & \cdot \\ \cdot & & 1.24 & & & & & \cdot \\ \cdot & & & 0.85 & & & & \cdot \\ \cdot & & & & 0.96 & & & \cdot \\ \cdot & & & & & 0.93 & & \cdot \\ \cdot & & & & & & 1.17 & \cdot \\ 0 & \cdot & \cdot & \cdot & \cdot & \cdot & \cdot & 0.77 \end{bmatrix}
$$

$$8 \times 8$$

$$
\begin{bmatrix} v_{x_1} \\ v_{x_2} \\ v_{x_3} \\ v_{x_4} \\ v_{x_5} \\ v_{x_6} \\ v_{x_7} \\ v_{x_8} \end{bmatrix} = \begin{bmatrix} +1 & 0 & 0 \\ -1 & 0 & +1 \\ 0 & 0 & -1 \\ 0 & +1 & -1 \\ 0 & -1 & 0 \\ -1 & +1 & 0 \\ 0 & -1 & 0 \\ +1 & 0 & 0 \end{bmatrix} \begin{bmatrix} \Delta X_E \\ \Delta X_F \\ \Delta X_G \end{bmatrix} + \begin{bmatrix} 0 \\ -0.22 \\ 0 \\ +0.25 \\ +0.17 \\ -0.01 \\ 0 \\ +0.15 \end{bmatrix} \qquad \mathbf{P}_x = \mathbf{P}_y
$$

$$
\mathbf{V}_x \qquad\qquad \mathbf{B} \qquad\qquad \mathbf{X}_x \qquad\qquad \mathbf{L}_x
$$

Two sets of normal equations are formed, one in Y and one in X. The procedure for the formation of the normal equations is that described in Section A-3. The coefficients $[\mathbf{B}^T\mathbf{PB}]$ in both sets of equations will be identical since $\mathbf{B}$ and $\mathbf{P}$ are identical for both sets. If the adjustment is being performed on a desk calculating machine, the values used in the solution of both sets of normal equations can be placed side by side because of the equality of the coefficients.

The proper values of ΔY are added to the approximate Y coordinates of the junction points, and the proper values of ΔX are added to the approximate X coordinates. These results are the adjusted coordinates. The standard errors of the ΣL's and ΣD's are determined as discussed in Section A-3 for determining the standard errors of the DEs.

After the adjusted coordinates of the junction points have been determined, the coordinates of the intermediate traverse stations in each section are made to be consistent with those of the junction points. This adjustment is an application of the compass rule, as outlined in Section 8-15 and as illustrated by the example in Section 8-19.

A-6 Adjustment of Intersection Observations

In the example of an intersection solution given in Section 9-13, the minimum of two measured angles α and β were used to compute the X and Y coordinates of P. If the third angle γ shown in Fig. 9-11 is also measured, the position of P can be determined by the method of variation of coordinates. The coordinates of known point C in Fig. 9-11 are $X_C = 1,465,872.89$ and $Y_C = 639,864.01$. Angle $\gamma = 87° 32' 02''$. Since an approximate position for P has already been computed in Example 9-3, the three lines AP, BP, and CP can be inversed to determine their preliminary lengths and azimuths as shown in Table A-6. The azimuths of AB and BA were computed in Example 9-3. The azimuth of BC is determined by inversing.

Using the scheme shown in Fig. A-2 to designate the *at*, *from*, and *to* stations, in forming the observation equation given by Eq. (A-34) for angle α, A is the *at* station, P is the *from* station, and B is the *to* station. The constant K_α is computed from Eq. (A-28b) and Table A-6.

$$\alpha'_{AB} = 108° 49' 36''$$
$$-\alpha'_{AP} = \underline{-34° 20' 20''}$$
$$74° 29' 16''$$

$$-\text{measured } \angle\alpha \quad \underline{-74° 29' 16''}$$
$$K_\alpha = 0$$

Then, since ΔX_A, ΔY_A, ΔX_B, and ΔY_B are all zero,

$$v_\alpha = -\frac{\cos \alpha'_{AP}}{AP \text{ arc } 1''} \Delta X_P + \frac{\sin \alpha'_{AP}}{AP \text{ arc } 1''} \Delta Y_P + 0$$

TABLE A-6

Line	Preliminary Length	Preliminary Azimuth	cos α'	sin α'
AB		108° 49' 36''		
AP	5886	34° 20' 20''	+ 0.82572	+ 0.56409
BP	9258	326° 36' 04''	+ 0.83486	− 0.55046
CP	8837	306° 37' 23''	+ 0.59655	− 0.80258
BC		39° 05' 15''		

For angle β in Eq. (A-34), B is the *at* station, A is the *from* station, and P is the *to* station. The constant K_β is computed

$$
\begin{aligned}
\alpha'_{BP} &= 326° \, 36' \, 04'' \\
-\alpha'_{BA} &= -288° \, 49' \, 36'' \\
\hline
&37° \, 46' \, 28'' \\
\end{aligned}
$$

$$
\begin{aligned}
-\text{measured } \angle\beta \quad & -37° \, 46' \, 28'' \\
\hline
& K_\beta = 0
\end{aligned}
$$

Then, since ΔX_A, ΔY_A, ΔX_B, ΔY_B are all zero,

$$
v_\beta = +\frac{\cos \alpha'_{BP}}{BP \text{ arc } 1''} \Delta X_P - \frac{\sin \alpha'_{BP}}{BP \text{ arc } 1''} \Delta Y_P + 0
$$

For angle γ in Eq. (A-34), C is the *at* station, B is the *from* station, and P is the *to* station. The constant K_γ is

$$
\begin{aligned}
\alpha'_{CP} &= 306° \, 37' \, 23'' \\
-\alpha'_{CB} &= -219° \, 05' \, 15'' \\
\hline
&87° \, 32' \, 08'' \\
\end{aligned}
$$

$$
\begin{aligned}
-\text{measured } \angle\gamma \quad & -87° \, 32' \, 02'' \\
\hline
& K_\gamma = +06''
\end{aligned}
$$

Then, since ΔX_B, ΔY_B, ΔX_C, and ΔY_C in Eq. (A-34) are all zero,

$$
v_\gamma = \frac{\cos \alpha'_{CP}}{CP \text{ arc } 1''} \Delta X_P - \frac{\sin \alpha'_{CP}}{CP \text{ arc } 1''} \Delta Y_P + 6''
$$

The three resulting observation equations are then

$$
\begin{aligned}
v_\alpha &= -28.94 \; \Delta X_P + 19.77 \; \Delta Y_P + 0 \\
v_\beta &= +18.60 \; \Delta X_P + 12.26 \; \Delta Y_P + 0 \\
v_\gamma &= +13.92 \; \Delta X_P + 18.73 \; \Delta Y_P + 6''
\end{aligned}
$$

$$
\mathbf{V} = \begin{bmatrix} v_\alpha \\ v_\beta \\ v_\gamma \end{bmatrix} \quad
\mathbf{B} = \begin{bmatrix} -28.94 & +19.77 \\ +18.60 & +12.26 \\ +13.92 & +18.73 \end{bmatrix} \quad
\mathbf{X} = \begin{bmatrix} \Delta X_P \\ \Delta Y_P \end{bmatrix} \quad
\mathbf{L} = \begin{bmatrix} 0 \\ 0 \\ +6'' \end{bmatrix}
$$

Assuming that all three angles have the same weight, P is the unit matrix. Then by Eq. (A-23),

$$
\mathbf{X} = -(\mathbf{B}^T\mathbf{B})^{-1}\mathbf{B}^T\mathbf{L}
$$

The solution of this equation gives $\Delta X_P = -0.07$ and $\Delta Y_P = -0.13$. The adjusted coordinates of P are then

$$
\begin{aligned}
X'_P &= 1,458,780.23 \qquad Y'_P = 645,135.92 \qquad \text{(from Example 9-3)} \\
& \underline{-0.07} \qquad \underline{-0.13} \\
X_P &= 1,458,780.16 \qquad Y_P = 645,135.79
\end{aligned}
$$

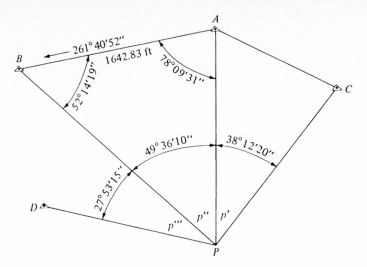

FIGURE A-7 Resection on four control points.

A-7 Adjustment of Conditioned Resection

In the example of a resection solution given in Section 9-14, the minimum of two measured angles p' and p'' were used to compute the X and Y coordinates of P. If a third angle p''' is also measured as shown in Fig. A-7, the position of P can be obtained by the method of variation of coordinates. Table A-7 lists the coordinates of the four control points in Fig. A-7.

The measured angles are shown in Fig. A-7. The length and azimuth of AB needed to compute the preliminary coordinates of P are 1642.83 ft and 261° 40′ 52″, respectively, obtained by inversing AB. The position computation for the preliminary position of P (see Section 9-8) using the results of triangle PBA of the example shown in Table 9-6 is laid out in Table A-8.

The lines needed for the adjustment are inversed to obtain the distances and preliminary azimuths of these lines. This is shown in Table A-9. Some of the lines are obtained from the triangle solution and the position given in Section 9-8 and Table A-8 but are shown in Table A-9 for convenience.

TABLE A-7

Station	X	Y
A	118,476.24	267,935.80
B	116,850.70	267,698.11
C	119,406.52	267,394.23
D	117,012.87	266,623.98

TABLE A-8

	azimuth AB =	261° 40′ 52″	azimuth BA =	+81° 40′ 52″
	$-\angle A$	$-78°$ 09′ 31″	$+\angle B$	+52° 14′ 19″
	azimuth AP =	183° 31′ 21″	azimuth BP =	133° 55′ 11″
		$-133°$ 55′ 11″		
	$\angle P$	49° 36′ 10″ (checks)		

Station	Length	Cos Azimuth	Sin Azimuth	Y	X
A				267,935.80	118,476.24
	1705.38	−0.998 11075	−0.061 44050	−1,702.16	−104.78
P				266,233.64	118,371.46
	2111.26	+0.693 64981	−0.720 31239	+1,464.48	−1,520.77
B				267,698.12	116,850.69
			fixed	267,698.11	116,850.70

TABLE A-9

Line	Preliminary Length	Preliminary Azimuth	Cos α'	Sin α'
PA	1705	3° 31′ 21″	+0.99811	+0.06144
PB	2111	313° 55′ 11″	+0.69365	−0.72031
PC	1555	41° 43′ 41″	+0.74631	+0.66559
PD	1414	286° 01′ 48″	+0.27614	−0.96112

In forming the observation equations given by Eq. (A-34), the unknown point P is the *at* station for each angle. With reference to Fig. A-7, for angle p''', D is the *from* station, and B is the *to* station. The constant $K_{p'''}$ is obtained by Eq. (A-28b).

$$\alpha'_{PB} = \quad 313° 55′ 11″$$
$$-\alpha'_{PD} = \underline{-286° 01′ 48″}$$
$$27° 53′ 23″$$

$$-\text{measured } \angle p''' \quad \underline{-27° 53′ 15″}$$
$$K_{p'''} = +8″$$

Then, since ΔX_B, ΔY_B, ΔX_D, and ΔY_D are all zero,

$$v_{p'''} = \frac{\cos \alpha'_{PD}}{PD \text{ arc } 1''} \Delta X_P - \frac{\sin \alpha'_{PD}}{PD \text{ arc } 1''} \Delta Y_P - \frac{\cos \alpha'_{PB}}{PB \text{ arc } 1''} \Delta X_P$$
$$+ \frac{\sin \alpha'_{PB}}{PB \text{ arc } 1''} \Delta Y_P + 8''$$

For angle p'', B is the *from* station and A is the *to* station. The constant $K_{p''}$ is computed

$$\alpha'_{PA} = 363°\ 31'\ 21''$$
$$-\alpha'_{PB} = -313°\ 55'\ 11''$$
$$\overline{49°\ 36'\ 10''}$$

$$-\text{measured } \angle p'' \quad \frac{-49°\ 36'\ 10''}{K_{p''} = 0}$$

Then, since ΔX_A, ΔY_A, ΔX_B, and ΔY_B are all zero,

$$v_{p''} = \frac{\cos \alpha'_{PB}}{PB \text{ arc } 1''}\Delta X_P - \frac{\sin \alpha'_{PB}}{PB \text{ arc } 1''}\Delta Y_P - \frac{\cos \alpha'_{PA}}{PA \text{ arc } 1''}\Delta X_P$$

$$+ \frac{\sin \alpha'_{PA}}{PA \text{ arc } 1''}\Delta Y_P + 0$$

For angle p', A is the *from* station and C is the *to* station.

$$\alpha'_{PC} = 41°\ 43'\ 41''$$
$$-\alpha'_{PA} = -3°\ 31'\ 21''$$
$$\overline{38°\ 12'\ 20''}$$

$$-\text{measured } \angle p' \quad \frac{-38°\ 12'\ 20''}{K'_p = 0}$$

Then, since ΔX_A, ΔY_A, ΔX_C, and ΔY_C are all zero,

$$v_{p'} = \frac{\cos \alpha'_{PA}}{PA \text{ arc } 1''}\Delta X_P - \frac{\sin \alpha'_{PA}}{PA \text{ arc } 1''}\Delta Y_P - \frac{\cos \alpha'_{PC}}{PC \text{ arc } 1''}\Delta X_P$$

$$+ \frac{\sin \alpha'_{PC}}{PC \text{ arc } 1''}\Delta Y_P + 0$$

The reduced observation equations are then

$$v_{p'''} = -27.50\ \Delta X_P + 69.83\ \Delta Y_P + 8''$$
$$v_{p''} = -52.97\ \Delta X_P + 77.81\ \Delta Y_P + 0$$
$$v_{p'} = +21.75\ \Delta X_P + 80.86\ \Delta Y_P + 0$$

$$\mathbf{V} = \begin{bmatrix} v_{p'''} \\ v_{p''} \\ v_{p'} \end{bmatrix} \quad \mathbf{B} = \begin{bmatrix} -27.50 & +69.83 \\ -52.97 & +77.81 \\ +21.75 & +80.86 \end{bmatrix} \quad \mathbf{X} = \begin{bmatrix} \Delta X_P \\ \Delta Y_P \end{bmatrix} \quad \mathbf{L} = \begin{bmatrix} 8'' \\ 0 \\ 0 \end{bmatrix}$$

Assuming all three angles have equal weights, $\mathbf{P}$ is the unit matrix. Then by Eq. (A-23),

$$\mathbf{X} = -(\mathbf{B}^T\mathbf{B})^{-1}\mathbf{B}^T\mathbf{L}$$

The solution of this equation gives $\Delta X_P = +0.03$ and $\Delta Y_P = -0.03$. The adjusted coordinates of P are then

$$X'_P = 118{,}371.46 \qquad Y'_P = 266{,}233.64 \qquad \text{(from Table A-8)}$$
$$\frac{+0.03}{X_P = 118{,}371.49} \qquad \frac{-0.03}{Y_P = 266{,}233.61}$$

A-8 Remarks on Adjustment of Observations

Since this appendix is intended only to introduce the reader to the concepts of least-squares adjustments, no attempt has been made to develop the subject in depth. This is the proper function of a separate textbook. A study of this appendix should raise several questions. For example, in the adjusted level net, how reliable are the adjusted elevations of the points? Or in a control net, how reliable are the adjusted plane coordinates? And do different parts of the net bear any relation to other parts of the net as far as dependability is concerned? Answers to these questions can be answered only by a rather thorough study of random-error theory and error propagation in a framework of modern statistics.

For a more complete study of the adjustment of surveying measurements, the readings listed in the Bibliography at the end of this appendix should be consulted.

The matrix formulation of the solution to the problem of the adjustment of weighted observations is the most efficient method because modern high-speed computers and programs for manipulating matrices are generally available. Before applying the matrix methods, however, a sound understanding is needed of what the least-squares adjustment does for a set of measurements.

BIBLIOGRAPHY

Adams, O. S., and Claire, C. N. *Manual of Plane Coordinate Computation.* Special Publication No. 193, U.S. Coast and Geodetic Survey, Washington, D.C.: U.S. Government Printing Office.

Brown, D. C. 1957. *A Treatment of the General Problem of Least Squares and the Associated Error Propagation.* App. A., R.C.A. Data Reduction Technical Report No. 39.

Curry, S., and Sawyer, R. 1989. STAR*NET:Rigorous 2D and 3D survey network adjustment program. *Tech. Papers Ann. Conv. Amer. Cong. Surv. & Mapp.* p. 269.

Deming, W. E. 1964. *Statistical Adjustment of Data.* New York: Dover.

Derenyi, E., and Chrzanowski, A. 1968. Preanalysis of trilateration nets for engineering surveys. *Surveying and Mapping* 28:615.

Meade, B. K. 1983. Least squares adjustment of a quadrilateral with six measured distances. *Surveying and Mapping* 43:385.

Mikhail, E. M. 1976. *Observations and Least Squares.* New York: Harper & Row.

——, and Gracie, G. 1981. *Analysis and Adjustment of Survey Measurements.* New York: Van Nostrand Reinhold.

Montag, R. E. 1982. Geometry as a basis for formulating the condition equation in trilateration adjustment. *Tech. Papers 42nd Ann. Meet. Amer. Cong. Surv. & Mapp.* p. 477.

Rainsford, H. F. 1957. *Survey Adjustments and Least Squares.* London: Constable.

Richardus, P. 1966. *Project Surveying.* New York: Wiley.

Stoughton, H. W. 1977. An algorithm to adjust a quadrilateral. *Surveying and Mapping* 37:201.

——. 1977. Introduction to least squares adjustment for the land surveyor. *Proc. Fall. Conv. Amer. Cong. Surv. & Mapp.* p. 375.

Veress, S. A. 1974. Selection of least-squares adjustment for control survey. *Proc. Ann. Meet. Amer. Cong. Surv. & Mapp.* p. 361.

Wolf, P. R. 1969. Horizontal position adjustment. *Surveying and Mapping* 29:635.

B The Adjustment of Instruments

B-1 Introduction

No matter how perfectly an instrument is adjusted when it leaves the instrument maker, it seldom remains in that condition for any considerable length of time, particularly when in daily use, subjected to rough handling and transported by car or truck over all kinds of roads. Although the field operations can be conducted so as to eliminate errors due to imperfect adjustment, this requires more time. For this reason every engineer and surveyor should be able to determine whether or not his instruments are in proper adjustment and, if they are not, how to adjust them.

Nearly all the adjustments of the transit and the level depend on the principle of reversal. By reversing the position of the instrument, the effect of the error is doubled. The adjustments are made by turning the capstan-headed screws or nuts that control the positions of the cross hairs and the bubble tubes. The beginner should be very careful not to put too much tension on these screws, as otherwise the threads may be stripped or the screw broken. The adjusting pins should be of the proper size to prevent damaging the holes in the screws and nuts.

When an instrument is badly out of adjustment, it may be necessary to repeat the adjustments several times. Time will usually be saved in such cases by not attempting to perfect each adjustment the first time, since the adjustment of the other parts of the instrument may affect the earlier adjustments.

Modern-day transportation provides the means to ship an instrument to an instrument maker for routine cleaning, adjusting, and lubrication, and to put it back into operation in a short period of time. Although the various adjustments to be discussed in this chapter can be performed by the engineer or surveyor, they should be performed by specialists. No attempt should be made by the surveyor to adjust an optical reading theodolite, a total station instrument, or a self-leveling level unless he has been trained to do so. These adjustments must be performed in a properly equipped shop or laboratory.

B-2   Cross Hairs of Dumpy Level

The first adjustment of the dumpy level is to make the horizontal cross hair truly horizontal when the instrument is leveled. With the level firmly set up, although not necessarily leveled, the clamp screw is clamped. A well-defined point that lies on, say, the left edge of the horizontal cross hair is selected. The telescope is then slowly rotated in azimuth about the vertical axis of the level by means of the tangent screw. If the point stays on the horizontal cross hair all the way over to the right edge, no adjustment is required. If the point departs from the horizontal cross hair, two adjacent capstan screws (shown in Fig. 3-9) are loosened, and the cross-hair ring is rotated to bring the point halfway back to the horizontal cross hair. The capstan screws are then tightened, and the test is repeated until the point stays on the cross hair.

B-3   Bubble Tube of Dumpy Level

The second adjustment of the dumpy level is to make the axis of the bubble tube perpendicular to the vertical axis. With the level firmly set up, it is leveled as best as can be using the leveling screws. The telescope is turned to bring the level bubble over one pair of leveling screws, and the bubble is carefully centered. The telescope is now rotated 180° in azimuth about the vertical axis. If the bubble returns to the center, the bubble tube is in adjustment. If the bubble does not return to the center, it is brought halfway back to center by raising or lowering the capstan end of the bubble tube.

Some level tubes contain two capstan-headed screws, one on top and one on the bottom. These are worked in opposition to one another; that is, one screw is loosened and the other one is tightened to raise or lower the end of the tube. Still other dumpy levels contain adjusting screws at both ends of the bubble tube. Either end can then be manipulated.

After the bubble has been brought halfway back to the center using the adjusting screws, it is then centered the remainder of the way by means of the leveling screws. Now again on rotation through 180°, the bubble should remain centered. If it does not, the adjustment is repeated until the bubble does remain centered.

B-4   Line of Sight of Dumpy Level

The final adjustment of the dumpy level is to make the line of sight parallel to the axis of the bubble tube, or to make the line of sight horizontal when the bubble is in the center of the tube. This is accomplished by the method referred to as the two-peg adjustment and is a form of reciprocal leveling described in Section 3-37. Two pegs perhaps 100 ft apart are set into the ground as shown in Fig. B-1. The level is first set up so close to point *A* that the eyepiece barely clears a leveling rod held on the point. By sighting through the wrong end of the telescope, the reading of a rod held on *A* can be obtained. The reading of a rod held on *B* is obtained in the usual way. If the level is in adjustment, the difference between these two readings will be the true difference in elevation. The level is next set up very close to *B*,

FIGURE B-1 Adjustment by two-peg method.

and readings are taken of rods held on B and A. If the difference in elevation obtained from this position of the level does not agree with the value obtained with the level near A, the instrument is out of adjustment. The true difference in elevation will be the mean of the two differences. From this value and from the rod reading on B, the correct rod reading on A can be calculated. The cross hair is brought to this calculated reading by moving the cross-hair ring vertically up or down by backing off on the top capstan screw and tightening the bottom screw, or vice versa. This must be done in small increments, as it is a very sensitive operation. After the adjustment has been made, the level bubble should be checked to make sure it is still centered. The reading at A should be then verified.

B-5 Cross Hair of Tilting Dumpy Level

The first adjustment of a tilting dumpy level is to make the horizontal cross hair truly horizontal when the bull's-eye bubble is centered. The test is the same as for the dumpy level. A point is chosen that coincides with one end of the cross hair. If the point does not remain on the hair as the telescope is slowly turned about its vertical axis, the adjustment is made by loosening two adjacent capstan screws and rotating the cross-hair ring.

B-6 Bull's-Eye Bubble of Tilting Dumpy Level

The second adjustment of the tilting level is that of making the circular bubble stay in the center as the telescope is rotated about the vertical axis. If the bubble moves off center when the telescope is turned 180° in azimuth, it is brought halfway back to center in both directions by raising or lowering the bubble mount by means of either capstan screws or spring screws, the method depending on the make of level, and the remaining distance with the leveling screws. This adjustment is not critical because of the tilting feature of the level.

B-7 Bubble Tube of Sensitive Level of Tilting Level

This adjustment gives a horizontal line of sight when the main sensitive bubble is centered. If the level bubble is of the coincidence type, the coincidence of the two ends of the bubble indicates centering. The two-peg method described in Section B-4 is used to make this adjustment. When the rod reading that will give a horizontal line of sight has been calculated, the telescope is tilted up or down by means of the tilting knob until the horizontal cross hair is on the correct reading. The main bubble is brought to center or to coincidence by raising or lowering one end of the bubble vial.

B-8   Plate Bubbles of Transit

The first adjustment of the transit is to make the axes of the plate levels perpendicular to the vertical axis of the instrument. The transit is set up and each bubble is brought to the center of its tube by means of the leveling screws. The instrument is then revolved on its vertical axis through 180°. If either bubble moves toward one end of the tube, it is brought halfway back by means of the capstan-headed screw at the end of the tube, and the remaining distance by means of the leveling screws. This operation is repeated until both bubbles remain in the centers of the tubes as the telescope is turned in azimuth.

B-9   Cross Hairs of Transit

The first adjustment of the cross hairs of a transit is to make the vertical cross hair perpendicular to the horizontal axis of the telescope, so that when the other adjustments have been made it will be truly vertical. This adjustment is made in a manner similar to the one used in making the horizontal cross hair in a level truly horizontal. One end of the vertical hair is brought to some well-defined point, and the telescope is revolved on its transverse axis so that the point appears to move along the hair. If it does not remain on the hair throughout the motion, two adjacent screws that control the cross-hair ring are loosened and the ring is rotated.

The purpose of the second adjustment of the vertical cross hair is to make the line of sight perpendicular to the horizontal axis. Whether or not an adjustment is necessary can be determined by prolonging a straight line by the method of double centering, as described in Section 6-6. If, as shown in Fig. B-2, the two points B and C do not coincide, the cross-hair ring should be shifted laterally until the vertical cross hair appears to strike D, which is at one-fourth of the distance from the second point C to the first point B. Since a double reversing is involved, the apparent error CB is four times the real error.

The third adjustment of the cross hairs is necessary when the transit is used as a level, or when it is used in measuring vertical angles. The purpose of the adjustment is to bring the horizontal hair into the plane of motion of the optical center of the object glass, so that the line of sight will be horizontal when the telescope bubble is in adjustment and the bubble is centered. In making the

FIGURE B-2 Second adjustment of vertical cross hair.

FIGURE B-3 Adjustment of horizontal cross hair.

adjustment the transit is set up at *A,* as shown in Fig. B-3, and two stakes are driven at *B* and *C* on the same straight line and at distances of about 30 and 300 ft from *A.* With the telescope clamped in the normal position and the line of sight represented by *cd,* rod readings *Be* and *Cd* are obtained by holding leveling rods on the two stakes. The telescope is then plunged and the line of sight *fg* is brought to the former rod reading *Be.* If the rod reading on *C* agrees with the reading previously obtained, no adjustment is necessary. If the two readings do not agree, the target is set at *Cb,* which is the mean of the two readings *Cd* and *Cg,* and the horizontal cross hair is brought to the position *a* where it appears to bisect the target, by shifting the cross-hair ring vertically by means of the capstan-headed screws on the top and bottom of the telescope. If the instrument is badly out of adjustment, the reading *Be* will be changed perceptibly as the cross-hair ring is shifted, and the adjustment may have to be repeated several times.

B-10   Standards

If the standards of the transit are in adjustment, the horizontal axis of the telescope is perpendicular to the vertical axis of the instrument. When the plates are leveled, the horizontal axis is truly horizontal, and the line of sight moves in a vertical plane as the telescope is raised and lowered. The test of the standards is made by sighting, with the telescope normal, to some well-defined point on a high object, such as a flagpole or church spire. In Fig. B-4, *A* is such a high point. With the horizontal motion clamped, the telescope is depressed and a point *B* is set on the ground in the line of sight. The telescope is then plunged, the lower clamp is released, and a sight is taken on *B.* With the lower clamp tight, the telescope is again elevated. If the line of sight strikes the high point *A,* no adjustment is necessary. If, as in Fig. B-4, the line of sight now strikes at *C,* one end of the horizontal axis must be raised or lowered to bring the line of sight midway between *C* and *A.* This is done by loosening the upper screws that hold the horizontal axis in place and then turning the small capstan-headed screw that controls the position of the block on which the horizontal axis rests. In the example shown in Fig. B-4, the left side should be raised or the right side lowered when the telescope is in the reversed position. When the axis has been properly adjusted, the upper screws are tightened just enough to prevent looseness of the bearing. If the line of sight is then lowered, it should strike point *B.*

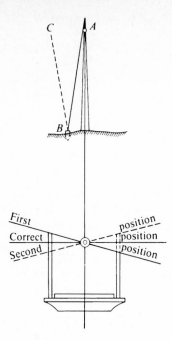

FIGURE B-4 Adjustment of standards.

B-11   Telescope Bubble

The purpose of the adjustment of the telescope bubble is to make the axis of its tube parallel to the line of sight. This adjustment is necessary whenever the transit is to be used as a level. When the direction of a horizontal line of sight has been determined by the peg method, as described in Section B-5, the line of sight is brought to the correct reading by raising or lowering the telescope with the vertical tangent screw. The bubble tube is then adjusted to bring the bubble to the center of the tube.

B-12   Vertical-Circle Vernier

The vertical circle should preferably read 0° when the telescope is horizontal. With the instrument carefully leveled and the telescope bubble in the center of the tube, the screws holding the vernier in place are loosened slightly, and the zeros are made to coincide by tapping the vernier lightly.

B-13   Centering the Eyepiece

If the intersection of the cross hairs appears to be out of the center of the field of view when the instrument is otherwise in adjustment, the end of the eyepiece must be out of position. Erecting eyepieces are usually held in place by a ring, similar to the cross-hair ring, which is controlled by a set of adjusting screws between the eyepiece and the cross-hair ring. By moving this ring, the intersection of the cross hairs can be brought to the center of the field. As the inverting eyepiece is very short, its inner end needs no support.

B-14   Protection and Cleaning of Instruments

The instrument man can do much to render unnecessary the frequent adjustment of any surveying instrument. The workmanship on a high-grade transit or level is comparable to that of the finest watch, and the instrument should be handled accordingly. When it is carried from point to point on the shoulder, the clamps should be tight enough to prevent needless wear, yet loose enough to yield if the parts are accidentally bumped. When the transit is not in use or is being carried, the needle should be raised. As the instrument is taken from the shoulder, it should be carefully set down. Otherwise a cross hair may be broken or the instrument thrown out of adjustment. When being transported by car or truck, it should be held in the lap, if possible. When being transported long distances, it should be packed in its box with crumpled newspapers used to hold it securely in place. The instrument box should then be packed in a box large enough to permit the use of several inches of excelsior padding on all sides.

It should be remembered that repairs to a damaged instrument are costly. In addition, it is almost impossible to restore a badly damaged instrument to its original condition. For these reasons a transit, level, or total station should not be left unguarded when a setup on a sidewalk or pavement is unavoidable. The tripod legs should be spread and pushed into the ground sufficiently to prevent its being blown or knocked over.

The instrument should be protected from the rain by a waterproof cover. If such a cover is not at hand, the dust cap should be placed on the objective and the object glass turned upward. When brought in wet from the field, the instrument should be thoroughly dried before it is replaced in the instrument box. The lenses should be wiped with a piece of the softest chamois skin, silk, or linen. If objects appear blurred or indistinct, the cause probably is a dirty eyepiece. Any dust should be removed with a fine camel's-hair brush. If very dirty, the eyepiece can be cleaned with alcohol and wiped with a clean soft cloth. The lenses should not be cleaned too frequently, as there is danger of their being scratched.

Extreme care should be used in wiping the vertical circle and vernier. The edges should not be rubbed, and the hands should be kept off the vertical circle to prevent tarnishing. If any of the graduated circle becomes tarnished, the dust should first be removed with a camel's-hair brush. A thin film of watch oil is then spread on the tarnished spots. The oil is allowed to remain for a few hours and is then wiped off with a soft piece of linen.

Although an instrument should not be taken apart unnecessarily, any part that does not work freely should not be forced, but cleaned. Old oil and accumulated dirt should be thoroughly removed with alcohol or, in the case of screws and screw holes, by the use of soap and water. No oil should be used on the exposed parts, as it soon accumulates enough dust to prevent the part from working freely. On interior bearings, only the finest grade of watch oil should be used, and this very sparingly since it tends to collect dust and in cold weather may become stiff and prevent the free movement of the part. Usually enough oil will remain if the oiled part is wiped dry with a soft cloth before it is replaced in the instrument.

C Tables

TABLE C-1   Lengths of Circular Arcs for Radius = 1

Deg.	Length	Deg.	Length	Deg.	Length	Deg.	Length
1	0.017 45 329	**31**	0.541 05 207	**61**	1.064 65 084	**91**	1.588 24 962
2	0.034 90 659	32	0.558 50 536	62	1.082 10 414	92	1.605 70 291
3	0.052 35 988	33	0.575 95 865	63	1.099 55 743	93	1.623 15 620
4	0.069 81 317	34	0.593 41 195	64	1.117 01 072	94	1.640 60 950
5	0.087 26 646	35	0.610 86 524	65	1.134 46 401	95	1.658 06 279
6	0.104 71 976	**36**	0.628 31 853	**66**	1.151 91 731	**96**	1.675 51 608
7	0.122 17 305	37	0.645 77 182	67	1.169 37 060	97	1.692 96 937
8	0.139 62 634	38	0.663 22 512	68	1.186 82 389	98	1.710 42 267
9	0.157 07 963	39	0.680 67 841	69	1.204 27 718	99	1.727 87 596
10	0.174 53 293	40	0.698 13 170	70	1.221 73 048	100	1.745 32 925
11	0.191 98 622	**41**	0.715 58 499	**71**	1.239 18 377	**101**	1.762 78 254
12	0.209 43 951	42	0.733 03 829	72	1.256 63 706	102	1.780 23 584
13	0.226 89 280	43	0.750 49 158	73	1.274 09 035	103	1.797 68 913
14	0.244 34 610	44	0.767 94 487	74	1.291 54 365	104	1.815 14 242
15	0.261 79 939	45	0.785 39 816	75	1.308 99 694	105	1.832 59 571
16	0.279 25 268	**46**	0.802 85 146	**76**	1.326 45 023	**106**	1.850 04 901
17	0.296 70 597	47	0.820 30 475	77	1.343 90 352	107	1.867 50 230
18	0.314 15 927	48	0.837 75 804	78	1.361 35 682	108	1.884 95 559
19	0.331 61 256	49	0.855 21 133	79	1.378 81 011	109	1.902 40 888
20	0.349 06 585	50	0.872 66 463	80	1.396 26 340	110	1.919 86 218
21	0.366 51 914	**51**	0.890 11 792	**81**	1.413 71 669	**111**	1.937 31 547
22	0.383 97 244	52	0.907 57 121	82	1.431 16 999	112	1.954 76 876
23	0.401 42 573	53	0.925 02 450	83	1.448 62 328	113	1.972 22 205
24	0.418 87 902	54	0.942 47 780	84	1.466 07 657	114	1.989 67 535
25	0.436 33 231	55	0.959 93 109	85	1.483 52 986	115	2.007 12 864
26	0.453 78 561	**56**	0.977 38 438	**86**	1.500 98 316	**116**	2.024 58 193
27	0.471 23 890	57	0.994 83 767	87	1.518 43 645	117	2.042 03 522
28	0.488 69 219	58	1.012 29 097	88	1.535 88 974	118	2.059 48 852
29	0.506 14 548	59	1.029 74 426	89	1.553 34 303	119	2.076 94 181
30	0.523 59 878	60	1.047 19 755	90	1.570 79 633	120	2.094 39 510

727

TABLE C-1 (continued)

Min.	Length	Min.	Length	Sec.	Length	Sec.	Length
1	0.000 29 089	**31**	0.009 01 753	**1**	0.000 00 485	**31**	0.000 15 029
2	0 58 178	32	9 30 842	2	00 970	32	15 514
3	0 87 266	33	9 59 931	3	01 454	33	15 999
4	1 16 355	34	9 89 020	4	01 939	34	16 484
5	1 45 444	35	10 18 109	5	02 424	35	16 969
6	0.001 74 533	**36**	0.010 47 198	**6**	0.000 02 909	**36**	0.000 17 453
7	2 03 622	37	10 76 286	7	03 394	37	17 938
8	2 32 711	38	11 05 375	8	03 879	38	18 423
9	2 61 799	39	11 34 464	9	04 363	39	18 908
10	2 90 888	40	11 63 553	10	04 848	40	19 393
11	0.003 19 977	**41**	0.011 92 642	**11**	0.000 05 333	**41**	0.000 19 877
12	3 49 066	42	12 21 730	12	05 818	42	20 362
13	3 78 155	43	12 50 819	13	06 303	43	20 847
14	4 07 243	44	12 79 908	14	06 787	44	21 332
15	4 36 332	45	13 08 997	15	07 272	45	21 817
16	0.004 65 421	**46**	0.013 38 086	**16**	0.000 07 757	**46**	0.000 22 301
17	4 94 510	47	13 67 175	17	08 242	47	22 786
18	5 23 599	48	13 96 263	18	08 727	48	23 271
19	5 52 688	49	14 25 352	19	09 211	49	23 756
20	5 81 776	50	14 54 441	20	09 696	50	24 241
21	0.006 10 865	**51**	0.014 83 530	**21**	0.000 10 181	**51**	0.000 24 726
22	6 39 954	52	15 12 619	22	10 666	52	25 210
23	6 69 043	53	15 41 708	23	11 151	53	25 695
24	6 98 132	54	15 70 796	24	11 636	54	26 180
25	7 27 221	55	15 99 885	25	12 120	55	26 665
26	0.007 56 309	**56**	0.016 28 974	**26**	0.000 12 605	**56**	0.000 27 150
27	7 85 398	57	16 58 063	27	13 090	57	27 634
28	8 14 487	58	16 87 152	28	13 575	58	28 119
29	8 43 576	59	17 16 240	29	14 060	59	28 604
30	8 72 665	60	17 45 329	30	14 544	60	29 089

TABLE C-2 Trigonometric Formulas for the Solution of Right Triangles

Let A = angle BAC = arc BF and let radius $AF = AB$ = $AH = 1$. Then

$\sin A = BC$	$\csc A = AG$
$\cos A = AC$	$\sec A = AD$
$\tan A = DF$	$\cot A = HG$
vers $A = CF = BE$	covers $A = BK = LH$
exsec $A = BD$	coexsec $A = BG$
chord $A = BF$	chord $2A = BI = 2BC$

In the right-angled triangle ABC, let $AB = c$, $BC = a$, $CA = b$. Then

1. $\sin A = \dfrac{a}{c}$

2. $\cos A = \dfrac{b}{c}$

3. $\tan A = \dfrac{a}{b}$

4. $\cot A = \dfrac{b}{a}$

5. $\sec A = \dfrac{c}{b}$

6. $\csc A = \dfrac{c}{a}$

7. vers $A = 1 - \cos A = \dfrac{c-b}{c} = $ covers B

8. exsec $A = \sec A - 1 = \dfrac{c-b}{b} = $ coexsec B

9. covers $A = \dfrac{c-a}{c} = $ vers B

10. coexsec $A = \dfrac{c-a}{a} = $ exsec B

11. $a = c \sin A = b \tan A$

12. $b = c \cos A = a \cot A$

13. $c = \dfrac{a}{\sin A} = \dfrac{b}{\cos A}$

14. $a = c \cos B = b \cot B$

15. $b = c \sin B = a \tan B$

16. $c = \dfrac{a}{\cos B} = \dfrac{b}{\sin B}$

17. $a = \sqrt{c^2 - b^2} = \sqrt{(c-b)(c+b)}$

18. $b = \sqrt{c^2 - a^2} = \sqrt{(c-a)(c+a)}$

19. $c = \sqrt{a^2 + b^2}$

20. $C = 90° = A + B$

21. area $= \dfrac{1}{2} ab$

TABLE C-3 Trigonometric Formulas for the Solution of Oblique Triangles

No.	Given	Sought	Formula
22	A, B, a	C, b, c	$C = 180° - (A + B)$
			$b = \dfrac{a}{\sin A} \times \sin B$
			$c = \dfrac{a}{\sin A} \times \sin (A + B) = \dfrac{a}{\sin A} \times \sin C$
		area	$\text{area} = \frac{1}{2} ab \sin C = \dfrac{a^2 \sin B \sin C}{2 \sin A}$
23	A, a, b	B, C, c	$\sin B = \dfrac{\sin A}{a} \times b$
			$C = 180° - (A + B)$
			$c = \dfrac{a}{\sin A} \times \sin C$
		area	$\text{area} = \frac{1}{2} ab \sin C$
24	C, a, b	c	$c = \sqrt{a^2 + b^2 - 2ab \cos C}$
25		$\frac{1}{2}(A + B)$	$\frac{1}{2}(A + B) = 90° - \frac{1}{2}C$
26		$\frac{1}{2}(A - B)$	$\tan \frac{1}{2}(A - B) = \dfrac{a - b}{a + b} \times \tan \frac{1}{2}(A + B)$
27		A, B	$A = \frac{1}{2}(A + B) + \frac{1}{2}(A - B)$
			$B = \frac{1}{2}(A + B) - \frac{1}{2}(A - B)$
28		c	$c = (a + b) \times \dfrac{\cos\frac{1}{2}(A + B)}{\cos\frac{1}{2}(A - B)} = (a - b) \times \dfrac{\sin\frac{1}{2}(A + B)}{\sin\frac{1}{2}(A - B)}$
29		area	$\text{area} = \frac{1}{2} ab \sin C$
30	a, b, c	A	$\text{Let } s = \dfrac{a + b + c}{2}$
31			$\sin \frac{1}{2}A = \sqrt{\dfrac{(s - b)(s - c)}{bc}}$
			$\cos \frac{1}{2}A = \sqrt{\dfrac{s(s - a)}{bc}}$

TABLE C-3   Trigonometric Formulas for the Solution of Oblique Triangles (Continued)

No.	Given	Sought	Formula
32			$\tan \frac{1}{2}A = \sqrt{\dfrac{(s-b)(s-c)}{s(s-a)}}$ $\sin A = \dfrac{2\sqrt{s(s-a)(s-b)(s-c)}}{bc}$ $\cos A = \dfrac{b^2 + c^2 - a^2}{2bc}$
33		area	$\text{area} = \sqrt{s(s-a)(s-b)(s-c)}$

Index